Microcomputer Repair
Fourth Edition

James L. Antonakos
Broome Community College

Upper Saddle River, New Jersey
Columbus, Ohio

For Michele, my wife, my love, my savior

Library of Congress Cataloging-in-Publication Data
Antonakos, James L.
    Microcomputer repair / James L. Antonakos. -- 4th ed.
        p. cm.
    Includes index.
    ISBN 0-13-019578-2
    1. Microcomputers -- Maintenance and repair.   I. Title.

TK7887.A58   2002
621.39'16'0288--dc21

Editor in Chief: Stephen Helba
Assistant Vice President and Publisher: Charles E. Stewart, Jr.
Production Editor: Alexandrina Benedicto Wolf
Production Coordination: Custom Editorial Productions, Inc.
Design Coordinator: Robin G. Chukes
Cover Designer: Tom Mack
Cover Art: Marjory Dressler
Production Manager: Matthew Ottenweller

This book was set in Times Roman by Custom Editorial Productions, Inc., and was printed and bound by Banta Book Group. The cover was printed by Phoenix Color Corp.

All character names (e.g., Joe Tekk and Ken Koder) are registered trademarks of Prentice Hall.

Prentice-Hall International (UK) Limited, *London*
Prentice-Hall of Australia Pty. Limited, *Sydney*
Prentice-Hall Canada, Inc., *Toronto*
Prentice-Hall Hispanoamericana, S.A., *Mexico*
Prentice-Hall of India Private Limited, *New Delhi*
Prentice-Hall of Japan, Inc., *Tokyo*
Prentice-Hall Singapore Pte. Ltd.
Editora Prentice-Hall do Brasil, Ltda., *Rio de Janeiro*

10 9 8 7 6 5 4 3 2 1
ISBN: 0-13-019578-2

Preface

This text/lab manual is designed to teach beginning electronics students the fundamental skills required for hardware and software servicing of microcomputer systems. No prior knowledge of electronics is required or assumed, and math and theory are held to a minimum. Instead, emphasis is placed on the development of essential troubleshooting and repair skills.

TEXT ORGANIZATION

Microcomputer Repair, Fourth Edition, is made up of 58 exercises, which are divided into six units. Each unit addresses a specific range of topics. This text organization gives instructors the flexibility to tailor their courses to their students' technical backgrounds and employment preparation goals. The six units are:

- Unit I—Introduction (Exercises 1–4)
  Introduces students to the use of this text, the laboratory, safety requirements, and basic tool usage. This unit may be omitted if students are already familiar with the electronics lab.
- Unit II—Basic Skills (Exercises 5–8)
  Familiarizes students with fundamentals of electronic hardware, basic soldering skills, and integrated circuit identification. This unit may be omitted if students have completed a laboratory course in basic electronics.
- Unit III—The Windows Operating System (Exercises 9–21)
  Describes the Microsoft Windows operating system in detail for the beginner. All aspects of computing under Windows are discussed.
- Unit IV—Computer Networks (Exercises 22–31)
  The important field of computer networking is explained in detail for the beginner from media to protocols. Networking under Windows is also explained in detail.
- Unit V—Microcomputer Hardware (Exercises 32–49)
  Gives students hands-on experience in the teardown and assembly of microcomputer systems. These exercises should be completed by all students. If the number of computers available in the lab is limited, these exercises may be given as laboratory/classroom demonstrations.
- Unit VI—Selected Topics (Exercises 50–58)
  Covers a wide range of microcomputer-related topics, from assembly language to computer networks and software viruses. These exercises are designed to introduce students to many advanced topics, so that they are more familiar with the complete world of microcomputer systems.

A detailed glossary of microcomputer-related terms is provided for quick reference.

EXERCISE ORGANIZATION

Each exercise in *Microcomputer Repair,* Fourth Edition (with the exception of the first, an introductory exercise) is organized into the following sections:

- **Introduction** sets the stage for the exercise by examining how Joe Tekk, a fictitious technician at a fictitious company (RWA Software), deals with the exercise topic.
- **Performance Objectives** state exactly what new skills the student will have at the completion of the exercise.
- **Key Terms** found in the exercise are listed for examination before reading the Background Information.
- **Background Information** presents all of the technical information and theory necessary to achieve the performance objectives.
- **Troubleshooting Techniques** are included to show real-world situations and their solutions.
- **Summary** section lists the main items covered in the exercise.
- **Self-Test** is provided for students to test their understanding of the Background Information. **Answers to Odd-Numbered Self-Test Questions** are found at the end of this book.
- **Familiarization Activity** requires students to actually perform and apply the information presented in the Background Information section of the exercise.
- **Questions/Activities** allow students to draw their own conclusions about the experience gained from the Familiarization Activities. Students may be required to do further research into a specific topic. This section is optional and may be assigned according to the instructor's course objectives.
- **Review Quiz** is given at the end of the exercise and outlines exactly what tasks students are expected to be able to complete and to what degree of accuracy, within a given amount of time, they will be expected to complete those tasks.

SUPPLEMENTS

Microcomputer Repair, Fourth Edition, is supported by the following materials:

- A CD-ROM that contains DOS-executable programs called WINQUIZ and NETQUIZ, which take the student through a series of true/false, multiple-choice, and fill-in-the-blank questions about Windows and networking, as well as many different files supporting the exercise material, including WINZIP and LanExplorer.
- A second CD-ROM containing McAfee Utilities is also included, providing a demonstration version of many important system utilities.
- An Instructor's Manual is available which contains suggestions for implementing each of the laboratory exercises and sample tests, as well as the even-numbered self-test answers.

CHANGES AND ADDITIONS TO THE FOURTH EDITION

Several changes and additions were made to create the fourth edition of *Microcomputer Repair.* These changes and additions are summarized as follows:

- Six new exercises increase coverage of Windows and networking. These are
  1. Exercise 11, An Introduction to Windows NT and Windows 2000
  2. Exercise 20, Windows CE
  3. Exercise 21, Other Network Operating Systems
  4. Exercise 30, Windows NT Domains
  5. Exercise 31, An Introduction to Telecommunications
  6. Exercise 57, Performance and Diagnostic Software
- Many new photographs of current computer hardware.
- Unit III from the third edition (The Disk Operating System) has been moved to Appendix B at the request of reviewers.
- A new Appendix C covers Fault Tolerance in Windows NT.
- A new Appendix D provides information on installing and upgrading Windows.
- Exercise 51, Computer Languages, now contains exposure to additional languages typically used on the PC, such as C++ and Visual BASIC.
- A new glossary contains detailed terms related to microcomputer systems.

Furthermore, each exercise now has Key Terms and Summary sections, additional information, and several new self-test questions and activities. The additional information includes Windows 2000, Windows CE, Windows ME, Netware, Linux, IEEE 802 standards, network access points, fiber optic cable, virtual private networking, e-mail, denial of service attacks, firewalls, switching power supply theory, motherboard manufacturers, Pentium III architecture, backside cache, USB, FireWire, the electrophotographic process, WINZIP, TFT displays, CLV and CAV, multimedia files types, C++ programming, FLASH BIOS, Macro viruses, PGP, Windows NT internal operation, benchmarks, fault tolerance, Windows installation and upgrading, plus many other details.

ACKNOWLEDGMENTS

I would like to thank my editor, Charles Stewart, and his assistant, Mayda Bosco, for all their help while I was putting this book together. I would also like to thank the many students and instructors who used the third edition, as well as my production supervisor at Custom Editorial Productions, Inc., Megan Smith-Creed, and my copyeditor, Julie Hotchkiss. More than anything, I appreciate the assistance and technical expertise I received from my friend and colleague, Kenneth C. Mansfield Jr.

These companies and the following individuals were especially helpful regarding permission requests and demonstration versions, and I deeply thank them.

- Netscape Communications Corporation, for allowing screen shots of Netscape Navigator. Screen shots copyright Netscape Communications Corporation, 1999. All rights reserved. Netscape, Netscape Navigator, and the Netscape N logo are registered trademarks of Netscape in the United States and other countries.
- Bong Sarmiento of Sunrise Telecom, Inc., for providing LanExplorer. LanExplorer is a registered trademark of Sunrise Telecom, Inc.
- Microsoft Corporation for allowing screen shots of their operating systems and software applications, particularly Microsoft Office. Screen shots reprinted by permission of Microsoft Corporation.
- Network Associates, Inc. and McAfee.com Corp., for allowing screen shots of VShield and VirusScan, and providing a demonstration version of McAfee Utilities.
- Power Quest Corp., for allowing screen shots of PartitionMagic.
- Symantec Corp., for allowing screen shots of Norton Utilities and pc ANYWHERE. © 1991, 1995, 1997 Symantec Corporation, 10201 Torre Avenue, Cupertino, CA 95014 USA. All rights reserved.
- Sara Rogers of WinZip Computing, Inc., for providing a demonstration version of WinZip. Copyright 1991–2000, WinZip Computing, Inc. WinZip is available from www.winzip.com. WinZip screen images reproduced with permission of WinZip Computing, Inc.

The following individuals provided many useful comments during the rewrite, and I am grateful for their advice: Paula M. Greenwald, Niagara County Community College, New York; Patrick R. Fontenot, Northwest Vista College, Texas; and Robert Smith, Texas State Technical College.

James L. Antonakos
antonakos_j@sunybroome.edu
http://www.sunybroome.edu/~antonakos_j

Contents

UNIT **I** Introduction

1 Using the Instructional System

INTRODUCTION

This exercise serves as an introduction to the entire book. Here you will see how all the remaining exercises are laid out and how best to use them in order to gain the maximum benefit.

Each exercise begins with a short example of why the exercise is applicable to micro-computer use. A fictitious employee is typically used to convey the problem or situation. Here is the first example:

Joe Tekk has just been hired by RWA Software as its new computer specialist. His manager was impressed during Joe's job interview by the fact that Joe had read RWA's company literature before the interview. When asked why, Joe said, "I like to be prepared, to know in advance what to expect of a situation."

PERFORMANCE OBJECTIVES

Every exercise starts with **performance objectives**. The performance objectives let you know what new skills and knowledge you will learn in the course of completing the exercise. The performance objectives also tell you how you can check to see if you have acquired these new skills and what degree of accuracy you can expect to attain.

Your instructor will usually administer the requirements of the performance objectives. You can think of these objectives as a test. However, unlike most tests, you know exactly what the test will be. Thus, you can practice the performance objectives as many times as you want before attempting them with your instructor. By making sure you can satisfy the objectives before being tested on them, you can ensure your success. And that is the whole idea behind performance objectives—making sure you know exactly how to apply your new skills and knowledge.

KEY TERMS

The **key terms** section lists many of the new terms encountered in the exercise. Looking at them before jumping into the background information gives you a quick peek at what is coming. You should be familiar with the meaning of all the key terms when you complete the exercise.

BACKGROUND INFORMATION

The **background information** section presents all the information you need in order to perform the exercise and pass the review quiz. The background information section, which is usually the longest section in the exercise, contains important and detailed information. Because of this, you may want to read through the section more than once. A good rule is to read through this section first, just to get an overview of what it covers. Then read through it again, this time much more slowly. Use a highlighter or marking pen to point out key areas for yourself. You may even want to jot notes in the margins. Remember, as a computer technician you will be using this book as a constant reference. The more meaningful the notes you place in it, the more useful it will become for you. Also, keeping notes in this lab manual will help you prepare for job interviews.

The background information section also contains a **troubleshooting** area. Tips, techniques, and actual problems and their solutions are presented.

SUMMARY

The last section before the self-test is the **summary**. Many of the main points or topics covered in the exercise are summarized. Any unfamiliarity with the summary indicates that certain sections should be reread.

SELF-TEST

True/False

Multiple Choice

Completion

Open-Ended

The next section you will encounter in each exercise is the **self-test**. The self-test is there to help you check your understanding of the material you just covered in the background information section.

As you can see, the self-test is divided into several types of test questions. This is done to make the test more interesting and more reflective of what you may need to review, and to help you get used to the different types of questions you may be asked during job interviews. In order to help you check your progress, answers to odd-numbered self-test questions are given at the end of the book. It's a good idea to try the self-test before going on to the next section of the exercise, where you will *apply* the information you learned in the background information section.

FAMILIARIZATION ACTIVITY

The next section in each exercise is the **familiarization activity**. It is here that you get "hands-on" applications of what you have just learned.

The familiarization activity is just that—an activity or series of activities the purpose of which is to familiarize you with the applications of what you have just learned.

You will usually perform the familiarization activity in the lab. However, there may be some exercises that your instructor will assign as outside work. This is usually the case for exercises on software. In these exercises, the familiarization activity usually consists of a series of software interactions with a computer. These interactions can be performed as a homework assignment or done in some other place such as a computer room that provides access to computers for all students. You can then complete the review quiz at another time under the direction of your instructor.

QUESTIONS/ACTIVITIES

The next section you will find in all the exercises is the **questions/activities** section. Here you will find questions about the familiarization activity you just completed. These questions are designed to help reinforce important concepts you should have picked up when you were doing the familiarization activity. There may be times when other activities are suggested, and your instructor may or may not assign them. These other activities usually include outside assignments and are selected to give you the opportunity to broaden your understanding of the subject of the exercise.

REVIEW QUIZ

The last section, the **review quiz**, restates the performance objectives. Thus you start the exercise knowing what you should learn from it and you end at the point at which you should be ready to quiz yourself. At this point in the exercise, you should check with your instructor to see when and how you will be quizzed on the stated performance objectives.

ANSWERS TO SELF-TEST

The **answers to the odd-numbered self-test questions** are given at the end of the book. They are placed there to keep you from getting distracted while you are trying the self-test. Keep in mind that the self-test is designed to help you. It is a personal self-check. The best way to benefit from it is to try all the questions first, writing down your answers as you go. Then, after completing the entire test, check your answers. Taking the test in this manner prevents you from seeing the answer to the next question, which will happen if you look at the answer each time you complete a question. The answers to the even-numbered self-test questions are provided in the instructor's manual.

You now have an overview of how each exercise is set up. You should also have an understanding of the purpose of each section in the exercise. Keep in mind that all the sections of each exercise are important. They are designed to be used as an integrated whole. By using them in the manner in which they were designed to be used, you can gain the skills needed to become a successful computer technician.

2 Laboratory Familiarization

INTRODUCTION

The technicians' lab at RWA Software has a small handwritten sign next to the main tool chest. The sign reads

> *"If it's not yours, don't take it.*
> *If you take it, bring it back."*

The sign was put up by the senior technician after he discovered that a cable-crimping tool was missing (borrowed for an off-site job). The sign was the first thing the senior technician showed to Joe Tekk his first day on the job.

PERFORMANCE OBJECTIVES

Upon completion of this exercise, you will be able to

1. Explain the rules for checking out hand tools and equipment.
2. Follow procedures for checking out parts and equipment for your own personal use.
3. Follow the policies and procedures for working on your own personal equipment in the lab.
4. State the times the lab is available for your use.
5. Follow the procedures for reporting damaged or improperly operating tools and equipment.

KEY TERMS

Safety   Policy
Respect   Procedure
Rule

BACKGROUND INFORMATION

Your microcomputer-repair lab is unique. Because you will be working with actual computer equipment, the rules and regulations may be different from those of other labs. The most important consideration in your lab (and any other lab) is *safety*—both your personal safety and the safety of others.

Always remember that you will be working with equipment that operates on electrical power. Electricity is potentially very dangerous and can cause severe electrical shock, burns, and other hazards that can do serious personal injury to you and others. The next lab exercise concerns the details of laboratory safety. For now, do not handle any electrical or mechanical equipment in the lab without the express permission of your instructor. If you see or suspect any potentially dangerous condition in the lab, report it to your instructor immediately.

When done properly and with safety in mind, maintenance and repair of microcomputer equipment can be a very rewarding and satisfying career. It is important that you approach this subject with a mature attitude, which includes respect for those with whom you will be working as well as for the tools and equipment you will be using. The lab is no place for horseplay or practical jokes. It is a serious environment for learning a very complex subject in a very practical manner. If you use the time spent here wisely, it will be a memorable experience that will stay with you throughout your professional career.

TROUBLESHOOTING TECHNIQUES

Proper setup and use of the lab equipment will help make the job of finding a faulty component or circuit easier for you. Oscilloscopes, DMMs, logic analyzers, and other types of test equipment must be used safely as well.

SUMMARY

In this exercise we discovered that

- Safety is the most important consideration in the laboratory environment.
- A mature attitude is required for work in the laboratory.

SELF-TEST

The following self-test is designed to help you check your knowledge of the rules and regulations regarding the microcomputer-repair lab. The information for this test may be supplied to you in writing or verbally by your lab instructor. This test may then be graded in class under the supervision of your lab instructor. Use this self-test as a learning tool. If you give a wrong answer, make sure that you correct it. Having all the right answers to this self-test will make it a useful document for review.

Open-Ended

Answer the following questions:

1. What is the room number of the lab?

2. State the days and times you are required to attend the lab.

3. How many instructors are available in the lab to assist you?

4. What are the names of your instructors?

5. Explain the lab attendance policy. As an example, will you be allowed to make up a regularly scheduled lab if you are absent?

6. How many lab partners will you have? How are they selected?

7. Must you keep the same group of lab partners throughout the course? Explain.

8. What days and times is the lab available for your use other than the required attendance times? Must an instructor be in the lab before you can use it?

9. If you know ahead of time that you will miss a required lab, what arrangements, if any, can you make with your instructor to make up that lab?

10. Is there a way you can contact your instructor when you are not at the school site? How?

11. Explain the procedure for checking out equipment.

12. Explain the procedure for returning equipment.

13. Explain the policy for checking out equipment to be used outside the lab.

14. Explain the policy for reporting damaged or missing parts and equipment.

15. May you bring in your own equipment to be repaired? Explain.

FAMILIARIZATION ACTIVITY

Perform the following activities as directed by your lab instructor. The activities may not be assigned in the same sequence, and some of the activities may be omitted.

1. In the given space, sketch the layout of your microcomputer-repair lab. Be sure to show the following:
   a. All exits   b. Emergency telephone   c. Circuit breakers
   d. First-aid kit   e. Your workstation

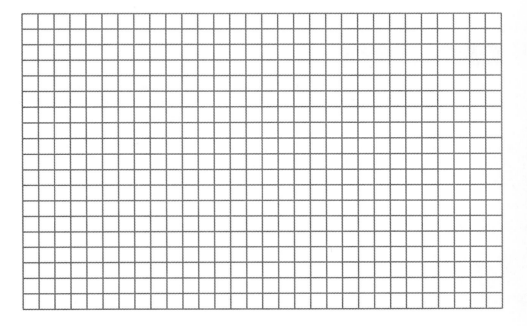

2. List the names of your lab partners and the phone numbers at which they can be contacted.

Name: _____   Phone: _____

Name: _____   Phone: _____

Name: _____   Phone: _____

Name: _____   Phone: _____

3. List any equipment that is permanently assigned to your workstation.

Name of Equipment   **Serial Number**

_____   _____

_____   _____

_____   _____

_____   _____

4. If you have a storage cabinet assigned to you, give its number and combination.

Number: _____

Combination: _____

5. Ask your instructor if there are any other activities to be done before the completion of this lab exercise.

Answer each of the following questions.

1. What is the policy for checking out parts from the lab stockroom?
2. What is the policy for checking out test equipment from the lab stockroom for use in the lab?
3. What is the policy for checking out parts or equipment for your own personal use outside the lab?
4. Are lab reports required? If so, when are they to be turned in? What is the required format for lab reports?
5. What is the policy for working on your own personal equipment in the lab?
6. What is the policy for making up a lab session that you have missed?

Under the supervision of your instructor

1. Explain the rules for checking out hand tools and equipment.
2. State the policy for checking out any parts or equipment for your own personal use.
3. State the policy for working on your own personal equipment in the lab.
4. State the times the lab is available for your use.
5. State the policy for reporting damaged or improperly operating tools and equipment.

3

Electrical and Mechanical Safety

At the end of his first week at RWA Software, Joe Tekk felt very comfortable in the computer lab. His first real troubleshooting experience had gone well. A customer's completely dead PC turned out to have a defective component in the power supply. Joe found the bad part and replaced it. The computer was still partially disassembled on a workbench, but it was plugged in and operational. The other technicians took Joe out to lunch to celebrate.

When Joe came back from lunch, he casually tossed his car keys onto the shelf above the workbench. The keys fell off and landed inside the exposed power supply unit of the customer's PC, causing a direct short between the AC power line and ground. The short drew so much current it blew the etched power line trace off the circuit board and tripped the circuit breaker on the workbench.

It took another week to repair the power supply again.

PERFORMANCE OBJECTIVES

Upon completion of this exercise, you will be able to

1. State verbally the safety rules of the microcomputer lab.
2. Discuss the methods for handling unexpected situations during lab sessions.

KEY TERMS

Reporting injuries
Safety glasses

Correct lifting techniques
Static discharge

BACKGROUND INFORMATION

Safety in the microcomputer laboratory requires observance of the commonsense safety rules that you should follow in any situation when working with or on electrical and mechanical equipment. The following safety rules should be observed at all times:

1. *Do not allow "horseplay" in the lab.* Many lab injuries are caused by students playing jokes or "booby-trapping" equipment. This practice can cause serious permanent injuries to yourself and others. This kind of behavior should not be tolerated in any laboratory situation.

2. *Always get instructor approval.* Your instructor is there to help. Always ask for instructor approval before starting any new task. Doing this can save valuable lab time and also help prevent injuries to yourself and/or damage to equipment.
3. *Report any injuries immediately.* Always report any injury to your lab instructor. You should do this no matter how small the injury. What may appear to be a small cut, mild shock, or minor bruise could lead to serious complications if not properly treated.
4. *Use **safety glasses**.* When any mechanical or electrical equipment is used, there is always the chance of sparks or particles being ejected. This can happen in an electrical circuit when a part such as an electrolytic capacitor has been installed incorrectly. Remember that it takes only a very small particle to cause permanent eye damage.
5. *Use tools correctly.* The improper use of tools can result in injuries to yourself or others as well as permanent damage to the tools. Never attempt to use a tool that is damaged. Never use a tool that you do not know how to use. Never use a tool for a purpose other than that for which it was designed.
6. *Use equipment correctly.* What applies to tools applies equally well to electrical or mechanical equipment. If you are not sure about the operation of any piece of equipment, ask your instructor before attempting to use it.
7. *Do not distract others.* Do not talk to or otherwise distract someone who is in the process of using electrical or mechanical equipment. Doing so could lead to personal injury and/or damage to the equipment.
8. *Use correct lifting techniques.* Always use the proper method for lifting or pushing heavy objects. Ask for help in lifting very heavy objects. If you cannot lift or move an object for any reason, let your instructor know.
9. *Remove jewelry.* Remove all rings, watches, chains, and other jewelry, which are all capable of conducting electricity and causing a shock.
10. *Avoid **static discharge**.* Follow the appropriate measures to avoid static damage of sensitive equipment and devices.

TROUBLESHOOTING TECHNIQUES

In addition to the physical kind of professionalism required of you in lab, you must also prepare yourself mentally for the lab environment. You are in laboratory to observe, to listen, to use all of your senses as an active participant. If you are required to write a report about the lab exercise, everything that occurs in lab from the moment you walk in the door is important. Spend a few minutes trying to be extra observant of what you see around you. You may find that there is a lot going on that you simply filter out without even thinking about it. Practice being a good observer. Your lab work will show the results.

SUMMARY

In this exercise we discovered that

- The laboratory is a serious place.
- The appropriate safety procedures should be followed at all times.
- Tools and equipment should be used properly.

SELF-TEST

The following self-test is designed to test your knowledge of the safety requirements of your microcomputer lab. Answers to odd-numbered questions are given at the end of the book.

Multiple Choice

Select the best answer.

1. When in doubt about operating a piece of equipment, you should
   a. Ask your lab partner for a demonstration.
   b. Try it yourself first so you don't appear stupid.
   c. Ask your instructor.
   d. Simply proceed. You are expected to know how to work all the equipment before starting any lab assignment.

2. If you accidentally cut your finger while using a small hand tool, you should
   a. Report the accident immediately to your instructor.
   b. Wait until after class and then report the accident to your instructor.
   c. Quietly leave the lab in order not to disturb anyone, and seek first aid.
   d. Ignore the incident, and continue with the lab experiment.
3. If you find that the tool you are using in your lab experiment is damaged, you should
   a. Try to repair it to save the school money.
   b. Not use it and let your instructor know that the tool is damaged.
   c. Use it so that you don't waste time in the lab.
   d. Put the damaged tool aside and borrow a similar tool from the lab group next to you.
4. When your lab partner is performing a complex measurement on a piece of equipment, it's best for you to
   a. Start a conversation to ease his or her nerves.
   b. Hum or whistle a soft tune.
   c. Shout at others to keep quiet so that he or she is not disturbed.
   d. Not talk or otherwise distract your partner.
5. If you are asked to lift something and you feel that it is too heavy for you to lift, you should
   a. Put your back into it and do the best you can.
   b. Let your instructor know and ask for assistance.
   c. Try a "test lift" first.
   d. Wait for someone else to do it.
6. If you find that you need a screwdriver but your lab group has not checked one out, you should
   a. Check out a screwdriver.
   b. Use a pocket knife to save lab time.
   c. Omit that part of the experiment and come back to it in the next lab session, when someone will remember to check out a screwdriver.
   d. Borrow a screwdriver from another lab group.
7. If a piece of lab equipment appears to be operating incorrectly, you should
   a. Get the service manual for that piece of equipment and attempt to repair it.
   b. Use it anyway to become accustomed to the kind of equipment you may find in the field.
   c. Wait until the lab group next to you is finished with their assignment and then use their piece of equipment.
   d. Immediately report the problem to your instructor.
8. Unknown to you, your lab partner, as a practical joke, rigs a piece of equipment in the lab in such a manner that it sparks when you attempt to use it.
   a. This kind of behavior shows the great creativity of your lab partner.
   b. Such behavior is very dangerous and is not allowed in any lab situation.
   c. This kind of behavior may cause your lab partner to be expelled from the lab.
   d. Both b and c are correct.

Open-Ended

Answer the following questions:

9. State the 10 safety rules presented in this exercise.
10. When should safety glasses be worn in the lab?

FAMILIARIZATION ACTIVITY

Your lab instructor may give a safety demonstration or lecture. After completing the self-test, check your answers against those at the end of the book. Your instructor may have an open class discussion concerning the answers to this self-test.

Answer the following questions:

1. Why is safety needed in the computer lab?
2. Who is responsible for safety in the computer lab?
3. What should you do when you are not sure of an assignment?
4. Name the potential safety hazards of the microcomputer lab.
5. List the 10 safety rules presented in this exercise.
6. What are the special safety precautions that should be observed in your lab situation?

Under the supervision of your instructor

1. State the potential safety hazards in the microcomputer lab.
2. Outline the 10 safety rules presented in this exercise.
3. Explain any special safety precautions that should be observed in your lab situation.

4 Hand Tool Identification and Use

INTRODUCTION

Joe Tekk was deeply involved in a repair job on an old high-speed line printer. Lying inside the unit, he was using his screwdriver to pry a metal support from the chassis. The metal support did not budge, no matter how hard Joe tugged on his screwdriver.

Giving up, Joe climbed out of the printer cabinet and set his screwdriver down on a bench. He went looking for a different tool. He did not notice he had bent the screwdriver shaft, ruining it.

PERFORMANCE OBJECTIVES

Upon completion of this exercise, you will be able to

1. Identify the hand tools used in the computer-repair lab.
2. Explain the correct use of hand tools found in the computer-repair lab.
3. Strip and cut electrical wire to specified lengths.

KEY TERMS

Hand tools

Diagonal cutters

Wire stripper cutters

Long-nose pliers

Flat-bladed screwdriver

Phillips screwdriver

IC extractor

Antistatic wrist strap

Tweezers

Nut driver

Pliers

BACKGROUND INFORMATION

GENERAL INFORMATION

Hand tools are instruments used with the hands to extend the hands' working capabilities. In the microcomputer lab, hand tools are used to aid in the disassembly and reassembly of microcomputer equipment and parts. They are not intended to serve as any kind of electrical testing devices; attempting to use them in this manner is very dangerous.

Figure 4.1 illustrates the hand tools that will be used in this exercise.

FIGURE 4.1 Hand tools for microcomputer repair

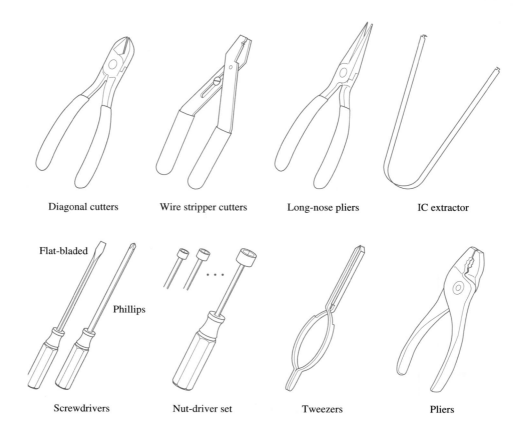

Diagonal cutters   Wire stripper cutters   Long-nose pliers   IC extractor

Flat-bladed   Phillips

Screwdrivers   Nut-driver set   Tweezers   Pliers

FIGURE 4.2 Insulated hand tools

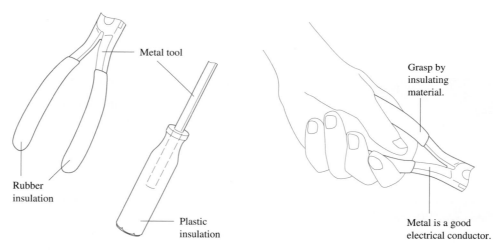

Metal tool

Rubber insulation

Plastic insulation

Grasp by insulating material.

Metal is a good electrical conductor.

All the hand tools shown in Figure 4.1 are made of metal. Because metal is a good conductor of electricity, such tools present a potential shock hazard when used in working with electrical equipment.

Properly made hand tools for working on electrical equipment have metal handles that are insulated with a rubber coating or cast in an insulating plastic. This is illustrated in Figure 4.2.

Note from Figure 4.2 that when using hand tools on electrical equipment, you do not touch the metal part of the hand tool with any part of your body. By grasping the tool only on the insulating material, you reduce your chance of electrical shock. Recall that an insulator is not a good conductor of electricity. Hand tools on which the insulation is frayed or has been removed should not be used; it is best to discard such tools for a new set.

FIGURE 4.3 Diagonal cutters

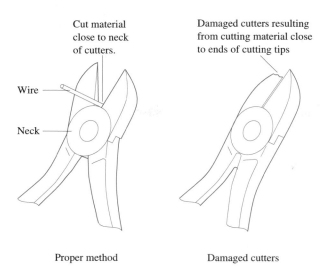

Proper method   Damaged cutters

FIGURE 4.4 Wire stripper cutters

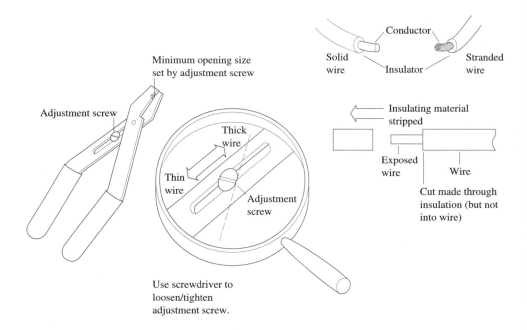

DIAGONAL CUTTERS

Figure 4.3 illustrates a typical set of ***diagonal cutters*** and the correct way to use them. As illustrated, diagonal cutters can be damaged by cutting thick material close to the ends of the cutting tips.

WIRE STRIPPER CUTTERS

Figure 4.4 illustrates a typical set of ***wire stripper cutters.*** As shown in the figure, these instruments are used to strip the insulation from wire to prepare the wire for use in an electrical connection.

Observe from Figure 4.4 that the minimum size of the wire stripper opening must be small enough to cut through the insulation of the wire completely, but not so small as to cut or nick the wire itself.

FIGURE 4.5 Long-nose pliers

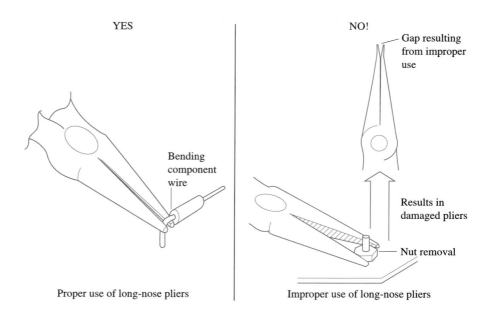

YES

NO!

Bending
component
wire

Gap resulting
from improper
use

Results in
damaged pliers

Nut removal

Proper use of long-nose pliers

Improper use of long-nose pliers

**FIGURE 4.6 Flat-bladed and
Phillips screwdrivers**

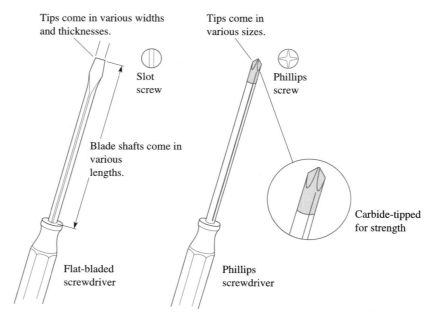

Tips come in various widths
and thicknesses.

Slot
screw

Blade shafts come in
various
lengths.

Flat-bladed
screwdriver

Tips come in
various sizes.

Phillips
screw

Carbide-tipped
for strength

Phillips
screwdriver

LONG-NOSE PLIERS

Figure 4.5 shows a typical set of *long-nose pliers*. As shown in the figure, these pliers are not intended for use in removing hardware such as nuts and bolts. Such misuse can result in permanent damage to these tools.

Long-nose pliers come in various sizes. Keep in mind that these are delicate instruments intended for delicate work, not for removing nuts and bolts.

FLAT-BLADED AND PHILLIPS SCREWDRIVERS

A typical set of *flat-bladed* and *Phillips screwdrivers* is shown in Figure 4.6. These screwdrivers come in various sizes and may use a carbide tip for strength. They are intended to be used for the insertion and removal of screws. Remember to use a screwdriver of the right size for the job at hand.

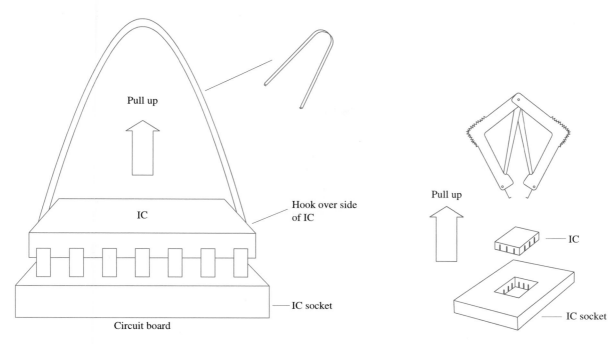

(a) DIP (Dual Inline Package) extractor

(b) PLCC (Plastic-Leaded Chip Carrier) extractor

FIGURE 4.7   IC extractors

IC EXTRACTORS

Figure 4.7 shows typical *IC* (integrated circuit) *extractors*. As shown in the figure, these instruments are intended for the removal of IC packages. It's important that all power be disconnected from the system before an IC is removed.

ANTISTATIC WRIST STRAP

Static electricity can damage many types of integrated circuits, including CPU and memory devices. To avoid static damage during handling, it is common to wear an *antistatic wrist strap*. A typical strap is shown in Figure 4.8(a). The strap wraps around your wrist and connects to a ground terminal via an attached cable, as indicated in Figure 4.8(b). The cable provides a path to ground for any static electricity encountered. A resistor built into the cable is typically used to reduce the current flow during a discharge.

Antistatic mats or pads may also be used. These are placed on top of the workbench and provide a safe surface for equipment and electronics.

TWEEZERS

A typical pair of *tweezers* used in microcomputer repair is shown in Figure 4.9. Tweezers may also be used as a soldering aid, as you will see in a later exercise. Be careful of static problems introduced by tweezers.

NUT DRIVERS

Figure 4.10 illustrates a typical *nut driver*. As shown in the figure, nut drivers are not adjustable and therefore come in various sizes to accommodate different-sized nuts. Note that

FIGURE 4.8 **(a) Antistatic wrist strap and (b) Using the wrist strap**

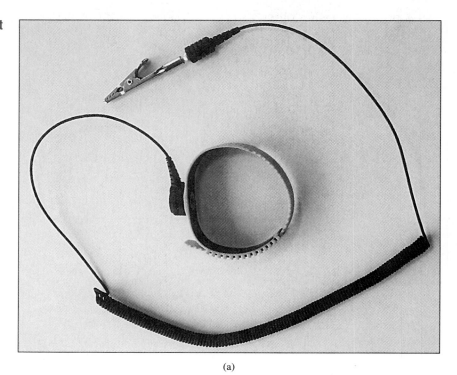

(a)

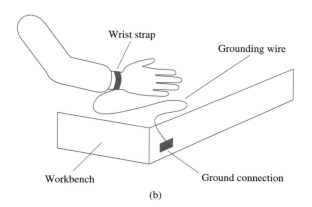

Wrist strap

Grounding wire

Workbench

Ground connection

(b)

FIGURE 4.9 **Tweezers**

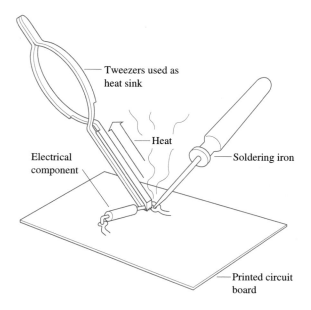

Tweezers used as heat sink

Heat

Electrical component

Soldering iron

Printed circuit board

FIGURE 4.10   Nut driver

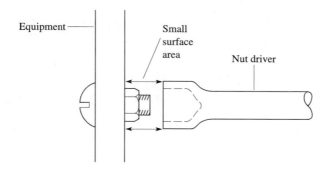

FIGURE 4.11   Common pliers

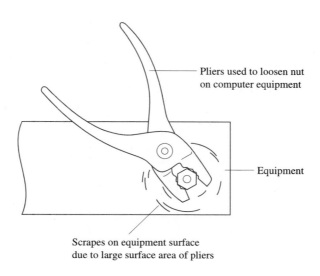

Pliers used to loosen nut on computer equipment

Equipment

Scrapes on equipment surface due to large surface area of pliers

these instruments present a small surface area, thus causing minimum marring of computer surfaces. Nut drivers are always the preferred instruments for loosening or tightening of nuts.

PLIERS

Common *pliers* are shown in Figure 4.11. Because pliers of this type can easily cause damage to computer surfaces, they should be the last instrument of choice for use on computer equipment.

TROUBLESHOOTING TECHNIQUES

A wise man once said, "Let the tool do the work for you." That is such good advice. Using the wrong tool can result in a damaged tool or piece of equipment and increase the risk of physical injury. It is well worth the investment to purchase a complete set of tools so that you will always have the right tool for the job. Figure 4.12 shows a tool set designed for electronic repairs.

SUMMARY

In this exercise we discovered that

- A large variety of tools is required for microcomputer repair. Special tool kits are available that contain the necessary tools.
- Hand tools must be used properly to ensure safety.
- Improper use of a hand tool can lead to injury or damage to the tool or part.
- Some tools come with carbide tips for strength.
- Antistatic precautions should be followed when working with sensitive parts.

FIGURE 4.12   Assortment of tools for electronic repair

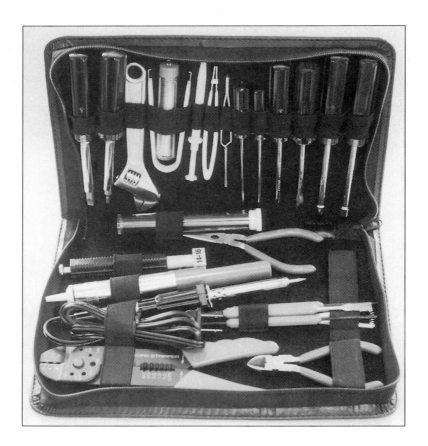

The following self-test is designed to test your knowledge of facts about hand tools used in the microcomputer-repair lab. Answers to the odd-numbered questions are presented at the end of the book.

True/False

Answer *true* or *false*.

1. Hand tools are instruments used to extend the capabilities of the hands.
2. Hand tools may be used to make electrical tests.
3. The metal parts of hand tools are good insulators.
4. All the hand tools presented in this exercise are made of metal.
5. A hand tool designed for working around electrical equipment will have insulating material around its handles.
6. Worn-out tools should be used as long as possible, since new tools are expensive.
7. An antistatic wrist strap or mat should be used when working with integrated circuits.

Multiple Choice

Select the best answer.

8. When using hand tools around electrical equipment, you should
   a. Touch only the metal parts of the tools.
   b. Avoid touching the metal parts of the tools.
   c. Not use hand tools around electrical equipment.
   d. None of the above.
9. Insulating material on hand tools is
   a. A good conductor of electricity.
   b. Used to prevent electrical shock.
   c. There to prevent the hand tool from rusting.
   d. Necessary in order to ground the tool properly.

10. Hand tools on which the insulation is frayed or missing
    a. Should be discarded and replaced by new hand tools.
    b. Are dangerous to use.
    c. Should not be used to repair computer equipment.
    d. All of the above.
11. Diagonal cutters are best used for
    a. Cutting wire and similar materials.
    b. Stripping insulation from wire.
    c. Removing nuts.
    d. Shortening small bolts.
12. Wire strippers should be adjusted so their opening
    a. Is small enough to cut through wire.
    b. Will leave a mark on the wire where insulation has been removed.
    c. Is large enough to leave some insulation on the wire.
    d. None of the above.
13. Using a carbide tip on a screwdriver increases its
    a. Length.
    b. Strength.
    c. Width.
    d. Resistance to static discharge.
14. Using the wrong tool can result in
    a. Damaged equipment.
    b. A damaged tool.
    c. Physical injury.
    d. All of the above.

Matching

Match each of the following uses with the proper tool in Figure 4.13.

15. Removing insulation from wire
16. Cutting electrical wire
17. Removing a nut
18. Removing an IC chip
19. Bending a wire

FIGURE 4.13 Tools for self-test questions 15–19

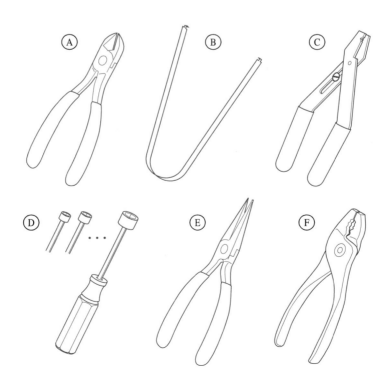

Completion

Fill in the blank or blanks with the best answers.

20. There are two basic kinds of screwdrivers; one is called a flat-bladed screwdriver, and the other is called a _____ screwdriver.
21. A pair of _____ may be used as a soldering aid.
22. A _____ _____ is the preferred instrument for removing nuts.
23. A pair of _____ should be the tool of last resort for removing a nut.
24. Never use a hand tool on electrical equipment when the hand tool does not have any _____.
25. Insulation is typically made of rubber or _____.

FAMILIARIZATION ACTIVITY

Using your wire stripper cutters, cut wires in the following lengths and strip $1/4$" of insulation from each end of each wire:

Three pieces 4" long
Two pieces 1" long
Five pieces 3" long

Exchange the wires you cut with the set cut by your lab partner. Check your lab partner's wires for the following:

1. Has each piece been cut to the correct length?
2. Has the insulation been stripped from both ends of each wire?
3. Has exactly $1/4$" of insulation been removed?
4. Are there any nicks in the wire resulting from incorrect adjustment of the wire strippers?

QUESTIONS/ACTIVITIES

Answer the following questions:

1. State the purpose of each hand tool you will be using in the lab.
2. Explain the proper use of
   a. Long-nose pliers
   b. Diagonal cutters
   c. Wire stripper cutters
3. State the electrical safety rules you should observe when using hand tools in the computer-repair lab.

REVIEW QUIZ

Under the supervision of your instructor

1. Identify the hand tools used in the computer-repair lab.
2. Explain the correct use of hand tools found in the computer-repair lab.
3. Strip and cut electrical wires to specified lengths.

UNIT ▐▐ Basic Skills

5

Microcomputer Familiarization

Don, the senior technician at RWA Software, asked Joe if he would come with him to pick up some new computer equipment. They drove to a local computer store and loaded the company van with several large boxes.

When they returned, Don asked Joe to unpack everything and set up the new computer. Thirty minutes later, Joe turned power on and the new computer booted up. He showed it to Don with admiration. "It's a real nice system, Don. It has everything: 1 GHz Pentium III, 128 megabytes of RAM, 3-D hardware acceleration, a 40-gigabyte hard drive, and lots of other goodies."

Don smiled at Joe. "I'm glad you like it. It's your new computer."

PERFORMANCE OBJECTIVES

Upon completion of this exercise, you will be able to

1. Discuss the history of the personal computer.
2. Identify the major parts of a microcomputer system.
3. Explain the purpose of each of these major parts.
4. Browse the World Wide Web.

KEY TERMS

Graphical user interface
Microcomputer
Peripheral device
Input device

Output device
I/O (input/output) device
Windows desktop
WYSIWYG

BACKGROUND INFORMATION

The personal computer has come a long way since its introduction in 1981. Back then, the 5 MHz 16-bit Intel 8086 microprocessor seemed fast, and 640K was plenty of memory. The first PC did not even have a hard drive. At power-on it loaded BASIC (a simple programming language) from on-board ROM or booted DOS from a 5.25-inch floppy drive.

Today Pentium III microprocessor clock speeds have hit 1000 MHz, making them two hundred times faster than the 8086. In addition, the Pentium III is a 64-bit microprocessor, with internal cache (high-speed memory), floating-point unit (high-speed math), and many other special hardware features designed to enhance its performance.

Hard drives for the PC have also evolved, from the initial 20-MB drives to the 10- (or more) GB drives in use today. The price of one bit of hard drive storage has gone from 4.29 cents per bit to less than 0.000001 cents per bit over the past 20 years. System memory has increased from 640K to 32 MB or 64 MB (minimum RAM for reasonable performance).

Along with the hardware evolution of the PC, there have been changes in the software used to control the PC (the operating system). The first operating system, DOS, had a text-based, command-line-oriented interface that required the user to remember and enter commands such as

```
DIR *.TXT /S /P
```

and

```
EDIT A:NAMES.LST
```

When Intel came out with more advanced microprocessors (the 80386, 80486, and finally the Pentium), DOS was unable to take advantage of the new features built in to each processor (through a special mode of operation called *protected mode*). This paved the way for a new operating system called Windows. The Windows operating system, unlike DOS, runs in protected mode and utilizes a **graphical user interface** (along with a mouse) to provide an easy-to-use environment.

Let's take a look at a typical PC.

A TYPICAL MICROCOMPUTER SYSTEM

Figure 5.1 illustrates the major parts of a **microcomputer** system. Table 5.1 lists the major parts shown in Figure 5.1, as well as the purpose of each part.

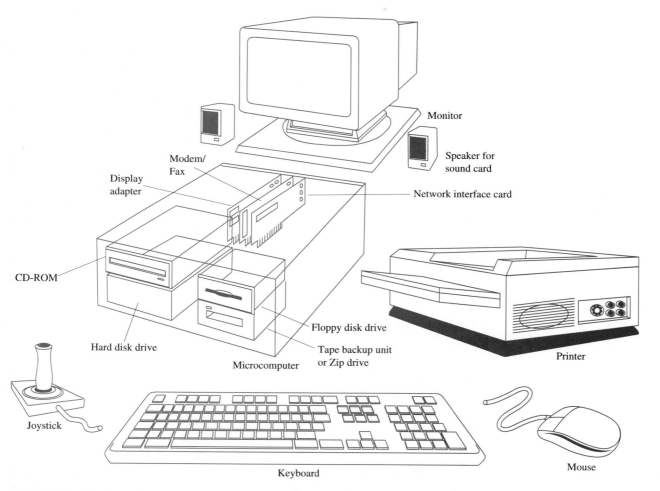

FIGURE 5.1   A microcomputer system

TABLE 5.1 Major parts of a microcomputer system

Part	Purpose
Microcomputer	Central component of the system. Performs all the calculations and logic functions. Also called the CPU (central processing unit).
Keyboard	Consists of miniature switches with alphanumeric and other labels. Allows the program user to enter information directly into the computer.
Monitor	Contains a viewing screen. Gives the program user temporary information useful in the operation of the microcomputer. Requires a display adapter card, such as a graphics accelerator.
Hard disk drive	Serves as a storage place for information. Consists of one or more rigid magnetic platters used to store programs and other items useful to the user. These platters cannot be changed by the user; however, the entire drive may be replaced.
Tape backup unit or Zip drive	Used to back up files to/from the hard drive. Tapes and cartridges are removable.
Floppy disk drive	Will copy information from or place information on small disks consisting of magnetic material. These disks are an easy and quick way of getting information into the microcomputer and can be changed by the user.
CD-ROM	Provides very large storage capability. Reads compact disks provided by the user. Rewritable CD-ROMs are also available.
Printer	Consists of a printing head and paper mechanism for the purpose of making permanent copies of useful information contained in the microcomputer.
Mouse	A small device moved by hand across a smooth surface. Used with information on the screen to control the microcomputer quickly and easily.
Joystick	Used for quick interaction with the monitor. Usually used for interacting with computer games.
Speakers	Provide left and right audio output from a sound board.
Telephone modem	A device for transferring information between computers by use of telephone lines.
Network interface card	Used to connect the computer to a network.

Each of these major parts is called a ***peripheral device*** because the device (such as the printer) is separate from the microcomputer.

Figure 5.2 shows the relationship of each peripheral device to the microcomputer. As shown in Figure 5.2, some of the peripheral devices serve only as input devices. An ***input device*** is one that can only input information to the microcomputer.

Other devices serve as output devices. An ***output device*** is one that can only get information from the microcomputer.

The third type of peripheral device is the kind that serves as both an input and an output device. These devices are capable of putting information into the microcomputer as well as getting information from the microcomputer. Peripheral devices that are capable of both inputting information and getting information from the microcomputer are called ***I/O (input/output) devices***.

FIGURE 5.2 Relationships of peripheral devices to the microcomputer

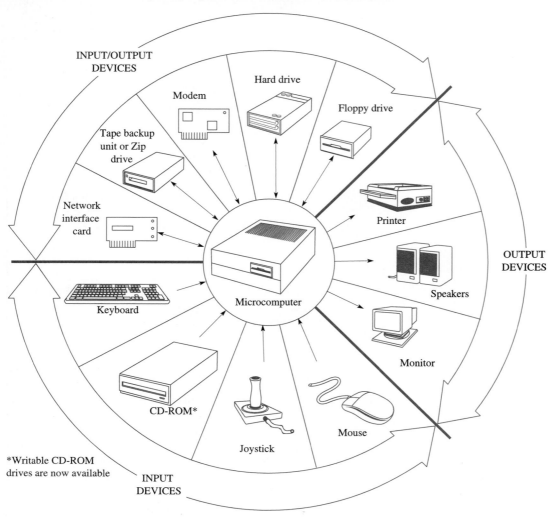

THE WINDOWS DESKTOP

As mentioned earlier, the operating system for the PC has evolved into a more powerful, graphical, point-and-click operating system that is much easier to use than DOS.

Figure 5.3 shows a screen shot of the *Windows desktop*. This is what you see after the machine boots up. Although the desktop is the subject of Exercise 12, a few points deserve mention now. Note that the time is displayed in the lower right corner. This is convenient and eliminates having to constantly enter the TIME command in DOS.

The Start button at the lower left is a very important starting point for many Windows operations. Left-clicking the Start button once will bring up its menu of items.

Along the left side of the screen are icons representing applications or folders (directories in DOS terminology) that can be started or opened using a left double-click on the mouse when the mouse pointer is over the icon. For example to start the MathPro application, place the mouse pointer (typically an arrow) over the icon that reads "mathpro" and left double-click.

If an application does not have an icon on the desktop you can still access it by using the Start button and selecting Programs from the menu. A list of all installed programs will appear, allowing you to select the MathPro application. This is typical of Windows, where one task can usually be performed several different ways.

Next, note the large square window covering most of the desktop. This is the Internet Explorer World Wide Web browser window. The Web site for the Yahoo search engine is

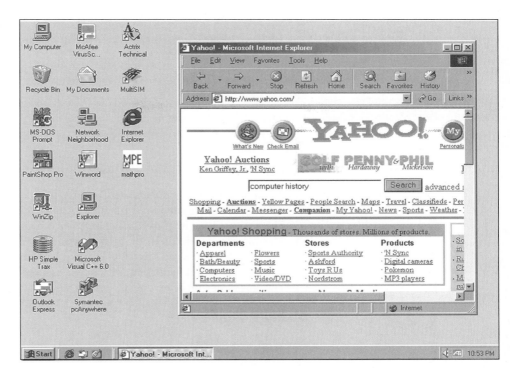

FIGURE 5.3 Windows desktop with open application (Reproduced with permission of YAHOO! Inc.© 2000 by YAHOO! Inc. YAHOO! and the YAHOO! logo are trademarks of YAHOO! Inc.)

currently being displayed. This was accomplished by entering "www.yahoo.com" in the Address field near the top of the browser window. The information contained in the Yahoo Web site includes images, text with different fonts and sizes, and interactive fields and buttons in a *WYSIWYG* format. WYSIWYG stands for what-you-see-is-what-you-get. This is a big improvement over the plain text interface used by DOS.

Last, notice that the user has entered the words "computer history" in the Search field of the Yahoo page. When the Search button is left-clicked, Yahoo will return a page of links to other Web pages that contain information about computer history. More information about the World Wide Web and the Internet can be found in Exercise 27.

The graphical interface of Windows, the existence of the World Wide Web, and the impressive power of today's computers have fundamentally changed the way we use our computers. Truly, we are in the middle of a PC technology revolution.

TROUBLESHOOTING TECHNIQUES

A brand-new computer fresh out of the box should assemble and boot up with a minimum of fuss. Many of the external connectors (keyboard, printer, mouse) only allow one way to plug in the appropriate cable, so there is no need to worry about accidentally plugging the printer cable into the mouse port.

If all external connections are correct and the machine will not boot, it could be the result of vibration damage that could have occurred during shipment. For example, an already-loose peripheral card may have simply popped out of its slot during transit. It is good to take a look inside the chassis (even before powering on for the first time, if necessary). A good visual inspection might turn up the problem.

SUMMARY

In this exercise we discovered that

- Computers have constantly evolved into more powerful machines.
- A microcomputer contains many different peripherals, such as a modem, NIC, and video display card.
- Peripherals come in three varieties: input (keyboard, mouse), output (speaker, display), or input/output (modem, NIC, disk drive).

- An operating system (such as Windows) is used to control the microcomputer system.
- A browser, such as Internet Explorer, displays Web pages in WYSIWYG format.

This self-test is designed to help you check your understanding of the background information presented in this exercise.

True/False

Answer *true* or *false*.

1. A complete microcomputer system consists of more than just the microcomputer itself.
2. All the calculations and logic functions are performed by the microcomputer.
3. The keyboard can be viewed as an example of an output device.
4. The monitor screen can be thought of as an input device.
5. The hard disk drive is a storage place for information.
6. The clock speed of the Pentium III is two hundred times faster than the 8086.
7. The cost of one bit of hard drive storage has constantly risen.

Multiple Choice

Select the best answer.

8. An output device
    a. Copies information from the microcomputer.
    b. Puts information into the microcomputer.
    c. Can copy information from or put information into the microcomputer.
    d. Is the keyboard.
9. An input device
    a. Copies information from the microcomputer.
    b. Puts information into the microcomputer.
    c. Can copy information from or put information into the microcomputer.
    d. Is the monitor.
10. An I/O device
    a. Copies information from the microcomputer.
    b. Puts information into the microcomputer.
    c. Can copy information from or put information into the microcomputer.
    d. Is the printer.
11. A joystick is classified as an
    a. Output device.
    b. Input device.
    c. I/O device.
    d. None of the above.
12. The monitor is used
    a. For storing permanent information.
    b. When the printer fails.
    c. For putting information into the computer.
    d. As a temporary storage for immediately useful information.
13. The Windows operating system uses
    a. A graphical user interface.
    b. Protected mode.
    c. A text-based user interface.
    d. Both a and b.
14. A browser is used to display
    a. The Windows desktop.
    b. The contents of a file.
    c. The contents of a Web page.
    d. All of the above.

Matching

Match each phrase on the left with the correct peripheral device or devices on the right.

15. Copies information from or places information into the microcomputer
16. Used with the monitor for quick control of the microcomputer
17. Can transfer information between computer systems
18. Makes a permanent copy of information in the microcomputer
19. Makes a temporary copy of information for immediate use

a. Monitor
b. Keyboard
c. Printer
d. Mouse
e. Joystick
f. Modem
g. Microcomputer
h. Hard drive
i. Floppy drive
j. CD-ROM
k. Zip drive

Completion

Fill in the blanks with the best answers.

20. Any device that copies information into the microcomputer is called a(n) _____ device.
21. Any device that copies information from the microcomputer is called a(n) _____ device.
22. I/O devices are capable of copying information _____ the microcomputer as well as putting information into it.
23. A(n) _____ is a small device moved by the hand across a smooth surface.
24. The disk drive that can have its disks changed by the user is called the _____ disk drive.
25. Yahoo is called a search _____.

FAMILIARIZATION
ACTIVITY

Your instructor may give a laboratory demonstration showing a complete microcomputer system, or your lab station may be equipped with a complete microcomputer system. In either case, you should know how to identify the following components of your system:

1. Microcomputer
2. Floppy, hard, and CD-ROM drives
3. Keyboard
4. Monitor
5. Printer
6. Modem
7. Network interface card
8. Mouse
9. Joystick
10. Tape backup unit or Zip drive

As an aid in familiarizing yourself with each of these peripheral devices, answer the following questions as they apply to the system used in the demonstration or at your lab station:

1. Who is the manufacturer of the microcomputer?
2. How many disk drives does the system contain?
3. Describe the monitor used in this system.
4. Who is the manufacturer of the printer?
5. How many keys are contained on the keyboard?
6. What is the capacity of the tape backup?
7. Sketch and examine the various connectors found on your computer. How many pins does each connector have? Is the connector male or female?

Using a current computer magazine, list two manufacturers of each of the following:

1. Microcomputer
2. Modem
3. Printer
4. Monitor
5. Mouse/joystick
6. Disk drive
7. CD-ROM
8. Tape backup unit

Using a current computer catalog, list the prices of each of the following:

1. Telephone modem
2. Printer
3. Monitor
4. Disk drive
5. Mouse/joystick
6. CD-ROM
7. Network interface card

Start up Internet Explorer, go to Yahoo, and search for "computer history." Look at five of the sites that show up on the search results.

Under the supervision of your instructor

1. Discuss the history of the personal computer.
2. Identify the major parts of a microcomputer system.
3. Explain the purpose of each of these major parts.
4. Browse the World Wide Web.

6

Electrical Component Identification

Joe looked over Don's shoulder. "What are you doing, Don?"

Don was examining a motherboard. "I'm giving this a visual." "A visual?" Joe wondered. "What are you looking for?"

Don shrugged. "I don't know. I'm just looking at everything, at all the components. Aside from some dust, I don't see . . . wait! There it is." Don pointed with his finger at a small resistor mounted near an expansion socket. "What do you see, Joe?"

Joe looked at the resistor. A small crack was visible near one end. "It looks broken. How would that happen?"

"Who knows?" Don replied. "Maybe they yanked real hard on the modem card that was plugged into that socket. Maybe they dropped the computer and didn't want to tell me."

Joe looked at the resistor again. "One resistor can stop the whole computer from working?"

Don laughed. "Sure, if it's the right one."

PERFORMANCE OBJECTIVE

Upon completion of this exercise, you will be able to

Identify, give the function of, and state any special considerations in regard to each of the following types of parts:

Resistors (fixed and variable)
Capacitors
Inductors
Diodes (including the LED)
DIP switches
Integrated circuits
IC sockets
Miniature relays

KEY TERMS

Resistor
Capacitor
Inductor
Diode
Rectification

Switch
Integrated circuit
IC socket
Relay

RESISTORS

Figure 6.1 shows several kinds of fixed *resistors*. As shown in the figure, resistors come in different sizes. The physical size of the resistor determines its wattage rating, which is a measure of how much electrical power it is capable of handling. The larger the wattage rating in watts, the more electrical power the resistor can dissipate. You should never replace a resistor with one that has a *lower* wattage rating.

The purpose of a resistor is to resist, or limit, the flow of current in a circuit. The value of a resistor is measured in ohms. The more ohms a resistor has, the more it will limit the flow of electrical current. The symbol for ohms is the Greek letter omega (Ω). Thus, 24 ohms is written as 24 Ω.

The value of a resistor in ohms is indicated by the colors of the bands on the resistor. Table 6.1 shows the meanings of the color-coded resistor bands. Figure 6.2 shows a diagram of the color-coded bands.

Figure 6.3 shows different variable resistors. As the name implies, a variable resistor (sometimes called a potentiometer) is capable of having its resistance value changed.

Resistor networks are shown in Figure 6.4. A resistor network consists of several resistors connected in a specified way. These networks are used to simplify the construction of digital circuits, so that only one network needs to be inserted into the circuit rather than several separate resistors.

FIGURE 6.1   Various fixed resistors (courtesy Stackpole Carbon Co.)

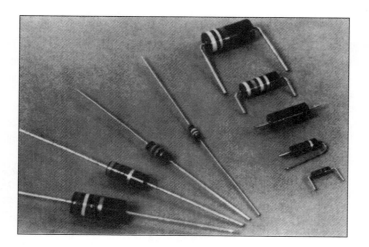

TABLE 6.1   Resistor color code

Digit* (Bands 1 and 2)	Color
0	Black
1	Brown
2	Red
3	Orange
4	Yellow
5	Green
6	Blue
7	Violet
8	Gray
9	White
Tolerance (Band 4)	**Color**
5%	Gold
10%	Silver
20%	No band

*Or multiplier for band 3.

FIGURE 6.2   Color-coded resistor bands

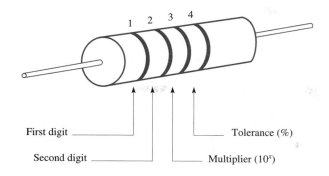

First digit

Second digit

Multiplier (10^x)

Tolerance (%)

FIGURE 6.3   Variable resistors [(a) courtesy Clarostat Mfg. Co., Inc. (b) courtesy Bourns, Inc.]

FIGURE 6.4   Resistor networks (courtesy Bourns, Inc.)

FIGURE 6.5 Resistor symbols and typical circuit connections

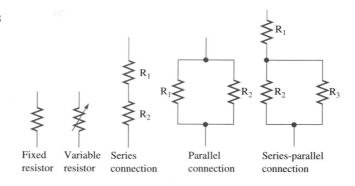

Fixed resistor   Variable resistor   Series connection   Parallel connection   Series-parallel connection

TABLE 6.2 Metric notation for resistors

Metric Prefix	Meaning	Examples
Kilo (K)	×1000	12 KΩ = 12 × 1000 = 12,000 Ω 4.3 KΩ = 4.3 × 1000 = 4300 Ω 280 KΩ = 280 × 1000 = 280,000 Ω
Mega (M)	×1,000,000	18 MΩ = 18 × 1,000,000 = 18,000,000 Ω 8.2 MΩ = 8.2 × 1,000,000 = 8,200,000 Ω 750 MΩ = 750 × 1,000,000 = 750,000,000 Ω

The schematic symbols for resistors are shown in Figure 6.5, which illustrates several different circuit connections of resistors. Note that a variable resistor has an arrow drawn through it to indicate that its value can be changed.

Most resistor values are quite large, usually in thousands or even millions of ohms. A notation called metric notation is used to represent these values. Table 6.2 lists the metric notation used for typical resistors.

CAPACITORS

As shown in Figure 6.6, **_capacitors_** come in a variety of shapes and sizes. The purpose of a capacitor is to store an electrical charge. Because of this capability, one of the uses of capacitors is to help maintain a steady supply of voltage to circuits within the computer.

FIGURE 6.6 Capacitors (courtesy KEMET Electronics, Corp.)

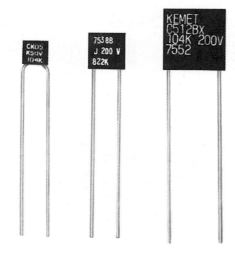

TABLE 6.3   Metric notation for capacitors

Metric Prefix	Meaning	Examples
Micro (μ)	×0.000001	12 μF = 0.000012 F 3.5 μF = 0.0000035 F 0.1 μF = 0.0000001 F
Pico (p)	×0.000000000001	12 pF = 0.000000000012 F 3.5 pF = 0.0000000000035 F 100 pF = 0.000000000100 F

The value of a capacitor is measured by farad (F). A capacitor also has a voltage rating. When replacing a capacitor, you should use a capacitor with the same value in farads and never one with a lower voltage rating.

Capacitor values are usually quite small. Typical values for a capacitor are in the range of millionths of a farad. As with resistors, metric notation is used to represent the value of a capacitor. Table 6.3 lists the metric notation used for typical capacitors.

The schematic symbol for and basic construction of a capacitor are shown in Figure 6.7. The basic construction of a capacitor involves two conductors separated by an insulator. The value of a capacitor depends on the area of the conductors facing each other and their separation. The value of a capacitor is also affected by the insulating material between the two conductors. With capacitors, this insulating material is called a *dielectric*. Some of the most common materials used for dielectrics are paper, mica, and ceramic. Table 6.4 lists some of the most common types of capacitors and their methods of construction. Figure 6.8 shows variable capacitors. Notice the adjustment screw used to change the capacitance.

FIGURE 6.7   Capacitor construction and schematic symbol

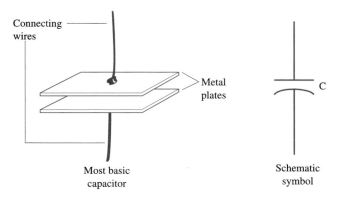

Connecting wires

Metal plates

Most basic capacitor

Schematic symbol

TABLE 6.4   Common types of capacitors

Capacitor Type	Comments
Mica capacitor	Uses mica as a dielectric. Values range from 1 pF to 0.1 μF, with voltage ratings from 100 to 2500 V.
Ceramic capacitor	Uses a ceramic material as the dielectric. Values range from 1 pF to 2.2 μF, with voltage ratings from 50 to 6000 V.
Paper/plastic capacitor	Uses a variety of materials for construction. Values of up to 100 μF can be achieved. Voltage ratings depend on the type of construction and specific material.
Electrolytic capacitor	Constructed in such a manner as to be *polarized*, which means that these capacitors must be put into the circuit in a certain way. They come in values up to 200,000 μF. Typical voltage ratings are around 400 V.
Variable	Constructed in such a manner that their capacitance values may be changed. The two most common types are air capacitors and trimmers and padders.

FIGURE 6.8   Variable capacitors (courtesy KSA Associates, Murata Erie North America, Inc.)

INDUCTORS

Figure 6.9 shows the basic construction of an ***inductor***. As shown in the figure, an inductor consists of a coil of wire. The center of this coil may consist of air or a magnetic material such as iron. Inductors are measured in henries (H). Values of inductors range from microhenries (μH) to several thousand millihenries (mH).

An inductor opposes a rapid change in current. One of the common uses of an inductor is to help maintain a steady source of current to supply electrical energy to computer circuits. Figure 6.10 shows some of the common types of inductors.

DIODES

A ***diode*** is an electrical device that allows current to flow in only one direction. Table 6.5 lists the most common types of diodes. The term *rectify* means to make straight. In electronics, ***rectification*** means the conversion of alternating current (current that repeatedly changes its direction) to direct current. Figures 6.11 through 6.15 show an assortment of typical diodes.

FIGURE 6.9   Basic construction of an inductor

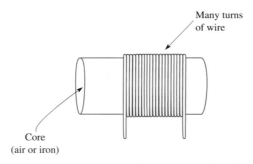

Many turns of wire

Core (air or iron)

FIGURE 6.10   Common types of inductors (courtesy Delevan/American Precision Industries): (a) fixed, (b) variable

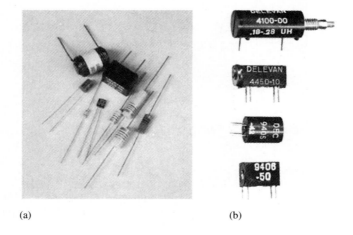

(a)   (b)

40

TABLE 6.5   Common types of diodes

Diode Type	Comments
Small-signal diode	Used to rectify small currents.
Power rectifier	Functions the same way as the small-signal diode. Used to rectify large currents.
Zener diode	Designed to have a specific breakdown voltage. This means that the diode can be used as a voltage regulator that can help maintain a constant voltage.
Light-emitting diode (LED)	Constructed in such a manner as to help you see the light emitted from it (a characteristic of *all* diodes).
Photodiode	Especially constructed so that it is sensitive to light levels (another characteristic of all diodes).

FIGURE 6.11   Small-signal diodes and schematic symbol

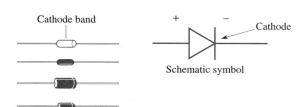

FIGURE 6.12   Power rectifiers

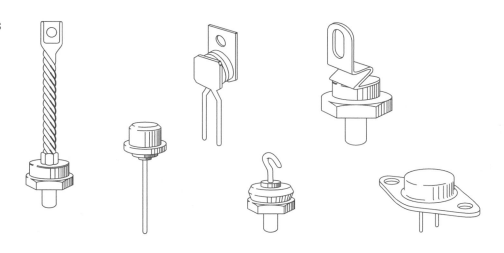

FIGURE 6.13   Zener diode schematic symbol

Schematic symbol

FIGURE 6.14   Light-emitting diode (LED) and schematic symbol

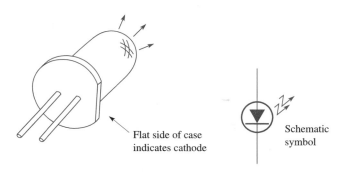

41

FIGURE 6.15 Photodiodes

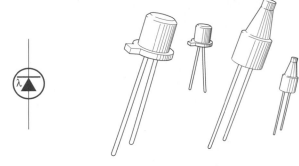

(a) Schematic symbol   (b) Typical devices

SWITCHES

The most common kind of *switch* used in microcomputer systems is the dual inline package (DIP) switch. Figure 6.16 shows typical DIP switches and the schematic diagram of a four-position DIP switch. Other types of switches are shown in Figure 6.17.

FIGURE 6.16   (a) Typical DIP switches and (b) schematic diagram of a four-position DIP switch (courtesy Grayhill, Inc.)

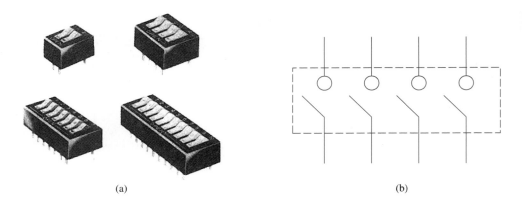

(a)   (b)

FIGURE 6.17   Other types of switches and terminology [(a), (b), and (c) courtesy Eaton Corp.; (d) courtesy Grayhill, Inc.]

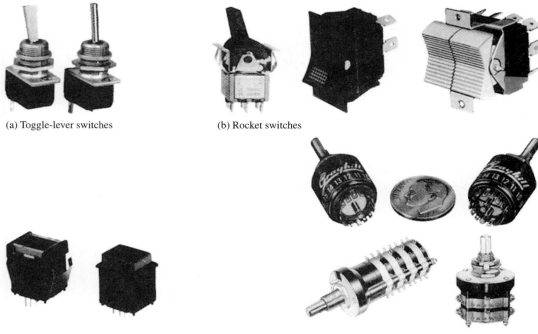

(a) Toggle-lever switches   (b) Rocket switches

(c) Push-button switches   (d) Rotary position switches

INTEGRATED CIRCUITS

Integrated circuits (ICs) come in a variety of sizes and types. Some of the most common types are shown in Figure 6.18. Integrated circuits are the workhorses of digital computers. You will spend considerable time working with these circuits as you progress through this book. What is important for now is that you recognize an IC and know how to identify the pin number for the circuit.

IC SOCKET

Integrated circuit *(IC) sockets* are used to provide solid mechanical and electrical connections for integrated circuits. Typical IC sockets are shown in Figure 6.19.

RELAY

A *relay* is an electrically operated switch. The schematic diagram and construction of a relay are shown in Figure 6.20. Relays are used to control high voltage/current by energizing the relay's coil.

FIGURE 6.18   Common types of integrated circuits

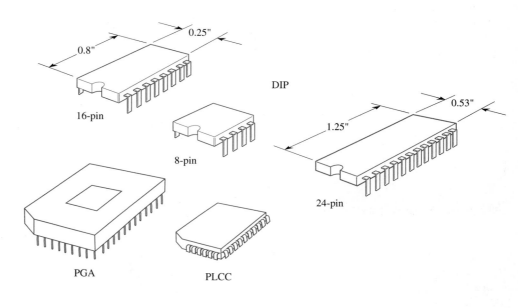

FIGURE 6.19   Typical IC sockets

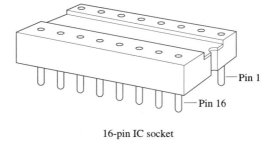

16-pin IC socket

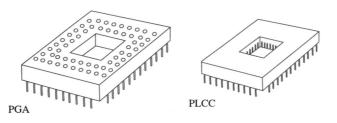

FIGURE 6.20 (a) Schematic and (b) construction of a typical miniature relay

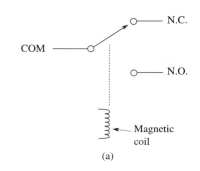

(a)

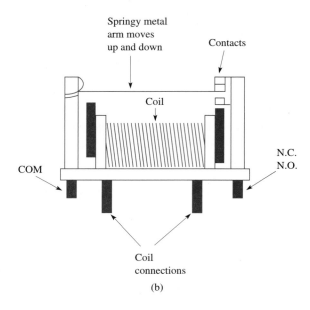

(b)

TROUBLESHOOTING TECHNIQUES

One of the first things on any hardware troubleshooting list should be a good visual inspection. The components we have just examined fail for different reasons and have their own special ways of introducing problems into a circuit.

Here is a short list of things we can check visually:

- Burned, cracked, or open resistors.
- Shorts between components (their leads are touching).
- Bad connections (connectors half out of their sockets).
- Components inserted incorrectly (ICs).
- Blown fuses.
- Foreign objects (a paper clip or staple that fell into the circuit).

You may think of other things to look for as well. A good visual is important and often leads to the source of the problem. This is especially true for equipment that was previously working. A circuit that never worked may have other problems, such as bad components or an incorrect design, and will probably require further troubleshooting.

REVIEW

Use Figure 6.21 as an aid to help you identify components used in computer circuits. The figure shows schematic symbols and other important facts about each component.

SUMMARY

In this exercise we discovered that

- Components such as resistors, capacitors, inductors, and diodes are used to limit and control current and store energy.

FIGURE 6.21 Common electrical components

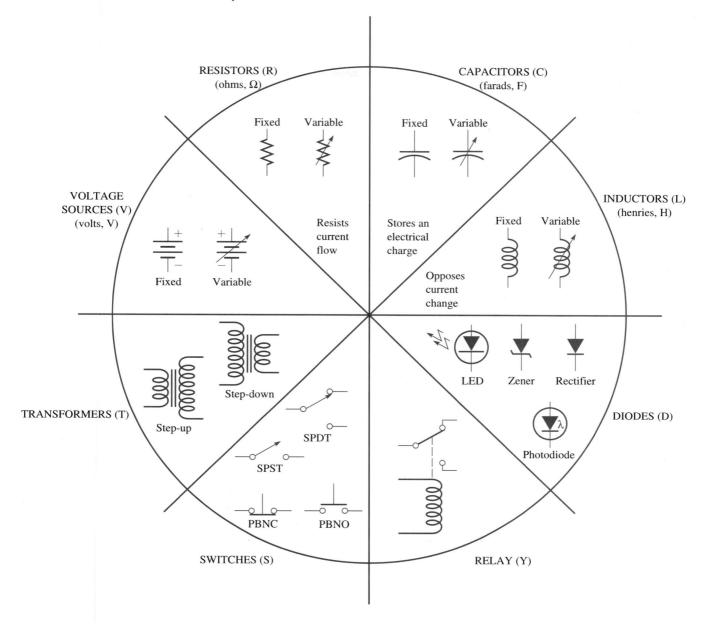

- Components come in many different values and varying units, such as 10 K ohms versus 22 microfarads.
- Components have distinctive markings to indicate such things as value and polarity. For example, the resistor color code uses three colored bands to indicate a resistor's value.
- Electromechanical components such as switches and relays are used to control current.

SELF-TEST

This self-test is designed to help you check your understanding of the material presented in this exercise.

True/False

Answer *true* or *false*.

1. A resistor opposes the flow of current.
2. The two major types of resistors are fixed resistors and variable resistors.

3. A resistor with a color code of *red-black-orange* would have a value of 2 KΩ.
4. The wattage of a resistor tells how much current it is capable of resisting.
5. A 10-KΩ resistor with a silver tolerance band would have a maximum allowable value of 11 KΩ.
6. A diode is used to rectify.
7. A capacitor stores an electrical charge.

Multiple Choice

Select the best answer.

8. The value of a capacitor is measured in
   a. Ohms.
   b. Henries.
   c. Farads.
   d. Volts.
9. A 120-µF capacitor has a capacitance value of
   a. 120 F.
   b. 0.000120 F.
   c. 12,000,000 F.
   d. 0.0000120 F.
10. The value of a 0.000000000015-F capacitor can be expressed as
    a. 15 pF.
    b. 1.5 pF.
    c. 15 µF.
    d. 1.5 µF.
11. The important consideration of electrolytic capacitors is that they
    a. Are polarized.
    b. Have small values.
    c. Should not be used in digital circuits.
    d. None of the above.
12. A capacitor will
    a. Store an electrical charge.
    b. Cause current to flow in one direction.
    c. Oppose a change in current.
    d. Convert AC to DC.
13. The unit of inductance is the
    a. Farad.
    b. Henry.
    c. Ohm.
    d. Watt.
14. This component uses a magnetic field to open/close a switch:
    a. Diode
    b. Inductor
    c. Relay
    d. DIP switch

Matching

Match each component name below with the correct symbol in Figure 6.22.

15. Diode
16. Inductor
17. LED
18. Variable resistor
19. Capacitor

FIGURE 6.22 Symbols for self-test questions 15–19

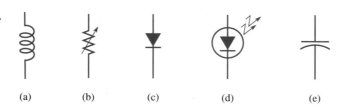

(a)   (b)   (c)   (d)   (e)

Completion

Fill in the blank or blanks with the best answers.

20. An inductor opposes a change in _____.
21. The term "DIP" stands for _____ _____ _____.
22. A diode is a device that causes current to flow in _____ _____.
23. A(n) _____ is constructed so that it is sensitive to light levels.
24. A(n) _____ diode can be used as a voltage regulator.
25. A(n) _____ capacitor is polarized.

FAMILIARIZATION ACTIVITY

1. In the spaces below, draw the schematic diagrams of the following components:
   a. Variable resistor   b. Inductor   c. Diode
   d. Eight-position DIP switch   e. LED   f. Zener diode

   a.   b.

   c.   d.

   e.   f.

2. Place each resistor your instructor gives you on a blank sheet of paper and, using the color code of each resistor, write the resistance value and wattage rating next to each resistor. Have your instructor check to make sure your values are correct.

   Instructor's OK: _____

3. Place each capacitor on a blank sheet of paper and write beside each one its value in farads, voltage rating (if marked), and type. Again, have your instructor check to make sure your values are correct.

   Instructor's OK: _____

4. With your lab partners, practice identifying components on the circuit board assigned to your lab station. Do this until you are confident that you will be successful in completing the performance objective for this lab.

Answer the following questions pertaining to the circuit board assigned to your lab station (this may be the motherboard of a microcomputer or other similar system board).

1. How many of the following does the board contain?
   a. Resistors
   b. Capacitors
   c. Integrated circuits
   d. IC sockets
   e. DIP switches
   f. Diodes
   g. LEDs
   h. Miniature relays
2. State the wattage value of each resistor on the circuit board.
3. If the circuit board contains any DIP switches, how many switch positions does each one have?
4. Determine what types of diodes are on the circuit board (small-signal, power, etc.).
5. If the circuit board contains any LEDs, can you tell which lead is the cathode?
6. Using the resistor color code and the physical size of the resistor, what is the largest-wattage resistor on the circuit board? The smallest? The largest resistance value? The smallest resistance value?
7. What is the largest-value capacitor on the circuit board?
8. Name the different types of capacitors.
9. What is the color code for a 10-KΩ resistor?
10. What types of capacitors have voltage polarity?
11. State the color code of a 52-Ω resistor with a 5% tolerance.
12. What is the difference between a small-signal diode and a power rectifier?
13. Briefly explain the operation of a miniature relay.

Under the supervision of your instructor

Identify, give the function of, and state any special considerations in regard to each of the following types of parts:

Resistors (fixed and variable)
Capacitors
Inductors
Diodes (including the LED)
DIP switches
Integrated circuits
IC sockets
Miniature relays

Once you have done this to the satisfaction of your instructor, you have completed the performance objective for this exercise.

7 Integrated Circuit Removal and Insertion

Don, the senior technician, was busy soldering IC sockets onto a printed circuit board. Joe Tekk asked him why he bothered with the sockets. "Wouldn't it be easier to just solder the IC in place?"

"No, not easier," Don replied. "It's still the same number of pins to solder. But there are advantages to having a socket. For example, it is easy to troubleshoot a circuit by replacing a socketed IC. Try doing that with an IC that is soldered in place."

Then Don winked at Joe. "I have another reason, too. If I'm looking for an IC I need, sometimes the only one I can find is sitting in a socket on a spare board. It saves me a trip to the store."

PERFORMANCE OBJECTIVES

Upon completion of this exercise, you will be able to

1. Examine a printed circuit board and locate
   a. All empty IC sockets.
   b. All IC sockets with mismatched ICs inserted in them.
   c. All ICs inserted with visible bent pins.
   d. All ICs incompletely inserted.
2. Correctly remove and reinsert at least one each of the following:
   a. A 14- or 16-pin IC
   b. A 24-pin IC
   c. An IC consisting of 30 or more pins
3. Point out any PGA or PLCC devices.
4. Determine the type of processor socket found on your lab computer's motherboard.

KEY TERMS

Insertion tool
Extraction tool
PGA (pin grid array)

PLCC (plastic leaded chip carrier)
ZIF (zero insertion force)

BACKGROUND INFORMATION

IDENTIFYING AN IC

An integrated circuit (IC) consists of many different electronic devices (such as transistors, resistors, capacitors, diodes, etc.), all connected together in a single small package. Figure 7.1 shows some typical integrated circuits.

FIGURE 7.1 Typical integrated circuits (photo copyright of Motorola, Inc.; used by permission)

FIGURE 7.2 Typical IC DIP sockets (courtesy Aries Electronics, Inc.)

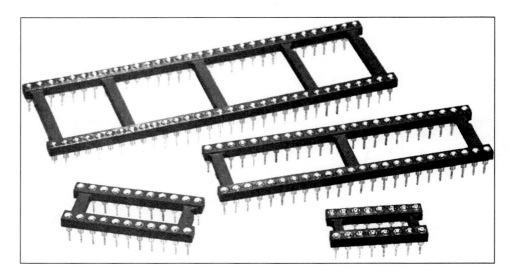

Note from Figure 7.1 that each IC contains several pins along its sides. These pins are used to make electrical connections to external circuits, usually through an IC socket. Figure 7.2 shows some typical IC sockets.

CONNECTING AN IC

Integrated circuits can be connected to external circuits by many methods. The following are three of the most common methods:

1. Soldering directly to the PC board (through-hole and surface-mount)
2. Insertion into an IC socket, where the socket itself is soldered directly to the PC board
3. Connection with wire, such as wire-wrapped connections

These three different methods of connecting integrated circuits are shown in Figure 7.3.
For this exercise, you will be working with ICs that are inserted into sockets.

PIN NUMBERING

One advantage of connecting an IC to a circuit through the use of sockets is that the IC may be removed easily for replacement. One disadvantage is that manufacturing costs are higher, since the socket must first be soldered to the board and then the IC inserted into the socket.

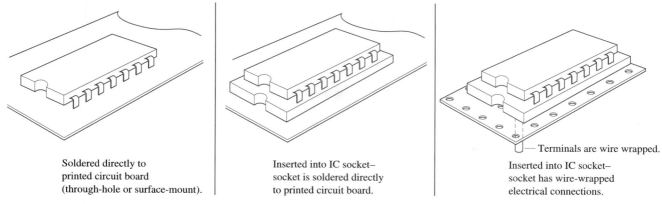

Soldered directly to
printed circuit board
(through-hole or surface-mount).

Inserted into IC socket–
socket is soldered directly
to printed circuit board.

—Terminals are wire wrapped.

Inserted into IC socket–
socket has wire-wrapped
electrical connections.

FIGURE 7.3   Three methods of connecting ICs

FIGURE 7.4   Pin numbers of various ICs

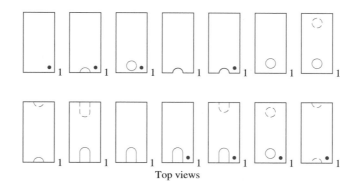

Top views

FIGURE 7.5   Pin numbers of IC sockets

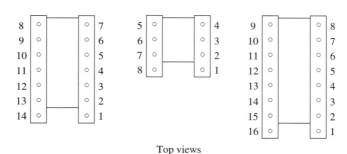

Top views

As a technician, you must ensure that you place an IC in the correct position when replacing it in a socket. All IC pins have distinct numbers, as do the corresponding sockets. The numbering of integrated circuits is shown in Figure 7.4. Figure 7.5 illustrates the numbering of IC sockets.

REASON FOR IC FAILURES

Integrated circuits are normally very reliable devices. One of the major causes of initial IC failure is improper insertion of the ICs into their sockets. Figure 7.6 shows the common insertion errors that may cause IC failures.

Because computer owners may have tried to repair their own computers before bringing them in for service, it is always good practice to give all printed circuit boards with IC sockets a complete visual inspection. Do this to ensure that the customer or another technician has not inserted an IC incorrectly into the board.

FIGURE 7.6 Common IC insertion errors

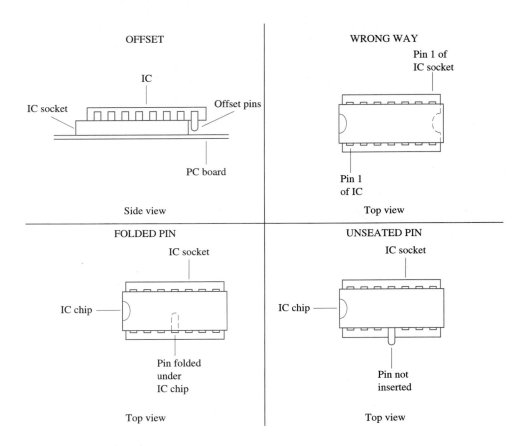

OFFSET

IC

IC socket

Offset pins

PC board

Side view

WRONG WAY

Pin 1 of
IC socket

Pin 1
of IC

Top view

FOLDED PIN

IC socket

IC chip

Pin folded
under
IC chip

Top view

UNSEATED PIN

IC socket

IC chip

Pin not
inserted

Top view

FIGURE 7.7 IC identification

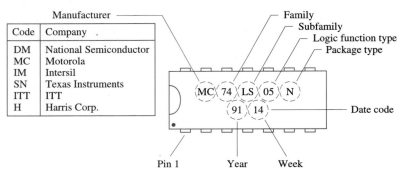

Manufacturer

Code	Company
DM	National Semiconductor
MC	Motorola
IM	Intersil
SN	Texas Instruments
ITT	ITT
H	Harris Corp.

Family
Subfamily
Logic function type
Package type

MC 74 LS 05 N

91 14

Date code

Pin 1   Year   Week

TYPES OF ICS

Figure 7.7 shows the nomenclature used to identify an IC. It is important that you always use an exact replacement, as recommended by the manufacturer.

Many different functions are performed by ICs in your computer. One common trouble-shooting technique is to replace a suspect IC with a known good one. It is important that you practice the technique of IC removal and insertion.

REMOVING AN IC

Figure 7.8 shows the proper procedure for removing an IC using an IC removal tool.

When removing an IC, it is important to pull it straight up in order to prevent bending the pins. There are times when the IC is so large that the removal tool will not work properly. If this is the case, use the method shown in Figure 7.9.

If you find it necessary to use the method shown in Figure 7.9, be sure to pry up, gently, first one end and then the other. Again, you are trying to pull the IC straight up in order to

FIGURE 7.8 Proper method of removing an IC

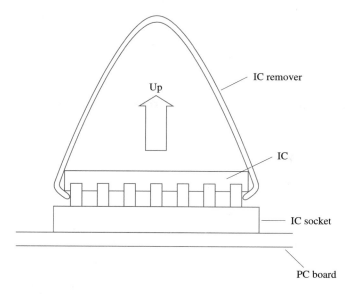

FIGURE 7.9 Removing a large IC

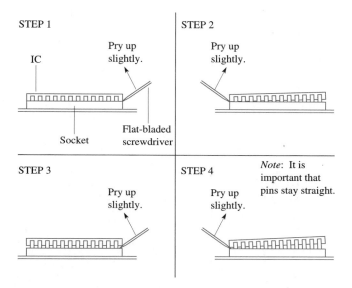

avoid bending the pins. Do not pull an IC out with your fingers. One slip, and the tiny metal pins of the IC can easily pierce your fingertips.

INSERTING AN IC

Figure 7.10 shows the correct method of inserting an IC into an IC socket. Note that the pins of an IC tend to spread out to the side.

The reason an IC comes from the factory with its pins spread out is to help form a better electrical connection within the IC socket. Because of the mechanical pressure caused by the spreading of the pins, a more reliable electrical connection is made between each IC pin and its corresponding socket connection.

Figure 7.11 shows how an ***insertion tool*** is used to insert an IC. The adapter holds the pins of the IC in grooves, which helps guide them into the socket holes.

It is important to note that when you are replacing an IC, you must always have the power off. If the power is on as you replace an IC, you could destroy the new IC. In a like manner, if power is still on when you remove an IC, you could destroy an otherwise good IC. Also, be sure to wear a protective wrist strap to avoid static damage to sensitive integrated circuits. Figure 7.12 shows an insertion tool and an ***extraction tool*** used to extract PLCCs, which are covered next.

FIGURE 7.10 Inserting an IC into its socket

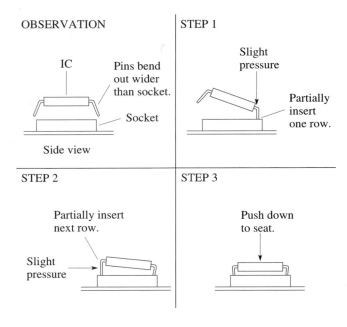

OBSERVATION

IC

Pins bend out wider than socket.

Socket

Side view

STEP 1

Slight pressure

Partially insert one row.

STEP 2

Partially insert next row.

Slight pressure

STEP 3

Push down to seat.

FIGURE 7.11 Using an insertion tool

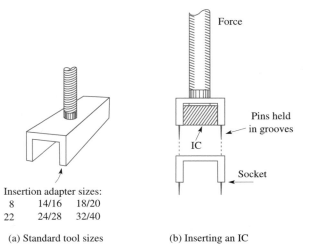

Force

Pins held in grooves

IC

Socket

Insertion adapter sizes:
8	14/16	18/20
22	24/28	32/40

(a) Standard tool sizes

(b) Inserting an IC

FIGURE 7.12 IC tools

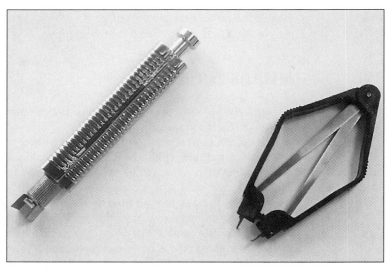

(a) DIP insertion tool

(b) PLCC extraction tool

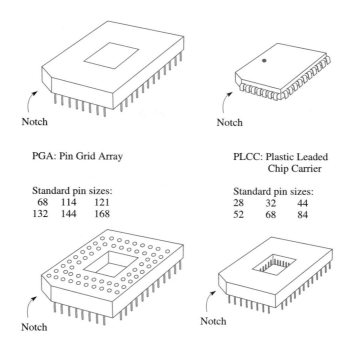

FIGURE 7.13 PGA and PLCC chips and sockets

Notch

Notch

PGA: Pin Grid Array

PLCC: Plastic Leaded
Chip Carrier

Standard pin sizes:
| 68 | 114 | 121 |
| 132 | 144 | 168 |

Standard pin sizes:
| 28 | 32 | 44 |
| 52 | 68 | 84 |

Notch

Notch

ADVANCED CHIPS AND PACKAGES

Early 16-bit microprocessors, such as Intel's 8088, were victims of the 40-pin DIP, which allowed the CPU to utilize only 40 external connections. This forced the designers to use eight pins for dual purposes (multiplexed address and data information) and reduced the performance of the processor. By the time the 64-pin DIP package became available, microprocessors had graduated to 32-bit data and 32-bit address buses and required sockets with even larger numbers of pins. Figure 7.13 shows two types of advanced packages—the *PGA (pin grid array)* and the *PLCC (plastic leaded chip carrier)*—that offer solutions to the connection problem. Both packages allow more pins to be connected in a smaller surface area of board space. In the PGA-style chip, all of the pins are arranged in a two-dimensional structure, protruding from the bottom of a square, ceramic chip housing, with up to 168 connections available. Considerable force is needed to push a PGA chip into its socket, making for very good connections.

The PLCC-style chip consists of springy pins mounted on all four sides of a square-shaped integrated circuit housing. Up to 84 pins are available, enough for a coprocessor chip or other suitable device. The PLCC is pushed straight down into its socket, with equal pressure applied to all four sides. The springy pins push against the connectors on the inside of the PLCC socket with plenty of force, resulting in good connections and the ability to lock the chip in place.

If you examine the motherboard of a newer microcomputer, the odds are good that you will find one, or both, of these types of sockets and chip styles.

**TROUBLESHOOTING
TECHNIQUES**

New motherboards are equipped with *ZIF (zero insertion force)* sockets that allow you to easily replace the processor. A small handle on the ZIF socket is lifted, releasing the processor by removing pressure on its pins (as indicated in Figure 7.14). The Pentium processors use ZIF sockets 5 and 7 (320 and 321 pins, respectively), the Pentium Pro uses the 387-pin socket 8, and the Pentium II/III uses a new connector design called *Slot 1*, a 242-contact rectangular cartridge that plugs into a slot on the motherboard.

Being familiar with the various processors and their sockets is a good beginning for the serious microcomputer repair enthusiast.

FIGURE 7.14   Operation of a ZIF socket

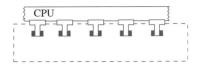

ZIF socket

(a) Lever down, pins locked in place

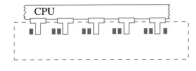

(b) Lever up, pins are released

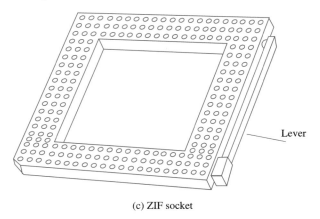

Lever

(c) ZIF socket

SUMMARY

In this exercise we discovered that

- There are many types of ICs and sockets, such as DIP, PGA, and PLCC.
- All ICs and sockets have distinct pin numbers.
- Information stamped onto an IC indicates the manufacturer, logic function, and date code.
- ICs are inserted/removed with special tools.
- A ZIF socket contains a lever to lock an IC into place.

SELF-TEST

This self-test is designed to help you check your understanding of the background information presented in this exercise.

True/False

Answer *true* or *false*.

1. The term "IC" stands for "integrated circuit."
2. An integrated circuit consists of many different devices, all connected together in a single, small package.
3. Integrated circuits are always inserted into IC sockets and are never soldered directly to the circuit board.
4. When an IC is inserted into a socket, it may be inserted in any direction and still operate properly.
5. To remove an IC, use a pair of long-nose pliers and remove one pin at a time.
6. Markings on the IC package identify pin 1.
7. In the date code 9530, the IC was manufactured in 1995.

Multiple Choice

Select the best answer.

8. An IC may be connected to a printed circuit board by
   a. Soldering it directly to the board.
   b. Inserting it into an IC socket, where the socket is soldered to the board.
   c. Using wire connections, such as wire-wrapping techniques.
   d. Any of the above.

FIGURE 7.15 IC pin numbers for self-test

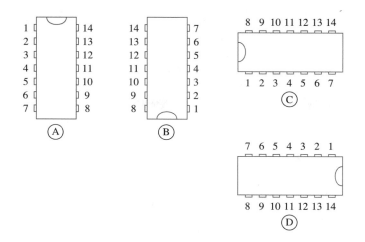

FIGURE 7.16 IC connections for self-test

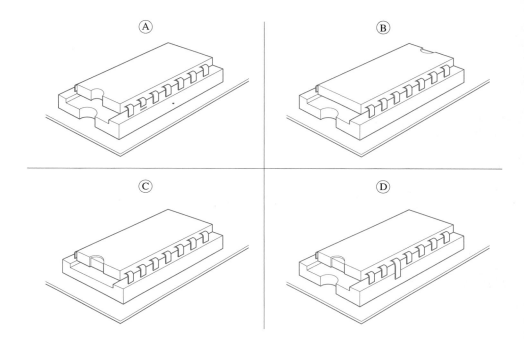

9. Figure 7.15 shows some ICs with corresponding pin numbers. The ICs on which the first halves of the pin numbers are correct are
   a. A, C, and D.
   b. B and C.
   c. A and D.
   d. All have the first half correct.

10. Referring again to Figure 7.15, the ICs on which all of the pin numbers are correct are
    a. A, C, and D.
    b. B and C.
    c. A and D.
    d. All have correct pin numbers.

11. Figure 7.16 shows ICs plugged into IC sockets. The connections that have mismatched pin numbers are
    a. A, C, and D.
    b. B and C.
    c. A and D.
    d. All have correctly matched pins.

FIGURE 7.17   IC chips for self-test

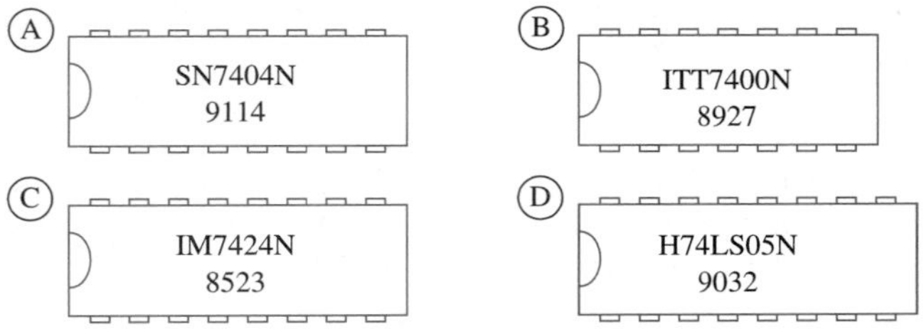

A. SN7404N
    9114

B. ITT7400N
    8927

C. IM7424N
    8523

D. H74LS05N
    9032

12. Referring again to Figure 7.16, the ICs that have one or more pins without proper connections are
    a. A, C, and D.
    b. B and C.
    c. A and D.
    d. All have improperly connected pins.
13. Which IC package has all of its pins coming out of the bottom?
    a. DIP
    b. PLCC
    c. PGA
    d. All of the above
14. ZIF stands for
    a. Zero insertion force.
    b. Z-axis integrated fixture.
    c. Zipped fixture.
    d. Zero insertion flange.

Matching

Match each of the following descriptions with the proper chip in Figure 7.17.

15. A 14-pin IC
16. Manufactured by Texas Instruments
17. Manufactured in 1990
18. Manufactured in 1985
19. Manufactured by ITT

Completion

Fill in the blanks with the best answers.

20. When removing an IC, it is important to pull it _____ _____ in order to prevent bending the pins.
21. The most common problem with IC removal is _____ pins.
22. A new IC will have its pins spread _____ than its package width.
23. Always replace a bad IC with a(n) _____ replacement.
24. When replacing an IC, be sure power is _____.
25. PGA stands for pin _____ array.

FAMILIARIZATION ACTIVITY

1. Using the proper procedures, remove five ICs from the printed circuit board assigned to your lab station.
2. Have your instructor check the removed ICs to make sure they were removed correctly.

   Instructor's OK: _____

3. Insert the five ICs you just removed into their proper sockets using the correct procedures.
4. Have your instructor check the replaced ICs to make sure they were replaced correctly.

   Instructor's OK: _____

Questions

1. Draw a sketch of a typical IC and explain how to identify the pin numbers.
2. Sketch a typical IC socket and state how to determine the pin numbers.
3. State the most common problems with IC removal.
4. State the most common problems with IC insertion.
5. Explain what is meant by a good visual inspection.

Activities

1. You will be tested on your ability to spot incorrectly inserted integrated circuits. Make sure you and your lab partners are proficient at doing this.
2. Practice removing and inserting integrated circuits until you are confident of your ability. You will be tested on IC removal and insertion.

Under the supervision of your instructor

1. Examine a printed circuit board and locate
   a. All empty IC sockets.
   b. All IC sockets with mismatched ICs inserted in them.
   c. All ICs inserted with visible bent pins.
   d. All ICs incompletely inserted.
2. Correctly remove and reinsert at least one each of the following:
   a. A 14- or 16-pin IC
   b. A 24-pin IC
   c. An IC consisting of 30 or more pins
3. Point out any PGA or PLCC devices.
4. Determine the type of processor socket found on your lab computer's motherboard.

8 Soldering and Desoldering Techniques

INTRODUCTION

Joe Tekk was buried in a sea of wires. Small plumes of smoke occasionally floated up from his work area, filling the room with the smell of hot solder. Joe was busy making up cables for some equipment being mounted in a rack cabinet. After soldering connectors onto both ends of a cable, he would then check the connections with an ohmmeter. On his workbench a set of data manuals lay open, showing pin assignments for various connectors and wiring diagrams for several different cables.

Don, the senior technician, complimented Joe on the quality of his soldering. "Where did you learn to solder like that?"

Joe replied that he and his friends like to spend their weekends installing car stereo systems, so he has had plenty of soldering practice.

PERFORMANCE OBJECTIVES

Upon completion of this exercise, you will be able to

1. Cut and tin both leads of a stranded wire.
2. Remove a given component (other than an IC) from a printed circuit board without damaging the board or component.
3. Solder a given component (other than an IC) to a printed circuit board, using proper heat sink techniques if required.
4. Desolder an integrated circuit from a printed circuit board without damaging the board or the circuit.
5. Solder an integrated circuit into a printed circuit board using proper heat sink techniques.

KEY TERMS

Soldering	Cold solder joint
Rosin	Solder bridge
Flux	Heat sink
Soldering iron	Desoldering
Tip	Surface-mount component
Tinning	Vacuum pickup

FIGURE 8.1   The process of soldering

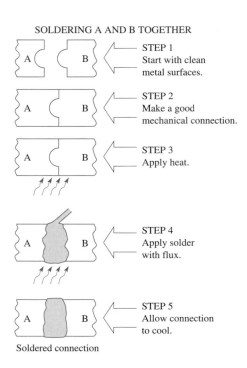

SOLDERING A AND B TOGETHER

STEP 1
Start with clean metal surfaces.

STEP 2
Make a good mechanical connection.

STEP 3
Apply heat.

STEP 4
Apply solder with flux.

STEP 5
Allow connection to cool.

Soldered connection

BACKGROUND INFORMATION

DEFINITION OF SOLDERING

Soldering is the process used to secure the wire connections of electrical components. Soldering is the least expensive, fastest, most reliable, and simplest method of making electrical connections between electronic components. The process requires three things:

1. A metal alloy, called solder
2. A material to clean the connection
3. A source of heat

Figure 8.1 illustrates the process of soldering.

WHY SOLDERING WORKS

The metal alloy called *solder* frequently consists of a combination of 60% lead and 40% tin, which is referred to as 60/40 solder. This combination, called an *alloy*, has a melting point, 370°F, that is lower than the melting point of either metal by itself. Solder has the ability to form an interface that embeds itself into the metals to be connected, forming an excellent electrical connection. This is shown in Figure 8.2.

FIGURE 8.2   The soldered electrical connection

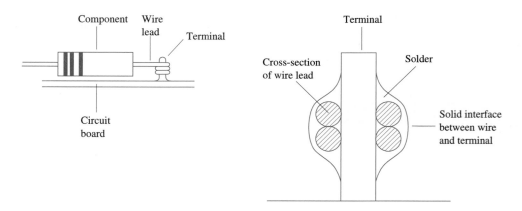

FIGURE 8.3   Three types of
solder

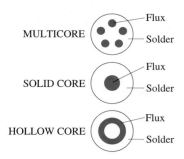

MULTICORE — Flux
— Solder

SOLID CORE — Flux
— Solder

HOLLOW CORE — Flux
— Solder

Because wire leads, which are made of copper, have a tendency to oxidize when heated, a method of keeping them clean is needed in order for the solder to make the required metal bonding. A chemical called a *rosin*, or *flux*, is used. The flux chemical comes contained within the core of the solder. Solder is available in a variety of shapes; the most common type is in the form of a long wire that is about $1/16"$ in diameter (good for most electrical work) or $1/32"$ in diameter (good for small, detailed work). Three types of solder are shown in Figure 8.3.

SOLDERING IRONS

The tool most commonly used as a source of heat for the melting and application of solder is a *soldering iron*. Figure 8.4(a) illustrates the basic construction of a soldering iron. Figure 8.4(b) shows an actual soldering station.

Soldering irons come in different wattage ratings, from 10 to 250 W. For most computer work, an iron with a wattage rating of between 10 and 50 W is used. In fact, the most common type used for computer work is an iron with a rating of 25 to 35 W. Soldering irons with higher wattage ratings get so hot that they can easily damage delicate components and printed circuit boards. These higher-wattage soldering irons are used only for larger and more rugged electrical soldering jobs, not for work on PCs.

The working part of a soldering iron is its *tip*. The tip is the part of the iron that melts the solder. Various soldering tips are shown in Figure 8.5.

TINNING AND CLEANING A SOLDERING TIP

The tip of a soldering iron must be capable of providing a concentrated area of heat. This is best achieved when the tip surface is bright and shiny. However, soldering tips will oxidize and turn black from the heat they generate, causing the amount of heat leaving the surface to be reduced, as shown in Figure 8.6.

FIGURE 8.4   (a) Basic
construction of a soldering iron

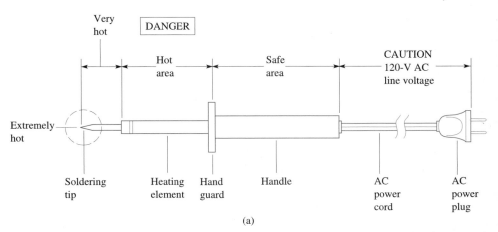

(a)

(continued on the next page)

FIGURE 8.4 *(continued)* (b) temperature-controlled soldering station

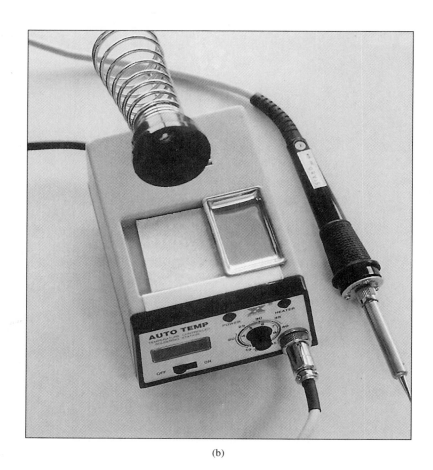

(b)

FIGURE 8.5 Various soldering tips

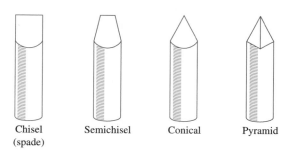

Chisel (spade) Semichisel Conical Pyramid

FIGURE 8.6 Heat on the surface of a soldering tip

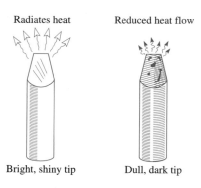

Radiates heat   Reduced heat flow

Bright, shiny tip   Dull, dark tip

In order to ensure that the tip of the soldering iron stays bright and shiny, a process called *tinning* is performed. Tinning is done by carefully melting solder on the tip of the soldering iron when the iron is first turned on for use. As soon as a small amount of solder melts on the tip, the tip is wiped clean with a small, damp cloth or damp sponge. The process of tinning the tip is shown in Figure 8.7.

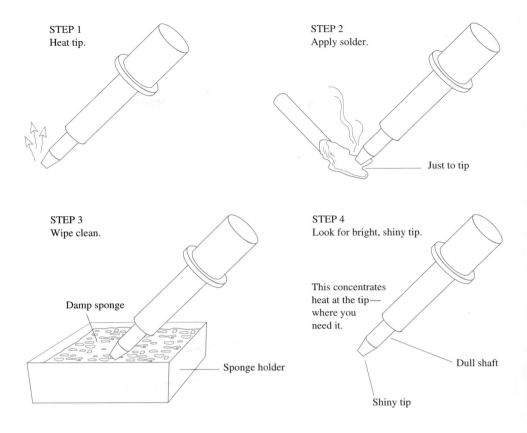

FIGURE 8.7 Tinning the tip of a soldering iron

STEP 1
Heat tip.

STEP 2
Apply solder.

Just to tip

STEP 3
Wipe clean.

Damp sponge

Sponge holder

STEP 4
Look for bright, shiny tip.

This concentrates heat at the tip— where you need it.

Dull shaft

Shiny tip

As you are using the soldering iron, you want to ensure that the tip is kept clean. This is achieved by wiping it quickly with a damp cloth or damp sponge. It is important that the tip remain clean throughout the entire soldering process.

HOW TO SOLDER

Soldering is a four-step process:

1. Make a good mechanical connection.
2. Heat both parts to be connected.
3. Apply solder to both parts at the same time.
4. Remove the iron and allow parts to cool slowly.

This four-step process is shown in Figure 8.8.

There are times when you may not have an actual physical connection on a printed circuit board to which to attach the lead of a component. In this case, the lead must be placed on the foil surface of the circuit board. When you must do this, be sure that the lead is absolutely still while the solder cools and hardens. This type of connection is shown in Figure 8.9.

COLD SOLDER JOINTS

A *cold solder joint* is an undesirable condition that results in an unreliable electrical connection. Cold solder joints are caused by

1. Dirty soldering tips.
2. Heat not uniformly distributed to both parts.
3. Movement of the connection during cooling.
4. Blowing on the connection or using an external source to cause rapid cooling of the connection.

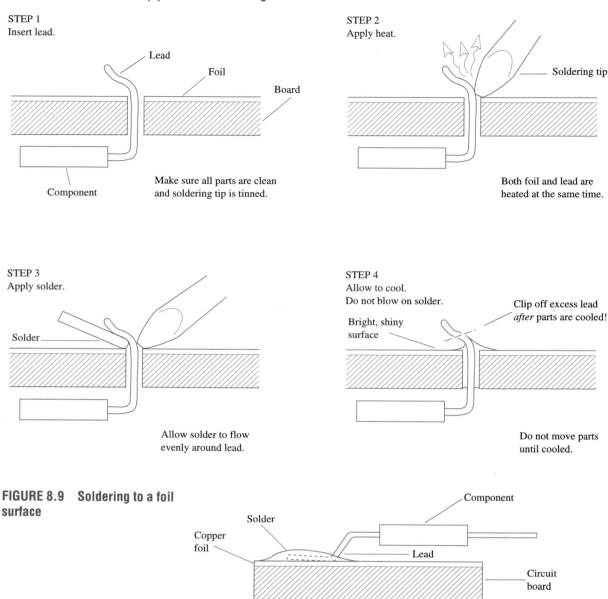

FIGURE 8.8   The four-step process of soldering

STEP 1
Insert lead.

Lead

Foil

Board

Component

Make sure all parts are clean
and soldering tip is tinned.

STEP 2
Apply heat.

Soldering tip

Both foil and lead are
heated at the same time.

STEP 3
Apply solder.

Solder

Allow solder to flow
evenly around lead.

STEP 4
Allow to cool.
Do not blow on solder.

Clip off excess lead
after parts are cooled!

Bright, shiny
surface

Do not move parts
until cooled.

**FIGURE 8.9   Soldering to a foil
surface**

Component

Solder

Copper
foil

Lead

Circuit
board

Some of the different types of cold solder joints and their causes are shown in Figure 8.10. A good soldered connection is smooth and shiny. A poor soldered connection (a cold solder joint) looks dull and pitted. If you find that you have created a cold solder joint, simply re-heat the joint to remove the old solder from both surfaces and, with a clean tip, resolder.

SOLDER BRIDGES

A **solder bridge** is an undesirable condition that can result in severe electrical damage to a computer system. Solder bridges are caused by solder accidentally allowing an electrical connection between two adjoining parts of a printed circuit board. A solder bridge is shown in Figure 8.11.

Always inspect your soldering work closely to ensure that no solder bridges have formed. If you find one, heat the solder carefully to remove it. The section on desoldering techniques will show you how to remove unwanted solder.

FIGURE 8.10   Cold solder joints and their causes

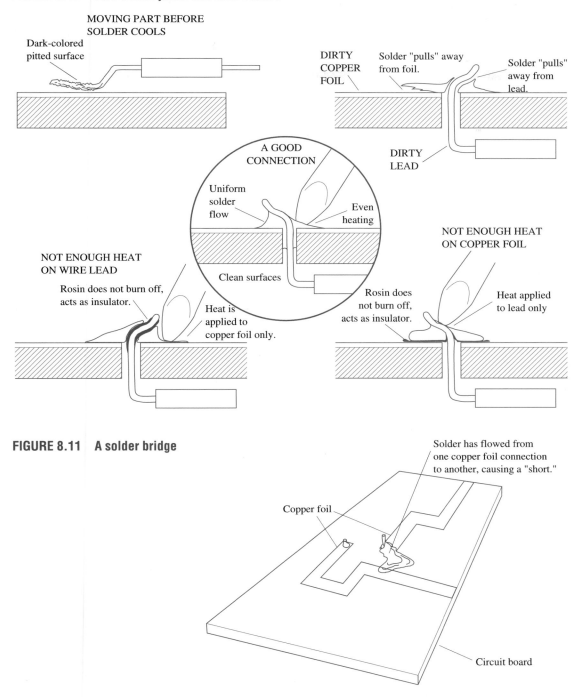

MOVING PART BEFORE
SOLDER COOLS

Dark-colored
pitted surface

DIRTY
COPPER
FOIL

Solder "pulls" away
from foil.

Solder "pulls"
away from
lead.

DIRTY
LEAD

A GOOD
CONNECTION

Uniform
solder
flow

Even
heating

Clean surfaces

NOT ENOUGH HEAT
ON WIRE LEAD

Rosin does not burn off,
acts as insulator.

Heat is
applied to
copper foil only.

NOT ENOUGH HEAT
ON COPPER FOIL

Rosin does
not burn off,
acts as insulator.

Heat applied
to lead only

FIGURE 8.11   A solder bridge

Solder has flowed from
one copper foil connection
to another, causing a "short."

Copper foil

Circuit board

TINNING WIRE

Insulated wire is either solid or stranded. Stranded wire is more flexible and not as break-able as solid wire. Figure 8.12 illustrates the two types of wire.

When stranded wire is used to make an electrical connection, the tips of the strands of wire should be soldered together. The process of doing this is called *tinning* the wire and is shown in Figure 8.13.

After the wire is properly tinned, it may then be used to make electrical connections. Tinning the wire helps make soldering it to other parts of the circuit an easy task, as shown in Figure 8.14.

FIGURE 8.12   Two types of wire

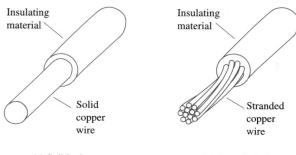

Insulating material

Solid copper wire

(a) Solid wire

Insulating material

Stranded copper wire

(b) Stranded wire

FIGURE 8.13   Process of tinning wire

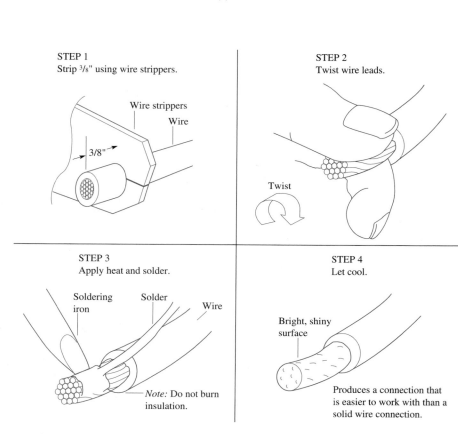

STEP 1
Strip ³/₈" using wire strippers.

Wire strippers

Wire

3/8"

STEP 2
Twist wire leads.

Twist

STEP 3
Apply heat and solder.

Soldering iron

Solder

Wire

Note: Do not burn insulation.

STEP 4
Let cool.

Bright, shiny surface

Produces a connection that is easier to work with than a solid wire connection.

FIGURE 8.14   Connecting tinned wire

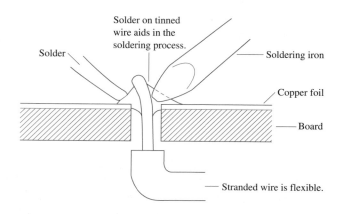

Solder on tinned wire aids in the soldering process.

Solder

Soldering iron

Copper foil

Board

Stranded wire is flexible.

FIGURE 8.15   The use of a heat sink during soldering

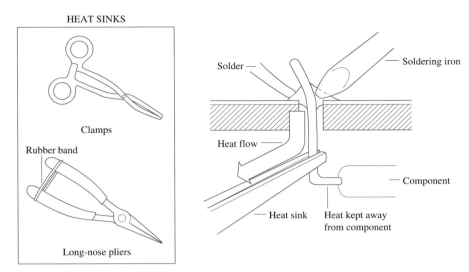

USING A HEAT SINK

Many of the components with which you will be working are very sensitive to heat. This is especially true of solid-state components such as integrated circuits, diodes, and transistors; other small components may also be sensitive. In order to help prevent heat from destroying these devices, a tool is placed between the heat source and the device to help conduct away the heat. Such a tool is called a *heat sink* and is illustrated in Figure 8.15.

It is important to use a heat sink to avoid damaging good electrical parts. If in doubt, use a heat sink; it can never hurt, but if you don't use it, you could wind up with a damaged part.

DESOLDERING TECHNIQUES

Desoldering is the process of removing a soldered component or simply removing unwanted solder (such as from a solder bridge).

The desoldering process consists of the following steps:

1. Heating the solder
2. Removing the solder
3. Removing the component
4. Cleaning the surface

The process of desoldering, along with various desoldering tools, is shown in Figure 8.16. In this exercise, you will have an opportunity to practice soldering and desoldering.

SURFACE-MOUNT COMPONENTS

Technological advances in printed circuit board design and fabrication now allow components (resistors, capacitors, integrated circuits) to be soldered directly onto the surface of the printed circuit board. No holes need to be drilled, since the connecting pins or pads of the component do not go through the board. Figure 8.17 illustrates the difference between a surface-mount capacitor and a disk capacitor. Note the size difference between the two capacitors, even though they both have ratings of 0.001 μF. *Surface-mount components* are much smaller than their through-hole-mounted counterparts. This allows more components to be placed on a board, or the same number of parts to be placed on a smaller board.

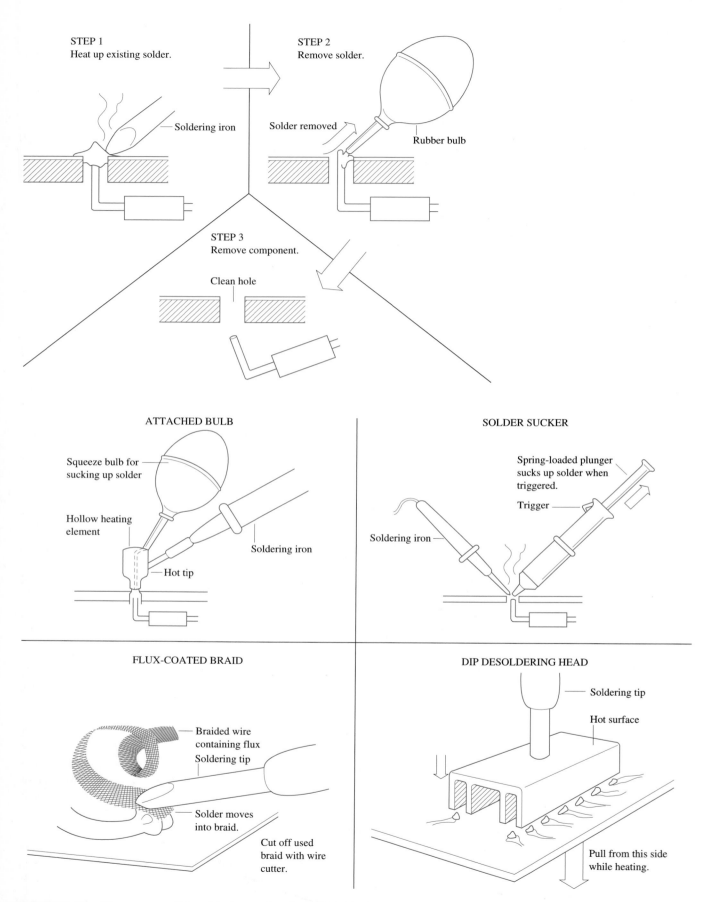

STEP 1
Heat up existing solder.

Soldering iron

STEP 2
Remove solder.

Solder removed

Rubber bulb

STEP 3
Remove component.

Clean hole

ATTACHED BULB

Squeeze bulb for
sucking up solder

Hollow heating
element

Hot tip

Soldering iron

SOLDER SUCKER

Spring-loaded plunger
sucks up solder when
triggered.

Trigger

Soldering iron

FLUX-COATED BRAID

Braided wire
containing flux
Soldering tip

Solder moves
into braid.

Cut off used
braid with wire
cutter.

DIP DESOLDERING HEAD

Soldering tip

Hot surface

Pull from this side
while heating.

FIGURE 8.16   The process of desoldering and various desoldering tools

FIGURE 8.17 Mounting capacitors two different ways

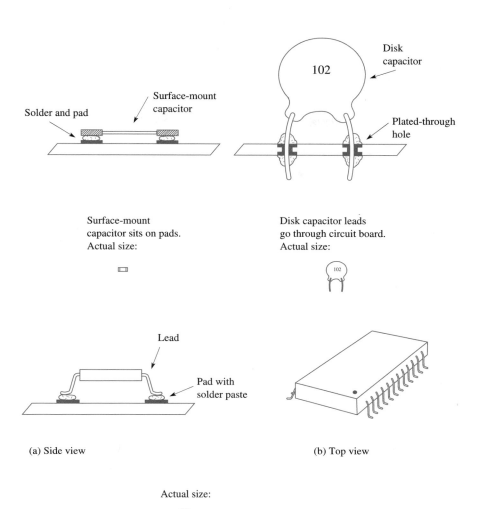

Solder and pad

Surface-mount capacitor

Disk capacitor

102

Plated-through hole

Surface-mount capacitor sits on pads. Actual size:

Disk capacitor leads go through circuit board. Actual size:

102

FIGURE 8.18 Surface-mount integrated circuit

Lead

Pad with solder paste

(a) Side view

(b) Top view

Actual size:

Surface mounting requires a solder paste to be applied to the pads on the surface of the printed circuit board. Then the component is carefully placed on the pads. There is no room for error when placing a component. If the component wiggles around on the pad, the solder paste smears, resulting in a bad connection when the board is heated up in a special machine used to melt the solder paste. This tight control of placement is especially important when dealing with a surface-mount integrated circuit, whose leads are very close together (as shown in Figure 8.18). For this reason, surface-mount components are rarely placed by human hands. Instead, large industrial pick-and-place machines are used to mechanically position each part on a surface-mount printed circuit board. These parts often need to be placed with an accuracy of one ten-thousandth of an inch!

If you are willing to spend a good amount of money, you can set up a surface-mount station that will allow you to experiment with the technology, rather than try to build 1000 motherboards a day. For a few thousand dollars you can buy a hot air rework system that holds printed circuit boards and uses hot air blown through a special tip to heat up the surface-mount pads when soldering/desoldering. For the same price, a manual placement machine allows precise placement (to 0.0005") of components onto a circuit board.

For several more thousand dollars a high-resolution vision system can be added, allowing visual inspection of surface-mount components and circuit boards.

On a smaller cost scale, many small tools and other items are available for surface-mount applications. These include *vacuum pickup* instruments (Figure 8.19), which can pick up small components using a suction cup and vacuum, pin straighteners, miniature soldering-iron tips, and surface-mount components of all varieties, from resistors to ICs, LEDs, and connectors.

FIGURE 8.19 Vacuum pickup
tool

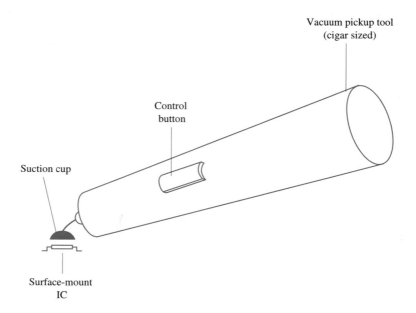

Control
button

Suction cup

Surface-mount
IC

**TROUBLESHOOTING
TECHNIQUES**

How exactly do you determine if a bad solder connection is causing your problem? Here is a short list of things to look for:

- The problem only shows up when the circuit is warmed up (or cooled down).
- The problem comes and goes when the circuit is moved or vibrated.
- Pushing on the circuit board in the right place causes intermittent operation.

SUMMARY

In this exercise we discovered that

- Soldering is used to make good electrical connections between metal connections.
- A chemical called rosin, or flux, is used to clean the metal to be soldered.
- Solder is composed of 60% lead and 40% tin.
- A soldering iron heats the metal and solder.
- A process called tinning keeps the soldering tip clean and hot.
- A cold solder joint is a poor electrical connection.
- A solder bridge is a short between two or more connections.
- Surface-mount components are soldered directly onto the surface of a circuit board.

SELF-TEST

This self-test is designed to help you check your understanding of the background information presented in this exercise.

True/False

Answer *true* or *false*.

1. Soldering is the process used to secure the wire connections of electrical components.
2. Solder is a metal alloy consisting of zinc and copper.
3. The material used to clean the connections to be soldered is called a flux.
4. Surfaces to be soldered will clean themselves because of the extreme heat produced by the soldering iron.
5. It is always best to use the highest-wattage iron available so that it takes less time to solder.
6. It is okay to blow on a solder connection to cool it quickly.
7. A dull, dark tip reduces heat flow from the soldering iron.

Multiple Choice

Select the best answer.

8. The process of soldering requires
    a. A metal alloy called the solder.
    b. A material to clean the connection.
    c. A source of heat.
    d. All of the above.
9. The metal alloy called solder consists of
    a. 50% zinc and 50% copper.
    b. 60% lead and 40% tin.
    c. 40% copper and 60% silver.
    d. None of the above.
10. Solder flux is necessary in order to
    a. Clean the metal leads to be soldered.
    b. Help increase the temperature of the solder.
    c. Prevent electrical shorts.
    d. Ensure a good mechanical connection.
11. The most common type of heat source for soldering in computer work is a
    a. High-wattage (250-W) soldering iron.
    b. Low-wattage (25-to-35-W) soldering iron.
    c. Variable-wattage soldering iron.
    d. Small butane flame tool.
12. When soldering, the proper process is to
    a. Heat both surfaces at the same time and apply the solder.
    b. Heat one surface at a time, applying solder to each.
    c. Heat the solder and let it drip onto the connection.
    d. Heat both surfaces, remove the source of heat, and then apply the solder before the surfaces cool.
13. A cold solder joint is
    a. The best type of connection.
    b. A poor, unreliable connection.
    c. Formed without a soldering iron.
    d. Bright and shiny.
14. A solder bridge is
    a. Unwanted.
    b. Used to conserve solder.
    c. The same as a cold solder joint.
    d. An open circuit between two conductors.

Matching

Match each phrase on the left with the correct term or phrase on the right.

15. The process of soldering the tips of braided wire
16. A dull and pitted solder surface
17. A bright and shiny solder surface
18. The purpose of applying solder to the tip of the iron
19. Material used to clean the surfaces to be soldered

a. Flux
b. Wire tinning
c. Wire stripping
d. Cold solder joint
e. Allows heat to leave the surface
f. A good solder joint
g. None of these

Completion

Fill in the blanks with the best answers.

20. A dirty soldering tip can cause a(n) _____ solder joint.
21. A solder _____ is caused by solder making an undesirable electrical connection on a printed circuit board.

22. When small, solid-state devices are soldered, a(n) _____ _____ should be used to prevent damage from heat.
23. Before soldering, make a good _____ connection where possible.
24. A cold solder joint can be caused by _____ of the soldered parts while they are cooling.
25. _____-mount components are soldered directly onto a circuit board.

TINNING THE TIP

Plug in your soldering iron and, after it heats, tin its tip using the proper tinning procedures. Have your instructor inspect the tip for proper tinning. Be very careful in moving the soldering iron around others, because it is very hot and can cause severe burns.

Instructor's OK: _____

Wire Tinning

Cut your 6" braided wire in half. Strip both ends of both wires back $1/4$" using your wire strippers. Using proper tinning techniques, tin each end of each stripped wire. Have your instructor inspect your work.

Instructor's OK: _____

Component Soldering

Using the printed circuit board assigned to you, solder one each of the following into the board at locations assigned by your instructor:

• A resistor
• A capacitor
• A diode or transistor
• An integrated circuit

   Be sure to inspect your work for cold solder joints and solder bridges. Have your instructor approve your work.

Instructor's OK: _____

Component Desoldering

Using the desoldering tool assigned to you, remove one each of the following components from your assigned printed circuit board. Your instructor will point out the components to be removed.

• A resistor
• A capacitor
• A diode or transistor (use heat sink techniques)
• An integrated circuit (use heat sink techniques)

Instructor's OK: _____

1. Explain what soldering does and why it works.
2. Use sketches to show and explain the difference between a good soldering connection and a bad soldering connection.
3. List the steps involved in the soldering process.
4. List at least three different kinds of soldering tips and their uses.
5. Use words and diagrams to describe the different types of desoldering tools and their uses.
6. Write a short paragraph, using original illustrations where necessary, describing how to desolder an integrated circuit from a printed circuit board.

Under the supervision of your instructor

1. Cut and tin both leads of a stranded wire.
2. Remove a given component (other than an IC) from a printed circuit board without damaging the board or component.
3. Solder a given component (other than an IC) to a printed circuit board using proper heat sink techniques if required.
4. Desolder an integrated circuit from a printed circuit board without damaging the board or the circuit.
5. Solder an integrated circuit into a printed circuit board using proper heat sink techniques.

The familiarization activity section of this exercise may be enough to satisfy the performance objectives. Check with your instructor.

UNIT ▐▐▐ The Windows Operating System

9

The Early Days of Windows

INTRODUCTION

Joe Tekk was working on a computer with a Windows problem. He opened and closed applications, resized several application windows, switched between different tasks, and tested a few screen savers. Then he just let the computer sit and run for a few hours.

Don, the senior technician, saw the computer later in the day and asked Joe what he was doing with it.

"It's an old 25-megahertz 80386 system one of our clients uses at his office. He says that Windows periodically locks up on him, usually when a screen saver is running."

"You mean Windows 3.1? Not Windows 98?"

"Yes," Joe answered, "The machine has only 4 Meg of RAM and a 100-Meg hard drive. Windows 98 would probably run very poorly on it, if it ran at all."

"Is it Windows 3.1 or Windows for Workgroups?"

"Windows for Workgroups. All I know so far is that the screen saver doesn't lock up when I turn off the network stuff."

Don was curious. "What does the network have to do with the screen saver?"

"Nothing, I think," Joe replied. "It might be a memory problem instead. I'm going to check lots of different things before I send it back."

PERFORMANCE OBJECTIVES

Upon completion of this exercise, you will be able to

1. Enter and leave Windows 3.x.
2. Select icons with the mouse.
3. Resize and move windows on the desktop.
4. Start up a Windows application.
5. Start up a DOS shell and run a DOS application from inside Windows.

KEY TERMS

Desktop
Program groups
Protected mode
Real mode

Dialog box
Button
DOS shell

TABLE 9.1 Timeline of Windows operating system releases

Date	Operating System
November 1985	Windows 1.0
April 1987	Windows 2.0, Windows /386
June 1988	Windows /286 and /386 ver. 2.1
May 1990	Windows 3.0
October 1991	Windows 3.0 with Multimedia Extensions
April 1992	Windows 3.1
October 1992	Windows for Workgroups 3.1
August 1993	Windows NT 3.1
February 1994	Windows for Workgroups 3.11
September 1994	Windows NT 3.5
June 1995	Windows NT 3.51
August 1995	Windows 95
August 1996	Windows NT 4.0
November 1996	Windows CE 1.0
November 1997	Windows CE 2.0
June 1998	Windows 98
July 1998	Windows CE 2.1
May 1999	Windows 98 SE
February 2000	Windows 2000
September 2000	Windows ME

BACKGROUND INFORMATION

In the early days of personal computers (before 1990), the prevailing operating system was DOS. But DOS was overtaken by a new type of operating system called Windows. Table 9.1 shows when all of the various versions of the Windows operating systems were released.

Note the sequence of Windows (from 1.0 through ME), Windows NT, and Windows CE releases. Windows 3.1 was by far the most popular of the early Windows releases, with Windows for Workgroups 3.11 adding the missing networking component.

Windows 95, the successor to Windows 3.11, provided enhanced networking, an improved graphical interface, and improved multitasking. Windows 95 and its follow-ups, Windows 98 and Windows ME, are introduced in Exercise 10.

Windows NT, designed from the beginning as a robust operating system, has evolved into an Internet-ready, secure, reliable operating system with support for multiple processors. Coverage of Windows NT and its successor, Windows 2000, is found in Exercises 11 and 30.

The growing market of handheld computing devices found an operating system tailored to its special needs: Windows CE. This operating system is a scaled-down version of Windows 95. Upgrades in Windows CE take advantage of improvements in handheld technology (for example, from B/W to color screens). Windows CE is covered in Exercise 20.

For the remainder of the exercise we will examine the features of Windows 3.x.

The Windows operating system allows you to control the computer in a totally different way than with DOS. For example, DOS commands (such as DIR and FORMAT) must all be entered from the keyboard, on a line-by-line basis. If your typing is not up to speed, it may be very inconvenient for you to continually enter DOS commands while servicing a computer or working on a project.

Windows practically eliminates the need to use the keyboard, getting all of its commands from the mouse. Clicking the mouse button once, or twice quickly, is all that is needed to generate a command. This is possible through the use of a *graphical interface*. In Windows, graphical symbols are used to represent application programs and groups of programs, as

FIGURE 9.1 Typical
Windows 3.x desktop

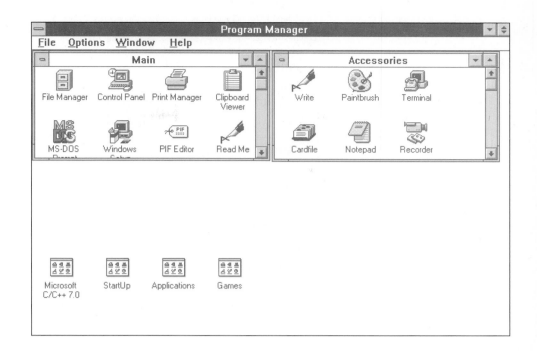

well as commands. This is illustrated in Figure 9.1, which shows a typical Windows 3.x start-up screen (called the **desktop**).

The four square icons at the bottom of the screen in Figure 9.1 represent **program groups**. Each program group may contain one or more programs, as indicated by the open windows for the Main and Accessories groups. To open up the Games window, all you need to do is position the mouse pointer over any portion of the Games icon and press the left mouse button twice quickly. This is roughly equivalent to changing directories in DOS (as in CD \GAMES).

In the next few sections, you will learn how to perform many useful functions inside Windows 3.x.

STARTING WINDOWS

All that is necessary to start Windows 3.x from the DOS prompt is the command

```
C:\> WIN
```

Windows does all the rest, eventually bringing up the Program Manager window shown in Figure 9.1. Depending on your processor's speed, total system RAM size, and hard disk access time, Windows will take different amounts of time to start up and perform other operations. Increasing the amount of RAM (adding 16MB, 32MB, or more) is the easiest way to make Windows run a little faster.

Windows is known as a **protected-mode** operating system, whereas DOS is a **real-mode** operating system. These two modes of operation are found on the 80386, 80486, and Pentium processors from Intel. All of these processors start up in the real mode, acting like really fast 8086 (or 8088) machines, the first machines DOS ran on. Windows switches the processor into the protected mode, in which the full 32-bit power of the processor is available.

LEAVING WINDOWS

At the end of a Windows 3.x session, you must follow a specific sequence to exit to DOS. Simply turning the power off may have drastic consequences for your Windows environment because of the way Windows uses the hard disk to support virtual memory.

FIGURE 9.2   File menu

New...	
Open	Enter
Move...	F7
Copy...	F8
Delete	Del
Properties...	Alt+Enter
Run...	
Exit Windows...	

FIGURE 9.3   Exit Windows dialog box

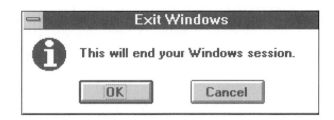

There are two ways to leave Windows 3.x. The first involves the use of the File menu. Examine Figure 9.1 again. Do you see the menu bar near the top of the screen? The four menu items are File, Options, Window, and Help. To choose the File menu, either press Alt+F on the keyboard or position the mouse cursor over the menu name (File) and click the left mouse button. You will get the File menu shown in Figure 9.2.

To select Exit Windows from the File menu, either press X on the keyboard or position the mouse pointer over the Exit Windows line and click the left mouse button again. This will cause the Exit Windows *dialog box* to appear, as shown in Figure 9.3. A dialog box usually contains a brief description of what Windows is doing, and one or more *buttons* to choose from. Notice that the two buttons in the Exit Windows dialog box provide for two choices. If you have changed your mind and are not yet ready to exit, use the Cancel button. Otherwise, to exit Windows, simply press Enter on the keyboard, or position the mouse pointer over the OK button and click the left mouse button. Windows will perform its shut-down sequence and return you to the DOS prompt.

The second way to exit Windows 3.x involves the use of the Program Manager window Close button. In all windows, there is a button in the upper left-hand corner that looks like the front of a file cabinet drawer. Clicking twice quickly (double-clicking) on this button in the Program Manager window brings up the Exit Windows dialog box. Clicking on this button in any other window simply closes the window.

OPENING A PROGRAM GROUP WINDOW

Recall from Figure 9.1 that there were four icons at the bottom of the Windows start-up screen that look identical except for their different names. These icons represent program groups. A program group window is opened by double-clicking with the mouse pointer on the desired icon. For example, double-clicking on the Games icon brings up the Games window shown in Figure 9.4.

Notice that there are seven game applications inside the Games window. Any of these game programs may be executed by double-clicking on the associated icon. Also notice that when the Games window opens, it overwrites portions of the Main and Accessories windows. This is similar to what happens when we place more than one folder on a desktop. Sometimes other folders get covered up. It is easy to view the covered (hidden) window again: simply click with the mouse pointer anywhere inside the visible portion of the covered window. Figure 9.5 shows what happens when the mouse is clicked inside the Main window.

FIGURE 9.4   Opening the Games window

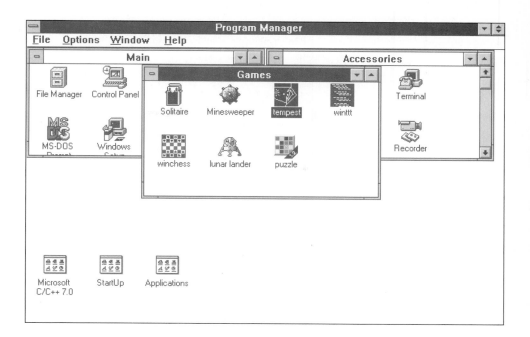

FIGURE 9.5   Reselecting the Main window

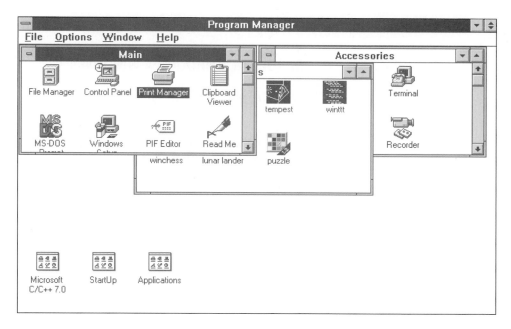

Windows provides an easy way for you to tell which window is the active window. Compare Figures 9.4 and 9.5. The bar containing the name of the Games window is highlighted in Figure 9.4, and in Figure 9.5 the bar for the Main window is highlighted.

MOVING A WINDOW

Refer again to Figure 9.1. There is a large empty portion on the lower right of the desktop. In order to see all three windows that are open (Main, Accessories, and Games), you can move one of the windows into this area. To move a window

1. Place the mouse pointer inside the name bar of the window.
2. Press *and hold* the left mouse button.
3. Drag the window to its new place on the desktop. The window will be outlined as it is being dragged.
4. Release the mouse button.

FIGURE 9.6   Moving the Games window

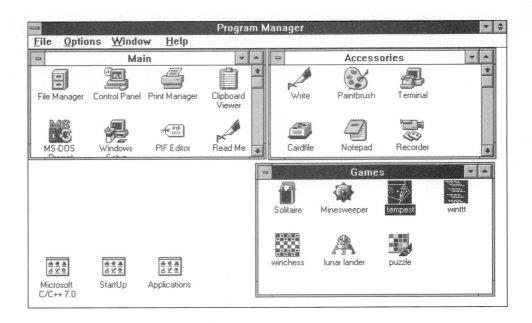

If you use these four steps to move the Games window, you end up with the desktop shown in Figure 9.6.

SCROLLING THROUGH A WINDOW

In Figure 9.6, there are differences between the Games window and the Main and Accessories windows. Notice the scroll bars at the right sides of the Main and Accessories windows. The scroll bar is present whenever there are more icons inside the window than the current window size can display. In the Main window, we cannot fully read the names of the first two icons in the bottom row. In the Accessories window, the scroll bar indicates that there are other icons present that are not shown at all.

One way to view hidden icons in a window is to use the scroll bar to move up/down (or left/right with a horizontal scroll bar) within the window. The up and down arrows shown in Figure 9.6 are used for this purpose. Clicking the down arrow inside the Main window results in the new window shown in Figure 9.7. Now the names of all four bottom-row icons are completely visible.

FIGURE 9.7   Seeing more of the Main window

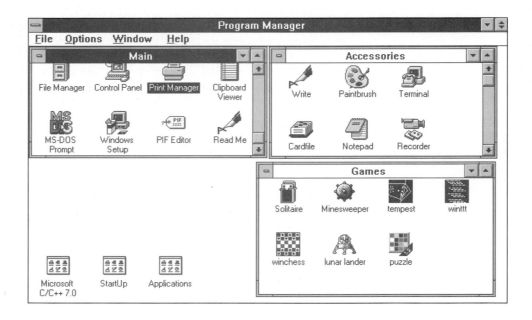

RESIZING A WINDOW

Instead of using the scroll bar, we can graphically resize the window, making it as large or as small as Windows will allow. This is accomplished in the following way:

1. Place the mouse pointer on an edge or corner of the window.
2. Press *and hold* the left mouse button.
3. Drag the window edge (or corner) until the window is the desired size.
4. Release the mouse button.

Using these steps on the Accessories window results in the desktop shown in Figure 9.8.

Since the scroll bar is still present in the window, there must be more icons that are still hidden. To make the window as large as the entire screen (and, hopefully, see the entire contents), simply double-click on the upward-pointing triangle at the top right-hand corner of the Accessories window. The result of this operation is shown in Figure 9.9. To get back to the original desktop, click on the button at the far right end of the menu bar (two triangles pointing in different directions).

FIGURE 9.8   Resizing the Accessories window

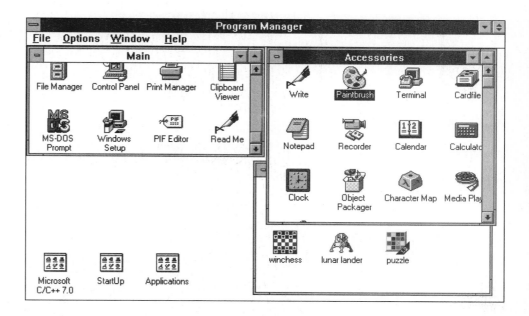

FIGURE 9.9   Resizing the Accessories window to full screen

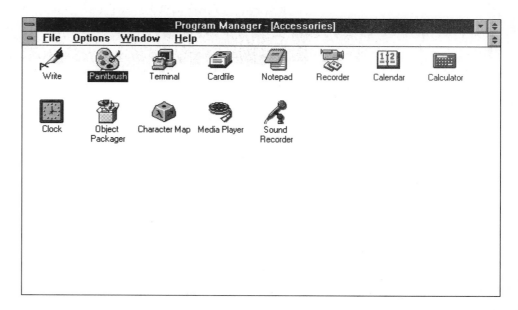

RUNNING A DOS APPLICATION FROM INSIDE WINDOWS

Refer again to Figure 9.8. The MS-DOS Prompt icon in the bottom left-hand corner of the Main window is used to start up a new **DOS shell**, a copy of the DOS environment maintained by Windows. Once inside a DOS shell, you can issue DOS commands as you normally would (DIR, for example). Depending on the settings for the MS-DOS Prompt icon, your DOS shell may run in a small window or may cover the full screen.

Windows requires that you enter the DOS command EXIT when you are through using the DOS shell. EXIT returns you to the Windows desktop.

TROUBLESHOOTING TECHNIQUES

Windows 3.1 users unlucky enough to get a general protection fault understand the frustration of having the operating system *lock up* during an unexpected problem. The advanced Windows operating systems (95, 98, 2000, and NT), as you will see in the remaining exercises of this unit, are much more *robust*, handling even unexpected errors gracefully, rarely getting hung up by a single application. For example, we can always press Ctrl+Alt+Del in Windows 98 to get the operating system's attention, but would only do so in Windows 3.1 under extreme circumstances.

This is just one of the major differences between Windows 3.1 and the advanced Windows operating systems. Read on to discover the others.

SUMMARY

In this exercise we discovered that

- Windows uses a graphical interface.
- Windows is started from DOS.
- Program group windows on the desktop can be resized, opened, closed, and moved.
- DOS applications can be run from inside the Windows operating system.

SELF-TEST

This self-test is designed to help you check your understanding of the background information presented in this exercise.

True/False

Answer *true* or *false*.

1. Windows and DOS are both real-mode operating systems.
2. Windows uses a graphical interface.
3. All icons in a window are displayed when the scroll bar is absent.
4. To exit Windows, you simply turn the computer off.
5. It is not possible to run a DOS program from inside Windows.
6. Program groups reside on the desktop.

Multiple Choice

Select the best answer.

7. To open a program group window, position the mouse pointer over the icon and
   a. Single-click the mouse button.
   b. Double-click the mouse button.
   c. Press and hold the mouse button.
8. To move a program group window, position the mouse pointer inside the name bar and
   a. Single-click the mouse button and drag.
   b. Double-click the mouse button and drag.
   c. Press and hold the mouse button and drag.

9. To start Windows from DOS, enter
   a. START
   b. START/WIN
   c. WIN
10. Windows can be controlled using the
    a. Mouse.
    b. Keyboard.
    c. Both a and b.

Completion

Fill in the blank or blanks with the best answers.

11. Windows is a(n) _____-mode operating system.
12. The main screen used by windows is called the _____.
13. When Windows is exited, a(n) _____ _____ appears with a message and two buttons.
14. An application program is executed by _____ _____ on its icon.
15. A window may contain horizontal or vertical scroll _____.
16. To leave a DOS shell and return to Windows, enter _____.

FAMILIARIZATION ACTIVITY

Be sure you have a floppy disk with some files on it.

1. Start Windows.
2. Open the Main program group by double-clicking on its icon.
3. Double-click on the File Manager icon.
4. Make sure drive A: is empty.
5. Click on the drive icon for drive A:. A dialog box should appear saying "There is no disk in drive A."
6. Place your disk in drive A: and click the Retry button. You should get a directory listing of your disk.
7. Click the drive C: icon. Your hard drive directory should now be displayed.
8. Leave File Manager.
9. Open the Accessories program group.
10. Double-click on the Clock icon.
11. Choose Settings from the Clock menu.
12. Change the clock setting to Analog (if not already set). What does the clock look like?
13. Change the clock setting to Digital. Now what does it look like?
14. Resize the clock to an appropriate size.
15. Exit Windows without closing the Clock window.
16. Restart Windows. Is the Clock still visible?
17. Double-click on the MS-DOS Prompt icon.
18. Enter the DOS command DIR.
19. When the DIR command completes, use EXIT to return to Windows.
20. Leave Windows by double-clicking the Program Manager Close button.

QUESTIONS/ACTIVITIES

1. What are the three main factors affecting the speed of Windows?
2. List some of the differences between the DOS environment and the Windows environment.
3. Explain why a program group is similar to a DOS directory.
4. Examine the files in your \WINDOWS directory. How many files are there? What are the more common extensions? Do you notice any relationship between the file names and the graphical screen information?

Under the supervision of your instructor

1. Enter and leave Windows 3.x.
2. Select icons with the mouse.
3. Resize and move windows on the desktop.
4. Start up a Windows application.
5. Start up a DOS shell and run a DOS application from inside Windows.

10 An Introduction to Windows 95 and Windows 98

INTRODUCTION

Joe Tekk was just about to leave for his lunch break when he was stopped by Don, the senior technician. "Hey, Joe, what happened with the laser printer?"

Joe replied that he had no luck trying to install the old drivers on the printer installation disks, so "I downloaded the new driver from the Web and installed it. Now the printer works fine."

"You did what?" Don asked, seeking more of an explanation.

"I brought up Netscape and used Yahoo to search for the printer manufacturer. Yahoo gave me a link to their home page. I went to it and found new drivers on their support page. I just had to click on the right one to download a copy to my hard drive."

Don was impressed. "Joe, you fixed a problem the customer has had for months with that laser printer. Good job!"

After Joe left, Don smiled to himself. "I would have searched with Alta Vista."

PERFORMANCE OBJECTIVES

Upon completion of this exercise, you will be able to

1. Explain the various items contained on the Windows 95 desktop.
2. Discuss the differences between Windows 3.x and Windows 95.
3. Briefly list the new features of Windows 95.
4. Identify the differences between Windows 95 and Windows 98.

KEY TERMS

Folder
Icon
Taskbar
Background image
Long file name
Context-sensitive menu

Registry
NetBEUI
Window
Preemptive multitasking
Recycle Bin
Channel bar

BACKGROUND INFORMATION

The Windows operating system has gone through many changes since it first appeared in the mid-1980s. It has evolved from a simple add-on to DOS to a multitasking, network-ready, object-oriented, user-friendly operating system. As the power of the underlying CPU running Windows has grown (from the initial 8086 and 8088 microprocessors through the Pentium 4, as well as other microprocessors), so too have the features of the Windows operating system. For users familiar with Windows 3.x, the good news is that many of the operating system features are still there. For example, a left double-click is still used to launch an application. The purpose of this exercise is to familiarize you with many of the features that are new in Windows 95. Where possible, comparisons will be made to Windows 3.x to help you gain an appreciation for how things have changed. In addition, many of the new features of Windows 98 will be introduced.

THE DESKTOP

Once Windows 95 has completely booted up, you may see a display screen similar to that shown in Figure 10.1. This graphical display is called the *desktop*, because it resembles the desktop in an office environment. The desktop may contain various **folders** and **icons**, a taskbar, the current time, and possibly open folders containing other folders and icons. A folder is more than just a subdirectory. A folder can be shared across a network, and it can be cut and pasted just like any other object. You can even e-mail a folder if you want.

Typically, the bottom of the display will contain the **taskbar**, which contains the Start button, icons for all applications currently running or suspended, open desktop folders, and the current time. You can hide, resize, or move the taskbar to adjust the display area for applications. Simply left-clicking on an application's icon in the taskbar makes it the current application.

A new application may be launched by left double-clicking its desktop icon. The desktop may contain a picture, centered or tiled, called the **background image**. The desktop itself is an object that has its own set of properties. For example, you can control how many colors are available to display the desktop. Everything is controlled through the use of easily navigated pop-up menus. The desktop is the subject of Exercise 12.

FIGURE 10.1   Windows 95 desktop

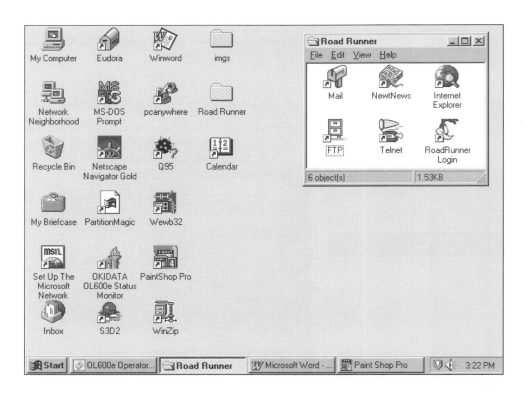

FIGURE 10.2   Two examples of long file names

```
Volume in drive D is FIREBALLXL5
Volume Serial Number is 245F-15E6
Directory of D:\repair3e\e15

.                  <DIR>          12-31-97 12:58a .
..                 <DIR>          12-31-97 12:58a ..
LIST     TXT            0  01-06-98  2:01p list.txt
E15      DOC       31,232  01-06-98  2:01p e15.doc
WOWTHI~1 DOC       20,480  01-06-98  1:38p Wow This is a LONG filename.doc
WOWTHI~2 DOC       20,992  01-06-98  2:01p Wow This is LONG too.doc
         4 file(s)         72,704 bytes
         2 dir(s)      23,674,880 bytes free
```

LONG FILE NAMES

File names in DOS were limited to eight characters with a three-letter extension (commonly called 8.3 notation). Because Windows 3.x ran on top of DOS, it, too, was limited to file names of the 8.3 variety, even though Program Manager allowed longer descriptive names on the program icons.

Windows 95 eliminates the short file name limitation by allowing up to 255 characters for a file name. As shown in Figure 10.2, a ***long file name*** has two representations. One is compatible with older DOS applications (the old 8.3 notation using all uppercase characters). The other, longer representation is stored exactly as it was entered, with uppercase and lowercase letters preserved. To be compatible with older DOS applications, Windows 95 uses the first six characters of a long file name, followed in most cases by ~1. When two or more long file names appear in the same directory, Windows 95 will enumerate them (~1, ~2, etc.), as you can see in the directory listing of Figure 10.2. To specify a long file name in a DOS command, use the abbreviated 8.3 notation, or enter the entire long file name surrounded by double quotation marks. For example, both of these DOS commands are identical in operation:

```
TYPE    WOWTHI~1.DOC
TYPE    "Wow This is a LONG filename.doc"
```

The great advantage of long file names is their ability to describe the contents of a file, without having to resort to cryptic abbreviations.

CONTEXT-SENSITIVE MENUS

In many instances, right-clicking on an object (a program icon, a random location on the desktop, the taskbar) will produce a ***context-sensitive menu*** for the item. For example, right-clicking on a blank portion of the desktop produces the menu shown in Figure 10.3. Right-clicking on the time in the lower corner of the desktop generates a different menu, as indicated by Figure 10.4. Note that the two example menus are different. This is what the

FIGURE 10.3   Context-sensitive desktop menu

FIGURE 10.4 Context-sensitive time/date menu

"context-sensitive" term is all about. Windows 95 provides a menu tailored to the object you right-click on. This is a great improvement over Windows 3.x, which rarely did anything after right-clicking.

IMPROVED HELP FACILITY

The built-in help available with Windows 95 is significantly different from that provided by Windows 3.x. To get help, go to the Start menu. Figure 10.5 shows the Help Topics window that is displayed when Help is selected.

Three tabs appear at the top of the display. The Index tab allows the user to enter keywords that might be found by looking in an index. As each letter is entered, the display is updated to show all matching items. Help for the highlighted selection (left-click to choose a different help topic) is then displayed by left-clicking the Display button or by double-clicking on the help topic.

The two other tabs, Contents and Find, provide additional support in the form of a guided tour of Windows 95, troubleshooting methods, tips, and alternate ways of finding specific help topics.

WINDOWS EXPLORER

Windows 3.x provided two main applications that made life bearable: Program Manager and File Manager. In Windows 95, the services provided by these two applications, as well as

FIGURE 10.5 Help Topics window

FIGURE 10.6   Windows
Explorer

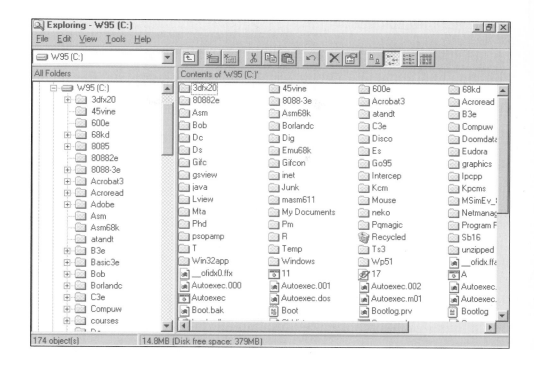

many new features, are found in the new Windows Explorer program. Figure 10.6 shows a typical Explorer window. Although we will cover Windows Explorer in detail in Exercise 14, it is worth taking a quick look at now. The small box in the upper left corner containing W95 (C:) indicates the current folder selected. Clicking the down arrow produces a list of folders to choose from. The two larger windows display, respectively, a directory tree of drive C: (folders only) and the contents of the currently selected folder (which also happens to be drive C:). Note the different icons associated with the files shown. Windows Explorer allows you to change the icon, or associate it with a different file. In general, as with Windows 3.x, double-clicking on a file or its icon opens the application associated with it.

Windows Explorer also lets you map network drives, search for a file or folder, and create new folders, among other things. It is truly one of the more important features of Windows 95.

THE REGISTRY

The *Registry* is the Windows 95 replacement for the SYSTEM.INI and WIN.INI configuration files used by Windows 3.x. The Registry is an internal operating system file maintained by Windows 95. As each application is installed, the installation program makes "calls" to the Registry to add configuration information, storing similar information to what was previously stored in the .INI files. In this way, the Registry is protected and therefore is harder to corrupt. The Registry is accessed using the REGEDIT program as illustrated in Figure 10.7. The Registry is nothing to fool around with simply because you feel like experimenting. A corrupt Registry can prevent Windows 95 from booting and could possibly require a complete reinstallation of Windows 95.

The Registry contains all the information Windows 95 knows about both the hardware and software installed on the computer.

NETWORKING

Windows 95 offers a major improvement in networking capabilities. Windows 3.1 had no built-in networking support and required network software to be loaded and maintained by DOS. This situation was improved slightly with the release of Windows for Workgroups 3.11, which provided limited networking via a network protocol called *NetBEUI* (NetBIOS Extended User Interface), which allows e-mail and file sharing in small peer-to-peer networks.

FIGURE 10.7 REGEDIT window

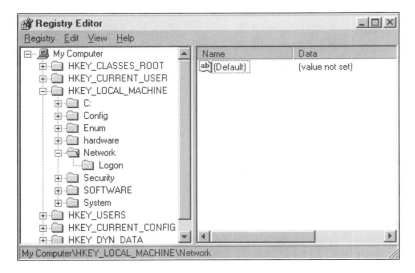

NetBEUI is still available in Windows 95 (providing the Network Neighborhood feature, network drives, shared printers), along with other additions. Two important protocols have been added, PPP (point-to-point protocol) and TCP/IP (transmission control protocol/Internet protocol). PPP is used with a serial connection (such as a modem) and is the basis for the dial-up networking provided by Windows 95. TCP/IP is the protocol used by the Internet. You can use TCP/IP applications such as Netscape or Internet Explorer to browse the World Wide Web, connect to a remote computer and share files, send and receive e-mail, and much more. Typically, TCP/IP is used in conjunction with a network interface card for fast data transmission, although it can also be used over a modem connection (by encapsulating it inside the PPP).

DOS

Yes, DOS is still a part of Windows. However, the role of DOS in Windows 95 is different in many ways from what was required under Windows 3.x. For example, Windows 3.x relied on the file system set up and maintained by DOS. This is commonly referred to as running "on-top" of DOS. Windows 95 does things differently, providing its own improved file system (long file names) and a *window* to run your DOS application in. You can open several DOS windows at the same time if necessary, and run different applications in each. Figure 10.8 shows a typical DOS window.

FIGURE 10.8 DOS window

```
E:\>dir

 Volume in drive E is EXTRA
 Volume Serial Number is 2F31-ODED
 Directory of E:\

QUAKE2       <DIR>        12-19-97 11:44p Quake2
MARIJUA1 GIF        4,391 11-23-97 12:02a Marijua1.gif
NETANT       <DIR>        12-03-97  3:42a NetAnt
NETXRAY      <DIR>        12-04-97 12:58a NetXRay
NET_TOOB     <DIR>        12-06-97 10:41p NET_TOOB
DOOM         <DIR>        12-09-97  5:58a Doom
        1 file(s)         4,391 bytes
        5 dir(s)    123,871,232 bytes free

E:\>ver

Windows 95. [Version 4.00.950]

E:\>_
```

FIGURE 10.9   Close window

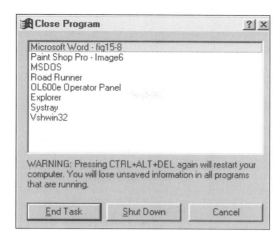

Furthermore, in Windows 3.x, it is possible for a wayward application running in a DOS shell to completely hang the system. Applications are not allowed that much control in Windows 95. If a DOS application (or any other, for that matter) hangs up, all you need to do is press Ctrl+Alt+Del (oddly enough) to bring up the Close window. Figure 10.9 illustrates the Close window.

It is important to note that Windows 95 is always in control somewhere in the background, monitoring everything. A DOS application cannot hang the system like its Windows 3.x counterpart.

PREEMPTIVE MULTITASKING

A significant change in the way multiple programs are executed was made when Windows 95 was developed. Windows 3.x used a technique called *cooperative multitasking* to run more than one 16-bit application at a time. Each application would get a slice of the processor's computation time, with periodic switches between applications. The problem with this form of multitasking is that one task can take over the system (by requesting all available memory or other resource) and prevent the other tasks from running.

Windows 95 uses a method called ***preemptive multitasking*** to run multiple 32-bit applications while still running the older 16-bit Windows 3.x applications in cooperative mode (in a single shared block of memory). Preemptive multitasking means that the current application can be interrupted and another application started or switched to. This provides a degree of *fairness* to the set of applications competing for processor time. For example, referring back to Figure 10.1, it is possible to quickly double-click on several icons (Netscape, Eudora, Winword) before the first application has completely come up. All applications clicked will start up, with the first one clicked getting the initial shot at the desktop.

In Windows 3.x, when you double-clicked an application to start it, you had to wait until the hourglass went away before you could do anything else. In Windows 95, the hourglass still indicates that the operating system is busy with a chore, but it may be preempted to begin a new task at any time. Once again, we see that both the user and the operating system have more control under Windows 95 than was possible with Windows 3.x.

RECYCLE BIN

The ***Recycle Bin*** shown in Figure 10.10(a) is a holding place for anything that is deleted in Windows 95. The nice thing about the Recycle Bin is that *you can get your files back* if you want to, using the Undelete option. Double-clicking on the Recycle Bin icon brings up the window shown in Figure 10.10(b). Any or all of the files shown in the window may be recovered. *Warning:* If you delete files while in DOS, they are not deposited in the Recycle Bin, and may be impossible to recover at a later time (with the old UNDELETE command).

FIGURE 10.10   Recycle Bin

(a) Program icon

Recycle Bin				_ □ ×
File Edit View Help				
Name	Original Location	Date Deleted	Type	Size
~$fig	D:\repair3e\e15	1/9/98 3:44 PM	Microsoft Word Doc...	1KB
~$list	D:\repair3e\e15	1/9/98 3:44 PM	Microsoft Word Doc...	1KB
ENGLISH.RES	C:\GRAVIS\GRAVU...	1/4/98 12:06 AM	Intermediate File	102KB
GRAVUTIL	C:\GRAVIS\GRAVU...	1/4/98 12:06 AM	Application	365KB
GRAVUTIL	C:\GRAVIS\GRAVU...	1/4/98 12:06 AM	Help File	9KB
GRAVUTIL	C:\GRAVIS\GRAVU...	1/4/98 12:06 AM	Shortcut to MS-DOS...	1KB
OKIDATA OL600e...	C:\WINDOWS\Des...	1/7/98 9:24 PM	Shortcut	1KB
7 object(s)		475KB		

(b) Window

OTHER NEW FEATURES

A summary of several other new features is included here to help you get a good idea of how much of an improvement Windows 95 is over Windows 3.x.

- *Remote Procedure Calls*   RPCs allow computers in a network to share their processing capabilities. For example, a 486 machine could issue an RPC to a Pentium-based machine to execute code *on that machine* and send the results back to the 486.
- *Support for OLE 2*   Object Linking and Embedding 2 (OLE 2) provides powerful features supporting a dynamic application environment. Cutting and pasting text, images, and other types of objects are just the beginning of OLE 2. Windows 95 provides a large set of enhanced OLE 2 features, such as drag-and-drop, nested objects, and optimized object storage.
- *Accessibility Options*   Many new features have been added to allow differently abled users to customize their computers to meet individual needs. The keyboard, display, sounds, and mouse can be set to a variety of combinations to suit most needs.
- *Microsoft Exchange*   E-mail is now a standard feature on the Windows 95 operating system.

WINDOWS 95 VERSION B

A number of bugs encountered in version A (the first release) of Windows 95 can be fixed by downloading Service Pack 1 from Microsoft's Web site (http://www.microsoft.com). The service pack updates Windows 95 by modifying various portions of the operating system (such as the kernel), and provides additional drivers and other improvements.

Version B of Windows 95 (called OEM Service Release 2) offers additional improvements, such as FAT32 (a better way of organizing files on your hard drive to reduce lost storage space), many new drivers, and other enhancements. Unfortunately, you cannot upgrade Windows 95 A to Windows 95 B.

THE WINDOWS 98 DESKTOP

Many changes in Windows 98 are visible right on the desktop. A ***channel bar*** provides one-click access to your favorite or often-used Internet links. A channel is a connection to a Web

FIGURE 10.11   Windows 98
desktop

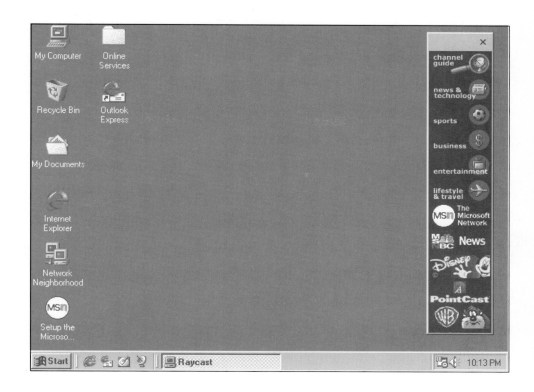

site that allows you to schedule automatic updates of the information you need. The desktop can be organized and operated as a Web page if you desire, even displaying a specific Web page as its background.

Figure 10.11 shows a sample Windows 98 desktop. The channel bar is to the right and contains several preassigned Internet links. The taskbar has a new area that holds often-used shortcuts. This is a nice feature because desktop shortcuts are hidden from view in Windows 95 whenever a window uses the full screen.

WHAT ELSE IS NEW IN WINDOWS 98?

The Welcome to Windows tour (Programs, Accessories, System Tools) contains details for users of Windows 3.x, Windows 95, and new users. All the new features of Windows 98 are highlighted and presented in an easy-to-view multimedia format. Figure 10.12 shows the main feature screen. Clicking on a feature opens up a short slide show with narration and a quick run-through of the menus required to use the feature. Let's take a brief look at each feature group.

Windows 98 Is Easier to Use

In addition to the Web-style features we will cover shortly, many other improvements are found in Windows 98. Two or more monitors (with associated display adapter cards) can be used at the same time. Special OnNow hardware can be managed to reduce power use. Advanced Plug & Play includes support for the Universal Serial Bus. A new utility called Microsoft Magnifier uses a portion of the desktop as a magnifying glass. The portion of the desktop closest to the mouse is displayed with an adjustable level of magnification. In addition, a substantial amount of online help is provided.

Windows 98 Is Enhanced for the Web

Windows 98 comes with Internet Explorer 4.0 included; a Connection Wizard to help make your connection to the Internet a smooth process; desktop Web page capabilities; Channels (favorite Internet links); Outlook Express, an all-purpose e-mail application; NetMeeting (use a digital whiteboard to sketch proposals, use Chat to confer with group members); and

FIGURE 10.12 New features in Windows 98

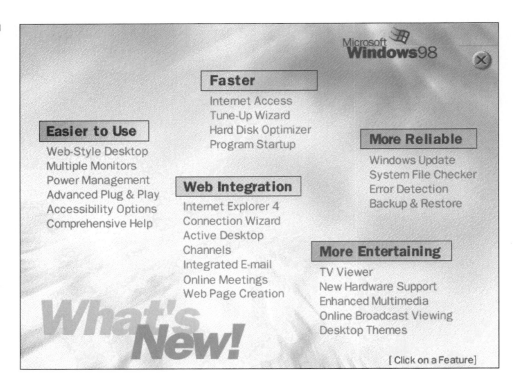

FIGURE 10.13 Web-style desktop

even FrontPage Express for creating your own Web pages. Figure 10.13 shows the Windows 98 desktop setup for Web-style appearance. If a Web page is used as the background, you can elect to hide the desktop icons to get a true browser appearance.

Windows 98 Is Faster and More Reliable

Microsoft has added a number of features to Windows 98 to enhance its operating speed and increase reliability. The Hard Disk Optimizer will convert FAT16 file systems into the newer FAT32 file system, which allows for larger drives and more efficient use of drive space. In addition, applications that are used frequently can be relocated to a special portion of the disk to allow quicker launching.

Windows 98 can be updated automatically over the Web. For reliability, a System File Checker application checks the integrity of the files required by the operating system. At boot time, ScanDisk runs automatically if the previous shutdown was not completed correctly.

Windows 98 Is More Entertaining

A number of multimedia improvements are found in Windows 98. With special hardware, Windows 98 allows you to view broadcast television from the airwaves or gather and present information from the Internet (using NetShow). The Universal Serial Bus is also supported; this allows devices such as DVD (digital versatile disk) players and *force-feedback* joysticks to be connected.

DirectX support is built-in, enabling real-time high-quality 3-D graphics for supported graphics adapters. There are also many new screen savers (which take advantage of DirectX for stunning visual effects).

Overall, Windows 98 is another big step forward, as Windows 95 was compared to Windows 3.x.

WINDOWS 98 SE (SECOND EDITION)

As with Windows 95, Microsoft released an updated version of Windows 98 called Windows 98 SE. Some of the significant improvements are

- Internet Connection Sharing
- Improved USB support
- IEEE 1394 FireWire support

Internet Connection Sharing allows the computers on a home network to share an Internet modem connection provided by one of the networked computers.

The IEEE 1394 FireWire standard is a new, high-speed serial bus that has become popular for connecting hard drives and capturing data from digital video recorders and other multimedia devices.

Many other improvements come with Windows 98 SE. Newer versions of Internet Explorer, Windows Media Player, and Net Meeting are available, as are bug fixes, additional hardware support, and online help.

WINDOWS ME (MILLENNIUM EDITION)

Windows ME is the next step forward for home operating system users (just as Windows 2000, covered in Exercise 11, is the next step for corporate and professional operating system users). Some of the major additions in Windows ME are

- System Restore
- System File Protection
- Easier network configuration
- Improved multimedia support

The System Restore feature identifies corrupted system files and automatically replaces them. System Restore can also be used to recover from a faulty software installation or after a catastrophic operating system failure. The System Restore will restore the system files to a *time* before the event that damaged the system. System File Protection helps prevent important system files from being improperly altered or updated.

Home network users get additional support from Windows ME, which automatically identifies shared resources and provides more assistance when adding a new machine to the network.

The emphasis on multimedia in Windows ME is evident. A new application called Movie Maker allows you to create your own movies captured from your camcorder or

VCR. A new version of Windows Media Player contains additional support for Internet audio and video streaming.

There are many other improvements, including new versions of Internet Explorer, Outlook Express, and numerous enhancements to Windows applications.

One change that is also significant is the way Windows ME treats DOS applications. With no real-mode DOS support from Windows ME, some older DOS applications may not run anymore. This is a point to consider for those users thinking of upgrading from Windows 95 or 98.

Visit http://www.zdnet.com for many useful articles on all the Windows operating systems.

TROUBLESHOOTING TECHNIQUES

There are so many new features in Windows 95/98 it is easy to lose track of them and not use one or more of them to make your life easier. For instance, rather than getting out a large, heavy Windows reference book, simply search the topics in Help for the answers you need.

Also, remember that it is usually okay to right-click on *anything*, any object (file, program icon, taskbar, display), and a context-sensitive menu will typically pop up. Sometimes you might get a What's This? button, which provides a short description of whatever you right-clicked on.

SUMMARY

In this exercise we discovered that

- Windows uses a graphical desktop containing folders and icons.
- Windows supports long file names of up to 255 characters.
- Windows provides context-sensitive menus.
- Windows contains support for networking and DOS.
- New Windows programs use preemptive multitasking.
- Deleted files go into the Recycle Bin, and you can get your files back from the Recycle Bin.

SELF-TEST

This self-test is designed to help you check your understanding of the background information presented in this exercise.

True/False

Answer *true* or *false*.

1. Long file names are limited to 500 characters.
2. NetBEUI is a networking protocol.
3. The Registry is just a database of .INI files.
4. You can open only one DOS window at a time.
5. Once a file enters the Recycle Bin, it is gone forever.
6. FAT32 is available on all Windows 95 computers.
7. Windows 98 contains a channel bar used to watch television.
8. The Windows 98 desktop can be organized and operated as a Web page.
9. The Windows 98 operating system supports two displays at the same time.

Multiple Choice

Select the best answer.

10. The desktop contains
    a. The taskbar, application icons, and folders.
    b. The taskbar, running applications, and a communications console.
    c. A set of folders for each hard drive.
    d. All of the above.

11. A context-sensitive menu is displayed when
    a. Double-clicking on a program icon.
    b. Pressing both mouse buttons at the same time.
    c. Right-clicking almost anything.
    d. None of the above.
12. Windows 95 supports
    a. 16-bit applications.
    b. 32-bit applications.
    c. Both a and b.
    d. DOS applications only.
13. Windows Explorer
    a. Replaces Program Manager and File Manager.
    b. Explores the Internet.
    c. Searches for viruses on the hard drive.
    d. None of the above.
14. Windows 98 includes Outlook Express, which provides
    a. A digital whiteboard and a chat feature.
    b. An HTML editor to create custom Web pages.
    c. An all-purpose e-mail application.
    d. Reduced system power consumption during peak system activity.
15. The new Windows 98 utility program Microsoft Magnifier is used to
    a. Show all channels on the desktop.
    b. Show a portion of the display as a magnifying glass.
    c. Send e-mail messages on the Internet.
    d. Access the devices on a Universal Serial Bus.
16. FAT32 is better than FAT16 because
    a. Hard drive space is used more efficiently.
    b. Memory resources are reduced and the system runs faster.
    c. It cannot be used on small disk drives.
    d. 32 is higher than 16 and it is a multiple of 2.

Completion

Fill in the blank or blanks with the best answers.

17. Two new Windows 95 protocols are PPP and _____.
18. Windows 95 runs 32-bit applications using _____ multitasking.
19. The _____ contains folders, icons, and the taskbar.
20. Windows Explorer can be used to map a(n) _____ drive.
21. A centralized location storing all configuration information about a Windows 95 system is called the _____.
22. Windows 98 can use multiple _____ at the same time.
23. ScanDisk runs automatically on Windows 98 if the previous _____ was not successful.

FAMILIARIZATION ACTIVITY

1. Boot up Windows 95/98.
2. On the taskbar, left-click the Start button.
3. Move the mouse pointer up to the Help icon and left-click it.
4. In the Help Topics window, left-click the Index tab.
5. Enter the word "network" in the text box (left-click inside the box if the cursor is not visible).
6. From the list of network topics, chose Dial-Up Networking by left-clicking on it.
7. Left-click the Display button.
8. Left-click the Cancel button.
9. Double left-click the Dial-Up Networking topic.
10. Left-click the Cancel button.

11. Left-click the Cancel button to return to the desktop.
12. Right-click on an empty portion of the desktop.
13. Move the mouse pointer to the New menu selection.
14. Move the mouse pointer to the Shortcut selection on the submenu.
15. Left-click on the Shortcut item.
16. Click the Browse button.
17. Locate the Windows folder icon (using the scroll bar).
18. Double left-click on the Windows folder icon.
19. Locate the Calendar icon (using the scroll bar).
20. Left-click on the Calendar icon.
21. Left-click the Open button.
22. Left-click on the Next button.
23. Click on the Finish button.
24. Double left-click on the Calendar icon.
25. Close the Calendar application.

QUESTIONS/ACTIVITIES

1. Does Windows 95 ever lose control of the system?
2. How are long file names backward compatible with the older DOS 8.3 notation?
3. What types of networking does Windows 95 provide?
4. Name five features found in Windows 98 that are not found in Windows 95.

REVIEW QUIZ

Under the supervision of your instructor

1. Explain the various items contained on the Windows 95 desktop.
2. Discuss the differences between Windows 3.x and Windows 95.
3. Briefly list the new features of Windows 95.
4. Identify the differences between Windows 95 and Windows 98.

11 An Introduction to Windows NT and Windows 2000

INTRODUCTION

Joe Tekk was examining the computer book section of a local bookstore. Shelf after shelf contained books about the Windows operating system. There were books about Windows 95, Windows 98, Windows NT, and Windows 2000.

Joe thought about where the future of the Windows operating system is headed. Just skimming through the new features of Windows 98 showed him that a major push toward integrating the Web into the operating system was undertaken.

Next, Joe turned his attention to the Windows NT operating system. With support for multiple processors, improved security and system management tools, and full 32-bit code utilization, Windows NT has significant differences from Windows 95/98, and is well suited for network server operation.

A voice from behind broke into Joe's thoughts. "I remember when the entire operating system for the PC fit into 32K of RAM."

Joe turned to see who had spoken to him. It was an old man, well into his eighties, hunched over and standing with the help of a cane. Joe noticed that the cane had a microchip encased in clear plastic on the handle.

"Now you need 16MB just to run the install program," Joe replied. He spent the next two hours talking to the old man about operating systems and learned a great deal from him.

PERFORMANCE OBJECTIVES

Upon completion of this exercise, you will be able to

1. Identify the differences between Windows 95/98 and Windows NT.
2. Identify the key features of Windows NT.
3. Identify any common features of Windows 95/98 and Windows NT.
4. Describe the features of Windows 2000.

KEY TERMS

NTFS   Network administrator
Domain

BACKGROUND INFORMATION

Windows NT is another operating system developed by Microsoft. It was developed to create a large, distributed, and secure network of computers for deployment in a large organization, company, or enterprise. Windows NT actually consists of two products: Windows

NT Server and Windows NT Workstation. The server product is used as the server in the client-server environment. Usually a server will contain more hardware than the regular desktop-type computer, such as extra disks and memory. The workstation product is designed to run on a regular desktop computer (consisting of an 80486 processor or better). Windows NT provides users a more stable and secure environment, offering many features not available in Windows 95/98 such as *NTFS*, a more advanced file system than FAT 16/32. The newest version of Windows NT, called Windows 2000, is covered at the end of this exercise.

We will use the Windows NT Server product to illustrate the user interface into the Windows NT environment. Let's begin by looking at the Windows NT login process.

WINDOWS NT OPERATING SYSTEM LOGON

One of the first things a new user will notice about the Windows NT environment is the method used to log in. The only way to initiate a log on is to press the Ctrl+Alt+Del keys simultaneously as shown in Figure 11.1. This, of course, is the method used to reboot a computer running DOS or Windows 95/98. Using Windows NT, the Ctrl+Alt+Del keys will no longer cause the computer to reboot, although it will get Windows NT's attention.

If the computer is not logged on, Windows NT displays the logon screen, requesting a user name and password. During the Windows NT installation process, the Administrator account is created. If the computer (Windows NT Server, Windows NT Workstation, Windows 95/98, or Windows for Workgroups 3.11) is configured to run on a network, the logon screen also requests the domain information. After a valid user name, such as Administrator, and the correct password are entered, the Windows NT desktop is displayed, as shown in Figure 11.2.

FIGURE 11.1   Windows NT Begin Logon window

FIGURE 11.2   Windows NT desktop

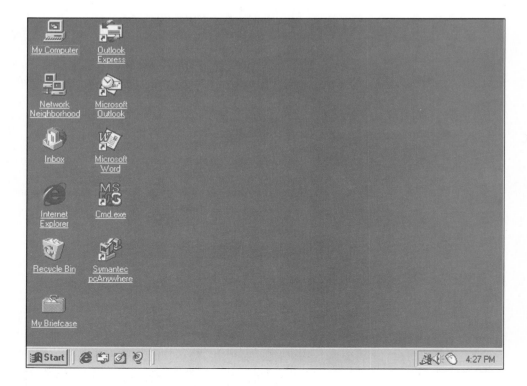

WINDOWS NT SECURITY MENU

After a Windows NT Server or Workstation is logged in, pressing the Ctrl+Alt+Del keys simultaneously results in the Windows NT Security menu being displayed as illustrated in Figure 11.3. From the Windows NT Security menu it is possible for the operator to select from several different options, including Cancel to return to Windows NT.

The Lock Workstation option is used to put the Windows NT Server or Workstation computer in a *locked* state. The locked state is usually used when the computer is left unattended, such as during lunch, dinner, nights, and weekends. When a computer is locked, the desktop is hidden and all applications continue to run. The display either enters the screen saver mode or displays a window requesting the password used to unlock the computer. The password is the same one used to log on.

The Logoff option is used to log off from the Windows NT computer. The logoff procedure also can be accessed from the shut-down menu by selecting the appropriate setting. The logoff procedure terminates all tasks associated with the user but continues running all system tasks. The system returns to the logon screen shown in Figure 11.1. The Shut Down option must be selected before system power can be turned off.

The Change Password option is used to change the password of the currently logged-in user.

The Task Manager option causes the Windows NT Task Manager window to be displayed. The Task Manager is responsible for running all the system applications, as indicated by Figure 11.4. Notice that individual applications may be created, selected, and

FIGURE 11.3   Windows NT Security menu

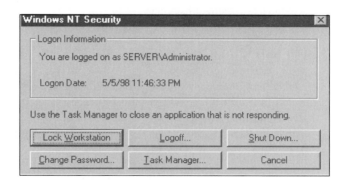

FIGURE 11.4   Task Manager applications

ended or switched by using the appropriate buttons. It is sometimes necessary to end tasks that are not functioning properly for some reason or another. In these cases, the status of the application is usually "not responding."

Each application controls processes that actually perform the required tasks. Figure 11.5 shows a number of processes being executed by the Task Manager. Applications may create as many processes as necessary. Extreme caution must be exercised when ending a process shown on the Processes display. The processes used to control Windows NT can also be ended, causing the computer to be left in an unknown state. If processes must be terminated, it is best to use the Applications tab.

The Task Manager can also display the system performance. Figure 11.6 shows a graphical display of current CPU and memory utilization. It also shows a numeric display of other critical information.

WINDOWS NT DESKTOP, TASKBAR, AND START MENU

The look and the feel of the Windows 95/98 desktop have been incorporated into the Windows NT desktop. At first glance, it might be hard to tell the difference between the two operating systems. Figure 11.7 shows the Start menu of a typical Windows NT desktop. Aside from the Windows NT text display along the left margin of the Start menu, it contains some of the same categories found on Windows 95/98 computers. As you will see, there are many similarities to Windows 95/98 (such as the desktop, Start menu, and the taskbar), and many differences.

WINDOWS NT CONTROL PANEL

Let's continue our investigation of Windows NT by looking in the Control Panel as illustrated in Figure 11.8. Within the Control Panel, there are several changes. For example, the System icon in Windows NT brings up a completely different display than Windows 95/98, as shown in Figure 11.9. This is no longer the place to go when examining system

FIGURE 11.5   Windows NT processes

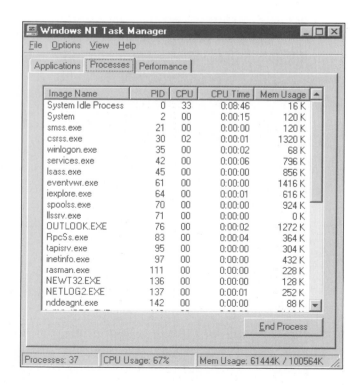

FIGURE 11.6   Task Manager Performance display

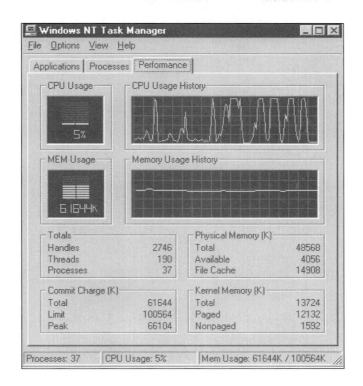

FIGURE 11.7   Windows NT Start menu

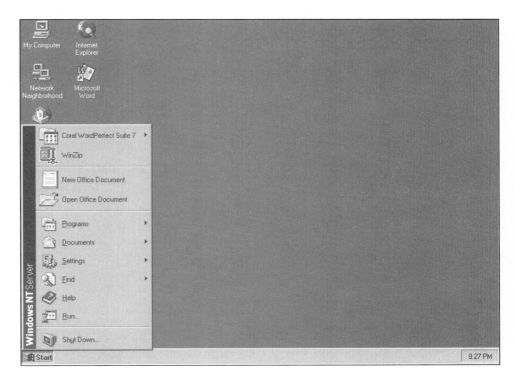

hardware components. The Devices icon provides that function in Windows NT. This is illustrated in Figure 11.10. Each individual device may be configured from this window.

The Services icon is used to control the software configuration on Windows NT. Usually, a service is started when the computer is booted, but it can also be started or stopped when a user logs in or out of Windows NT. Note that the Administrator account must be used when making changes to the Windows NT service configuration. A typical

FIGURE 11.8 Items in the
Windows NT Control Panel

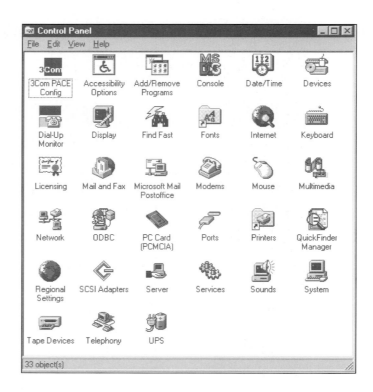

FIGURE 11.9 Startup and
shutdown system properties

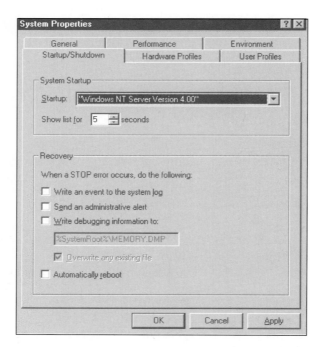

Services window is shown in Figure 11.11. You are encouraged to explore the different
icons in the Windows NT Control Panel.

WINDOWS NT DOMAINS

Windows NT computers usually belong to a computer network called a ***domain***. The domain will collectively contain most of the resources available to members of the domain. Computers running Windows NT Server software offer their resources to the network clients

FIGURE 11.10   A list of devices

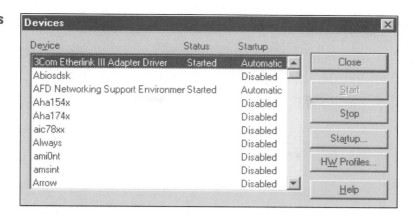

FIGURE 11.11   A list of services

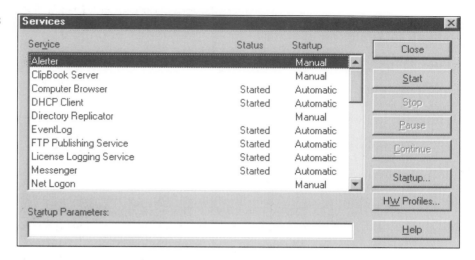

(Windows NT Workstations, Windows 95/98, and Windows for Workgroups 3.11). For example, during the logon process, a Windows NT Server responsible for controlling a domain will verify the user information (a user name and password) before access to the computer is allowed. The Network Neighborhood allows access to the resources available on other computers in the domain. Figure 11.12 shows the Network menu, where network components are configured.

The *network administrator* determines how the network is set up and how each of the components is configured. It is always a good idea to know whom to contact when information about a network is required. If the setting is not correct, unpredictable events may occur on the network, creating the potential for problems.

WINDOWS 2000

At first glance, Windows 2000 looks similar to the older versions of Windows NT with many subtle and not so subtle changes. The Windows 2000 operating system consists of two general versions (similar to Windows NT 4.0), the Windows 2000 Server and Windows 2000 Professional. The Windows 2000 Server product is further broken down into three products: the basic Windows 2000 Server and two additional products called Windows 2000 Advanced Server and Windows 2000 Datacenter Server. With all of these server products to choose from, a Windows 2000 Server solution is available for every size business or organization.

Windows 2000 Professional is designed for the client computers in any size business. Windows 2000 Professional extends the security, reliability, and manageability available in

FIGURE 11.12 Currently installed network protocols

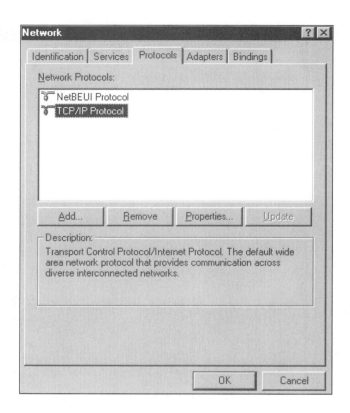

NT 4.0 Workstation but also includes many of the latest features and enhancements available in Windows 98. The Windows 2000 Server products are designed for file, print, application, and Web servers. The Windows 2000 Server also supports the creation of domains, allowing for central administration of all Windows 2000 services. The Windows 2000 Advanced Server and Windows 2000 Datacenter Server offer increased capabilities for enterprise applications and data warehousing. We will briefly examine some of the features and benefits of these new Windows operating systems.

Windows 2000 Server

Let's begin by looking at the desktop view of Windows 2000 Server illustrated in Figure 11.13, which shows the Administrative Tools portion of the Start menu. First of all, you may notice many differences in the list of Administrative Tools. In addition, you should notice that Windows NT Explorer is not listed in the Programs menu as in the previous versions of the operating system. In Windows 2000 Server, Windows NT Explorer has been renamed Windows Explorer and has been moved to the Accessories submenu. An examination of Windows Explorer shows a few more of these subtle differences, as illustrated in Figure 11.14.

Notice that the Folders list shown in the left pane of the Explorer display now includes an entry for My Documents, and the Network Neighborhood has been replaced by My Network Places. Another important change is that additional help information is easily accessible on the Explorer display. Virtually every commonly used application has been enhanced in the Windows 2000 upgrade.

From a system administration point of view, Windows 2000 supports all of the existing applications such as the BackOffice application suite. In many cases, new versions of the applications programs are available. There are a few major differences that help to simplify the administration of a Windows 2000 Server computer. A new application called Windows 2000 Configure Your Server provides a quick method to access the most common applications, as shown in Figure 11.15.

In addition to the obvious differences between Windows 2000 and its predecessors, you will also notice changes in many application names and/or the locations where they are stored. For example, the functionality of Disk Administrator program is now part of the Computer Management application, as shown in Figure 11.16.

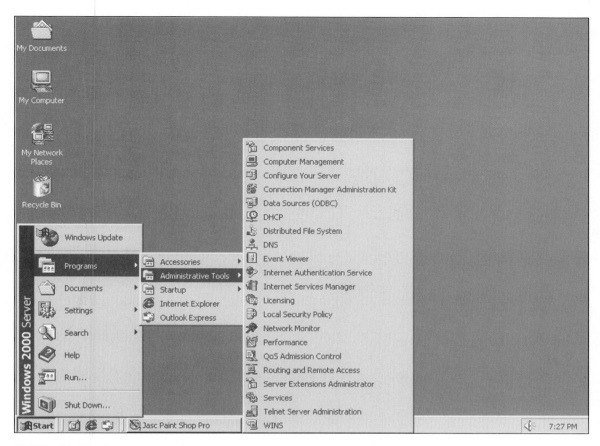

FIGURE 11.13 Windows 2000 Server desktop

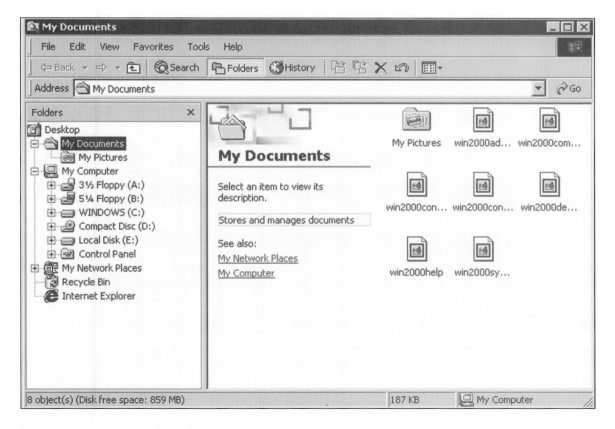

FIGURE 11.14 Windows 2000 Server Explorer

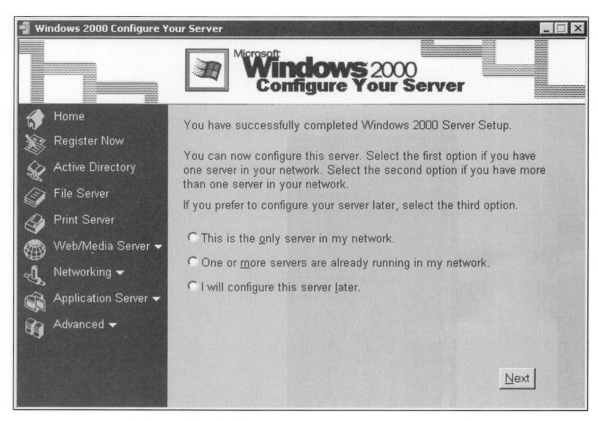

FIGURE 11.15 New Windows 2000 configuration tool

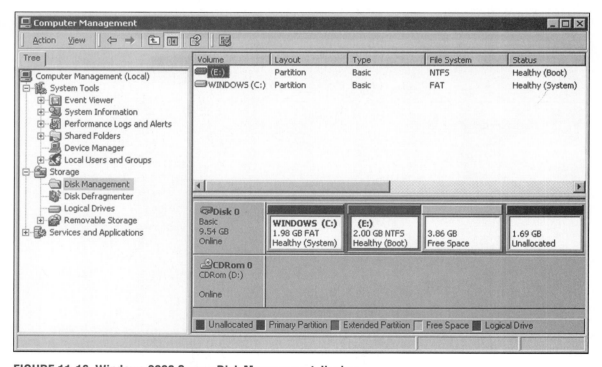

FIGURE 11.16 Windows 2000 Server Disk Management display

Generally, all the old features of Disk Administrator are supported, but in addition, new features have also been added, such as a built-in defragmenter. Some of the most notable features of Windows 2000 Server are shown in Table 11.1. You are encouraged to explore these features using the updated Windows 2000 Server help system that is displayed in Figure 11.17.

TABLE 11.1 A list of built-in features of Windows 2000 Server

Windows 2000 Features	Description
Supports the latest Windows technologies	Active directory interfaces Group policies NTFS 5.0 Plug-and-play support Resources Reservation Protocol support
Internet Server applications	Domain Name Services (DNS) Dynamic Host Configuration Protocol (DHCP) Lightweight Directory Access Protocol support Internet Printing Internet Connection sharing Network Address Translation (NAT) Streaming media server
Safe mode start-up	System Troubleshooting Tool
IntelliMirror	Windows 2000 Professional client support
File server	Folder sharing
Message queuing	Support for distributed applications
Public Key and Certificate Management	Support for Virutal Private Networks Kerberos V support
Disk and File Management	Disk Quotas Distributed File System Distributed Link Training Distributed Authoring and Versioning Content Indexing Encrypting File Systems (EFS)

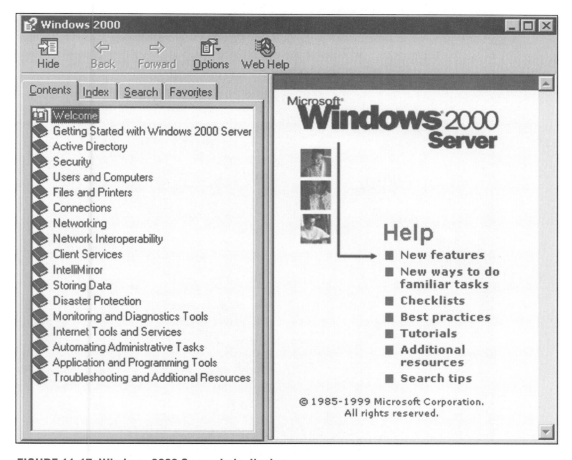

FIGURE 11.17 Windows 2000 Server help display

After a careful review, a System Administrator will realize that there are many reasons to upgrade to Windows 2000 Server.

Windows 2000 Professional

The Windows 2000 Professional operating system can be used in combination with Windows 2000 Server and offers many features to the user. The Windows 2000 Professional operating system is easier to use, manage, and troubleshoot compared to Windows NT Workstation. There is additional support for adding new hardware, including an Add/Remove Hardware Wizard, plug-and-play support, and additional power-saving options. Windows 2000 Professional also provides additional support for mobile users. These features include built-in support for Virtual Private Networks, Internet Printing, and offline folders that, when used with a Synchronization Manager, are designed to keep everything current with a minimum of effort. Additional features allow for remote administration and installation.

Note that the Windows 2000 Professional operating system is not able to run *any* of the Windows 2000 Server applications. For example, a Windows 2000 Professional computer cannot offer Directory Services or function as a DNS, DHCP, or a WINS server. If any of the server-based applications are required, it is necessary to install at least one copy of Windows 2000 Server. With these exceptions, Windows 2000 Professional shares the same user interface enhancements and many of the general user application programs as Windows 2000 Server. You are encouraged to explore the features of all the Windows 2000 operating systems.

TROUBLESHOOTING TECHNIQUES

Windows NT provides many different troubleshooting aids to tackle a wide variety of common problems. The Administrative Tools menu shown in Figure 11.18 shows a list of common applications designed to properly configure a Windows NT computer. One of the most common administrative tools is the Windows NT Diagnostics application shown in Figure 11.19. The Diagnostics menu contains several different tabs, each showing a specific area of the system. When experiencing problems, it is always a good idea to examine all the information shown on the diagnostics window.

FIGURE 11.18 Administrative Tools menu for Windows NT

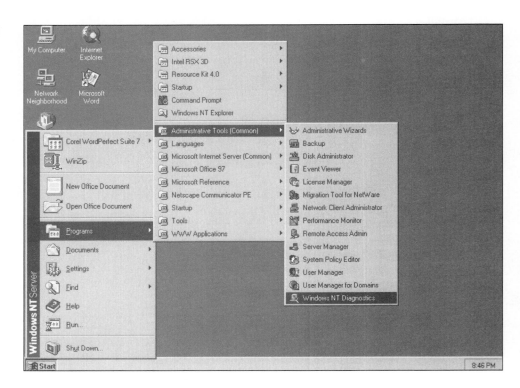

FIGURE 11.19   Windows NT Diagnostics System information

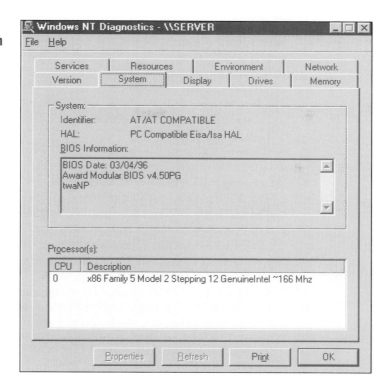

Within the figure:

Windows NT Diagnostics - \\SERVER

File   Help

| Services | Resources | Environment | Network |
| Version | System | Display | Drives | Memory |

System:
Identifier: AT/AT COMPATIBLE
HAL: PC Compatible Eisa/Isa HAL
BIOS Information:

BIOS Date: 03/04/96
Award Modular BIOS v4.50PG
twaNP

Processor(s):

| CPU | Description |
| 0 | x86 Family 5 Model 2 Stepping 12 GenuineIntel ~166 Mhz |

Properties   Refresh   Print   OK

SUMMARY

In this exercise we discovered that

- Windows NT comes in server and workstation versions.
- Ctrl+Alt+Del is used to initiate a logon sequence.
- Windows NT Server manages groups of computers in a domain.
- Windows 2000 extends the capabilities of the Windows network operating system and offers enhanced applications, new features, and an updated help system that can simplify network administration.

SELF-TEST

This self-test is designed to help you check your understanding of the background information presented in this exercise.

True/False

Answer *true* or *false*.

1. Logoff can be selected from the Windows NT Security menu.
2. A task that is not responding must be stopped manually.
3. The Windows NT operating system has the look and feel of a Windows 98 computer.
4. The Windows NT operating system requires an 80486 processor or better.
5. The Devices icon in the Windows NT Control Panel replaces the Add New Hardware icon in Windows 98.
6. There is only one version of Windows 2000.
7. Windows 2000 Professional is a server operating system.

Multiple Choice

Select the best answer.

8. Windows NT consists of two products:
   a. Client and server.
   b. Client and workstation.
   c. Server and workstation.
   d. Server and client workstation.

9. Windows NT uses the
   a. FAT32 file system.
   b. NTFS file system.
   c. RBFS file system.
   d. All of the above.
10. The Task Manager is used to
   a. Put the computer in a locked state.
   b. Log off from Windows NT.
   c. Create new system applications.
   d. None of the above.
11. When a computer is in a locked state,
   a. The desktop is hidden and applications are paused.
   b. The desktop is hidden and applications continue to run.
   c. The desktop is not hidden and applications are paused.
   d. The desktop is not hidden and applications continue to run.
12. The Windows NT Security menu is displayed when the
   a. Computer is booted the first time.
   b. Computer is accidentally turned off and then turned back on.
   c. User presses the Ctrl+Alt+Del keys simultaneously.
   d. User presses Ctrl+C or Ctrl+Break repeatedly.
13. Which is not a Windows 2000 operating system?
   a. Windows 2000 Server.
   b. Windows 2000 Light.
   c. Windows 2000 Datacenter Server.
14. In Windows 2000, the Network Neighborhood has been replaced by
   a. My Network Places.
   b. My Network Sites.
   c. Network Locations.

Completion

Fill in the blank or blanks with the best answers.

15. The _____ option must be used before powering down a Windows NT computer.
16. The Windows NT task can be stopped when a process is _____ _____.
17. If a Windows NT task must be stopped, it is best to use the _____ tab.
18. Disk _____ is an example of an administrative tool.
19. When experiencing problems, it is a good idea to examine the _____ window.
20. Windows 2000 Professional contains support for _____ Private Networks.
21. Windows 2000 extends the capabilities of the Windows _____ operating system.

FAMILIARIZATION ACTIVITY

Using a Windows NT computer, perform the following activities:

1. Open the Control Panel.
2. Run each of the Control Panel applications to determine what type of settings may be controlled.
3. Close the Control Panel.
4. Examine the administrative tools.
5. Review online help information.

QUESTIONS/ACTIVITIES

1. Create a new Windows NT user using the User Manager administrative tool.
2. Look at the Windows NT Event Viewer application.

Under the supervision of your instructor

1. Identify the differences between Windows 95/98 and Windows NT.
2. Identify the key features of Windows NT.
3. Identify any common features of Windows 95/98 and Windows NT.
4. Describe the features of Windows 2000.

12 The Desktop

INTRODUCTION

Jeff Page, who was in charge of all Web development at RWA Software, ran into Joe Tekk's office. He was covered with sweat. "Joe, you've got to come and help me. My machine is trashed."

Joe followed as Jeff ran back to his office. "I've rebooted my machine five times and get the same result each time. My desktop is completely blank. There are no icons for anything."

Joe looked at the display. The taskbar was visible at the bottom of the display screen, but the rest of the screen was a large empty patch of green background. Joe tried to reassure Jeff. "Believe it or not, Jeff, I've seen something like this before. I think I might know what is wrong."

Joe sat down and grabbed the mouse. Before long, he had clicked his way into the Recycle Bin, which contained, among other things, all of the missing desktop icons. Within moments the icons were restored.

Jeff wondered how Joe had identified the problem so quickly. Joe explained, "Last week one of the network guys brought his daughter with him while he did some work. She ran over to an open laptop and deleted every icon on the desktop. I think she figured out what happens when you right-click, and learned how to delete things to make them invisible. Apparently she knows just enough about Windows to cause trouble with it."

"Do you think she's still here?" Jeff asked, looking worried. "I have to leave for a basketball game in 20 minutes and I need to leave my machine on."

Joe laughed and shook his head. "Password protect your screen saver. That will keep her out."

PERFORMANCE OBJECTIVES

Upon completion of this exercise, you will be able to

1. Demonstrate the different features of the Start menu.
2. Identify items on the taskbar.
3. Change the appearance and properties of the desktop.
4. Create and use shortcuts.
5. Explain the function of the standard desktop icons.

KEY TERMS

Wallpaper	Resolution
Shortcut	My Computer
System settings area	Network Neighborhood
Start button	Attachment
Bitmap file	

Since the desktop is common to all four advanced Windows operating systems (95, 98, 2000, NT), we will refer to it simply as the Windows desktop. Desktop features unique to a single operating system will be discussed as necessary. The desktop is the centerpiece of the Windows operating system. The desktop typically contains a number of standard icons provided by the operating system as well as user-defined icons. As with the Windows 3.x operating system, an application is launched by double-clicking the program icon, although single-clicking also may be used in Web-style desktop settings.

The desktop also contains the taskbar, a system tray, any open folders and applications, and an optional background image called a *wallpaper*. Figure 12.1 shows the contents of a typical desktop. All the icons down the left-hand side of the desktop are automatically created and placed on the desktop when Windows is installed. These icons are entry points to the many built-in features of Windows. All the other icons are *shortcuts*, user-defined links to programs or other files that are used frequently.

In the upper right corner of the desktop shown in Figure 12.1 is an open *folder* called "Road Runner," which contains icons for six objects. Recall that Windows is an object-oriented operating system. Double-clicking on any of the icons in the Road Runner folder will start their associated applications. Note that the taskbar at the bottom of the desktop contains an entry for the Road Runner folder.

The lower right-hand corner of the desktop contains the *system settings area*, where several background tasks (virus protection, speaker volume, and the time-of-day clock) are represented by small icons. The icons may be hidden if necessary, to make more room on the taskbar.

Finally, the background of the desktop itself is an object that has properties that we can alter, such as what type of image to display, the overall display resolution, and the number of colors available. In the following sections we will discuss each of the many desktop components in greater detail.

THE START BUTTON

Microsoft made it easy to begin using Windows by placing the *Start button* at the lower left-hand corner. Everything it is possible to do in Windows is accessible through the Start button. Left-clicking on the Start button generates the menu shown in Figure 12.2. The Shut Down, Run, and Help items are selected by left-clicking on them. The other four items

FIGURE 12.1 Sample Windows desktop

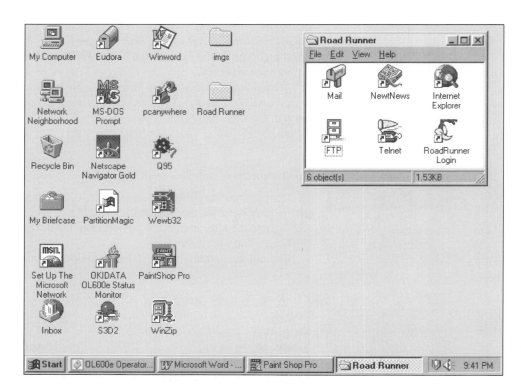

FIGURE 12.2   Start menu

FIGURE 12.3   Windows 95 Shut Down menu

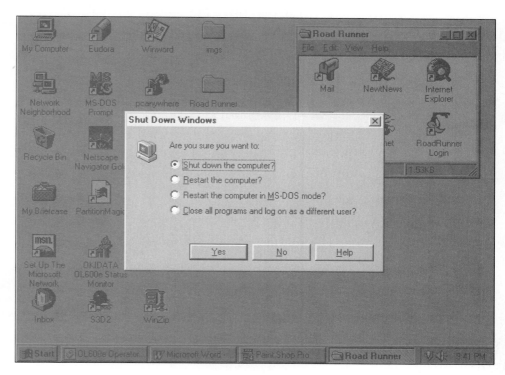

produce submenus when selected by the mouse (which only needs to pass over the menu item to create the associated submenu). Let's look at each menu item and its operation.

The Shut Down item, when left-clicked, produces the display shown in Figure 12.3. The display is darkened, except for a bright Shut Down menu. You may back out of Shut Down by left-clicking the No button (or the Close box). You may also choose to restart the computer in the MS-DOS mode of operation.

The Run menu item allows a program to be executed by typing the file name in the text box or by selecting it from a graphical file menu accessed through the Browse button. Figure 12.4 shows a Run dialog box with the file name "scandisk" entered in the text box. The Run menu gives you one method to access programs that are not contained on the desktop.

The next Start menu item, Help, provides access to the help facility, which is vastly improved over that of Windows 3.x. The Help menu is shown in Figure 12.5. Do not let the short list of topics mislead you. There is so much help available that you could spend days reading the various topics provided. There is help for Dial-Up Networking, managing printers, customizing your desktop and Windows environment, and much more. You are encouraged to spend some time looking through the Help information; you may pick up some valuable tips along the way.

FIGURE 12.4   Run dialog box

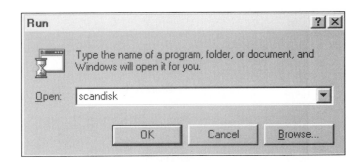

FIGURE 12.5   Windows 95 Help menu

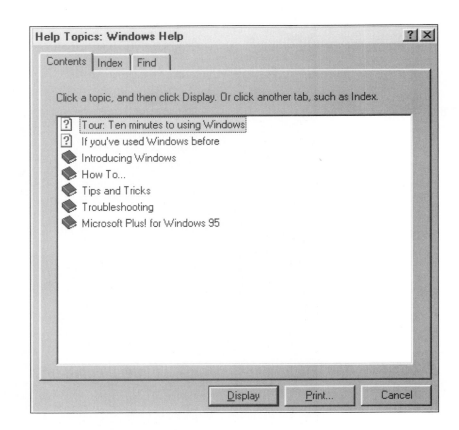

FIGURE 12.6   Windows 95 Settings submenu

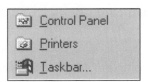

The Settings menu item opens up a submenu to three other applications, as illustrated in Figure 12.6. The Control Panel is a powerful set of utilities that are used to control how the hardware and software work under Windows. The Control Panel is the subject of Exercise 13. The Printers item brings up the Printers folder, which we will examine in Exercise 15. The third item, Taskbar, opens up the Taskbar Properties window, which we will look at in the next section. The contents of the Start menu are controlled using the Taskbar option from the Settings submenu.

The Documents menu item produces a submenu containing links to the last 15 documents opened. As shown in Figure 12.7, files such as Microsoft Word documents appear on the Documents submenu, placed there automatically when opened by the user.

FIGURE 12.7 Documents submenu

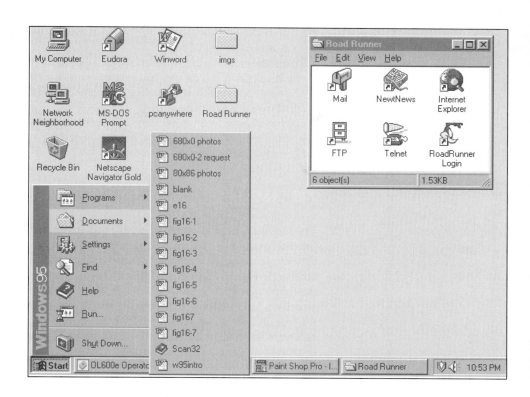

FIGURE 12.8 Programs submenu

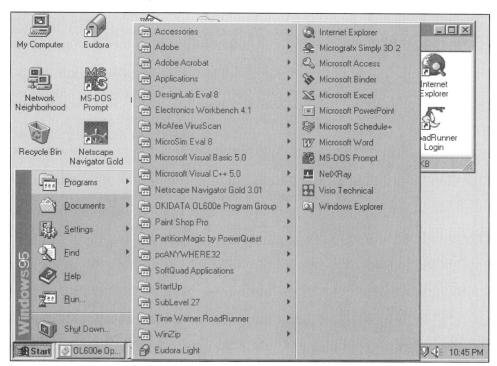

The last item on the Start menu is the Programs menu item, the starting point when starting a new application. Placing the mouse pointer over the Programs menu item generates a submenu similar to that illustrated in Figure 12.8. The Programs submenu consists of folders and program icons. Each folder icon and associated folder name shown in the list of applications indicates that a submenu exists for that particular item. Also note the small black triangles that point to the right on many of the menu items. These indicate that submenus exist for the menu items. As the mouse pointer is dragged over this type of menu item, a new submenu will appear.

When the mouse pointer is dragged over a program icon (NetXRay, for example), the icon becomes highlighted, and its associated application can be launched by simply left-clicking on it.

As new applications are installed, they are automatically added to the list of applications on the Programs submenu.

If you change your mind and do not want to run an application from the Programs submenu, simply left-click anywhere on the desktop outside the submenu area or choose a different menu item from the Start menu.

THE TASKBAR

The *taskbar* is used to switch between tasks running on the desktop. If no tasks are running, the taskbar contains only the Start button and the System Settings area. When other applications are running, or folders are open on the desktop, their icons will be displayed on the taskbar. In Figure 12.1 the taskbar contains four entries. The first three are for applications, the fourth is for the open Road Runner folder. Program and folder entries in the taskbar appear from left to right in the order they were started or opened. The taskbar can be resized, moved, and set up to automatically hide itself when not in use. Figure 12.9(a) shows a taskbar with so many applications displayed on it, it is impossible to read the names of any of them. In Figure 12.9(b) the same taskbar has been resized to allow more space for each application to be identified. To resize the taskbar, move the mouse near the top of the taskbar. It will change into an up/down arrow. Left-click and hold to drag the taskbar to a larger or smaller size, and then release.

The taskbar can also be moved to any of the four sides of the display. Figure 12.10 shows the taskbar moved to the right-hand side of the display. To move the taskbar, left-click and hold on any blank portion of the taskbar, then drag the mouse to the desired side of the display and release.

The taskbar has its own set of properties (such as its Auto hide feature) that can be adjusted. Recall that the Settings menu item of the Start menu displays a submenu containing a taskbar item. This is the entry point to the Taskbar Properties window. You can also right-click on a blank area of the taskbar and get a context-sensitive menu containing a Properties item you can click. Figure 12.11 shows the contents of the Taskbar Properties window.

You should experiment with each of the settings to see what they do and to settle on a format for the taskbar that you find pleasing. For example, if the Always on top box is unchecked, it is possible for an application to completely cover up the taskbar when it is opened up to full screen size. The only way to get the taskbar back is to minimize all applications or move applications around on the desktop to uncover the taskbar. Checking the Always on top box allows Windows to do the housecleaning for you and keep the taskbar visible no matter how many applications are open (unless the Auto hide feature is enabled).

THE BACKGROUND

The look of the desktop background can be adjusted for a pleasing appearance. Right-clicking anywhere on an empty portion of the desktop brings up the Display Properties

FIGURE 12.9 Two views of the taskbar

(a) The taskbar with many applications displayed on one row

(b) The same taskbar resized to two rows

FIGURE 12.10   A different location for the taskbar

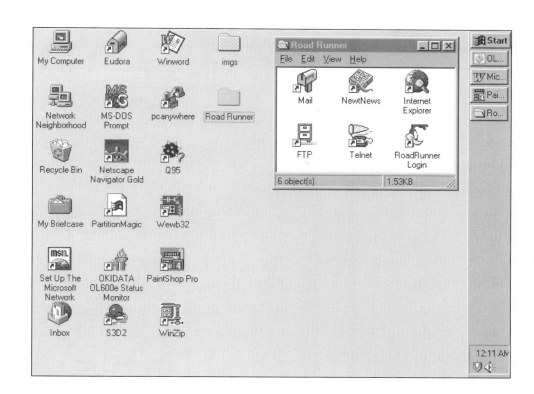

FIGURE 12.11   Taskbar Properties window

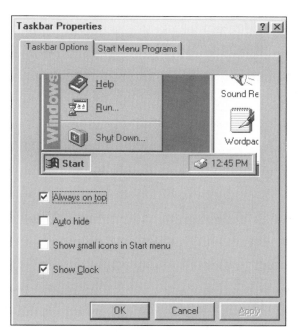

window shown in Figure 12.12. The Background controls allow you to choose a pattern or wallpaper to paint over the background. Both may be selected by double-clicking on their entry. Custom wallpapers can be selected using the Browse button, which allows selection of a *bitmap file* (.BMP extension), a standard Windows file format used for graphic images. The bitmap files displayed in the list are located in the directory in which the Windows operating system is installed.

When a pattern or wallpaper type is selected with a single click, the display icon shows an example of what you will see. This is illustrated in Figure 12.13. Note that the pattern chosen will completely cover the background of the desktop. If a wallpaper is

FIGURE 12.12   Display Properties window showing Background controls

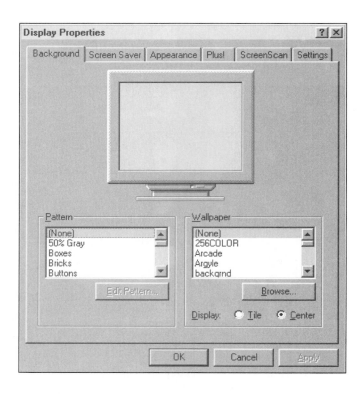

FIGURE 12.13   Choosing the Boxes pattern

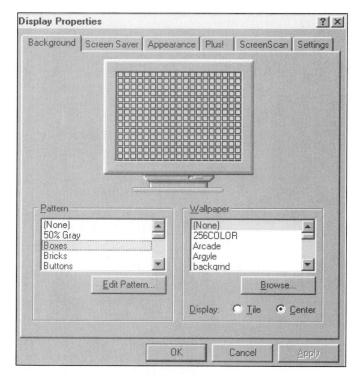

selected instead, you may choose to tile the wallpaper (repeat the wallpaper pattern over and over to cover the entire desktop) or simply center the wallpaper pattern, leaving the rest of the desktop filled with a blank background or a patterned background. All of these options are shown in Figure 12.14.

In Figure 12.14(a), notice that icons are drawn on top of the background pattern in a special way so that you can still read their names. In Figure 12.14(b), icons and folders are drawn on top of the background wallpaper, which is centered. The desktop area not

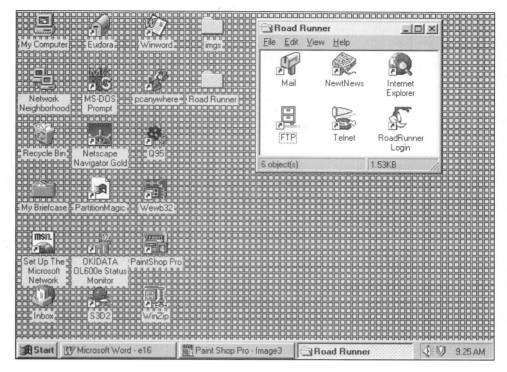

(a)

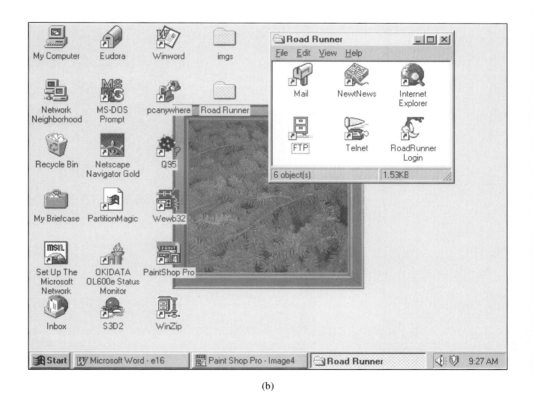

(b)

covered by a small wallpaper image can be patterned, as shown in Figure 12.14(c). If the wallpaper is tiled, the entire surface of the desktop is covered by repeating the wallpaper image horizontally and vertically as many times as necessary. This last option is shown in Figure 12.14(d).

FIGURE 12.14 *(continued)*
(c) Bricks pattern and Forest wallpaper (centered), and (d) Forest wallpaper (tiled)

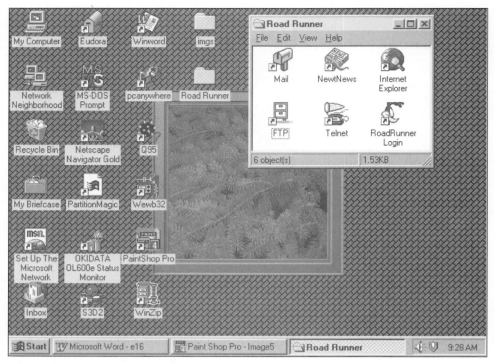

(c)

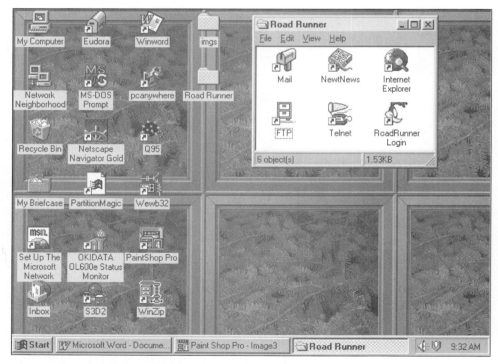

(d)

THE SYSTEM SETTINGS AREA

The system settings area of the taskbar typically contains a speaker icon (for controlling the speaker volume) and a 24-hour clock. Just placing the mouse pointer near the time display causes a small pop-up window to appear with the full date displayed. This is illustrated in Figure 12.15.

FIGURE 12.15 Pop-up date window

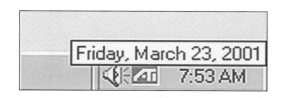

FIGURE 12.16 Date/Time Properties window

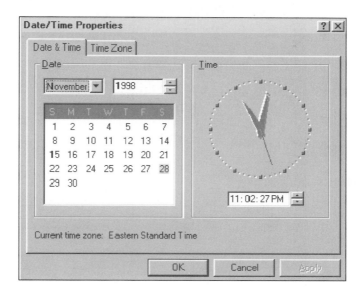

Double-clicking on the time display brings up the Date/Time Properties window illustrated in Figure 12.16. It is easy to change any part of the time or date with a few left-clicks or keystrokes. Windows will even adjust its clock automatically twice a year to properly "spring ahead" or "fall back" for daylight savings time as required.

Some applications place their icons in the system area displayed to the left of the time. By placing the mouse pointer over these icons, their current status is displayed. The applications are accessed by single-clicking or double-clicking them. We will examine these items in detail when we discuss accessories in Exercise 16.

CREATING AND USING SHORTCUTS

Some applications add their shortcut icons to the desktop automatically during installation. A shortcut to any application installed on a computer can be added to the desktop by simply right-clicking on an empty portion of the desktop and selecting Shortcut from the New submenu. The Create Shortcut dialog window is displayed, as shown in Figure 12.17.

Simply key the name of the file to be added or click the Browse box to locate the desired file. When the Browse box is selected, the Browse window is displayed. Figure 12.18 shows a sample Browse window for the C: drive. From this dialog box, any program located on the local hard drive or network hard drive can be selected. For example, suppose we want to add a shortcut to the Calculator program. Search for the application using the Browse window. After the Calculator program is located and selected, simply click the Open button to return to the Create Shortcut dialog box. Notice that the correct path and program name are displayed in the Command line text box shown in Figure 12.19.

By selecting the Next button, Windows provides an opportunity to enter a name for the icon (which is displayed under the icon on the desktop). The name of the file is used by default but can be changed by simply keying a new name in the text box, as illustrated in Figure 12.20.

When the Finish button is selected, Windows creates the shortcut and adds the program icon to the desktop. Figure 12.21 shows the desktop with the new Calculator icon.

FIGURE 12.17 Create Shortcut dialog window

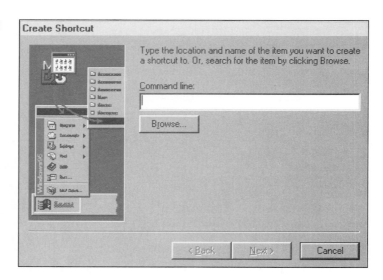

FIGURE 12.18 Browse window

FIGURE 12.19 Create Shortcut window with file selected

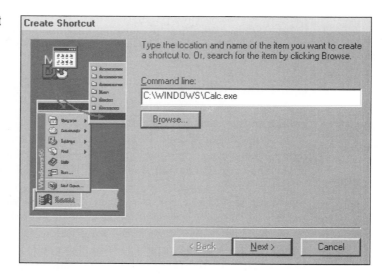

CHANGING OTHER DESKTOP PROPERTIES

Several other desktop properties you may want to experiment with are the Screen Saver, Appearance, and Settings items. Screen savers are programs that run in the background, only performing their screen saving when there has been no mouse or keyboard activity for a predetermined time period. A typical screen saver draws an interesting shape on the

FIGURE 12.20 Naming the shortcut

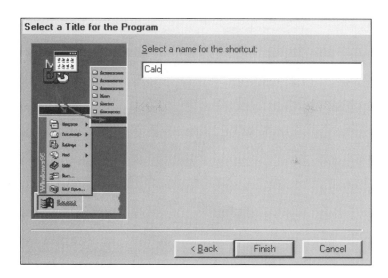

FIGURE 12.21 The desktop with the new shortcut

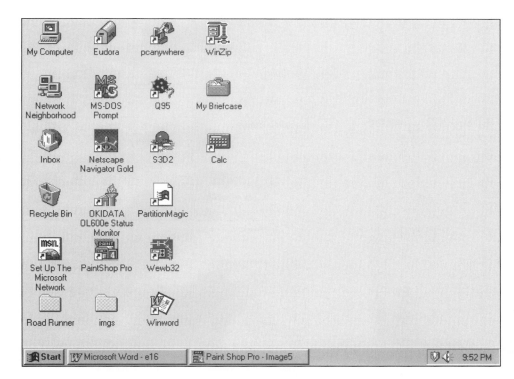

screen or performs some other graphical trick, keeping most of the screen blank. This is done to prevent an image from being burned into the phosphor coating on the display tube. When a key is pressed, or the mouse is moved, the screen saver restores the original display screen and goes back into the background.

Screen savers can also be password protected, to prevent unauthorized access to your desktop if you are away from your computer. Figure 12.22 shows the Screen Saver control window, with a small example of the selected screen saver being displayed.

The appearance of the desktop, from the color used in the title bar of a window, to the font of the desktop text, can be adjusted using the Appearance controls in the Display Properties window. Practically everything you see in the Appearance window illustrated in Figure 12.23 is clickable, including the scroll bar shown in the active window, which allows its width to be set. If the defaults for each item are not acceptable, spend some time tailoring them to suit your needs.

The Settings tab on the Display Properties window provides access to another important set of controls. As shown in Figure 12.24, the Settings controls allow you to change the size

FIGURE 12.22 Selecting a screen saver

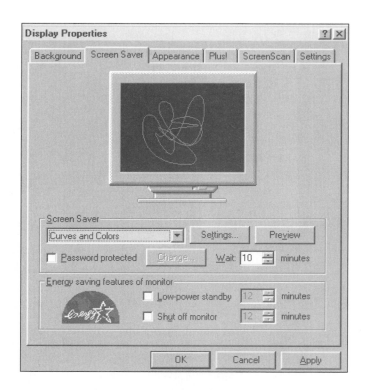

FIGURE 12.23 Desktop Appearance controls

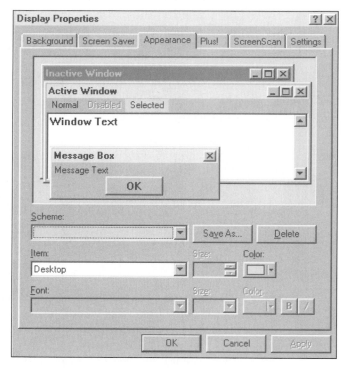

(also called *resolution*) of the desktop, the number of colors available, and the type of display connected to the computer. A small sample of the desired desktop is displayed to assist you when adjusting the settings.

MANAGING THE DESKTOP CONTENTS

The degree of control you have over the appearance of the desktop can be adjusted in many ways. For example, if you like to place new desktop icons (shortcuts or newly installed

FIGURE 12.24 The Settings tab of the Display Properties window

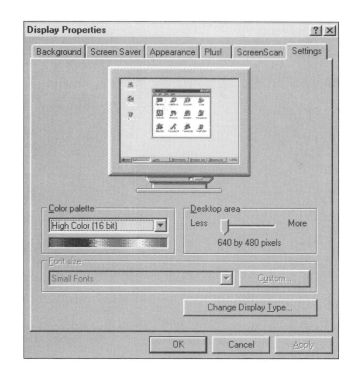

FIGURE 12.25 Arrange Icons submenu

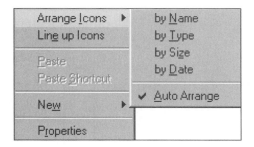

application folders) in specific locations, you may do so by left-clicking on them and holding the mouse button down as you drag the icons to their new positions on the desktop (drop them by releasing the mouse button). If you want Windows to manage your icons for you, or if you need to prevent inexperienced users from messing up the appearance of the desktop, select the Auto Arrange option from the Arrange Icons submenu shown in Figure 12.25. Right-clicking on a blank portion of the desktop brings up the context-sensitive menu.

When Auto Arrange is enabled, icons dragged from their positions will snap back in place when released. The desktop icons will remain in an orderly arrangement.

Many applications install folders on the desktop (such as the Road Runner folder shown in many of the figures in this exercise). The desktop itself is actually a special folder, and you can view its contents using Windows Explorer. This is demonstrated in Figure 12.26. In Exercise 14 you will see exactly how to do this, and many other useful operations, such as creating folders and moving folders to different locations.

MY COMPUTER

Double-clicking the My Computer icon opens up the folder shown in Figure 12.27. The contents will vary from machine to machine, depending on the actual hardware installed on each one. Double-clicking on any of the drive icons will open a directory window for the selected drive. The properties of a disk can be displayed by right-clicking on any of the drive icons and selecting Properties from the pop-up menu. The Disk Properties window shown in Figure 12.28 shows the disk label and current usage. The used space, free space, and capacity are displayed in bytes.

FIGURE 12.26 Viewing the desktop with Windows Explorer

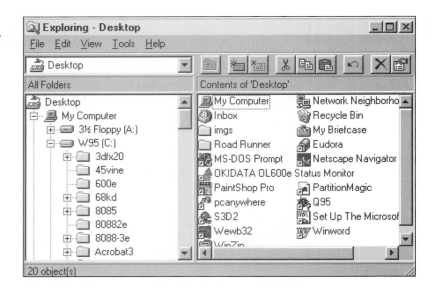

FIGURE 12.27 Contents of My Computer

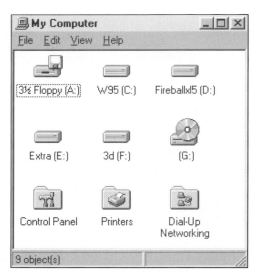

FIGURE 12.28 Disk Properties window

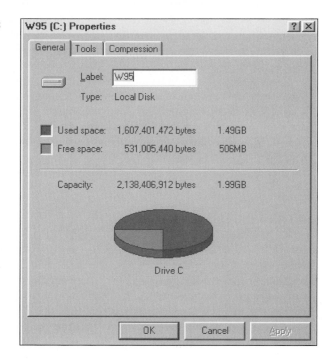

FIGURE 12.29   Disk Tools

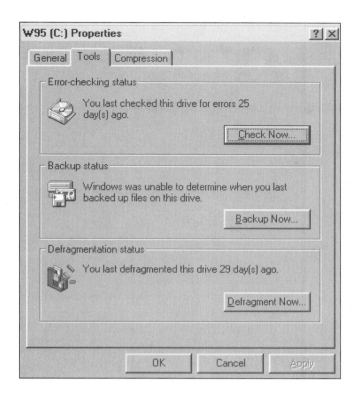

The Tools tab illustrated in Figure 12.29 shows the current status of each of the built-in disk tools provided by Windows. Each tool displays how long it has been since the utility has been run.

The Compression tab controls disk compression options. These utility programs will be examined in Exercise 16. The other system features are accessed and controlled through the Control Panel, Printers, and Dial-Up Networking folders, all of which we will cover in detail in Exercises 13, 15, and 17, respectively.

THE NETWORK NEIGHBORHOOD

If your computer is connected to a network, via a network interface card or a serial PPP (point-to-point protocol) connection, the Network Neighborhood icon, when clicked, allows you to examine other machines connected to the same network. Note that the other machines must be running a protocol called NetBEUI for them to be included in the network neighborhood. NetBEUI is part of Windows for Workgroups 3.11 and one of many protocols used with Windows.

As Figure 12.30 shows, the Network Neighborhood looks similar to a directory tree. In fact, files and printers may be shared among computers participating in the Network Neighborhood. Thus, one laser printer can serve the needs of a small laboratory or office.

THE RECYCLE BIN

Whenever anything is deleted in Windows, from executable programs to text files, images, and desktop icons, it is not yet gone for good. The exception to this rule are files deleted while running inside a DOS window. They are simply gone, unless the old UNDELETE command is still active. When you delete an item, Windows prompts you with a question, allowing you to change your mind, if necessary (as indicated in Figure 12.31).

The first destination of a deleted item is the Recycle Bin. Double-clicking the Recycle Bin icon on the desktop brings up the Recycle Bin window, which lists all files, if any, that have been deleted. Figure 12.32 shows the Recycle Bin window containing several deleted

FIGURE 12.30 Network Neighborhood

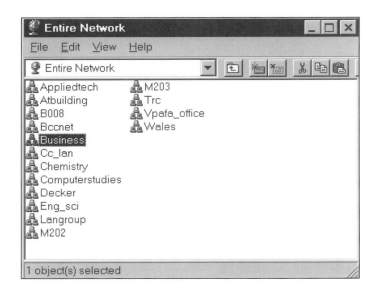

FIGURE 12.31 One of the Windows deletion safeguards

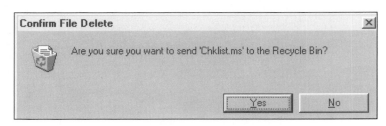

FIGURE 12.32 Contents of the Recycle Bin window

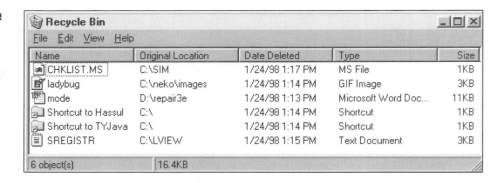

files. Under the File menu, the Empty Recycle Bin option is used to finally delete the files from the hard drive. Windows provides one final confirmation window, to make sure you really want to delete the items in the Recycle Bin. To get a file back, simply select it and choose the Undo Delete option from the Edit menu (or Restore from the File menu).

One thing to keep in mind is that the amount of free disk space does not change until the Recycle Bin is emptied. This is because a copy of the deleted file is stored in the Recycle Bin, which stores its contents on the hard drive.

MY BRIEFCASE

When you need to work on one or more files on two different computers (home and office machines), the My Briefcase utility can be used to organize and properly update the files. Files are copied into the Briefcase by dragging them onto the My Briefcase icon while in Windows Explorer. The entire Briefcase is copied onto a floppy disk by dragging its icon onto the icon for the floppy drive.

After the Briefcase files have been modified on a different computer, it is necessary to update the original files. Dragging the Briefcase icon from the floppy drive back to the desktop will cause Windows to evaluate the contents of the Briefcase and determine which files need updating.

THE INBOX

The Inbox is the central location for all e-mail activity. The first time the Inbox is selected, the Inbox Setup Wizard will begin the configuration process. The Microsoft Network, Microsoft Mail, and Microsoft Fax selections are available. Figure 12.33 shows the Inbox Setup Wizard window.

When the Inbox is selected after the services have been configured, the Microsoft Exchange application will start and begin to perform all e-mail and fax services. Figure 12.34 shows a Microsoft Exchange mail window with several e-mail messages.

E-mail is more than just letters and numbers. You can send and receive all types of files, from graphic images to executable programs. Typically, these types of files are called ***attachments***. The paper clip on the fourth e-mail message in Figure 12.34 indicates an attachment exists for that message. In general, double-clicking the attachment brings up the application associated with the attachment's file type. Or, if the attachment is an executable program, it is executed when double-clicked.

All in all, e-mail provides services that are essential to modern-day computing, business, and education. Spend some time learning to use this feature of the Windows operating system.

FIGURE 12.33   Inbox Setup Wizard window

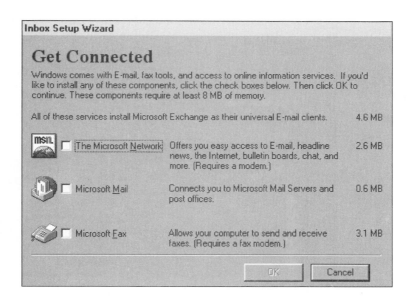

FIGURE 12.34   Microsoft Exchange mail window

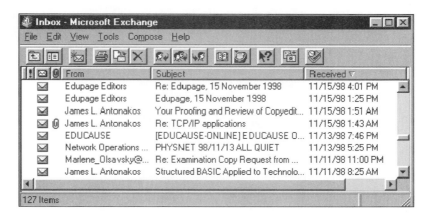

GETTING BACK TO THE DESKTOP FROM DOS

If you are running in a full-screen MS-DOS window, you can exit back to the desktop by entering EXIT at the DOS prompt. If you want to keep the DOS session active, but also access the desktop, pressing Alt+Enter will automatically resize your DOS environment to a window and bring back the desktop.

WINDOWS 98 HELP

One of the first differences between Windows 95 Help and Windows 98 Help is the appearance of the Help window. As shown in Figure 12.35, the Windows 98 Help window displays the Windows 98 logo alongside the Contents window. When a Help item is clicked in the Contents window, the associated Help information replaces the Windows 98 logo. This is a nice change from Windows 95 Help, which opened a new Help-specific window, making it awkward to return to the Contents window.

WINDOWS 98 SETTINGS MENU

Figure 12.36 shows the Settings menu found in the Windows 98 Start menu. Three new selections are available: Folder Options, Active Desktop, and Windows Update. The Folder Options selection allows you to choose how folders appear and function (single-click to open versus double-click), and handles file associations. Additional desktop properties can be examined or modified using the Active Desktop selection.

Selecting Windows Update automatically connects your computer to Microsoft's Windows update site on the Web. Once connected, you can pick and choose the updates

FIGURE 12.35   Windows 98 Help window

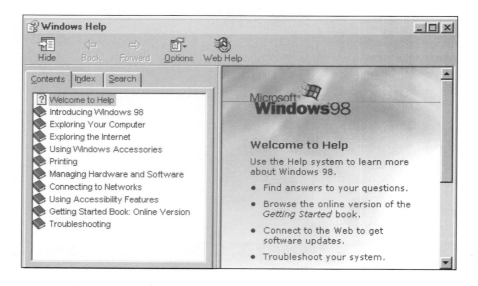

FIGURE 12.36   Windows 98 Settings menu

your Windows system requires. Software patches are downloaded and executed to make the necessary changes.

WINDOWS 98 EXPLORER

Under Windows 98, the Windows Explorer has been updated to work with the Web in addition to the desktop. Figure 12.37 illustrates how Explorer integrates the Web as just another item on the desktop. In addition to displaying lists of files and folders, Explorer is also designed to display Web content. Look along the left-hand side of the Explorer screen to see that the Voltage Regulator HTML document displayed is found in the Internet Explorer folder. The address of the HTML document is shown in the Address field.

WINDOWS 98 WEB-STYLE DESKTOP

To further integrate the Web into the Windows 98 computing environment, you have the option of configuring your desktop to act like a Web page. Folders can be opened or applications launched with a single left-click, instead of the usual double-click. When applications are launched in this way, the application is selected by simply moving the mouse pointer over the desktop icon. As Figure 12.38 shows, the names under each icon have the familiar underline associated with clickable links on a Web page. The Web-style desktop can also be configured so that the underline is displayed when the icon is selected. With all the available choices, no two desktops look the same.

WINDOWS NT DESKTOP EVOLUTION

The Windows NT desktop has evolved along the same path as the Windows desktop. The first version, Windows NT 3.5, took on the appearance of the Windows 3.1 desktop. This is because the first version of Windows NT was an industrialized version of Windows 3.1.

The desktop of the second version of Windows NT, version 4.0, looks like the Windows 95 desktop.

FIGURE 12.37   Windows 98 Explorer window displaying a Web page

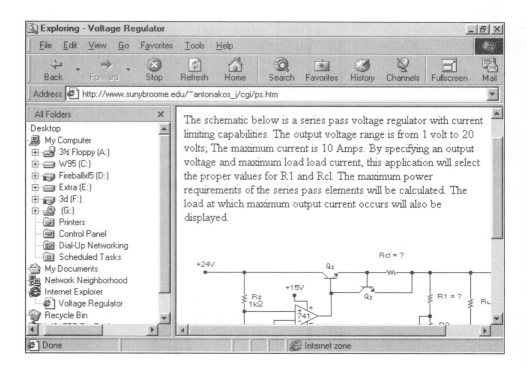

FIGURE 12.38 Windows 98 Web-style desktop

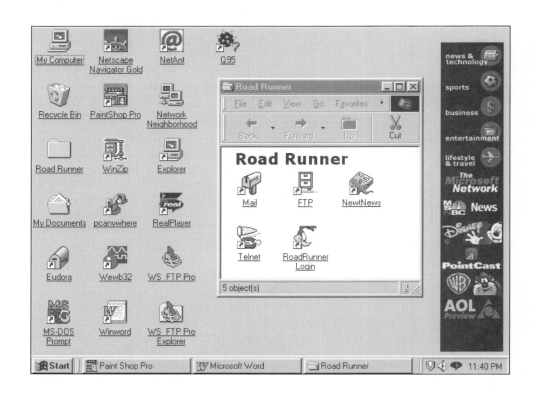

The release of Windows Explorer Version 4.0 offered the Windows 98 look and feel to the Windows 95 and Windows NT desktops. To determine the current operating system, look at the Start menu. Along the left-hand side is a graphic indicating the operating system version.

TROUBLESHOOTING TECHNIQUES

It is important to be patient with Windows. For example, suppose you have a number of applications open, and have just clicked the Close button on one of them, expecting to see the application close instantly; instead, you see nothing happen for a long time.

An impatient user may decide that Windows has died and simply turn the computer off. This is not the recommended approach (there may be important information that needs to be backed up to the hard drive). In this case, if Windows has "gone away" for a long time, attempt to switch to another application on the taskbar (or press Ctrl+Alt+Del to bring up the Close Program window). Just seeing the mouse pointer move when you use the mouse is a good sign that the operating system is still listening.

You may find that, eventually, Windows finishes closing the "dead" application, or reports an error message like the one shown in Figure 12.39. There is not really much that can be done for the application if an illegal operation is detected, but at least Windows catches these problems and does not crash because of them.

FIGURE 12.39 Illegal operation detected

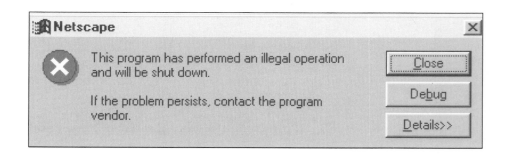

In this exercise we discovered that

- The desktop contains many powerful features (folders, icons, wallpaper).
- The taskbar contains many features (Start button, system settings area).
- The appearance of the taskbar and desktop can be adjusted.
- Shortcuts provide quick access to frequently used items.

SELF-TEST

This self-test is designed to help you check your understanding of the background information presented in this exercise.

True/False

Answer *true* or *false.*

1. The taskbar contains only the Start button, running applications, and the system settings area.
2. A shortcut for any application can be added to the desktop.
3. Context-sensitive help is displayed whenever the mouse is simply moved on top of any desktop icon.
4. The Documents menu item on the Start menu contains important help information for Windows.
5. When looking for help, enter HELP on the keyboard.
6. Ctrl+Alt+Del instantly reboots the computer.
7. The background may contain a pattern and a wallpaper.
8. The size of the desktop is fixed.
9. Right-clicking a desktop icon will cause its application to start up.
10. The name and position of each desktop icon may be changed.
11. The size of the taskbar is fixed.
12. The taskbar can be hidden (not shown on the desktop).

Multiple Choice

Select the best answer.

13. The Windows desktop is
    a. Part of the installation process.
    b. An essential component of the operating system.
    c. An optional application program.
14. When files are deleted in a DOS window, they are
    a. Gone for good.
    b. Sent to a temporary directory.
    c. Sent to the Recycle Bin.
15. When files are deleted in Windows, they are
    a. Gone for good.
    b. Sent to a temporary directory.
    c. Sent to the Recycle Bin.
16. The My Computer icon displays information about
    a. Each disk drive.
    b. Folders for the printer and control panel.
    c. Both a and b.
17. Wallpaper files are image files called
    a. Pixmaps.
    b. Bitmaps.
    c. Pic files.
18. To quickly bring up a context-sensitive display menu
    a. Left-click anywhere on the desktop.
    b. Right-click anywhere on the desktop.
    c. Press M on the keyboard.

19. Left-double-clicking a disk icon in My Computer
    a. Formats the disk.
    b. Displays the disk directory.
    c. Displays the disk properties.
20. If Windows goes away for a long time
    a. Just turn the computer off.
    b. Press every key on the keyboard and move the mouse.
    c. Press Ctrl+Alt+Del and see what happens.
21. When Auto Arrange is enabled, the desktop icons
    a. Periodically move themselves to new locations.
    b. Snap back into place if moved.
    c. Converge at the center of the screen.
22. Windows help is
    a. Minimal due to lack of disk space.
    b. Only available on CD-ROM.
    c. Online and significantly better than Windows 3.x.
23. The file extension .BMP stands for
    a. Binary map.
    b. Bitmap.
    c. Bi-modal program.
24. The Network Neighborhood shows
    a. All the LANs connected to a computer.
    b. All the computers on a LAN.
    c. All the computers on the Internet.

Completion

Fill in the blank or blanks with the best answers.

25. The Start button is located on the _____.
26. The desktop contains the _____ Neighborhood icon.
27. Wallpaper may be centered or _____.
28. The Programs submenu is located on the _____ menu.
29. The _____ is used to maintain a set of files that are used on more than one computer.
30. When a file is deleted, it goes to the _____ _____.
31. A(n) _____ takes over the display screen when there is no user activity for a length of time.
32. Electronic mail and fax services are provided by selecting the _____ icon.
33. The size of the display is set in the Display Properties _____ menu.
34. To exit from a DOS shell but keep the session running, press the _____ and _____ keys.
35. To shut down the computer, begin with the _____ button.
36. Network Neighborhood utilizes the _____ protocol.

FAMILIARIZATION ACTIVITY

1. Create a shortcut on the desktop to the Calendar application in the Windows directory and demonstrate its use.
2. Experiment with the desktop appearance by choosing five different patterns and five different wallpapers.
3. Compare the centered and tiled options for three different wallpapers.
4. Move the taskbar to all four corners of the display and vary the size. Explain your preference for the normal position of your taskbar.
5. Create several small text files with a text editor or word processor. Put a copy of each file into the Briefcase. Drag the My Briefcase icon onto the icon for your floppy drive. Take your floppy to a different computer and drag the Briefcase from the floppy onto

the desktop. Edit the files in the Briefcase. Drag the Briefcase back to the floppy. Put the floppy in the original computer and drag the Briefcase back onto the desktop. Double-click the My Briefcase icon and examine its contents. Update the files in any manner you choose.

6. Click the time display on the taskbar. Use the up and down arrows to determine the first and last year recognized by Windows. Can you grab the minute hand and move it around the clock?

7. Enable the Auto hide feature of the taskbar and experiment with the mouse to determine how and when the taskbar will reappear. Resize the taskbar. How large and small can it be?

8. Create a folder on the desktop and name it Temp. Drag some of the other desktop icons into it. Open the Temp folder and drag the icons back to the desktop one at a time. Do the icons snap into place or can you put them anywhere you choose?

9. Open the Recycle Bin. If there are already files in it, determine when they were put there and whether it is safe to delete them. Open a DOS window. Do a directory listing of the current drive and make note of the free space on the drive. Now empty the Recycle Bin. Has the free space changed?

10. Slowly move your mouse pointer over the items in the Programs submenu as if choosing an application to start up. Do you encounter any difficulty trying to select any of the applications?

QUESTIONS/ACTIVITIES

1. If you were a Windows 3.x user, how would you explain the similarities between Program Manager and the Windows desktop?
2. Why is it good to be patient with Windows?
3. If the operating system looks dead, what might you do to check for signs of life?
4. Does the taskbar have to be at the bottom of the screen? Does it have to be on the screen at all?
5. What is the Network Neighborhood?
6. Explain how to produce the Display Properties menu.

REVIEW QUIZ

Under the supervision of your instructor

1. Demonstrate the different features of the Start menu.
2. Identify items on the taskbar.
3. Change the appearance and properties of the desktop.
4. Create and use shortcuts.
5. Explain the function of the standard desktop icons.

13

The Control Panel

INTRODUCTION

Joe Tekk was sitting at his desk, eating lunch in front of his computer. His Windows 98 desktop showed the open Control Panel folder. Joe selected an icon from the bottom row, ODBC, and examined all of its various submenus and options, never actually selecting any of them.

Don, the senior technician, saw Joe's screen and was concerned. "What are you doing in Control Panel? Is there something wrong with your system?"

"No, Don," Joe answered. "I'm just checking out some of the operations I never use. Every now and then I discover one that has some new feature I am looking for."

PERFORMANCE OBJECTIVES

Upon completion of this exercise, you will be able to

1. Identify several of the many useful icons in the Control Panel and explain their functions.
2. Use the Control Panel to modify system properties such as time/date, sounds, the display format, and keyboard/mouse functionality.
3. Show how printers and modems are configured.

KEY TERMS

Add New Hardware Wizard
Add/Remove Programs Properties menu
Date/Time menu
Display menu
Modem

Mouse
Network menu
Password
Printers window
System Properties menu

BACKGROUND INFORMATION

The Control Panel is one of the most important tools included in Windows. Just like the Control Panel included with Windows 3.x, the Windows Control Panel allows you to fine-tune and customize the hardware and software of your system. The degree of control and the number of features available are much improved over Windows 3.x. Figure 13.1 shows three views of the Control Panel. Note the similarities in all three Control Panels. Windows 95 and Windows 98 are the most similar. The Windows NT 4.0 Control Panel shares many of the same icons, but has several icons that are unique to the Windows NT 4.0 environment, such as Devices, Licensing, and Services. As indicated in Figure 13.1, nearly every aspect of the operating system can be controlled, examined, or configured through one of the many Control Panel icons. Let's examine the operation of several important icons.

145

FIGURE 13.1 (a) Windows 95 Control Panel, (b) Windows 98 Control Panel, and (c) Windows NT 4.0 Control Panel (Web-style format)

(a)

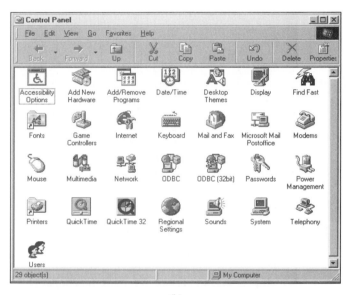

(b)

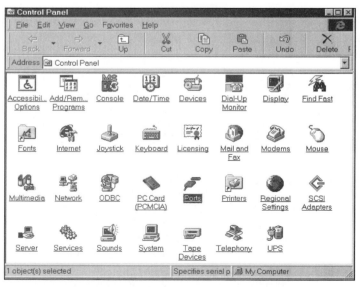

(c)

FIGURE 13.2 Accessibility Properties menu

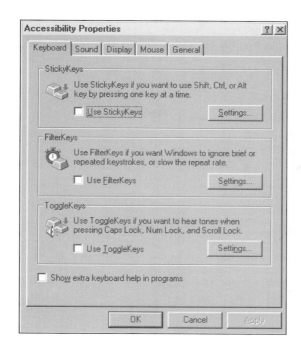

ACCESSIBILITY

Figure 13.2 shows the Accessibility Properties menu. For users who want to customize how the keyboard, sound, display, and mouse operate, this menu provides the way. As indicated in Figure 13.2, a lot of attention has been paid to how each key on the keyboard behaves. Other accessibility options involve generation of short tones when Windows performs an operation, helpful pop-up messages, using the arrow keys to control the mouse, and changing the display colors to high-contrast mode for easier viewing.

ADD NEW HARDWARE

The process used to add new hardware to Windows 95 or 98 usually involves two steps. First, the new hardware device must be physically added to the computer. This requires removing the cover from the computer, identifying the proper location to install the new hardware, and then performing the actual installation procedure. The second step involves configuring the software to properly communicate with the new device. This is accomplished using the *Add New Hardware Wizard* shown in Figure 13.3. This process is discussed in detail in Exercise 19. Adding new hardware in Windows NT is different and will also be discussed in Exercise 19.

FIGURE 13.3 Add New Hardware Wizard window

FIGURE 13.4   Add/Remove Programs Properties menu

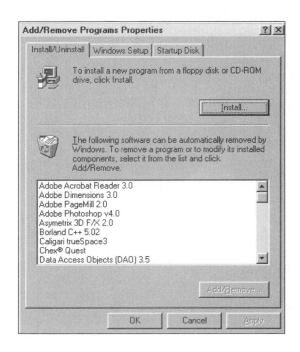

ADD/REMOVE PROGRAMS

When installing software applications in Windows, many software vendors require the installation to be performed from the Add/Remove Programs Properties menu illustrated in Figure 13.4. When the software is installed, it is registered with Windows. If at some point it is necessary to remove an application from the system, the user can simply return to the menu, select the application, and press the Add/Remove button. In addition to user application programs, the Windows operating system can also be modified by adding or removing components. This topic is discussed in Exercise 18.

DATE/TIME

This menu can be opened from inside Control Panel or by double-clicking the time display in the taskbar. The Date/Time Properties menu is shown in Figure 13.5. The user can easily change the time or date with a few left clicks. The Time Zone submenu allows you to select the time zone your computer is located in. Windows will then automatically adjust the clock for daylight savings time when required.

FIGURE 13.5   Date/Time Properties menu

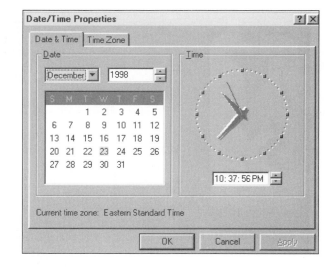

**FIGURE 13.6   Display
Properties menu**

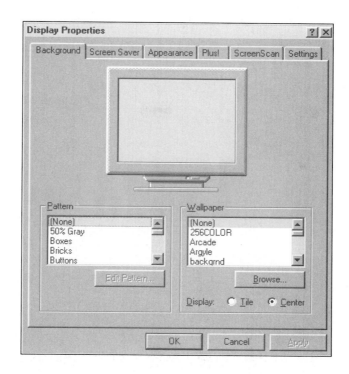

DISPLAY

The Display Properties menu is shown in Figure 13.6. Recall from Exercise 12 all of the various ways we can adjust the display. The menu in Figure 13.6 is also reachable by right-clicking on a blank area of the desktop and selecting Properties.

KEYBOARD

The Keyboard Properties menu is displayed in Figure 13.7. The Speed tab is used to set the repeat delay and repeat rate when a character is pressed and held down on the keyboard and

**FIGURE 13.7   Keyboard
Properties menu**

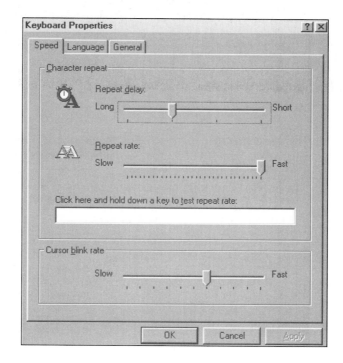

FIGURE 13.8 Modems
Properties menu

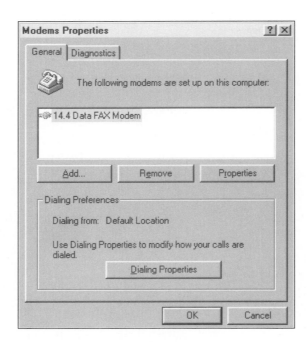

the rate at which the cursor blinks in a data entry field. Many people need to change these settings because of individual keyboarding styles. We will investigate the consequences of changing these settings during the activities at the end of this exercise.

The Language tab is used to indicate the keyboard language, and the General tab is used to identify the specific type of keyboard used, such as the Microsoft Natural keyboard.

MODEMS

Modems are very easy devices to work with in Windows. The Modems Properties menu (Figure 13.8) identifies the specific type of modem installed in your computer.

Windows can be set up to dial from many different dialing locations. For example, many offices may require a 9 to be pressed to access an outside line; other dialing locations may not require a 9. All the places frequently called can be given a name and can be selected very easily. Even specific communication details, such as the number of data bits and the type of parity that is used, can be adjusted.

The Diagnostics submenu is useful for interrogating the modem and examining its response to commonly used modem commands.

MOUSE

This menu provides all the functional control over the *mouse*. The mouse can be set up for left- or right-handed operation, its double-click speed adjusted, mouse trails enabled, and appearance changed by choosing one of several scenes (3-D pointer, for example). The initial Mouse Properties menu is shown in Figure 13.9.

The mouse is a significant component of the Windows operating system. The mouse properties you choose for yourself can make your system easier to use.

MULTIMEDIA

The Multimedia Properties menu is used to control the multimedia hardware and software installed on the system. Windows comes equipped with the ability to display MPEG (Motion Pictures Experts Group) video in real time and provides virtual device drivers for

FIGURE 13.9   Mouse Properties menu

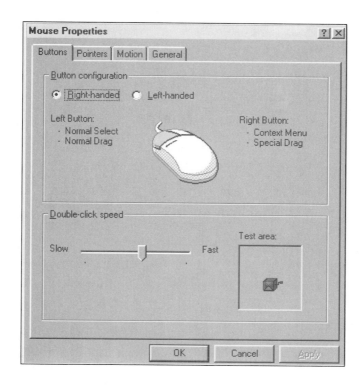

FIGURE 13.10   Multimedia Properties menu

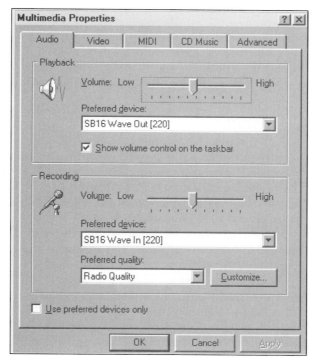

many popular sound cards and graphics accelerators. Windows can also take advantage of the new MMX technology available in the Pentium family of microprocessors. Figure 13.10 shows the initial Multimedia Properties menu.

NETWORK

The Network menu allows you to add, modify, or remove various networking components, such as protocols (NetBEUI, TCP/IP), drivers for network interface cards, and Dial-Up

FIGURE 13.11   Network menu

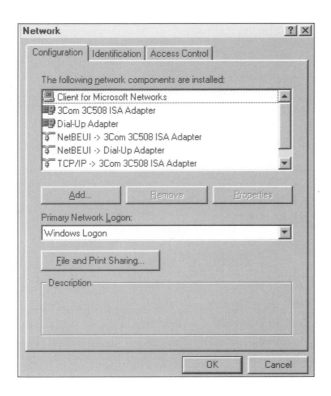

Networking utilities. You can also specify the way your machine is identified on the network, as well as various options involving file and printer sharing, and protection. Figure 13.11 shows a sample Network menu. Selecting any of the network components allows its properties to be examined.

Networking will be covered in detail in Exercise 17 as well as in Unit IV.

PASSWORDS

The Passwords Properties menu found in Windows 95 and Windows 98 is used to maintain information about *passwords* and security when a computer is shared among different people. Passwords are maintained in files with a .PWL extension if the computer does not participate in a Windows NT domain. The file name portion consists of the first eight letters of the user name entered at the Windows logon screen. If you are required to supply passwords for services used on the network, the passwords may be saved in a .PWL file. Unfortunately, you are allowed to delete any .PWL file you want, effectively removing all password protection for the affected user. This may be a factor in a security-conscious network.

Windows 95, Windows 98, and Windows NT can be configured to allow each computer user to maintain individual desktop settings plus other related preferences all protected by an initial password. Figure 13.12 illustrates the Passwords Properties menu. Windows NT Server can be configured as a domain controller responsible for maintaining passwords for all computers in the network.

PRINTERS

The Printers window shown in Figure 13.13 shows all the printers currently installed on a particular computer. The Add Printer icon is used to add a brand-new local or network printer to Windows. Although printers are covered in detail in Exercise 15, it is important to note the starting point for printer operations. The system tray portion of the taskbar will indicate when a printer is in use and the status of the current print job. Right-clicking on an installed printer will allow you to bring up the Properties window and change printer parameters as necessary.

FIGURE 13.12   Passwords Properties menu

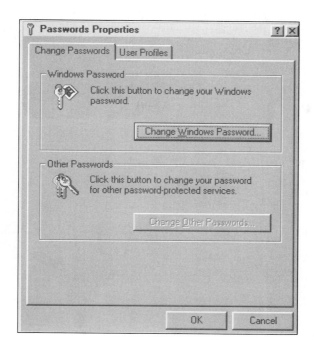

FIGURE 13.13   Printers window

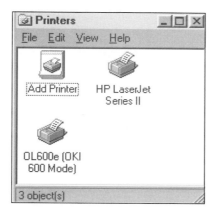

REGIONAL SETTINGS

The Regional Settings menu is used to configure Windows to conform to the many different international standards. For example, the display format for numbers, currency, date, and time can all be modified. Figure 13.14 illustrates the initial Regional Settings menu. To make the job simple, simply select the specific region, and all the individual items are automatically configured. If you travel internationally, your system can automatically adjust to the new region with a few clicks of the mouse.

SOUNDS

Windows uses sounds to accompany many typical operations, such as closing an application or shutting down the computer. The sound, if any, that is played for an event is specified using the Sounds Properties menu illustrated in Figure 13.15.

Sounds are stored in .WAV files in the main Windows subdirectory. Many different sound editors are available that allow you to create or modify sounds to fully customize your Windows environment.

153

FIGURE 13.14 Regional Settings Properties menu

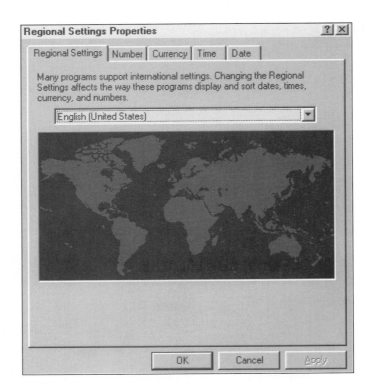

FIGURE 13.15 Sounds Properties menu

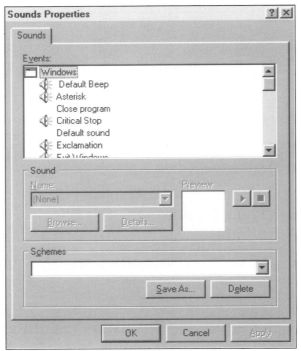

SYSTEM

The System Properties menu is the central location for all system-related information. The initial properties window for Windows 98 and Windows NT is shown in Figure 13.16. The Windows 95 menu is virtually the same as the Windows 98 menu. The operating system version, processor type, and amount of RAM are some of the system properties displayed.

The other submenus (Device Manager, Hardware Profiles, and Performance, to name a few) provide detailed access to the inner workings of the installed hardware as

FIGURE 13.16 (a) Windows 98 System Properties menu and (b) Windows NT System Properties menu

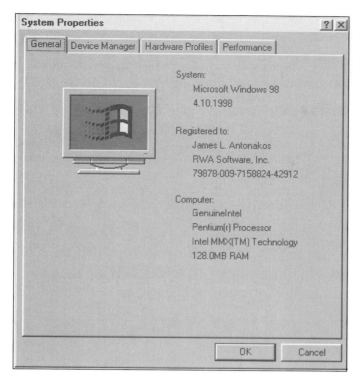

(a)

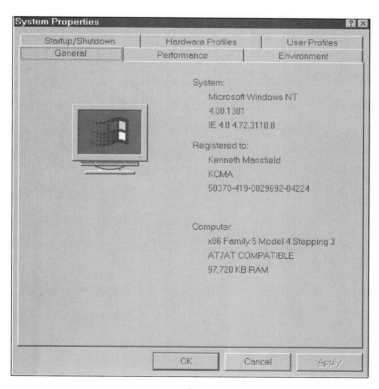

(b)

well as critical Windows variables, such as virtual memory settings and file system configuration.

There are many icons yet to be examined. You are encouraged to examine their properties on your own.

When troubleshooting on a Windows 95 or 98 computer, the System Properties menu can be used to investigate and diagnose many types of problems. The Device Manager submenu provides a single location to examine all the hardware components installed on a system. After installing a new piece of hardware, it is good practice to view the new hardware properties and make note of any configuration parameters.

On some occasions, by viewing these screens, problems may be discovered. This is indicated in Figure 13.17. When reviewing the System devices category, the exclamation point next to the Plug and Play BIOS item indicates that a problem exists.

By double-clicking the Plug and Play BIOS item, the associated Properties menu is displayed, as shown in Figure 13.18. The device status portion of the menu indicates that the device drivers for the Plug and Play BIOS have not been installed correctly and then goes on to suggest that the user click on the Drivers submenu tab to change the drivers. When the problem is resolved, the error indication will be removed from the Device Manager display.

FIGURE 13.17   System Properties reports an error

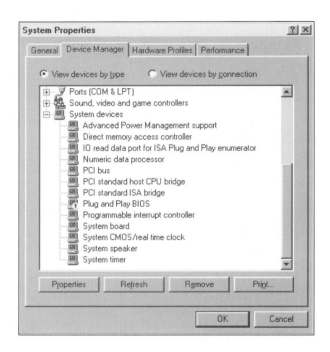

FIGURE 13.18   A course of action is suggested

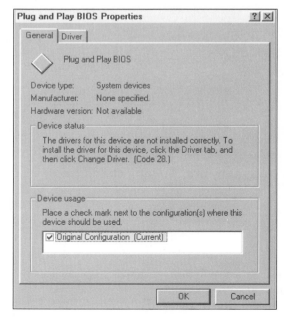

It pays to become familiar with the System Properties menu when troubleshooting problems using Windows 95 or 98.

In Windows NT, hardware errors are written to a log file, which can be examined for details concerning the nature of the problem.

SUMMARY

In this exercise we discovered that

- The Control Panel provides access to hardware and software tools used to manage your computer.
- The System Properties menu provides important diagnostic information and assistance.

SELF-TEST

This self-test is designed to help you check your understanding of the background information presented in this exercise.

True/False

Answer *true* or *false*.

1. The Control Panel is used to control hardware only.
2. Only one language can be used with the keyboard.
3. Windows takes advantage of MMX technology.
4. Each user of a computer can maintain his or her own desktop.
5. The Printers window shows all active printers.
6. The keyboard and mouse cannot be adjusted through Control Panel.
7. Sounds are stored in .WAV files.

Multiple Choice

Select the best answer.

8. Passwords are stored in
    a. .PWL files.
    b. .PAS files.
    c. .DAT files.
9. Changing the way numbers are displayed is performed in
    a. Multimedia Properties.
    b. Display Properties.
    c. Regional Settings.
10. Control Panel icons
    a. Are identical in all versions of Windows.
    b. Are only found in Windows NT.
    c. Establish control over almost all hardware and software properties.
11. The Systems Properties menu shows
    a. Information about the system printers.
    b. The version of the operating system.
    c. Information about system performance.
    d. All of the above.
12. The Accessibility options are used to
    a. Network several Windows machines together.
    b. Customize how the system and user interact.
    c. Control how the hard drive is accessed.
13. Which icon would be used when working with an MPEG file?
    a. Mouse.
    b. Modem.
    c. Multimedia.

14. Clicking the Printers icon shows
    a. All printers installed on a computer.
    b. All printers installed on a network.
    c. The printer status.

Completion

Fill in the blank or blanks with the best answers.

15. Windows can automatically play _____ video files.
16. To connect to a remote computer, we may use the _____ or _____.
17. The Network menu is used to change the _____ for the network interface card.
18. The _____ _____ submenu is used to display information about all the hardware components installed in a Windows 95 or 98 system.
19. Sounds are stored in _____ files.
20. Use Add New _____ to install a new video card.
21. NetBEUI and TCP/IP can be found in the _____ menu.

FAMILIARIZATION ACTIVITY

1. Go through each of the Control Panel operations covered in this exercise. Be sure to click on every tab to see the full extent of control you have over the operating system and the computer.
2. Briefly examine the remaining operations provided by the Control Panel. Determine what ODBC stands for. Find out how to send a fax.
3. Make a Windows start-up disk using Add/Remove Software. Windows NT calls this disk an emergency repair disk.

QUESTIONS/ACTIVITIES

1. How must Windows be configured to allow each user to maintain his or her own desktop settings?
2. Why is it necessary to have different dialing locations maintained inside of Dial-Up Networking?
3. How does Windows indicate problems with hardware settings?

REVIEW QUIZ

Under the supervision of your instructor

1. Identify several of the many useful icons in the Control Panel and explain their functions.
2. Use the Control Panel to modify system properties such as time/date, sounds, the display format, and keyboard/mouse functionality.
3. Show how printers and modems are configured.

14 Windows Explorer

Joe Tekk was busy using Windows Explorer to organize his hard drive. He moved directories, renamed them, created a folder of his most often used shortcuts, and copied several files from his hard drive to a floppy disk. Then he installed an application from a shared CD-ROM on RWA Software's Pentium III file server.

Don, the senior technician, was watching. "Joe, do you ever close Explorer?" Don asked.

Joe laughed. "Sometimes, Don, but usually I leave it minimized on the taskbar, just in case I need it for something."

PERFORMANCE OBJECTIVES

Upon completion of this exercise, you will be able to

1. Explain the basic features of Windows Explorer.
2. Show how to start an application.
3. Create a shortcut, move, find, and delete files and folders.
4. Map a network drive.

KEY TERMS

View menu
Cut
Copy

Paste
Undo
Network drive

BACKGROUND INFORMATION

Windows Explorer can be thought of as the *command post* of Windows, performing duties similar to the Program Manager and File Manager utilities in Windows 3.x.

Figure 14.1 shows a typical Explorer display of the folders on drive C:. Although there are indeed similarities between Windows 95 Explorer and Windows 98 Explorer (shown, respectively, in Figures 14.1(a) and 14.1(b) for comparison), Windows 98 Explorer offers additional features because Internet Explorer 4.0 has been integrated into the desktop. For example, Web pages can be viewed in the rightmost panel, without having to first open a browser. We will examine the new Windows 98 Explorer features at the end of this exercise. Bear in mind that much of what follows regarding Windows 95 Explorer operation also applies to Windows 98 Explorer. Furthermore, Windows NT Explorer is practically

FIGURE 14.1 (a) Windows 95 Explorer window and (b) Windows 98 Explorer Window

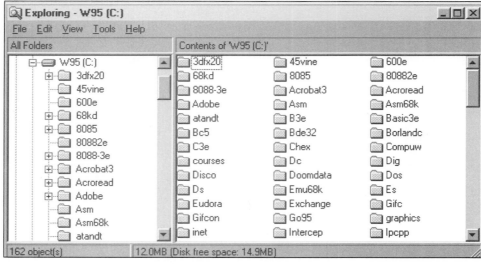

(a)

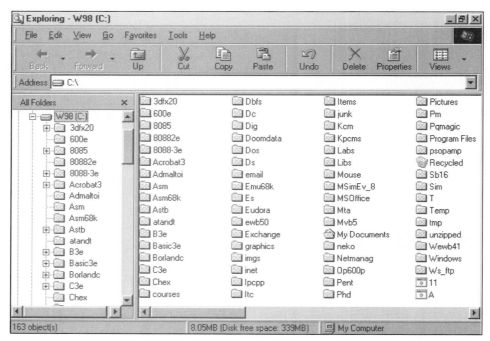

(b)

identical to Windows 95 Explorer, although you may install Internet Explorer 4.0 and gain the enhancements for your Windows NT environment. For now, let's take a detailed look at Windows 95 Explorer.

The Windows 95 Explorer window can be broken down into four areas: the pull-down menus, the folder display window, the file display window, and the status bar. The pull-down menus provide a way to access the features of Explorer, to configure it, and do many other useful things, such as map network drives. The folder display window allows the selection of any resource on the computer. This includes all the drives (floppy, hard, CD-ROM, network), special folders (Control Panel, Printers, Dial-Up Networking), and other system and user items.

The file display window shows a list of the folders and files in the currently selected location. For example, in Figure 14.1(a), the file display window shows the contents of the root directory of drive C:. Note the drive label "W95" at the top of the window. The status bar, located at the bottom of the Explorer window, shows the number of objects in the file display window as well as the amount of disk space used by the objects and the disk free space. Single-clicking on an item in the folder display window will show the contents of the item in the file display window.

CHANGING THE VIEW

The Windows 95 Explorer interface can be customized in many ways. First, let's add a graphical toolbar to the Explorer window. This is accomplished by selecting the Toolbar option in the View menu. The new Explorer display window should look similar to Figure 14.2. Many of the pull-down menu items are represented in the graphical toolbar. Holding the mouse still over a toolbar button will produce a small pop-up description window, such as "Delete" or "Map Network Drive." The function of each button is shown in Figure 14.3.

Another way to change Explorer's view is to adjust the method used to show files and folders. These methods are as follows:

- By large icon
- By small icon
- As a list
- With details

Figure 14.4 shows the large icon format in the file display window. The files displayed in the window scroll vertically and are listed in alphabetical order with folders first.

FIGURE 14.2 Explorer display with graphical toolbar

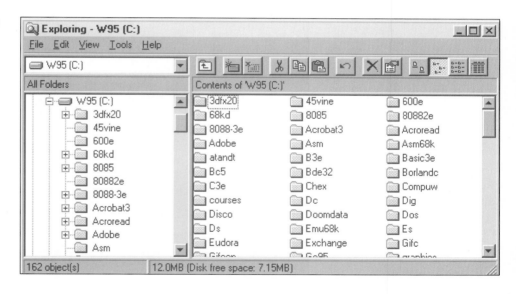

FIGURE 14.3 Explorer toolbar functions

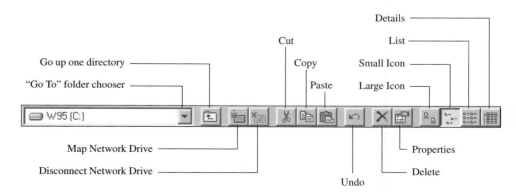

FIGURE 14.4 Explorer display window showing large icons

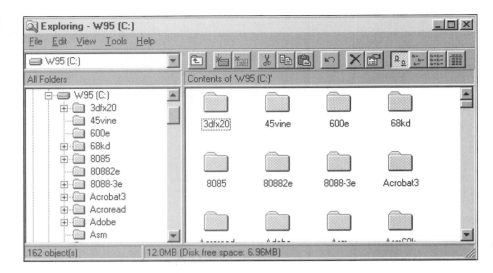

FIGURE 14.5 Explorer display window showing small icons

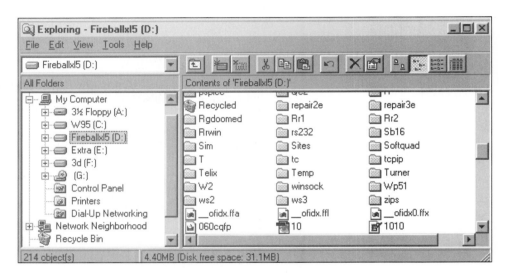

FIGURE 14.6 Explorer display window showing icon list

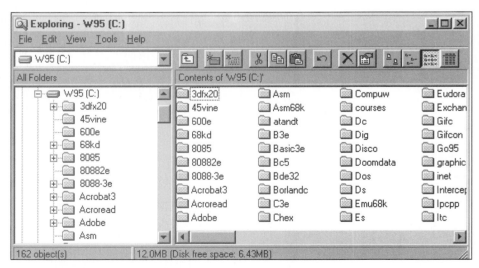

Figure 14.5 shows an example of small icon format. This display mode also scrolls vertically.

When files and folders are displayed using the list option, the display looks similar to that shown in Figure 14.6. Notice that the list scrolls *horizontally*, not vertically, and uses the small icon format.

FIGURE 14.7 Explorer display window showing folder details

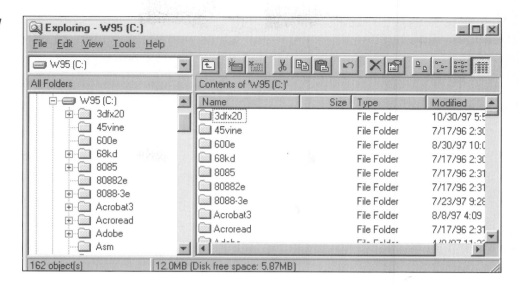

FIGURE 14.8 Explorer display window showing file details

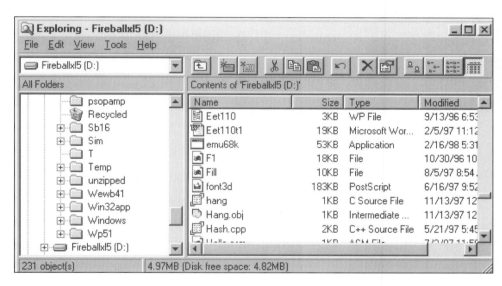

It may be necessary to view the details of each file. This includes the type of file or folder, its size, and last modification date. Figures 14.7 and 14.8 show examples of this display mode.

The arrangement of files and folders in the file display window is flexible (by name, date, size, and so on) and can be limited to certain types if desired. The Options selection in the View menu is used to adjust these features.

CREATING NEW FOLDERS

To help organize files, it is often necessary to create folders. By selecting File, New, Folder, Windows 95 automatically creates the new folder in the current directory (the directory shown in the file display window). You are prompted to enter a name for the new folder.

The destination of the new folder is important. For example, if you want the folder to be in the root directory of drive E:, you must make this directory the current directory. This is easily done by left-clicking on the icon for drive E: in the folder display window. Figure 14.9 shows a new folder being created. Windows 95 automatically names the folder New Folder and gives you the opportunity to rename it by keying in a new name. Figure 14.10 shows the new folder renamed Ken's Stuff.

To create a new folder inside an existing folder, left-click on the existing folder to select it and then create the folder as you normally would.

163

FIGURE 14.9   Creating a new folder

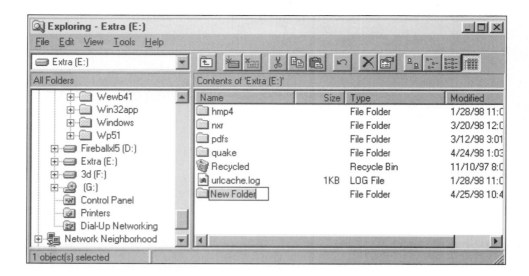

FIGURE 14.10   Naming the new folder

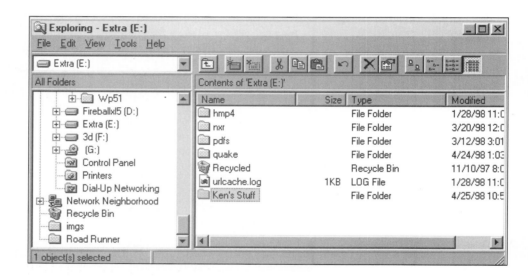

DELETING A FOLDER

Occasionally it becomes necessary to delete the contents of a folder. This is typically a result of the hard disk running out of free space. By selecting the folder and then selecting Delete, Windows 95 will prepare to delete the contents of the folder. First, as Figure 14.11 illustrates, Windows 95 will issue a confirmation message to make sure you really want to move the folder to the Recycle Bin. This is because you might have selected the wrong menu item or pressed a wrong button by accident. Or, being human, you simply may have changed your mind.

Windows 95 makes doubly sure that a file you have chosen for deletion should really be deleted. Figure 14.12 shows the warning window that is displayed when a program is deleted. Other selected file types, such as drivers or fonts, may require confirmation as well. In addition, Windows 95 usually knows when a file you try to delete is in use by an application (or itself) and may disallow the delete operation.

While files are being moved into the Recycle Bin, Windows 95 displays an animation of trash being thrown into a wastebasket. This is shown in Figure 14.13.

The entire folder can be recovered from the Recycle Bin at a later time, if necessary, by clicking File . . . Restore inside the Recycle Bin.

FIGURE 14.11   Confirmation window

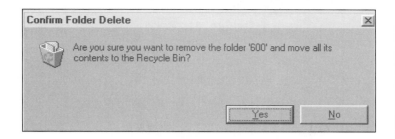

FIGURE 14.12   Confirming deletion of a program

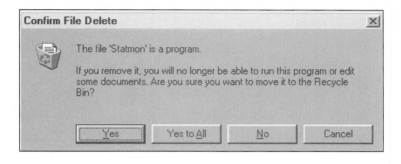

FIGURE 14.13   Deleting the files in a folder

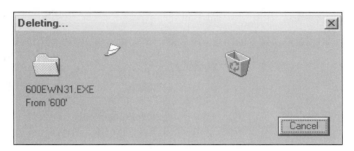

CREATING SHORTCUTS

A *shortcut* is an icon that you can double-click to start an application, rather than navigating to the application using Windows Explorer and double-clicking it there. For ease of use, shortcuts to often-used applications are usually placed on the desktop, so you can instantly access the applications without opening Explorer or using the Run or Programs menu.

To create a shortcut, use Explorer to navigate to the folder where the application is stored. Select the application by left-clicking it. Then choose File, Create Shortcut to create the shortcut icon, which is placed into the current folder. The name of the shortcut is automatically "Shortcut to . . . ," although you may rename it if you want. Drag and drop the shortcut onto the desktop for easy access to it.

By right-clicking the shortcut icon and selecting Properties from the menu that appears, you can examine or set its properties, such as the type of window it starts the application in (minimized, maximized), its attributes (hidden, read-only), and the working directory. Figure 14.14 shows the initial Properties display for a shortcut to an MS-DOS application called ASM. The Program tab displays the properties shown in Figure 14.15. Note that the shortcut can be started with the same command line parameters you would use while in DOS, just by entering them in the Cmd line box.

For appearance, you may want to change the icon used by a shortcut. Left-clicking the Change Icon button brings up the window shown in Figure 14.16. The new icon can be chosen from the group provided, or you can use Browse to select an icon from a different location.

The Font and Screen tabs control the appearance of the window the shortcut application runs in, and the Memory tab allows you to adjust the way memory is allocated to

FIGURE 14.14 Initial shortcut properties screen

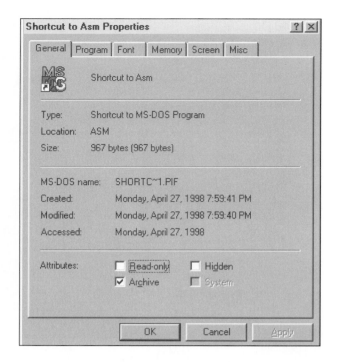

FIGURE 14.15 Program properties for shortcut

FIGURE 14.16 Choosing a shortcut icon

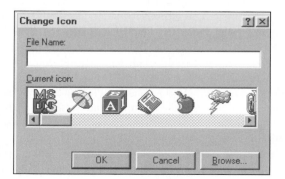

the application. Last, the Misc tab provides control over other important properties, such as the ability to use a screen saver with the application and how the application may be terminated.

Note: If you right-click a non-DOS application's shortcut icon, such as Word, and then select Properties, you get a screen with only two tabs (General and Shortcut)—not six tabs as in Figure 14.14. This is further proof that menus are context sensitive.

CHECKING/SETTING PROPERTIES

All the items in the Folder and File display windows can be examined in great detail by looking at the item properties. For example, by right-clicking a disk drive and selecting Properties from the menu, the hard disk Properties window is displayed, as illustrated in Figure 14.17.

From the disk Properties menu, any of the hard disk properties can be modified or disk tools run. Similarly, if we right-click on a folder and select Properties from the menu, folder properties are displayed. Examine Figure 14.18 to view the properties of a folder. The folder properties include items such as the file name, creation date, file location, size of all the files stored inside the folder, and file attributes. The file attributes can be changed as necessary.

Many property windows have an Apply button, which causes the changes made in the Properties window to take effect. Properties are typically set or cleared by clicking check boxes or radio buttons, or by entering data into a text box.

EDITING FEATURES

Explorer provides four essential editing features: *Cut*, *Copy*, *Paste*, and *Undo*. These operations are available in the Edit pull-down menu and as buttons on the toolbar. To cut a file or folder, left-click it and then select Edit, Cut, or click the Cut button. Cutting is typically used when you want to move a file or folder to another location. Unlike Delete, nothing is

FIGURE 14.17 Disk Properties window

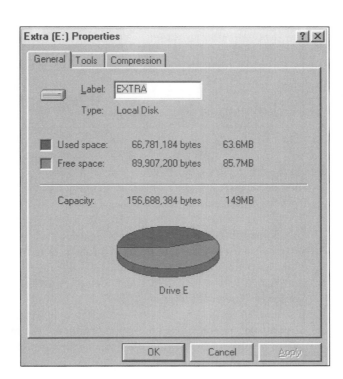

167

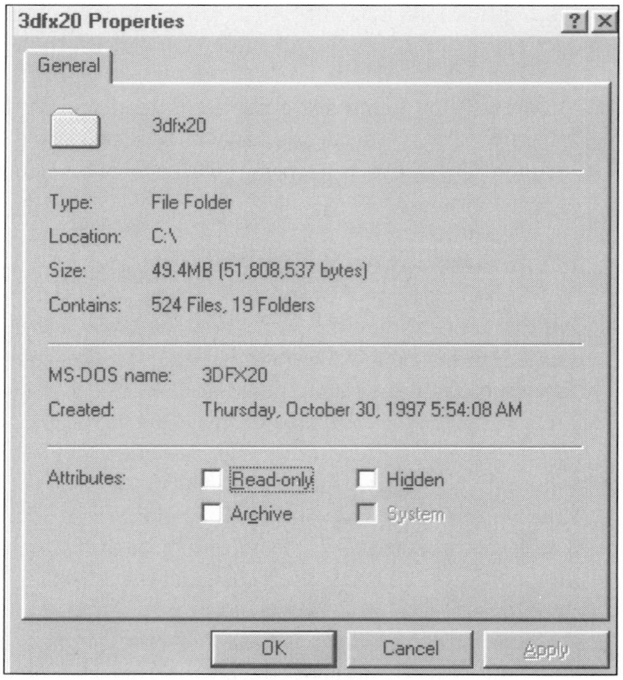

FIGURE 14.18   A Folder Properties menu

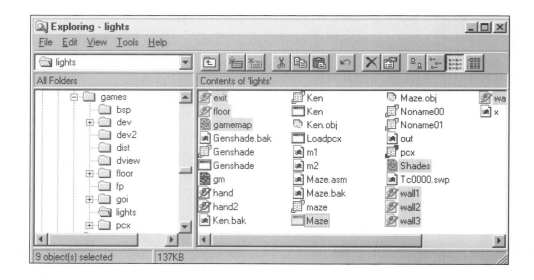

FIGURE 14.19   Selecting multiple files

placed into the Recycle Bin when you cut an object. To move the object that has been cut to another location, navigate to the destination, and click the Paste button (or choose Paste from the Edit menu).

Clicking the Copy button (or selecting Copy from the Edit menu) instead of Cut leaves the original file or folder in place, storing a copy when Paste is clicked.

To select more than one file from a folder, or multiple folders, press and hold the Ctrl (Control) button on the keyboard, and left-click on each file or folder you want to select. If you change your mind, simply left-click anywhere in a blank portion of the file display window to deselect everything.

Figure 14.19 shows an example of selecting multiple files in a folder.

The Undo button is used to back up and erase the results of the last operation. For example, if you accidentally move a folder to the wrong location, Undo will move it back for you. You may need to refresh the display to see the results.

FINDING THINGS

Windows Explorer has many options to assist you in finding a file or set of files on your machine. You can search subfolders, search by the date of creation or last access, or search for files of a certain size. You can even search for files containing specific text strings.

Clicking on the Find option under the Tools menu allows you to choose Files or Folders or Computer as the search location. Choosing Files or Folders produces the search window shown in Figure 14.20. The file ASM.C on drive C: is the subject of the search. You may enter any legal file name in the Named box, including file names with wild card characters. Clicking on Find Now begins the search, which can be stopped at any time by clicking Stop. Figure 14.21 shows the results of the search. As indicated, two copies of ASM.C were found, each in a different directory on drive C:.

Clicking the Date Modified tab brings up the search options shown in Figure 14.22. Notice that you can search for a recently created or modified file by selecting a *time frame* for the search.

Advanced search options, illustrated in Figure 14.23, include searching files for a string of text (the word "microprocessor" in Figure 14.23), searching for files of a particular size, or searching files associated with a specific application (illustrated in Figure 14.24).

Combinations of each search option may also be used to further restrict the scope of the search.

The second option on the Tools . . . Find . . . menu is Computer. Instead of searching for a file or folder, you can search the network your computer is connected to (even if you

FIGURE 14.20   Finding a file

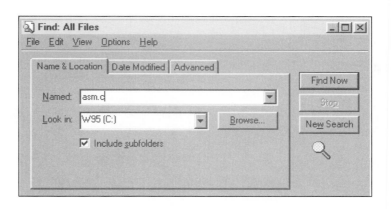

FIGURE 14.21   Search results

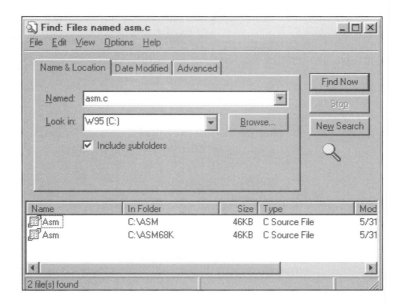

FIGURE 14.22 Using the date to limit the search

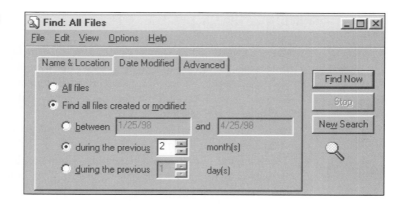

FIGURE 14.23 Searching for text

FIGURE 14.24 Selecting a file type to search for

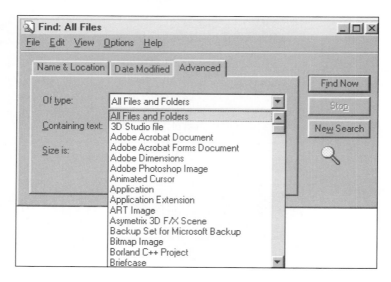

are using Dial-Up Networking) for a specific machine. In Figure 14.25 a machine called "Waveguide" is found using this search method.

There are many ways to search for an object using Windows.

WORKING WITH NETWORK DRIVES

If you have a connection to a network (dial-up PPP or network interface card), you can use Explorer to *map* a network drive to your machine. This is done by selecting Map Network Drive on the Tools menu. Figure 14.26 shows the menu window used to map a ***network drive***.

The computer automatically picks the first free drive letter (you can pick a different one) and requires a path to the network drive. In Figure 14.26 the path is \\SBCCAA\ ANTONAKOS_J. The general format is \\machine-name\user-name.

FIGURE 14.25   Searching for a computer on a network

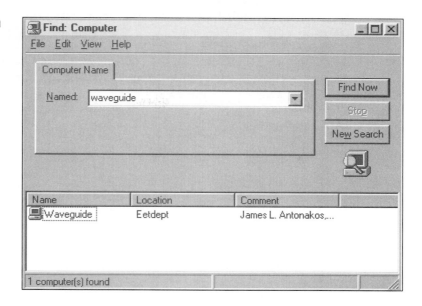

FIGURE 14.26   Mapping a network drive

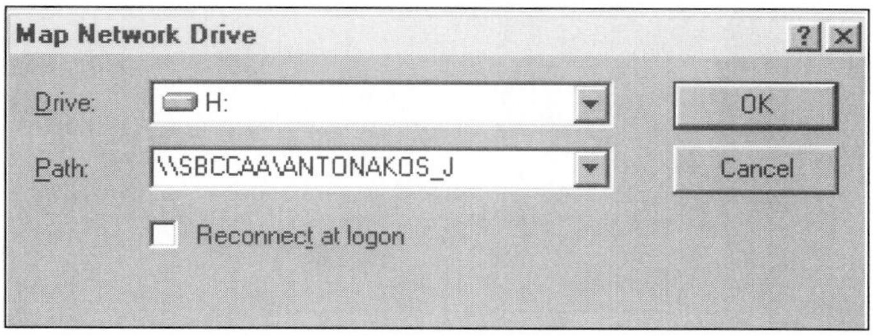

FIGURE 14.27   Supplying a network password

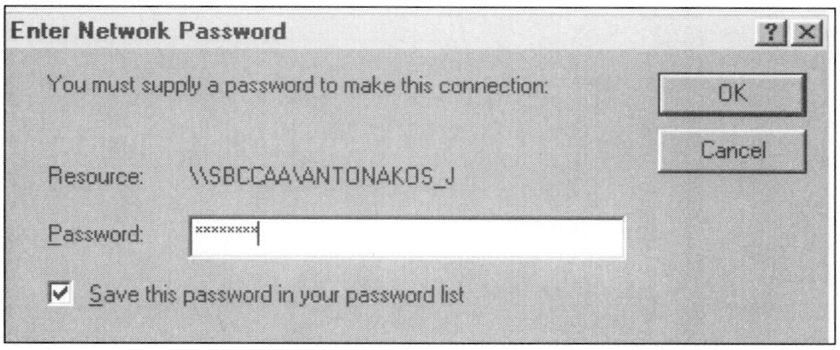

Access to the network drive may require a password, as indicated in Figure 14.27. If an invalid password is entered, the drive is not mapped.

If the drive is successfully mapped, it will show up in Explorer's folder display window. Figure 14.28 shows the contents of the mapped drive. Note that drive H: has a different icon from the other hard drives.

When you have finished using the network drive, you can disconnect it (via the Tools menu). This is illustrated in Figure 14.29.

USING GO TO

When the toolbar is turned on, the Go To folder chooser (above the folder display window) is available; this provides access to all components of the computer. Figure 14.30 shows many of the typical items found in the Go To list.

Notice that the first item in the list is the Desktop. Built-in desktop components such as My Computer, Network Neighborhood, and the Recycle Bin are in the Go To list, as well as

FIGURE 14.28 Contents of network drive H:

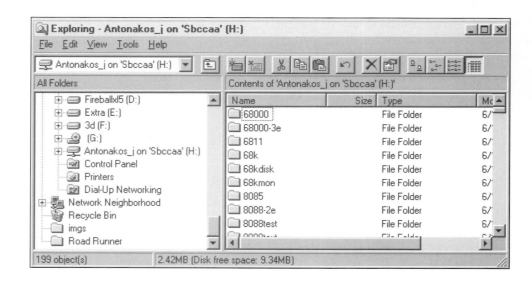

FIGURE 14.29 Disconnecting a network drive

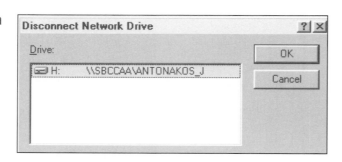

FIGURE 14.30 Using the Go To menu to select an item

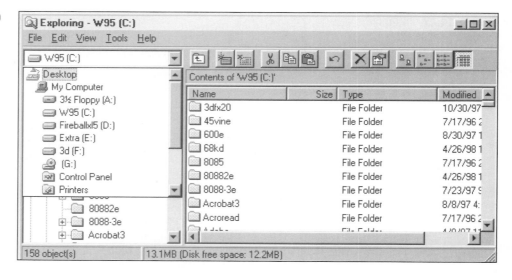

other desktop items added by the user. This is illustrated in Figure 14.31, which shows the remaining items in the Go To list. The imgs and Road Runner folders, which reside on the desktop, are part of the Go To list as well.

Left-clicking on an item (such as My Computer) in the Go To list displays its contents in the file display window. This is illustrated in Figure 14.32.

So almost anywhere you want to go on your machine is only a click away with the Go To chooser.

FIGURE 14.31 More Go To items

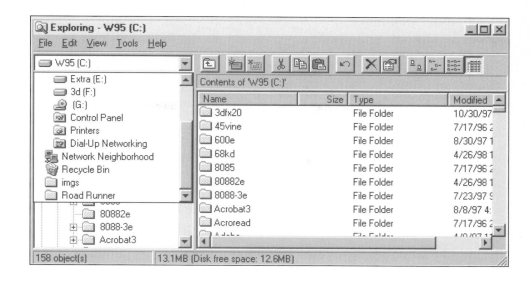

FIGURE 14.32 The contents of My Computer

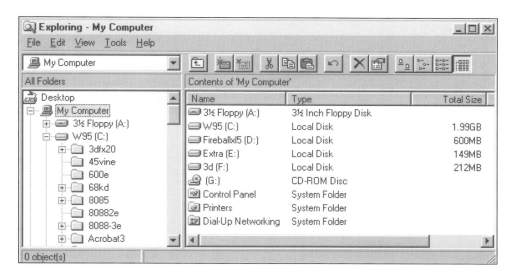

HELP

Help on almost any Windows 95 topic is provided through the Help menu (previously examined in Exercise 10). This is the same Help that is found on the Start menu. The initial Help screen is shown in Figure 14.33.

In addition, selecting the About Windows 95 item on the Help menu brings up the information window shown in Figure 14.34.

The Help menu completes a set of Explorer menus that provide a large amount of control over the Windows 95 environment. Spend some time getting familiar with all the features of Explorer.

WINDOWS 98 EXPLORER

As previously mentioned, many of the features found in Windows 95 Explorer are also found in Windows 98 Explorer. You can still map drives, navigate around directories, and change views. However, many new features are provided to improve the Windows 98 experience. One enhancement is illustrated in Figure 14.35, in which a Web page is being

FIGURE 14.33   Initial Help screen

FIGURE 14.34   The About Windows 95 Help window

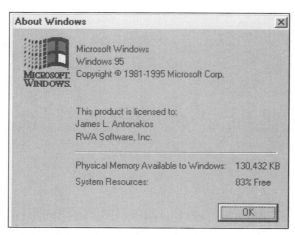

viewed in the rightmost panel. Compare this with Figure 14.1(b), which displays files and folders in the same panel.

In general, the features of Windows 98 Explorer are provided by Internet Explorer 4.0. The toolbar is context dependent with larger buttons that are more visible and convenient. The Address Bar eliminates the need for the Go To option in Windows 95 Explorer. The path to a directory folder or file, or the WWW address of a Web page, can be entered in the Address Bar (with Explorer 4.0 automatically completing the entry if it has been entered before). Furthermore, you may easily add a file, folder, or Web page to your list of favorite locations (with Web pages automatically updated in a background process as frequently as desired).

The context-sensitive aspects of Explorer 4.0 can be seen by comparing the graphical toolbars in Figure 14.35 and 14.1(b). In Figure 14.35 the buttons provide options normally available when browsing the Web, such as Stop, Refresh, and Home. The graphical toolbar

FIGURE 14.35 Windows 98 Explorer showing a Web page

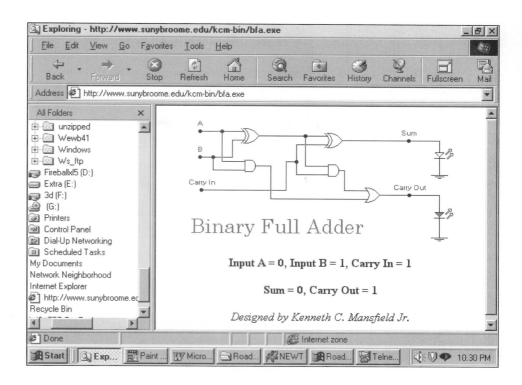

in Figure 14.1(b) provides the typical editing functions necessary when working with folders and files, such as Cut, Copy, and Paste.

WINDOWS NT

The Windows NT 4.0 operating system comes standard with the same interface as the Windows 95 operating system. All the discussion about Windows 95 Explorer earlier in this exercise is applicable.

Fortunately, the newer features available in Windows 98 Explorer can also be added to Windows NT 4.0 by upgrading to Windows Explorer 4.0. The upgrade can be downloaded directly from Microsoft or when installing other Microsoft software (such as Visual C++ 6.0). In either case, it is easy to get the new software and take advantage of the new features.

TROUBLESHOOTING TECHNIQUES

Although the graphical toolbar contains buttons for many useful operations, it is important to remember that some operations are missing. For example, you may get frustrated searching the toolbar for the Create Folder button, only to find that there is no button for the operation, which is available only as a pull-down menu item. It pays to periodically reexamine the meaning of each toolbar button. You may find that you have been using the Cut and Paste operations from the pull-down menu quite frequently, instead of their built-in toolbar buttons.

SUMMARY

In this exercise we discovered that

- The viewing format of files and folders can be adjusted.
- Explorer can be used to cut, copy, and paste files and folders.
- Network drives can be mapped/unmapped using Explorer.
- Explorer can be used to delete files and folders or check their properties.

This self-test is designed to help you check your understanding of the background information presented in this exercise.

True/False

Answer *true* or *false*.

1. The Windows Explorer window contains three separate areas.
2. The graphical toolbar contains a limited set of the available features from the pull-down menus.
3. Using the List option scrolls the files vertically.
4. Deleting a file consists of moving the file contents into the Recycle Bin.
5. New folders are always created in the currently selected drive's root directory.
6. There is only one way to view a list of files.
7. To make a copy of a file, use Copy and Paste.

Multiple Choice

Select the best answer.

8. The file display shows
   a. The number of objects in the window and the amount of disk space used.
   b. The graphical toolbar.
   c. A list of files and folders in the currently selected location.
   d. Help topics.
9. New folders are created in
   a. The root directory on the C: drive.
   b. The Windows subdirectory on the D: drive.
   c. The currently selected drive's root directory.
   d. The currently selected location.
10. When deleting a folder, Explorer
    a. Moves the folder into the Recycle Bin immediately.
    b. Prompts two times to be sure all files are to be deleted.
    c. Prompts one time to be sure all files are to be deleted.
    d. Prompts multiple times to be sure all files are to be deleted.
11. Explorer performs four essential editing features:
    a. Cut, Copy, Paste, Redo.
    b. Cut, Edit, Move, Rename.
    c. Cut, Paste, Undo, Redo.
    d. Cut, Copy, Paste, Undo.
12. When mapping a network drive
    a. The computer automatically picks the last free drive letter.
    b. The disk shows up in the Explorer display just like the local drives.
    c. Use of the Go To menu item is required to identify the specific network resources.
    d. None of the above.
13. You can select to view
    a. Files and folders.
    b. Different disk drives.
    c. Both a and b.
14. Files deleted by Explorer
    a. Are gone forever.
    b. End up in the Recycle Bin.
    c. Go to the TEMP directory.

Completion

Fill in the blank or blanks with the best answers.

15. The list of items on the desktop contains the built-in desktop components _____, _____, and _____.
16. _____-clicking on an item displays its contents in the file display window.

17. Many of the _____ menu items are located in the graphical toolbar.
18. _____ properties include file name, creation date, file location, and total size.
19. The _____ _____ button begins the search when finding files.
20. In \\eelab\bigcpu, eelab is the _____ name.
21. Explorer allows you to map a _____ drive.

FAMILIARIZATION ACTIVITY

1. Practice using all the display modes of Explorer to determine which one you prefer.
2. Use Explorer to map a network drive. What kinds of software are available on the network drive?
3. Practice creating and renaming files and folders, moving them to different locations, copying them, and navigating through the drives on your system.
4. Create shortcuts to your frequently used applications.

QUESTIONS/ACTIVITIES

1. What is the value of a mapped network drive?
2. Make a list of 10 things you normally do while using your computer (for example, rename or copy files, run applications). Explain how these chores can be performed by Windows Explorer (if possible).

REVIEW QUIZ

Under the supervision of your instructor

1. Explain the basic features of Windows Explorer.
2. Show how to start an application.
3. Create a shortcut, move, find, and delete files and folders.
4. Map a network drive.

15

Managing Printers

<table>
<tr><td>**INTRODUCTION**</td><td>*Joe Tekk lugged a heavy laser printer into his office and set it down on his desk. He plugged the power cable in, loaded paper into the paper tray, and connected the printer to his Windows 98 machine. The printer was an old HP LaserJet Series II that Joe had bought at a hamfest for $25.*</td></tr>
</table>

Joe Tekk lugged a heavy laser printer into his office and set it down on his desk. He plugged the power cable in, loaded paper into the paper tray, and connected the printer to his Windows 98 machine. The printer was an old HP LaserJet Series II that Joe had bought at a hamfest for $25.

Joe ran the Add Printer Wizard and loaded the drivers for the laser printer. However, when he printed a test page, nothing happened. After several minutes of troubleshooting, Joe found the problem: a bent pin on the printer connector. He straightened the pin with a set of needlenose pliers and tried the test page again.

His $25 laser printer worked just fine.

PERFORMANCE OBJECTIVES

Upon completion of this exercise, you will be able to

1. Check the properties of any available printers.
2. Install a new printer.
3. Connect to a network printer.

KEY TERMS

ECP printer port
Print spooling
Dithering

Local printer
Network printer

BACKGROUND INFORMATION

Windows has many built-in features to assist you in using your printer, from installing it to sharing the printer on a network. Figure 15.1 shows the starting point for many printer operations, the Printers folder. You can open the Printers folder by left-clicking Start, Settings, and Printer. Or, from inside the Control Panel, you can double-click the Printers folder. The printers installed on the system appear in the Printers folder as separate printer icons. Figure 15.1 shows just one installed printer, the Okidata OL-400e.

FIGURE 15.1   Printers folder

PRINTER PROPERTIES

Printer properties can be examined by right-clicking the desired Printer icon. Doing so opens up a context-sensitive menu that provides some basic printer control (pause/purge print jobs, set default printer) and a properties section. The Properties window for the Okidata laser printer is shown in Figure 15.2(a).

Figure 15.2(a) illustrates the Windows 95/98 Properties window for the Okidata printer. Note that the format of the tabs displayed along the top of the Properties window may contain a Sharing tab if printer sharing is enabled. Figure 15.2(b) shows the Windows NT Properties menu. Although much of the same information is available on Windows NT, it is located on different tabs. You are encouraged to familiarize yourself with these screens.

If the printer has been installed correctly, left-clicking the Print Test Page button will cause the printer to print a test page. The test page contains a graphical Windows logo and information about the printer and its various drivers. A dialog box appears asking whether the test page printed correctly. If the answer is no, Windows starts a printer help session. Figure 15.3 shows the initial Help window.

Windows will ask several printer-related questions to help determine why the printer is not working. The causes are different for network printers, so Windows provides two different troubleshooting paths (network versus local).

FIGURE 15.2   (a) Initial Windows 95/98 printer Properties window

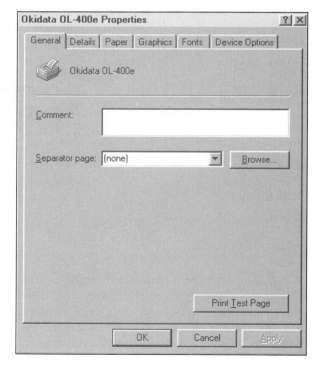

(a)

FIGURE 15.2 *(continued)* (b) Windows NT printer Properties menu

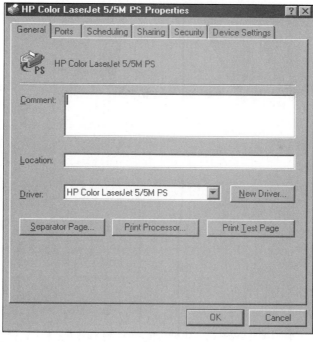

(b)

FIGURE 15.3 Built-in printer help

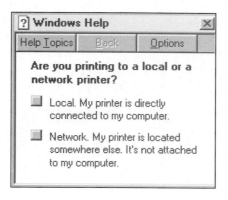

In Windows 95/98 the Details tab brings up the window illustrated in Figure 15.4(a). Many of the software and hardware settings are accessed from this window. Notice that LPT1 is set up as an ***ECP printer port***. ECP stands for *extended capabilities port*, a parallel port standard that allows 8-bit bidirectional data flow. This allows ECP to support other peripherals over the parallel printer port, such as external CD-ROM drives or scanners.

The Ports tab shown in Figure 15.4(b) shows a comparable Windows NT menu that indicates an HP Color LaserJet printer is attached on LPT2. The Enable printer pooling feature in Windows NT allows for several printers to work together to handle a heavy print workload. Print jobs in the queue are assigned to available printers in the pool so that each of the printers in the pool can be used to service many different print jobs simultaneously.

Left-clicking the Spool Settings button (or the Scheduling tab in Windows NT) allows you to check/set the spool settings for the printer. These settings are indicated in Figure 15.5. ***Print spooling*** is a technique that speeds up the time required by an application to send data to the printer. The hard disk is used as a temporary holding place for the print job. The application prints everything very quickly to a spooling file on the hard drive.

FIGURE 15.4 (a) Windows 95/98 printer Details menu and (b) Windows NT printer Ports menu

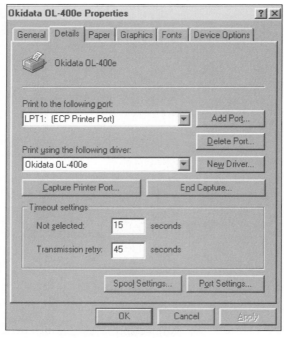

(a)

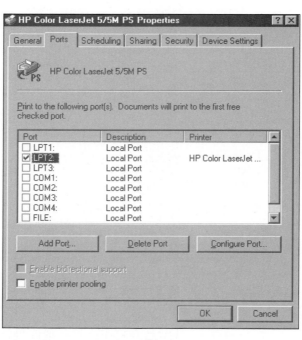

(b)

Windows then prints the spooled file in the background, while the user continues working on other things. The application does not have to wait for each page to be printed before returning control back to the user. This process is diagrammed in Figure 15.6.

Notice that in Figure 15.5(b), other useful information is available on the Windows NT Scheduling menu in addition to the spool settings. The times that the printer is available, priority for the print jobs, and other various settings can be modified as required.

The Paper tab in the Windows 95/98 printer Properties window brings up the window shown in Figure 15.7(a). Here the size and orientation of the paper are selected, as well as

FIGURE 15.5 (a) Windows 95/98 Spool Settings window and (b) Windows NT printer Scheduling menu

Spool Settings   ? X

⊙ Spool print jobs so program finishes printing faster
   ○ Start printing after last page is spooled
   ⊙ Start printing after first page is spooled
○ Print directly to the printer

Spool data format: [EMF   ▼]

○ Enable bi-directional support for this printer
○ Disable bi-directional support for this printer

[OK]   [Cancel]   [Restore Defaults]

(a)

HP Color LaserJet 5/5M PS Properties   ? X

| General | Ports | Scheduling | Sharing | Security | Device Settings |

Available: ⊙ Always
             ○ From [12:00 AM ⬍]   To [12:00 AM ⬍]

Priority
  Lowest [▐——————————————] Highest
  Current Priority: 1

⊙ Spool print documents so program finishes printing faster
   ○ Start printing after last page is spooled
   ⊙ Start printing immediately
○ Print directly to the printer

☐ Hold mismatched documents
☐ Print spooled documents first
☐ Keep documents after they have printed

[OK]   [Cancel]

(b)

FIGURE 15.6 Operation of a print spooler

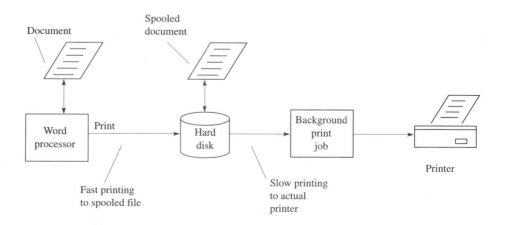

FIGURE 15.7 (a) Windows 95/98 Paper settings menu and (b) Windows NT printer Device Settings menu

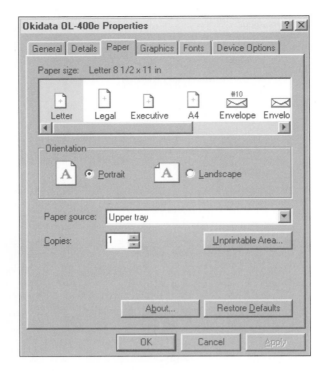

(a)

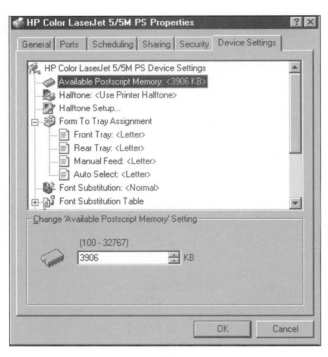

(b)

the default number of copies. Normally, pages are printed in Portrait mode. Selecting Landscape turns the printed document 90 degrees, which is useful for printing wide documents such as spreadsheets. In Figure 15.7(b) the Device Settings tab shows the paper settings as well as the Available Postscript Memory setting for the HP Color LaserJet printer running under Windows NT.

The Graphics settings menu illustrated in Figure 15.8 allows the resolution, *dithering*, and intensity of the printer to be adjusted. Dithering is a method used to represent a particular color

FIGURE 15.8 Windows 95/98 Graphics settings menu

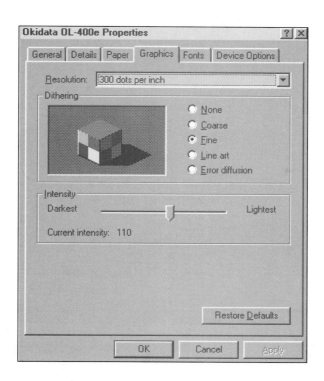

by using one or more colors that are similar. This is a necessary step when printing graphical documents that may contain many more colors (or shades of gray) than the printer supports.

You should experiment with the graphics settings until you find the right combination for your printer. Note that the printer properties window does not contain a Graphics tab in Windows NT, although you are allowed to adjust the resolution using the Advanced option under Printing Preferences.

ADDING A NEW PRINTER

To add a new printer, double-click the Add Printer icon in the Printers folder shown in Figure 15.1. This will start up the Add Printer Wizard, an automated process that guides you through the installation process.

The first choice you must make is shown in Figure 15.9. A *local printer* is local to your machine. Only your machine can print to your printer, even if your computer is networked. A *network printer* can be printed to by anyone on the network who has made a connection to that printer. A network printer is also a local printer to the machine that hosts it. If you are installing a network printer, the next window will look like that shown in Figure 15.10. The

FIGURE 15.9 Choosing local/ network printing

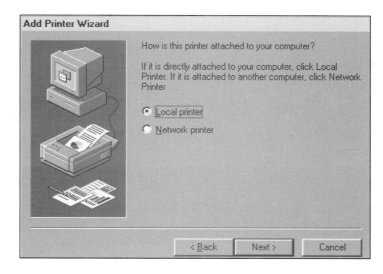

FIGURE 15.10 Mapping a
network printer

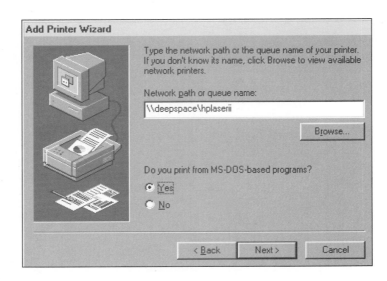

FIGURE 15.11 Choosing a
printer manufacturer/model

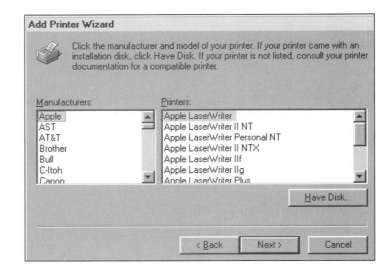

printer being mapped is an HP LaserJet II (named "hplaserii" on the network) connected to the machine "deepspace." You can also browse the Network Neighborhood to select a network printer. DOS accessibility to the network printer is controlled from this window as well.

Next, the manufacturer and model of your printer must be chosen. Windows has a large database of printers to choose from. Figure 15.11 shows the initial set of choices. If your printer is not on the list, you must insert a disk with the appropriate drivers (usually supplied by the printer manufacturer).

Once the printer has been selected, the last step is to name it (as in the network printer "hplaserii").

If only one printer is installed, it is automatically the default printer for Windows. For two or more printers (including network printers), one must be set as the default. This can be done by right-clicking on the Printers icon and selecting Set As Default. You can also access printer properties and change the default printer from inside the printer status window, using the Printer pull-down menu.

CHECKING THE PRINTER STATUS

To check the status of the current printer, double-click the Printer icon in the system settings area of the taskbar (next to the time display) or double-click the desired Printers icon in the Printers folder. The printer status window is shown in Figure 15.12. The printer status indicates that a document is being printed by the printer.

FIGURE 15.12 **Printer status window**

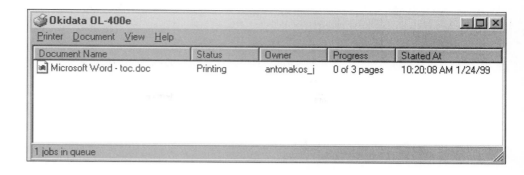

FIGURE 15.13 **Printer error window**

If Windows detects a problem with the printer (out of paper, offline, powered off), it will report the error without having to bring up the status window. Figure 15.13 shows a typical printer error window.

If a document cannot be printed for some reason before the system is shut down, Windows will save the print job and try to print it the next time the system is booted.

PAUSING A PRINT JOB

To pause a print job (possibly to add more paper), select Pause Printing from the Printer or Document pull-down menus in the printer status window. This will shut off the flow of new data to the printer. If several pages have already been sent, they will be printed before the pause takes effect.

RESUMING A PRINT JOB

To resume a paused print job, select Pause Printing again (it should have a black check mark next to it when paused).

DELETING A PRINT JOB

Individual print jobs can be deleted by selecting them with a single left-click and then choosing Cancel Printing from the Document pull-down menu. *All* the jobs in the printer can be deleted at the same time by selecting Purge Print Jobs from the Printer pull-down menu.

NETWORK PRINTING

To make the printer on your machine a network printer, you need to double-click the Network icon in Control Panel and then left-click the File and Print Sharing button. This opens up the window shown in Figure 15.14. The second box must be checked to allow network access to your printer.

After a network printer connection has been established, you may use it like an ordinary local printer. Windows communicates with the network printer's host machine using NetBEUI. What this means is that jobs sent to a network printer are sent in small bursts

FIGURE 15.14 Giving network
access to your printer

(packets) and typically require additional time to print due to the network overhead. In a busy environment, such as an office or college laboratory, printer packets compete with all the other data flying around on the network, and thus take longer to transmit than data traveling over a simple parallel connection between the computer and the printer.

WINDOWS NT PRINTER SECURITY

The Security tab in Windows NT printer properties offers several options not available in Windows 95/98. Figure 15.15 shows the window that allows access to view or change the Permissions, Auditing, and Ownership properties of a Windows NT printer. The system user can assign the printer permissions so that individuals, groups, or everyone in the entire domain may access a particular printer. The type of access (No Access, Print, Manage Documents, or Full Control) is specified for each user or group.

Windows NT printer auditing can be configured to retain certain events. These events (Print, Full Control, Delete, Change Permission, and Take Ownership) are grouped into two categories: success and failure. In general, auditing involves tracking certain users or examining the data to determine a trend such as paper usage, job length, or who prints the most.

The ownership of a printer can be changed by selecting the Ownership menu button. The owner of a printer can modify permission settings and grant permission to other users. The type of access to the printer determines if ownership of a printer can be taken.

By setting the Permissions, Auditing, and Ownership options appropriately, the system administrator can manage every aspect of the printing process for every computer user in

**FIGURE 15.15   Windows NT
printer Security menu**

FIGURE 15.16   Downloading a printer driver

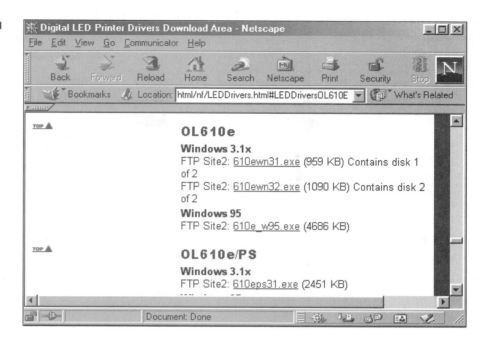

the Windows NT domain. You are encouraged to examine each of these menus in detail to determine what types of settings are currently selected on your computer.

TROUBLESHOOTING TECHNIQUES

You may run into a situation in which you need new drivers for your printer, but you do not have a disk from the manufacturer. One of the easiest solutions is to search the Web for the printer manufacturer and look for a driver download page for your printer. Figure 15.16 shows a Netscape window of a portion of Okidata's driver page. The printer drivers are contained in self-extracting executables. Download the one you need and run it. Be sure to read the README file or other preliminary information. Some printer manufacturers require the old drivers to be completely removed before beginning a new installation or update.

SUMMARY

In this exercise we discovered that

- Printers can be shared among networked computers.
- A print spooler is used to print a document in the background.
- Multiple printers may be installed on a system. Only one is selected as the default printer at any time.
- Windows NT provides security for printing operations.

SELF-TEST

This self-test is designed to help you check your understanding of the background information presented in this exercise.

True/False

Answer *true* or *false*.

1. Only one printer may be installed at a time.
2. Print spooling refers to the coiling of the printer cable around a spool.
3. One printer driver will work for many different printers.
4. Print jobs can be paused.

5. ECP stands for easily connected printer.
6. Windows NT has no security advantages over Windows 95/98.
7. There can be two default printers selected at the same time.
8. Dithering is only used on color printers.

Multiple Choice

Select the best answer.

9. A print job is deleted by
   a. Dragging it to the Recycle Bin.
   b. Selecting it and choosing Cancel Printing.
   c. Turning the printer off.
10. When Purge Print Jobs is selected,
    a. The current job is purged.
    b. Only network printer jobs are purged.
    c. All print jobs are purged.
11. A network print job typically takes
    a. Less time than a local print job.
    b. The same time as a local print job.
    c. More time than a local print job.
12. When you pause a print job,
    a. Paper immediately stops coming out of the printer.
    b. The current page finishes printing.
    c. All complete pages sent to the printer finish printing.
13. In the network printer path \\waveguide\riko, the name of the printer is
    a. Waveguide.
    b. Riko.
    c. Neither.
14. The Windows NT printer Security menu shows
    a. Administrator, Access, and Permission.
    b. Owner, Permissions, and Manager.
    c. Permissions, Auditing, and Ownership.
15. Data on the ECP printer port is
    a. Bidirectional.
    b. One way (output only).
    c. One way (input only).
16. When a document is spooled for printing, a copy is placed
    a. In the printer.
    b. On the hard drive.
    c. In the SPOOL folder.

Completion

Fill in the blank or blanks with the best answers.

17. A technique that attempts to match colors and gray levels during printing is called _____.
18. Data is sent to network printers using the _____ protocol.
19. The two types of printer connections are network and _____.
20. Saving a print job on the hard drive and printing it in the background is called _____.
21. One printer is always set as the _____ printer.
22. Windows NT printer auditing groups events into two categories: _____ and _____.
23. Network printing is accomplished using the _____ protocol.
24. Documents are printed in Portrait or _____.

1. Examine the properties of all the printers connected to your computer, including any network printers.
2. Install a new local printer.
3. Install a new network printer.
4. Connect to a network printer.
5. Send a 10-page document containing text and graphics to a network printer. Keep track of how long it takes to finish printing. Repeat the print job three more times and compare the timing results.
6. Download a new printer driver from the Web and install it.

1. Make a list of all the printers in the lab. How many of them can be found in the installation list of manufacturers and models (Figure 15.11)?
2. Go to a local computer store and check the prices of various printers. What is available for less than $200?

Under the supervision of your instructor

1. Check the properties of any available printers.
2. Install a new printer.
3. Connect to a network printer.

16

Accessories

INTRODUCTION

Joe Tekk stopped in his office for a few minutes on the way to a meeting. He sat down at his computer and checked his schedule with the Calendar utility, opened WordPad to write a quick note to a friend, and examined a graphical display of network traffic using System Monitor. Then his telephone rang. It was Jeff Page, wondering what a $550 motherboard would cost with a 17% discount. Joe brought up the Calculator utility, performed the calculation, and gave Jeff the answer.

As he closed the Calculator window, Joe wondered what he would do without all the utilities provided by Windows.

PERFORMANCE OBJECTIVES

Upon completion of this exercise, you will be able to

1. Describe the basic contents of the Accessories folder.
2. Demonstrate the operation of several utilities, such as the Calculator, Notepad, and Paint.
3. Explain the importance of the System Tools and Administrative Tools folders.

KEY TERMS

Browser
Internet tools
System tools
Fragmented
ScanDisk
Calculator

Clipboard
Dial-Up Networking
Direct cable connection
Hyperterminal
Notepad

BACKGROUND INFORMATION

Windows provides a large number of accessory applications designed to help perform all the small chores we might require when using a computer to simplify our lives. Figure 16.1 shows the Accessories menu, which is full of useful applications. Notice that there are several folders in the submenu as well.

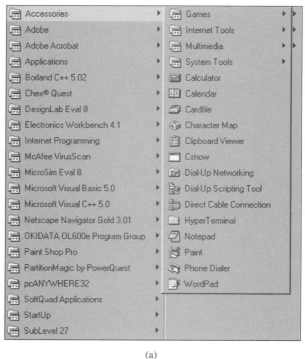

(a)

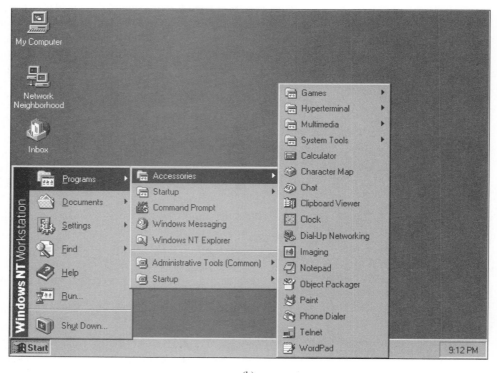

(b)

Notice the similarities between the Windows 95/98 Accessories menu in Figure 16.1(a) and the Windows NT Accessories menu in Figure 16.1(b). Many of the discussions that follow apply to Accessories applications found in both operating systems.

In this exercise we will take a brief look at many of the accessory applications. You are encouraged to spend additional time learning how to use any applications that are useful to you or recommended by your instructor.

FIGURE 16.2   Solitaire game window

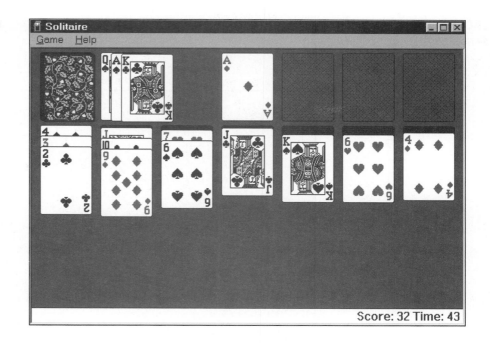

GAMES

Windows comes equipped with several games, just as Windows 3.x did. These games can be useful tools when introducing a new user to the Windows environment. The Solitaire game shown in Figure 16.2 can be used to introduce a user to the mouse. Experienced Windows users rely on the mouse regularly, so why not get used to using the mouse in a fun way?

INTERNET TOOLS

The Internet (covered in depth in Exercise 27) is a worldwide collection of networks that all communicate with each other, sharing files and data. One very popular form of posting and sharing information is through a Web page. A Web page is written in a language called HTML (hypertext markup language) and is viewed using a ***browser***, an application capable of interpreting HTML and displaying a graphical page layout, as illustrated in Figure 16.3. The browser being used in Figure 16.3 is Internet Explorer, which can be downloaded for free over the Web and is an automatic part of Windows.

MULTIMEDIA

There are several multimedia applications included in Windows. They provide features such as a CD player and a sound recorder, to name just a few. Figure 16.4 shows the contents of the Multimedia folder in the Accessories menu.

SYSTEM TOOLS

The system tools provide many important features for controlling, monitoring, and tuning your Windows system. Note that many of the Windows 95/98 tools are also found in Windows NT. Exceptions to this rule will be noted accordingly. Figure 16.5 shows the selections available in the System Tools menu. Let's look at several of these applications.

Windows 95/98 Disk Defragmenter

When a hard drive is used for a long period of time, many files are created, modified, and deleted. The nature of the file allocation mechanism eventually causes the files of the hard

FIGURE 16.3   Internet Explorer window

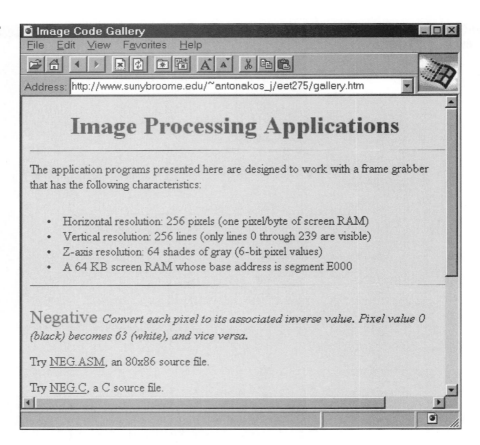

FIGURE 16.4   Multimedia applications

FIGURE 16.5   System Tools menu

FIGURE 16.6 Selecting a drive to defragment

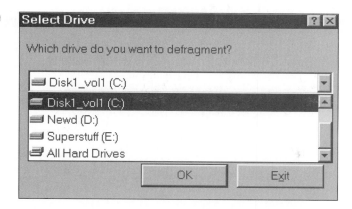

FIGURE 16.7 Defragmentation in progress

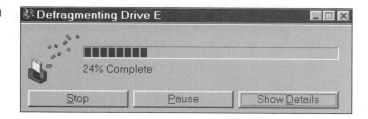

drive (or even a floppy) to become ***fragmented***. A fragmented file is spread out over many different areas of the disk, rather than being stored in one big block. This fragmentation increases the time required to read or write to the file and can lower performance significantly if a large portion of the disk becomes fragmented.

Fortunately for Windows 95/98 users the Disk Defragmenter application is included in the system tools to automatically defragment a disk drive. The files are read one by one and written back unfragmented (space is made available before writing the file back). Figure 16.6 shows the initial window for the Disk Defragmenter. Any or all of the disk drives may be selected. Depending on the size of the drive and the amount of fragmentation, the defragmentation process can be quite time consuming. A simplified status window reports the current progress, as shown in Figure 16.7. Clicking Show Details brings up a multicolored graphical display of the drive being defragmented, with different colors used to identify files being moved, files that cannot be moved, and free space. Stopping the process does not destroy any data on the drive.

Windows NT does not come with a defragmenting program of its own, but you may install a third-party product for disk maintenance, such as Diskeeper, from Executive Software (www.execsoft.com). Diskeeper works with NTFS, the native file system structure used by Windows NT.

The old-fashioned method of defragmentation was to back up the hard disk to tape, format the hard disk, and restore from the backup. Cautious users may still prefer this method, the reward being a *permanent* copy of the hard drive data.

Resource Meter

Three important resources used by Windows 95/98 are system, user, and GDI. The amount of each resource available determines how well Windows 95/98 can handle a new event, such as the launch of a new application. The Resource Meter, shown in Figure 16.8, displays the free percentage of each resource in a bar-graph display. The display is updated as resources are allocated and deallocated.

ScanDisk

Occasionally, a file or directory may develop a problem that prevents Windows 95/98 from reading or writing the file, or accessing the directory. This could be the result of having shut the computer down improperly, or having an application run amok and cause some damage. Or you may accidentally have bumped the computer, causing a head crash in the hard drive.

FIGURE 16.8   Windows 95/98 Resource Meter display

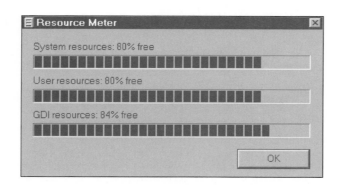

FIGURE 16.9   (a) ScanDisk control window and (b) Sample Chkdsk execution in Windows NT

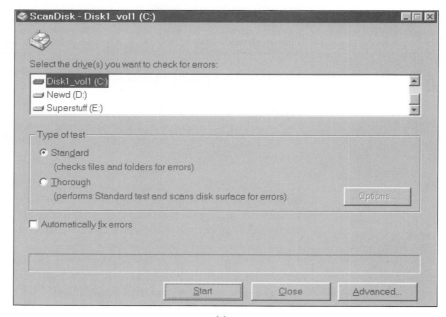

(a)

(b)

No matter what the cause, it is always a good idea to run the ***ScanDisk*** utility to check the integrity of a disk. Figure 16.9(a) shows the initial ScanDisk window. ScanDisk checks the organization of the FAT (file allocation table), the directory areas, and every file on the disk. The entire surface of the disk can be searched for bad sectors as well. Windows 95 version B and Windows 98 automatically run ScanDisk if they detect that the computer was not properly shut down the last time it was used.

In Windows NT, Chkdsk, not ScanDisk, is used to check the structure of a disk drive. Figure 16.9(b) shows the results of running Chkdsk on a Windows NT disk.

System Information

For a quick look at the information particular to your computer, use the System Information utility. Figure 16.10 shows the first screen of information, which gives a good summary of the pertinent data regarding the overall system. It would be worthwhile to spend some time looking at the items available in the System Information window. You might be surprised at what you find, such as applications running in the background that you did not even know were there.

System Monitor

To really monitor the performance of your computer, use the System Monitor utility. With this tool, you can display the *history* of usage for one or more resources. As Figure 16.11 shows, the percentage of processor usage and the amount of allocated memory are being

FIGURE 16.10   System Information window

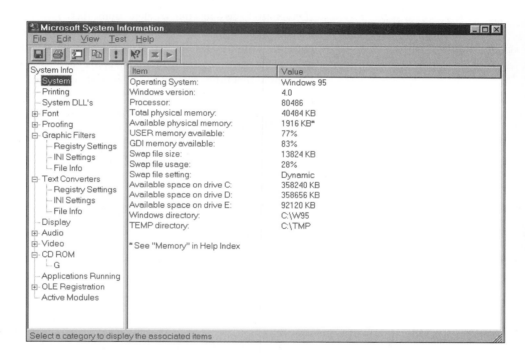

FIGURE 16.11   Monitoring system resources

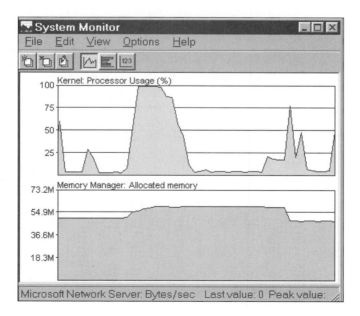

tracked by the system monitor. The Edit menu allows you to add items to track, choosing from four different areas: file system, kernel, memory manager, and network.

Compression Agent and DriveSpace

There are other tools in the System Tools menu that you may want to examine. One of these is Compression Agent. The Compression Agent utility is used to compress files on a *DriveSpace* compressed drive. DriveSpace is the technology used by Windows 95/98 to compress files on a hard drive or floppy disk, to increase the amount of storage capacity. There is a slight overhead required for decompression when a file is accessed, which can lower performance. Only compress your drive if you have run out of space and cannot add a new drive to your system.

CALCULATOR

If you need to do a few quick calculations, there is no need to go looking for your calculator. Windows has a built-in calculator that performs basic math functions in *standard mode* (shown in Figure 16.12). In *scientific mode*, the calculator has many additional features, such as the use of scientific numbers, conversion between different bases, and several transcendental functions. Figure 16.13 illustrates the scientific calculator.

CALENDAR

The Calendar application program is used to maintain a listing of important events or appointments. Each day is displayed showing time increments along the left-hand margin and can be modified very easily by simply keying information at the appropriate time. The calendar can also be set to generate an alarm up to 10 minutes before each appointment or event. Figure 16.14 shows a typical day in the calendar application.

When many computers are networked together, electronic calendars are used to simplify the scheduling process for groups of people. This may be one reason why use of electronic calendars is becoming very widespread.

CARDFILE

The Cardfile utility is an electronic form of the popular 3-by-5-inch index card. You can label a card with a heading, put other information on the body of the card, and save groups of cards in their own file. Figure 16.15 shows a sample set of cards.

FIGURE 16.12 Standard Calculator window

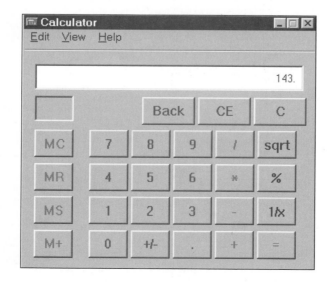

FIGURE 16.13 **Scientific Calculator window**

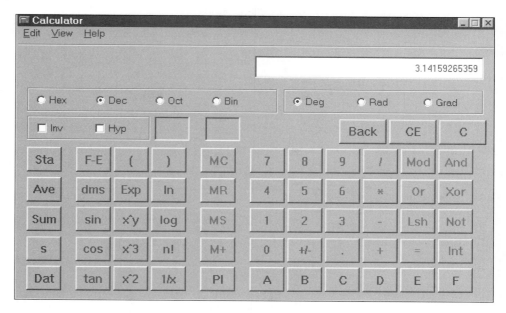

FIGURE 16.14 **Friday's schedule**

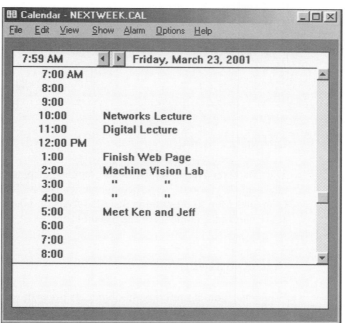

FIGURE 16.15 **Sample Cardfile contents**

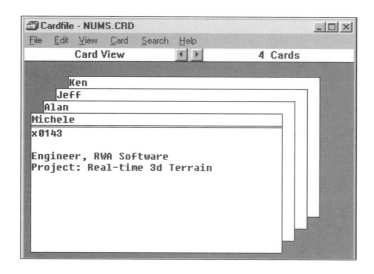

FIGURE 16.16   Character Map window

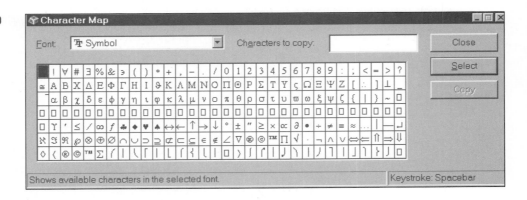

FIGURE 16.17   An image saved on the Clipboard

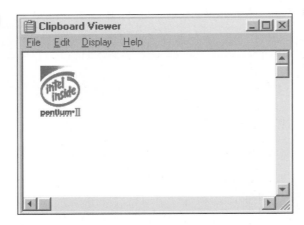

CHARACTER MAP

The Character Map tool provides a graphical display of each character available in a particular font. If you require a special character to include in a report or project, you can use Character Map to browse the installed fonts on your system and find the character. Figure 16.16 shows the contents of the Symbol font.

CLIPBOARD VIEWER

Whenever you cut or copy an object, or press the Print Screen button, data is placed on the Windows *Clipboard*. The Clipboard is a temporary holding place for information that you may want to exchange between applications. For example, when developing a Web page, you may use the Clipboard to cut and paste text and images from several different applications into your Web design tool. Figure 16.17 shows the Clipboard Viewer window, which contains a small graphic image placed on the Clipboard by a paint program.

MISSING SHORTCUTS

Occasionally you may come across an icon that does not have an application associated with it anymore (the application was deleted or moved to another folder or drive). In this case Windows will display a small searchlight while it attempts to locate the missing application. Figure 16.18 demonstrates this process.

A new application can be assigned by using the Browse feature. If there is no need for the shortcut any longer, it can be deleted by navigating to the Accessories folder using Windows Explorer and then right-clicking on the shortcut and choosing the Delete operation.

FIGURE 16.18   Missing Shortcut window

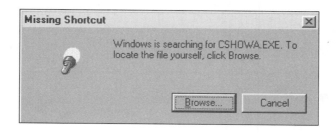

FIGURE 16.19   Dial-Up Networking window

FIGURE 16.20   Initial Direct Cable Connection window

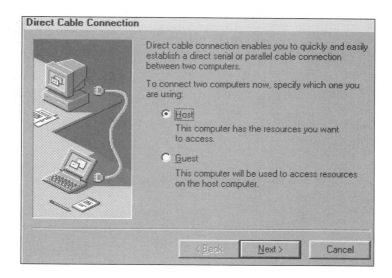

DIAL-UP NETWORKING

Dial-Up Networking (covered in detail in Exercises 17 and 30) allows your computer to use its modem to connect to a computer network (using a special protocol called PPP, for point-to-point protocol). Figure 16.19 shows the initial dial-up window. Double-clicking the My Office icon automatically dials the network mainframe computer and establishes the connection. Once connected, your computer can share files as if it were part of the actual network.

DIRECT CABLE CONNECTION

Two Windows machines can communicate with each other directly by using a ***direct cable connection*** between the serial or parallel ports of each machine. The initial window (shown in Figure 16.20) allows the user to specify Host or Guest mode. The type of connection (parallel versus serial) and port are specified next. More information about using a direct cable connection is presented in Exercises 17 and 30.

FIGURE 16.21 Using HyperTerminal to communicate with a mainframe

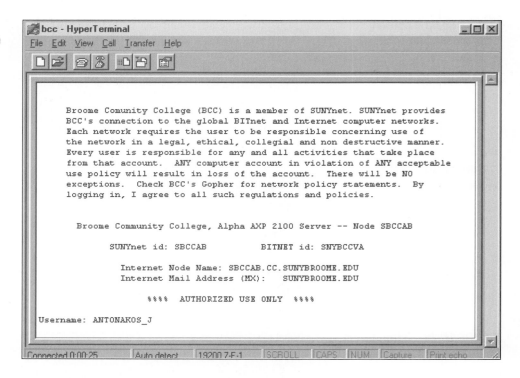

FIGURE 16.22 Notepad window

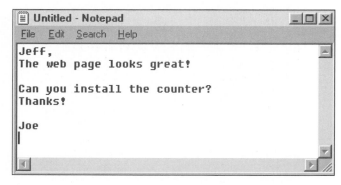

HYPERTERMINAL

If you do not own a modem communications program, Windows comes equipped with the HyperTerminal application, which emulates an ASCII data terminal and controls the modem at the same time. Figure 16.21 shows the start of a HyperTerminal session, with the user ANTONAKOS_J logging into the Broome Community College, Alpha server. HyperTerminal will capture its screen to a file, transfer files between computers, and emulate several different types of terminals.

NOTEPAD

The Notepad is a screen-based text editor. You can paste information into the notepad from the Clipboard or enter it directly from the keyboard. Figure 16.22 shows a quick note being entered. Notepad is useful for small text files (it can search for text, print, and perform simple editing chores), basically replacing the EDIT utility from DOS.

PAINT

The Paint utility allows you to create and edit graphical files (which may include text, graphics, or images). Figure 16.23 shows a simple schematic being edited. Paint saves files

204

FIGURE 16.23   Using Paint to create a graphics file

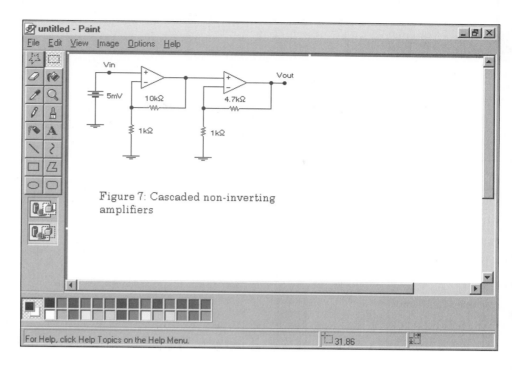

FIGURE 16.24   Phone Dialer window

in one format: bitmap. Bitmap files are native to the Windows operating system and have a special binary format that makes them compatible with the desktop display.

PHONE DIALER

Dialing a touch-tone number is as easy as a single click when you use Phone Dialer. Numbers may be saved for speed dialing or entered from a graphical keypad (or manually from the keyboard). The Phone Dialer window is shown in Figure 16.24. After the number has been dialed (via the internal modem), Phone Dialer instructs you to lift the handset and begin speaking.

WORDPAD

WordPad has many of the features of a full-fledged word-processing application. These include insertion of objects, expanded editing functions, and more control over the printer

FIGURE 16.25 Editing a document using WordPad

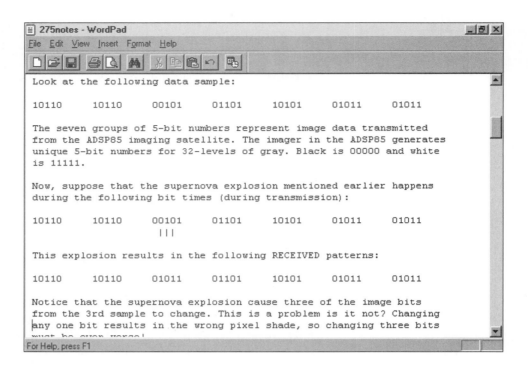

than in Notepad. Figure 16.25 shows a sample document being edited in WordPad. Together, WordPad and Paint provide you with a significant amount of text and graphical file processing, all included automatically.

WINDOWS NT ADMINISTRATIVE TOOLS

Windows NT provides additional tools for system administration. The very names of the tools suggest a different atmosphere from the Windows 95/98 environment. In fact, Windows NT allows users to have their own accounts on the system, with varying degrees of privilege and control over the system. Figure 16.26 shows the Administrative Tools

FIGURE 16.26 Windows NT Administrative Tools menu

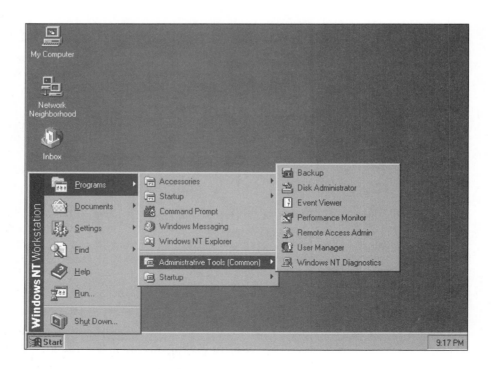

FIGURE 16.27 Event Viewer
display

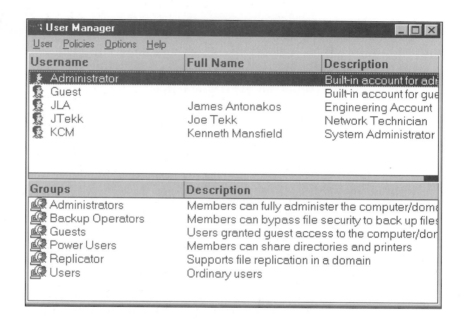

FIGURE 16.28 User Manager
display

(Common) menu. The applications listed in the menu are designed to monitor and configure many different aspects of the Windows NT operating system. Let's briefly examine three of these applications. You are encouraged to spend some time exploring these tools in the Familiarization Activity section.

Event Viewer

The Event Viewer tool, shown in Figure 16.27, keeps a running log of various system events. The historical information in the log may prove helpful when troubleshooting problems.

User Manager

The User Manager tool allows user accounts to be created, modified, and deleted (if your own account has sufficient privilege). Figure 16.28 illustrates the user base of a small system. Users who are not displayed in the list cannot log on to the Windows NT system. When Windows NT is first installed, an Administrator account is created automatically that has all available privileges, including the ability to begin creating other accounts on the new system.

FIGURE 16.29   Windows NT Memory Diagnostics display

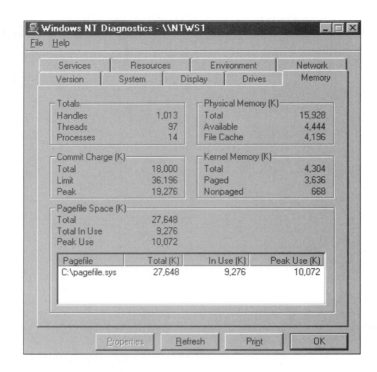

Windows NT Diagnostics

Windows NT keeps track of a significant amount of system parameters. Viewing these parameters and statistics is useful when fine-tuning the operating system for peak performance or locating a bottleneck. This activity is typically done by an experienced system administrator. The presentation of the system data is interesting in its own right, so a tour of the individual diagnostic tabs would be worthwhile. For example, the Memory Diagnostics tab display shown in Figure 16.29 uses nomenclature that is not familiar to the typical system user.

TROUBLESHOOTING TECHNIQUES

It is easy to forget that the Accessories folder contains so many useful utilities. It would be good to create shortcuts to the utilities you use the most, placing them on the desktop for quick access.

You may want to periodically review the contents of the Accessories folder. You may re-encounter a utility that you had forgotten about and now could use.

SUMMARY

In this exercise we discovered that

- A large number of useful applications are contained in Accessories.
- Accessories include system tools, Internet tools, multimedia, games, communication utilities, text and graphical viewers, and other helpful utilities.

SELF-TEST

This self-test is designed to help you check your understanding of the background information presented in this exercise.

True/False

Answer *true* or *false*.

1. Games have no value for the beginning user.
2. HTML is a program that executes when browsing the Web.
3. The Character Map tool provides directions on where to download fonts on the Internet.

4. A modem is required for Dial-Up Networking.
5. A direct cable connection can be used to connect two Windows machines.
6. HyperTerminal is the only utility that can dial a number with the modem.
7. Any user can create, modify, and delete accounts in Windows NT.
8. It is good to have a lot of fragmentation on a disk.
9. The Calculator only has one mode of operation.

Multiple Choice

Select the best answer.

10. A list of names might be stored using the
    a. Notepad.
    b. Cardfile.
    c. Calendar.
    d. Clipboard Viewer.
11. An image can be transferred between applications using the
    a. Notepad.
    b. Cardfile.
    c. Calendar.
    d. Clipboard.
12. The time/date of an important meeting can be stored using the
    a. Notepad.
    b. Cardfile.
    c. Calendar.
    d. Clipboard.
13. Windows NT systems events are viewed with the
    a. Notepad.
    b. Event Viewer.
    c. Calendar.
    d. Clipboard Viewer.
14. A direct cable connection is used to
    a. Connect a printer to a computer.
    b. Connect two computers together.
    c. Connect a modem to the printer port.
15. To view a Web page, you use a(n)
    a. Browser.
    b. HTML generator.
    c. Modem.

Matching

Match each utility on the right with each item on the left.

16. Used to create an image.
17. Connects with a mainframe.
18. Used to create a text document.
19. Performs numeric operations.

a. Calculator
b. Notepad
c. Paint
d. HyperTerminal

Completion

Fill in the blank or blanks with the best answers.

20. HTML stands for _____ _____ _____.
21. A disk that has each file stored in different locations, rather than one long block, is said to be _____.
22. The _____ utility checks a disk for errors (such as bad FAT entries).
23. Two editors capable of creating a text file are Notepad and _____.
24. Windows NT administrative accounts have all available _____.
25. Files transfer at a _____ rate when a disk is fragmented.
26. The _____ utility checks a disk for errors.

1. Use Netscape or Internet Explorer to browse the Web. Try www.windows95.com or www.winfiles.com, two very good sites for Windows users.
2. Check the amount of fragmentation for each drive on your lab system. If you have a floppy disk, check it also.
3. If approved by your instructor, defragment each drive. Make note of how long it takes to do a single drive and the size of the drive.
4. Determine the amount of available resources (system, user, GDI).
5. Use the System Monitor to watch at least four parameters (processor usage, etc.) for 10 minutes. Perform other work on the system as you watch the display. What do you notice?
6. Use the Calculator to determine how long, in minutes, a photon of light takes to travel from the Sun to Earth. The distance is 93 million miles and the speed of light is 186,000 miles/second.
7. Make a cardfile of your favorite television shows. Explain how you organized them.
8. Open the Clipboard Viewer. Is there anything on it? If so, where do you think the data came from?
9. Use Notepad to write a short letter requesting a sample catalog from a computer manufacturer. Print out the final version.
10. Open the letter from step 9 using WordPad. Use the text editing features to make some words bold, others italic, and others a different size. Center and right justify some of the text. Print out the final version.
11. Explore each of the Windows NT Administrative Tools on the Administrative Tools display menu.

1. What other utilities might be useful in the Windows environment?
2. Search the Web for the utilities you listed in your answer to question 1. With approval, download one of your choices and install it. How well does it meet your expectations?

Under the supervision of your instructor

1. Describe the basic contents of the Accessories folder.
2. Demonstrate the operation of several utilities, such as the Calculator, Notepad, and Paint.
3. Explain the importance of the System Tools and Administrative Tools folders.

17 An Introduction to Networking with Windows

INTRODUCTION

"That's the computer right there."

Those were the first words Joe Tekk heard when he entered a high school laboratory maintained under contract by RWA Software. "Pardon me?" Joe asked.

The laboratory technician was a senior, ready to graduate in a few months, with little patience for computers that did not work.

"It's that one right there. It won't connect to the network." He pointed at the computer until Joe got to it. Joe walked around to the back of the computer, pulled the T-connector off the back of the network card, and looked at it closely.

"Here's your problem," he said, to the surprise of the student. "The metal pin is missing from the center of the connector."

The student looked at the connector and then back at Joe. "How did you know to look for that?"

"I always pull the connector out first. I've seen this happen before. Now, it's just a habit."

PERFORMANCE OBJECTIVES

Upon completion of this exercise, you will be able to

1. Identify hard disk resources available on a network computer.
2. Identify printer resources available on a network computer.
3. Create a Dial-Up Networking connection.

KEY TERMS

Protocol
ISO/OSI network model
NetBIOS
Workgroup

Sharing files
Internet service provider (ISP)
Protocol analyzer
LanExplorer

BACKGROUND INFORMATION

Windows offers many different ways to connect your machine to one or more computers and plenty of applications to assist you with your networking needs. In this exercise we will briefly examine the basics of networking in Windows.

TABLE 17.1 ISO/OSI network model

ISO/OSI Layer	Operation (purpose)
7 = Application	Use network services via an established protocol (TCP/IP, NetBEUI).
6 = Presentation	Format data for proper display and interpretation.
5 = Session	Establish, maintain, and teardown session between both networked computers.
4 = Transport	Break application data into network-sized packets.
3 = Network	Handle network addressing.
2 = Data Link	Flow control, reliable transfer of data.
1 = Physical	All hardware required to make the connection (NIC, cabling, etc.) and transmit/receive 0s and 1s.

MICROSOFT NETWORKING

Although Windows supports many different types of common networking protocols, the backbone of its network operations is NetBEUI (NetBIOS Extended User Interface), a specialized Microsoft protocol used in Windows for Workgroups, Windows 95/98, and Windows NT. NetBEUI allows small networks of users to share resources (files and printers).

Whenever you use a network to transfer data, an entire set of *protocols* is used to set up and maintain reliable data transfer between the two network stations. These protocols are used for many different purposes. Some report errors, some contain control information, others carry data meant for applications. One widely accepted standard that defines the layering of the network (and thus a division of protocols) is the *ISO/OSI network model*, which is described in Table 17.1.

The ISO/OSI model breaks network communication into seven layers, each with its own responsibility for one part of the communication system. As indicated, the Physical layer is responsible only for sending and receiving digital data, nothing else. Any errors that show up in the data are handled by the next higher layer, Data Link. As we will see in this exercise, each layer has its own set of protocols.

THE NETBEUI PROTOCOL

The Network BIOS Extended User Interface (NetBEUI) protocol is the backbone of Windows for Workgroups, Windows 95/98, and Windows NT networking. File and printer sharing between these network operating systems is accomplished through the use of NetBEUI.

NetBEUI provides the means to gather information about the Network Neighborhood. Table 17.2 shows the advantages and disadvantages of the NetBEUI protocol. One of the main disadvantages with NetBEUI is that it is a nonroutable protocol. This means that a NetBEUI message cannot be routed across two different networks. It was designed to support small networks (up to 200 nodes) and becomes inefficient in larger installations.

TABLE 17.2 Advantages and disadvantages of NetBEUI

Advantages	Disadvantages
Easy to implement	Not routable
Good performance	Few support tools
Low memory requirements	Proprietary
Self-tuning efficiency	

FIGURE 17.1   Network Neighborhood window

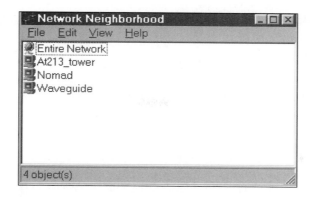

FIGURE 17.2   Items shared by Waveguide

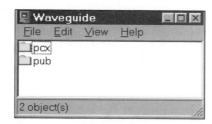

THE NETWORK NEIGHBORHOOD

The Network Neighborhood is a hierarchical collection of the machines capable of communicating with each other over a Windows network. Note that systems running Windows for Workgroups have the ability to connect to the network as well.

Figure 17.1 shows a typical Network Neighborhood. The three small PC icons named At213_tower, Nomad, and Waveguide all represent different machines connected to the network. Each machine is also a member of a **workgroup**, or *domain* of computers that share a common set of properties.

Double-clicking on Waveguide brings up the items being shared by Waveguide. As indicated in Figure 17.2, Waveguide is sharing two folders: pcx and pub.

The Network Neighborhood gives you a way to graphically navigate to shared resources (files, CD-ROM drives, printers).

NETWORK PRINTING

A network printer is a printer that a user has decided to share. For the user's machine it is a local printer. But other users on the network can map to the network printer and use it as if it were their own printer. Figure 17.3 shows a shared printer offered by a computer named Nomad. Nomad is offering an hp 890c.

It is necessary to install the printer on your machine before you can begin using it over the network.

FIGURE 17.3   A printer shared by Nomad

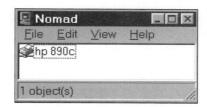

SHARING FILES OVER A NETWORK

A computer can share its disks with the network and allow remote users to map them for use as an available drive on a remote computer. The first time a disk is shared and a connection is established, it is necessary to provide a password to gain access to the data. The password is typically provided by the network administrator. This password is usually stored in the password file for subsequent access to the disk if it is reconnected after a reboot. Figure 17.4 shows the contents of My Computer. The small hand holding drive D: (Fireballxl5) indicates the drive is shared.

The user sharing the drive controls the access others will have to it over the network. Figure 17.5 shows the sharing properties for drive D: (right-click on the drive icon and select Properties). Clearly, the user has a good deal of control over how sharing takes place.

DIAL-UP NETWORKING

Dial-Up Networking is designed to provide reliable data connections using a modem and a telephone line. Figure 17.6 shows two icons in the Dial-Up Networking folder (found in

FIGURE 17.4　Indicating a shared drive

FIGURE 17.5　Sharing Properties window for drive D:

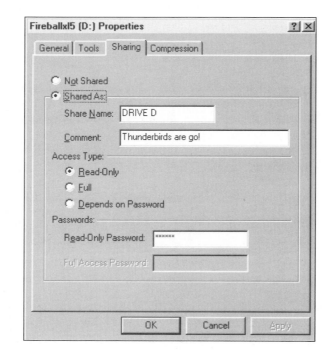

FIGURE 17.6   Dial-Up Networking icons

FIGURE 17.7   Make a New Connection window

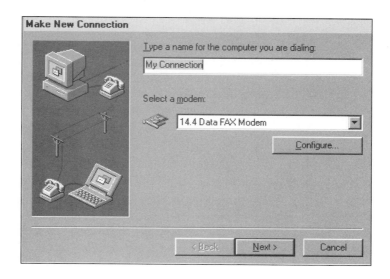

FIGURE 17.8   Information required to access host

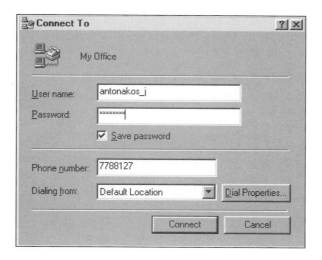

Accessories on the Start menu). Double-clicking the Make New Connection icon will start the process to make a new connection, as shown in Figure 17.7. The name of the connection and the modem for the connection are specified.

It is also necessary to provide an area code and telephone number during the configuration process. This number must be for a machine capable of supporting a PPP (point-to-point protocol) connection.

Once the connection has been created, it is activated by double-clicking it. To connect to a remote host, it is necessary to supply a user name and a password. This can be done automatically by the Dial-Up Networking software. Figure 17.8 shows the connection window for the My Office icon.

FIGURE 17.9   Active Dial-Up
Networking connection

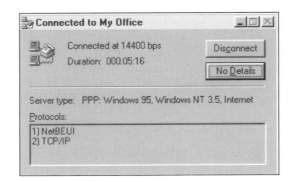

After the information has been entered, the Connect button is used to start up a connection. When the connection has been established, Windows displays a status window showing the current duration of the connection and the active protocols. Figure 17.9 shows the status for the My Office connection. Left-clicking the Disconnect button shuts the connection down and hangs up the modem.

CONNECTING TO THE INTERNET

Besides a modem or a network interface card and the associated software, one more piece is needed to complete the networking picture: the *Internet service provider (ISP)*. An ISP is any facility that contains its own direct connection to the Internet. For example, many schools and businesses now have their own dedicated high-speed connection (typically a T1 line, which provides data transfers of more than 1.5 million bits/second).

Many users sign up with a company (such as AOL or MSN) and then dial in to these companies' computers, which themselves provide the Internet connection. The company is the ISP in this case.

Even the local cable company is an ISP now, offering high-speed cable modems that use unassigned television channels for Internet data. The cable modem is many times faster than the fastest telephone modems on the market.

Once you have an ISP, the rest is up to you. You may design your own Web page (many ISPs host Web pages for their customers), use e-mail, browse the Web, Telnet to your school's mainframe and work on an assignment, or download a cool game from an ftp site.

TROUBLESHOOTING TECHNIQUES

Troubleshooting a network connection requires familiarity with several levels of operation. At the hardware level, the physical connection (parallel cable, modem, network interface card) must be working properly. A noisy phone line, the wrong interrupt selected during setup for the network interface card, incompatible parallel ports, and many other types of hardware glitches can prevent a good network connection.

At the software level there are two areas of concern: the network operating system software and the application software. For example, if Internet Explorer will not open any Web pages, is the cause of the problem Internet Explorer or the underlying TCP/IP protocol software?

Even with all of the built-in functions Windows automatically performs, there is still a need for human intervention to get things up and running in the world of networking. Let's take a look at a software tool designed to work with many aspects of the computer network.

LANEXPLORER

Computer networks play a large role in society today. Often it is necessary for a network engineer or a technology student to troubleshoot problems that arise on the network using a software tool called a ***protocol analyzer***. Using a protocol analyzer, the network interface card in a computer is put into promiscuous mode, allowing it to see all the traffic that is transmitted on the local network. This potentially causes a network security risk because it is possible to capture data that is considered to be confidential, such as passwords, social security numbers, and salaries. Therefore, extreme caution and good judgment should be exercised when using a protocol analyzer.

Traffic Monitoring

A typical use for a protocol analyzer is to collect baseline network traffic data. The baseline historical data is then used to compare against network data collected at a different time. The baseline data can be used as an early warning detection system because it is possible to identify several possible harmful situations. These situations include identification of network capacity issues, DHCP errors, duplicate IP assignments, and network utilization trends. In some situations, it is possible to identify a piece of faulty network equipment that has been causing excessive collisions. The process of investigating a network begins with monitoring all the network traffic to develop the baseline data. The baseline may be taken over the course of several hours, days, or weeks using a product such as Sunrise Telecom's LanExplorer (www.intellimax.com).

LanExplorer is one of the most popular protocol analyzer packages available. Note that a demonstration copy of LanExplorer is included on the companion CD. The first time LanExplorer is started, it is necessary to select the default network adapter. Since there is usually only *one* network card in a PC, there is only *one* choice available. When more than one network card is present, one of them must be selected. The first screen displayed by LanExplorer after selecting a default adapter is shown in Figure 17.10. The LanExplorer screen is divided into three areas. First is the Task Panel, which contains two items, the Traffic option and the Statistics option. The Traffic option allows for quick access to built-in displays. The Statistics option provides access to Distribution and Rate information, which is automatically set. Selections that can be made in the Traffic Task Panel are displays of the Matrix Table, Host Table, Matrix Chart, Host Chart, and Alarm Log. As the items are selected, the corresponding data is displayed in the panel to the right of the Task Panel. At the bottom of the LanExplorer display is the Console Panel, which is used to display a breakdown of the monitored network traffic. Note that as soon as LanExplorer is

FIGURE 17.10 Initial LanExplorer window

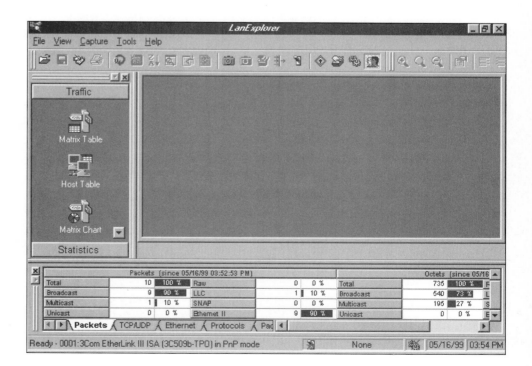

started, it begins to monitor all the traffic that is present on the network. Let's begin by discussing the various traffic display options.

When the Matrix Table display option is selected by left-clicking on the icon, the panel on the right displays a line of data for each packet that is examined. The information available for examination that is displayed on this screen include the following:

- Address (host name or host number)
- Octet ratio (bytes of data)
- Total octets
- Total packets
- Duration of the network activity
- Octets in, packets in
- Octets out, packets out
- Broadcast and multicast message counts
- IP packet type
- Time stamp information

Figure 17.11 shows a typical Matrix display containing Internet Protocol information. By using the scroll bars, all of the various data elements may be examined.

Moving down the list of Traffic Task Panel options, the Host Table is displayed, as shown in Figure 17.12. As the traffic is monitored, statistics are gathered for each of the hosts that is transmitting data on the network. The graphical bar chart displayed in Figure 17.12 indicates the hosts with the highest ratio of traffic.

The Traffic Matrix Chart shows the same information that was listed in the table format, but now is displayed graphically using a pie chart as illustrated in Figure 17.13. The graphical display shows a breakdown of the traffic data, by both a count and a percentage. The names of the hosts with the highest traffic are displayed in ascending order as space permits. As the traffic patterns change, the pie chart are automatically updated.

The remaining item in the Traffic Task panel is the Alarm Log. The Alarm Log is used to keep track of all security-related issues such as plain-text password transmissions, DHCP

FIGURE 17.11   LanExplorer Traffic Matrix display

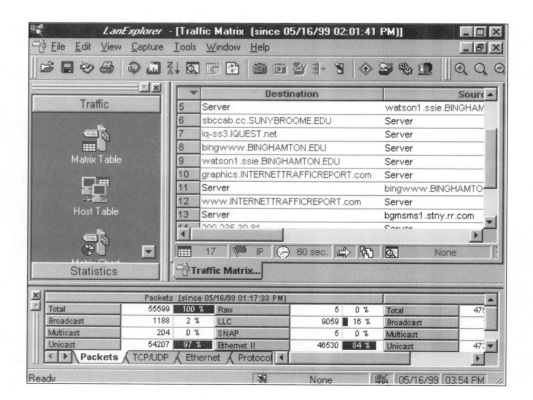

FIGURE 17.12   LanExplorer Host Table display

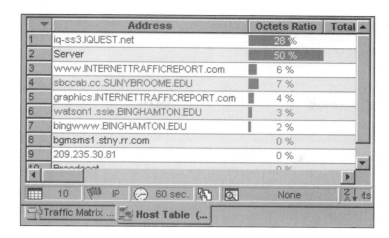

	Address	Octets Ratio	Total ▲
1	iq-ss3.IQUEST.net	28 %	
2	Server	50 %	
3	www.INTERNETTRAFFICREPORT.com	6 %	
4	sbccab.cc.SUNYBROOME.EDU	7 %	
5	graphics.INTERNETTRAFFICREPORT.com	4 %	
6	watson1.ssie.BINGHAMTON.EDU	3 %	
7	bingwww.BINGHAMTON.EDU	2 %	
8	bgmsms1.stny.rr.com	0 %	
9	209.235.30.81	0 %	
10	Broadcast	0 %	

☐ 10   ✋ IP   🕐 60 sec. 📇 🔍   None   ↕ ts

⊟ Traffic Matrix ... 📇 Host Table (...

FIGURE 17.13   LanExplorer Traffic Matrix Chart

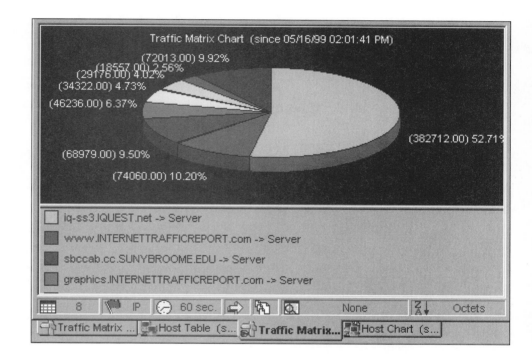

address issues, duplicate IP network assignments, and many other important items. You are encouraged to check the items on the Alarm Log frequently because many potential security issues may be identified.

The LanExplorer Console window located at the bottom of the display contains information about all the traffic being monitored. Figure 17.14 shows the various information displayed in the Console window. The first tab contains information about the Packet Statistics, which are displayed in Figure 17.14(a). The total number of packets as well as the packet type is identified. Figure 17.14(b) shows a breakdown of the TCP/UDP Internet Protocol packets. Figure 17.14(c) shows all of the Ethernet statistics. Figure 17.14(d) shows what type of protocols have been monitored on the network, and Figure 17.14(e) completes the list of available information by providing a breakdown of the various sizes of the packets that have been transmitted.

Each of the categories presented in the Console window display can be very important when developing a baseline of activity for a network or when performing troubleshooting.

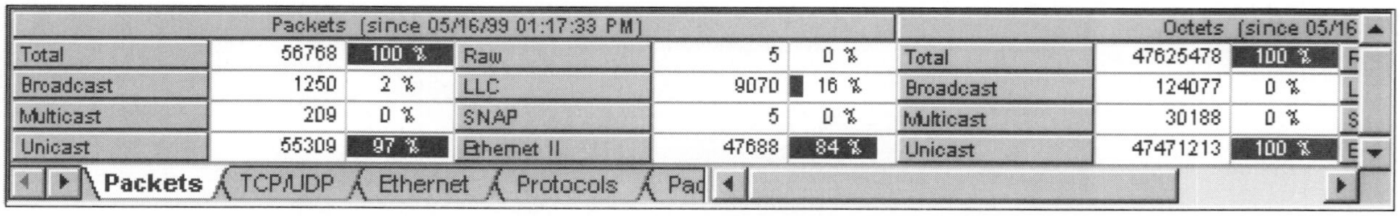

(a) Packet statistics

Packets (since 05/16/99 01:17:33 PM)						Octets (since 05/16
Total	56768	100 %	Raw	5	0 %	Total	47625478	100 %
Broadcast	1250	2 %	LLC	9070	16 %	Broadcast	124077	0 %
Multicast	209	0 %	SNAP	5	0 %	Multicast	30188	0 %
Unicast	55309	97 %	Ethernet II	47688	84 %	Unicast	47471213	100 %

◄ ► \ **Packets** \ TCP/UDP \ Ethernet \ Protocols \ Pac ◄

(a) Packet statistics

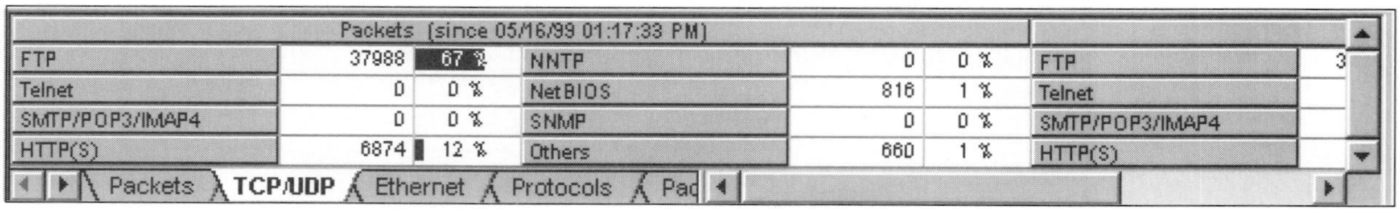

(b) TCP/UDP statistics

Packets (since 05/16/99 01:17:33 PM)						
FTP	37988	67 %	NNTP	0	0 %	FTP	3
Telnet	0	0 %	NetBIOS	816	1 %	Telnet	
SMTP/POP3/IMAP4	0	0 %	SNMP	0	0 %	SMTP/POP3/IMAP4	
HTTP(S)	6874	12 %	Others	660	1 %	HTTP(S)	

◄ ► \ Packets \ **TCP/UDP** \ Ethernet \ Protocols \ Pac ◄

(b) TCP/UDP statistics

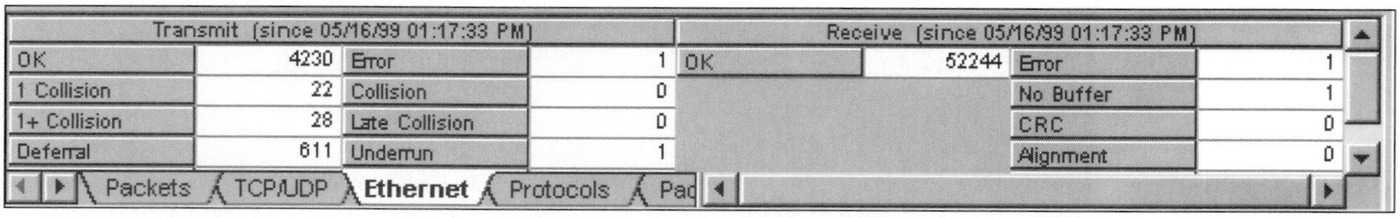

(c) Ethernet statistics

Transmit (since 05/16/99 01:17:33 PM)				Receive (since 05/16/99 01:17:33 PM)			
OK	4230	Error	1	OK	52244	Error	1
1 Collision	22	Collision	0			No Buffer	1
1+ Collision	28	Late Collision	0			CRC	0
Deferral	611	Underrun	1			Alignment	0

◄ ► \ Packets \ TCP/UDP \ **Ethernet** \ Protocols \ Pac ◄

(c) Ethernet statistics

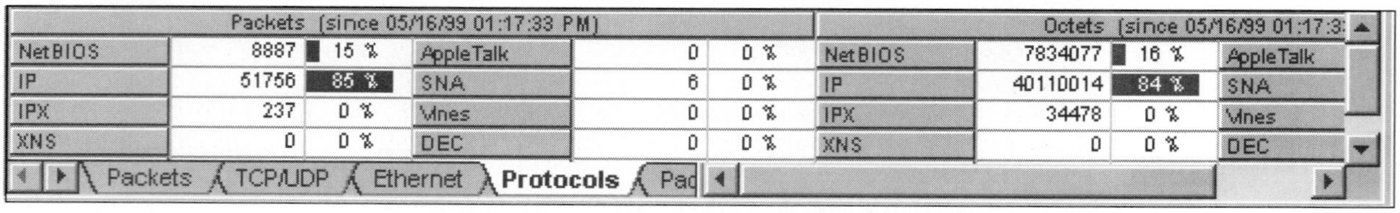

(d) Protocol statistics

Packets (since 05/16/99 01:17:33 PM)						Octets (since 05/16/99 01:17:3
NetBIOS	8887	15 %	AppleTalk	0	0 %	NetBIOS	7834077	16 %	AppleTalk
IP	51756	85 %	SNA	6	0 %	IP	40110014	84 %	SNA
IPX	237	0 %	Vines	0	0 %	IPX	34478	0 %	Vines
XNS	0	0 %	DEC	0	0 %	XNS	0	0 %	DEC

◄ ► \ Packets \ TCP/UDP \ Ethernet \ **Protocols** \ Pac ◄

(d) Protocol statistics

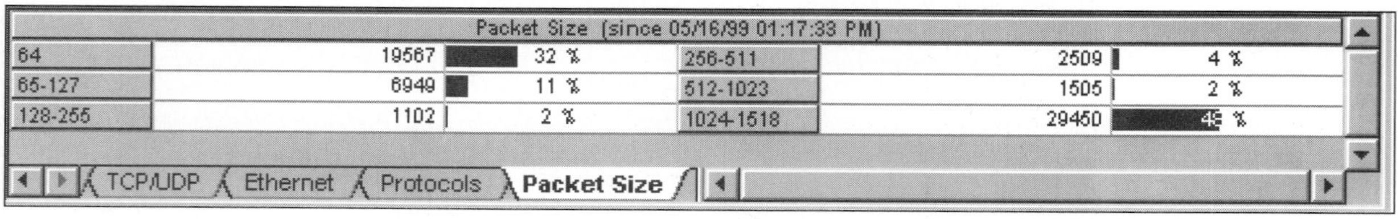

(e) Packet size statistics

Packet Size (since 05/16/99 01:17:33 PM)					
64	19567	32 %	256-511	2509	4 %
65-127	6949	11 %	512-1023	1505	2 %
128-255	1102	2 %	1024-1518	29450	45 %

◄ ► \ TCP/UDP \ Ethernet \ Protocols \ **Packet Size** \ ◄

(e) Packet size statistics

FIGURE 17.14   LanExplorer Console window display information

Packet Capture

Aside from the process of passively monitoring the network traffic, LanExplorer can also capture network traffic. Simply by selecting the Start option from the Capture pull-down menu or using the toolbar icon, LanExplorer will keep a copy of each packet of information transmitted. When the capture process is "in progress," the screen shown in Figure 17.15 is displayed. The number of packets captured, number of octets captured, packets seen, octets seen, elapsed time, filter information, and buffer usage is all displayed in real time.

When the buffer is full, or the Stop Capture command is executed, the user is presented with a screen similar to Figure 17.16. Although it looks very much like the Traffic Matrix display discussed earlier, packets that are captured may be decoded. To decode a packet,

FIGURE 17.15   LanExplorer packet capture status display

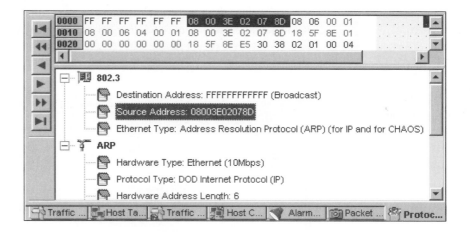

Packet capture in progress... To view captured data, click here or select the 'Stop Capture' command	
Packets Captured	4
Octets Captured	360
Packets Seen	4
Octets Seen	360
Elapse Time	00:00:02.650
Filter Applied	Default
Buffer Size	256 KB
Buffer Usage	0 %
Buffer Status	Stop capture after buffer is full

Traffic ...   Packet ...   Host Ta...   Traffic ...   Host C...   Alarm...   **Packet ...**

FIGURE 17.16   LanExplorer captured packet display

Packet Capture 1

	Destination	Source	Protocol
1	209.235.30.81	Server	NETBIOS Name Service
2	Server	209.235.30.81	ICMP
3	209.235.30.81	Server	NETBIOS Name Service
4	Server	209.235.30.81	ICMP
5	Server	bgmsms1.stny.rr.com	Domain Name Server
6	bgmsms1.stny.rr.com	Server	Domain Name Server
7	209.235.30.81	Server	NETBIOS Name Service
8	Server	209.235.30.81	ICMP
9	209.235.30.81	Server	NETBIOS Name Service

Default   2570   None   2570

Traffic Ma...   Host Tabl...   Traffic Ma...   Host Char...   Alarm Log   **Packet C...**

FIGURE 17.17   LanExplorer Protocol Decode display window

```
0000  FF FF FF FF FF FF 08 00 3E 02 07 8D 08 06 00 01    . . . . . . . .
0010  08 00 06 04 00 01 08 00 3E 02 07 8D 18 5F 8E 01    . . . . . . . .
0020  00 00 00 00 00 00 18 5F 8E E5 30 38 02 01 00 04    . . . . . . . .
```

- 802.3
  - Destination Address: FFFFFFFFFFFF (Broadcast)
  - Source Address: 08003E02078D
  - Ethernet Type: Address Resolution Protocol (ARP) (for IP and for CHAOS)
- ARP
  - Hardware Type: Ethernet (10Mbps)
  - Protocol Type: DOD Internet Protocol (IP)
  - Hardware Address Length: 6

Traffic ...   Host Ta...   Traffic ...   Host C...   Alarm...   Packet ...   **Protoc...**

the program user simply double-clicks the specific line in the display. This causes the Protocol Decode window to automatically be displayed as shown in Figure 17.17. The Protocol Decode window contains two areas. First, the raw data window is displayed at the top of the screen. Second, the protocol-specific breakdown is displayed at the bottom. The Source Address that has been selected in the Protocol area has a corresponding selection in the raw data display. Notice that the same source address is highlighted in both areas. Choosing the Save option creates a copy of this decoded information in a text file. Figure 17.18 shows the data that is written to the text file.

FIGURE 17.18   Decoded LanExplorer packet saved in text format

```
Capture 1:Packet 88

Destination Source       Protocol      Summary  Size    Time Tick
------------------------------------------------------------------
24.95.142.229  24.95.142.1   ARP    Request   60   05/16/99 15:27:33.767

Addr.  Hex. Data                                      ASCII
0000:  FF FF FF FF FF FF 08 00 3E 02 07 8D 08 06 00 01   ........>.......
0010:  08 00 06 04 00 01 08 00 3E 02 07 8D 18 5F 8E 01   ........>...._..
0020:  00 00 00 00 00 00 18 5F 8E E5 30 38 02 01 00 04   ......._..08....
0030:  18 5F 8E F8 03 03 A9 90 00 00 00 00               ._..........

802.3 [0000:000D]
  0000:0005  Destination Address: FFFFFFFFFFFF (Broadcast)
  0006:000B  Source Address: 08003E02078D
  000C:000D  Ethernet Type: Address Resolution Protocol (ARP) (for IP and
for CHAOS)
ARP [000E:0029]
  000E:000F  Hardware Type: Ethernet (10Mbps)
  0010:0011  Protocol Type: DOD Internet Protocol (IP)
  0012:0012  Hardware Address Length: 6
  0013:0013  Protocol Address Length: 4
  0014:0015  Opcode: Request
  0016:001B  Source HW Address: 08003E02078D
  001C:001F  Source IP Address: 24.95.142.1
  0020:0025  Destination HW Address: 000000000000
  0026:0029  Destination IP Address: 24.95.142.229
```

Packet Capture Filters

Many times when monitoring or troubleshooting a network, it is necessary to monitor only a small portion of the data being transmitted. For example, LanExplorer can be used to look for network traffic from a specific host computer or look for a specific type of protocol. This is accomplished by setting up a packet capture *filter*. Packet filter selections are made from either the network layer or address. Figure 17.19(a) shows the check box options available when selecting Layer 3+ IP/ARP and TCP/UDP categories. Each item that contains a check mark is identified for capture. Figure 17.19(b) shows the Address packet capture filter. Notice that the address may be a MAC address or an IP address. Addresses in the Known Addresses box may be dragged down to the Address Filter List as required. The Filter Mode is used to specify whether the addresses listed in the address filter list are inclusive or exclusive, allowing for unlimited filter choices.

FIGURE 17.19   Packet Capture Filter configuration options

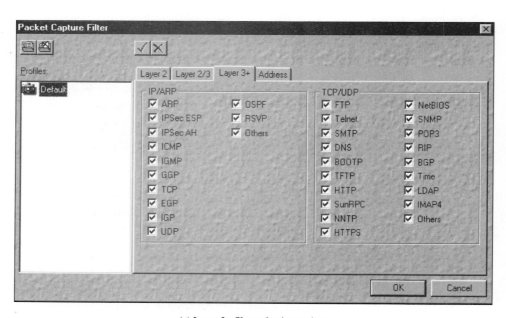

(a) Layer 3+ filter selection options

FIGURE 17.19   *(continued)*

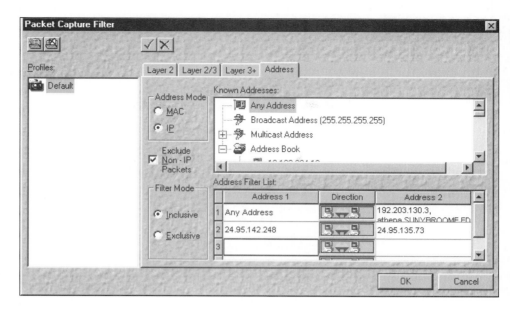

(b) Address filter selection options

By using a filter, much of the networking traffic is eliminated from the buffer, thereby saving only the traffic that is desired. Note that it is also possible to set a trigger event, which will cause LanExplorer to begin the packet capture. This helps guarantee that the network traffic that is captured follows the triggering event. This makes investigating network problems much easier by isolating the information. Notice from Figure 17.20 that it is possible to start or stop capturing network traffic using trigger events based on a date and/or time, a specific network event, or by the existence of a specific file.

Rather than pay thousands of dollars for a hardware-based protocol analyzer, spend a fraction of the amount on a software-based protocol analyzer such as LanExplorer. Your networking knowledge and experience will increase rapidly.

FIGURE 17.20   Packet Capture Trigger events

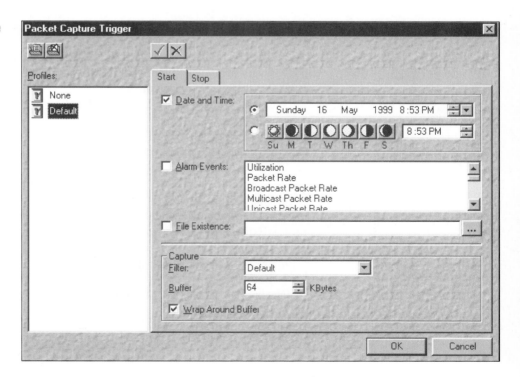

In this exercise we discovered that

- NetBEUI is the basic protocol in a Windows network.
- The ISO/OSI network model contains seven layers.
- In a Windows network, you can share files and printers.
- Dial-Up Networking provides a PPP connection over the telephone (using a modem).
- LanExplorer is a packet capture/display tool.

SELF-TEST

This self-test is designed to help you check your understanding of the background information presented in this exercise.

True/False

Answer *true* or *false*.

1. NetBEUI is a protocol only used by Windows NT.
2. Dial-up connections work with ordinary dial-up phone numbers.
3. The cable company is an example of an ISP.
4. Network printers can be used as soon as you map them.
5. Anyone who wants to can delete all the files on a shared drive.
6. The ISO/OSI network model contains eight layers.
7. The bottom ISO/OSI network model layer is the Physical layer.

Multiple Choice

Select the best answer.

8. NetBEUI is a protocol used
    a. Only for network printers.
    b. Only for file sharing.
    c. For sharing files and printers.
9. A workgroup is a set of users that
    a. Share common properties.
    b. Use the same printer.
    c. Work as a team on projects.
10. The Network Neighborhood shows
    a. The networked computers within 20 meters of your machine.
    b. Every computer on the entire network.
    c. Machines sharing resources.
11. What is required for Dial-Up Networking?
    a. A modem.
    b. A network interface card.
    c. A direct cable connection.
12. The Network Neighborhood shows
    a. Every computer on the network at the same time.
    b. Hierarchical groups of networked computers.
    c. All the computers on the Internet.
13. LanExplorer
    a. Captures and displays packets.
    b. Decodes NetBIOS encrypted passwords.
    c. Both a and b.
14. The "NetB" in NetBEUI stands for
    a. Network Backbone.
    b. Network Broadcast.
    c. Network BIOS.

Completion

Fill in the blank or blanks with the best answers.

15. NetBEUI stands for NetBIOS _____ _____ _____.
16. Another term for workgroup is _____.
17. Dial-Up Networking is accessed via the _____ folder in the Start menu.
18. The Dial-Up Networking connection uses the _____ protocol.
19. ISP stands for _____ _____ _____.
20. LanExplorer puts the network interface card into _____ mode to capture packets.
21. The highest ISO/OSI network model layer is the _____ layer.

FAMILIARIZATION ACTIVITIES

Do one or both of the following activities:

Activity 1

1. Set up a new modem connection in Dial-Up Networking. Your instructor will supply the Dial-Up Networking number and other parameters.
2. Begin a Dial-Up Networking session. How long does it take to connect?
3. Use the Network Neighborhood to view the machines reachable over your connection.
4. Try to find three machines sharing files or printers.

Activity 2

1. Establish a network connection using the network interface card. This may be automatically done at boot time.
2. If possible, determine the number of machines on your Windows network by counting the icons in the Network Neighborhood windows.
3. Use LanExplorer to monitor and examine network traffic at your school or business.

QUESTIONS/ACTIVITIES

1. Go to a local business that advertises on the Web. Ask them to describe their network connection. Do they have dial-up service? What is the cost? Who maintains their systems?
2. Search the Web for information on how satellites are used in Internet connections.

REVIEW QUIZ

Under the supervision of your instructor

1. Identify hard disk resources available on a network computer.
2. Identify printer resources available on a network computer.
3. Create a Dial-Up Networking connection.

18

Installing New Software

INTRODUCTION

Joe Tekk was very excited. He was downloading a demo version of a game over the Internet. He had considered buying the game, but he did not really know if it was worth the money.

Joe finished downloading a 12-MB self-exploding .EXE file that contained all the installation files in a compressed format. He planned to install the software on his D: drive, since the C: drive was quite low in disk space.

However, as he ran the installation procedure, it failed, stating the disk was out of room. Joe noticed the files were being extracted to the TEMP directory on the C: drive during the installation process.

Joe ended up having to reclaim some space on the C: drive so he could install the program. As it turned out, the 12MB of compressed files grew to more than 40MB during the installation process. When the C: drive finally contained enough free space, the installation was successful.

Joe had never run into this problem before, but it was one he would not soon forget, since he had wasted quite a bit of time.

PERFORMANCE OBJECTIVES

Upon completion of this exercise, you will be able to

1. Determine which Windows applications are currently installed.
2. Locate the updates for the Windows operating system on the Microsoft Web site.
3. Discuss the general steps required to perform an application installation and deinstallation.

KEY TERMS

.CAB file   Uninstall
Setup Wizard

BACKGROUND INFORMATION

The process of installing software involves moving files from a floppy disk or CD-ROM to a suitable hard disk location (local or network). Every single file located on a computer disk has been installed, one way or another. The files actually consist of executable images, data files, initialization files, dynamic link libraries, and other custom files necessary for the computer or application to run. The files are placed in a directory structure determined by an application's installation program.

What types of software do we install? The operating system itself, plus every single application. Note that many applications register themselves with the operating system during the installation process. Specific application settings are stored in the Registry.

EXISTING WINDOWS 3.X SOFTWARE

Most data files and application programs are upward compatible. This means that Windows 3.x applications can be installed on Windows 95, 98, and NT. Unfortunately, Windows 95, 98, and NT applications cannot be installed on a Windows 3.x computer. When a Windows 3.x computer is upgraded to Windows 95 or 98, all installed applications are also upgraded, by placing information from the old .INI files to the new system Registry. There is no upgrade path to Windows NT from any of the other Windows operating systems.

INSTALLING NEW SOFTWARE FROM THE WINDOWS CD-ROM

The Windows operating system can be configured in many different ways. There are likely to be many files that were not installed during the initial Windows installation. To see the current system configuration, double-click the Add/Remove Programs icon in the system Control Panel. Then select the Windows Setup tab as shown in Figure 18.1. Just by looking at the window, it is apparent what components are installed.

Check boxes along the left margin of the component window identify three situations. First, if a box is checked and the inside of the box is white, all components of the category have been installed. If a box is checked and the inside of the box is gray, only some of the components from that particular category are installed. A box with no check indicates that no components are installed for that category. The Details button is used to show the specific status of each component. If the Windows operating system components are changed, the computer will request the Windows CD-ROM or floppy installation media to copy the additional files.

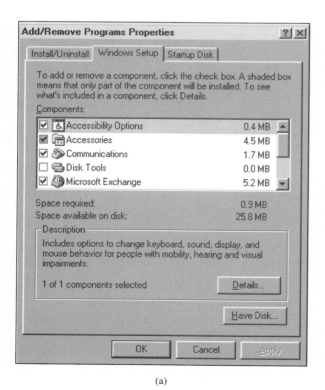
(a)

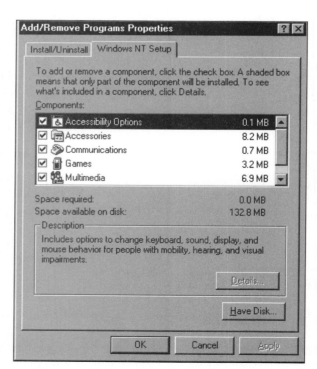
(b)

FIGURE 18.1   (a) Windows 95/98 Setup menu and (b) Windows NT Setup menu

The Windows files are stored in *.CAB files* on the Windows CD-ROM. These files contain the Windows operating system components in compressed form for distribution. Each file is extracted from the .CAB file using a special procedure built into the system.

GETTING THE LATEST UPDATES FROM MICROSOFT

As improvements are made in the Windows operating system, they are posted on the World Wide Web. Figure 18.2 shows the downloads page for support drivers, patches, and service packs from Microsoft. As you can see, each category contains specific applications such as Word, Exchange, and the different Microsoft operating systems.

INSTALLING THIRD-PARTY SOFTWARE

Application software is usually installed with a custom software installation wizard. Let's examine an installation process under Windows 95. Windows 98 and NT operate in a similar manner. The first step in performing a software installation is identifying which installation program to run. Figure 18.3 shows the Run window with the path specified to a setup file. When the OK button is selected, the *Setup Wizard* begins the installation process. Figure 18.4 shows the Norton Utilities For Windows 95 Setup Wizard installation screen. The installation program asks for the user name and company. This information is usually displayed by the application each time it is run. Each window has option

FIGURE 18.2 Operating system update from the Web

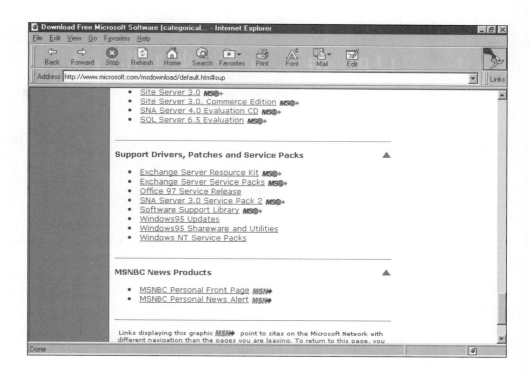

FIGURE 18.3 Location and name of the installation file

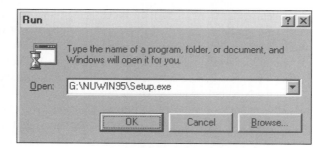

**FIGURE 18.4   Installation
Setup Wizard**

**FIGURE 18.5   License
agreement**

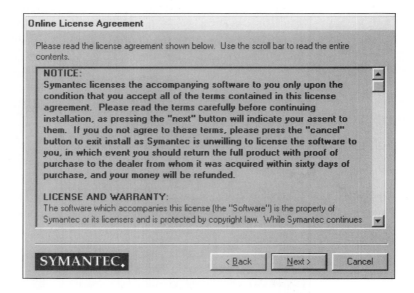

buttons that allow the program user to move back and forward through the screens presented during the installation.

The software producers do not actually sell their code, or executables—they license them. The license allows the user to run the product within the scope of the license agreement. It is impossible to install the software without agreeing with the license terms, as shown in Figure 18.5.

During an installation, the user is given choices about how the Setup Wizard is to install or reconfigure the application software. A complete installation usually installs all components of an application, and a custom installation may allow for one or more components of a product to be installed individually. Figure 18.6 shows a typical Setup option window indicating that a complete installation is to be performed. The installation proceeds by clicking the Next button.

Some applications will search for a previous installation of the application to identify which particular components are installed. When the user is prompted to confirm the program location, it will display the current installation locations. Figure 18.7 shows how a custom location can be selected as the installation directory. The directory will be created when the Next button is clicked.

The installation process continues with the Setup Wizard asking about how the new application software is to be configured, such as adding special features to the Recycle Bin, as shown in Figure 18.8 and whether to run the System Doctor when the system boots, as

FIGURE 18.6   Installation options

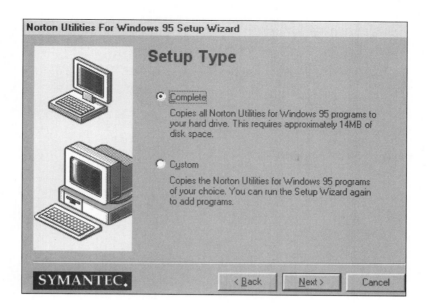

FIGURE 18.7   Selecting a custom storage location

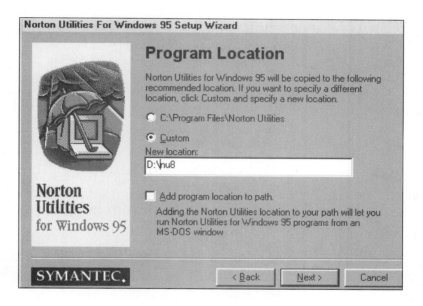

FIGURE 18.8   More installation options

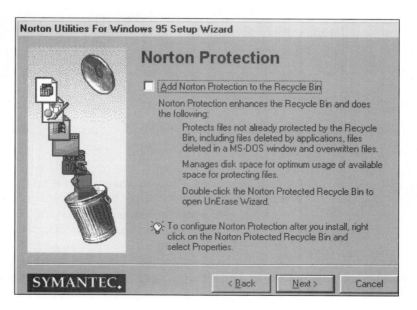

FIGURE 18.9   More installation options

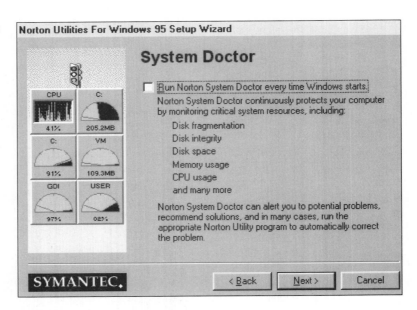

FIGURE 18.10   Files being copied to the hard disk

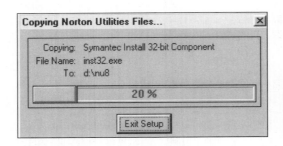

FIGURE 18.11   Installation process requires a system reboot

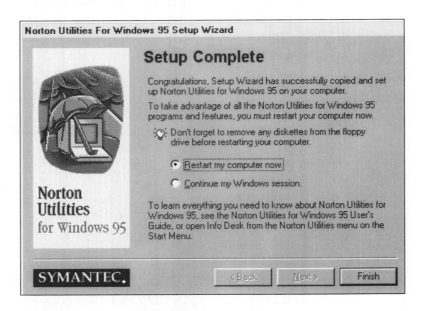

shown in Figure 18.9. These options are usually set to a default answer. When in doubt, use the default responses.

Eventually, the Setup Wizard begins the process of copying files from the installation media to the selected installation location. The progress meter shown in Figure 18.10 shows a percentage of the how close the installation is from being complete and the current file being processed.

When the installation process completes, it may require the system to be rebooted. This allows for new drivers to be loaded during system initialization. Figure 18.11 shows a congratulatory message for successfully completing the installation.

REMOVING APPLICATION SOFTWARE

Many situations may require a software application to be removed from a computer. Because an installation program may install application programs in many different directories on a disk, an application cannot be deleted by simply deleting the files in the installation directory. This action deletes many files and will remove the ability to run the program but leaves behind a trail of other files, which remain on the hard disk for no purpose.

To solve this dilemma, newer Windows applications provide an *uninstall* feature. Using the Add/Remove Programs icon in the Control Panel, the Install/Uninstall tab shows all installed products, as shown in Figure 18.12. These are the applications that can be uninstalled. To uninstall any product, simply select it from the list and press the Add/Remove button. Confirmation prompts will double-check to make sure the choice to remove a product is correct. With positive acknowledgment, the item is removed from the computer. Figure 18.13

FIGURE 18.12   Application selection window for Install/Uninstall feature

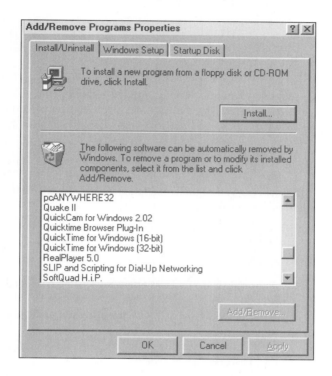

FIGURE 18.13   Progress display during removal of software

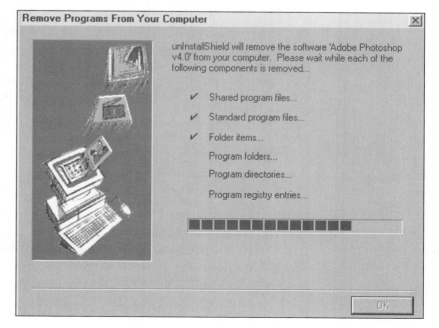

shows the current step being performed by the uninstall process and a progress display indicating the status of the procedure. Exercise extreme caution when uninstalling software. Unless the files have been backed up, it may be impossible to retrieve deleted data files.

TROUBLESHOOTING TECHNIQUES

When a software installation fails, it can usually be attributed to a lack of available resources on the computer, such as disk space or memory. Refer to the product literature to determine the application requirements and be sure your system meets the requirements before you begin an installation.

Be cautious when trying to make room for new applications. Do not delete directories manually or run an uninstall program without making quite sure that any files that may be deleted can be replaced if necessary. Remember, if you delete the wrong files, your system can be rendered unusable.

SUMMARY

In this exercise we discovered that

- Windows components are stored in compressed .CAB files.
- A Setup Wizard typically guides a software installation.
- You may need to reboot your computer after an installation.

SELF-TEST

This self-test is designed to help you check your understanding of the background information presented in this exercise.

True/False

Answer *true* or *false*.

1. All the optional Windows operating system components are installed when Windows is installed.
2. Due to software licensing issues, program updates cannot be made available over the Web.
3. The Windows operating system is stored in compressed form on the Windows installation CD-ROM.
4. Every installation wizard must do the same thing, regardless of the application being installed.
5. A checked box with a white background on the Windows Setup menu indicates only a portion of a component is installed.
6. Windows 3.x may be upgraded to Windows NT.
7. An application may register itself in the Registry during installation.
8. You can ignore the licensing agreement in a software installation.

Multiple Choice

Select the best answer.

9. Windows updates can be retrieved from the
    a. World Wide Web.
    b. Electronic mail attachments.
    c. Newest version of the Windows CD-ROM.
10. When installing software, the Setup Wizard
    a. Must reboot the system before the installation process can proceed.
    b. Prepares the files on the CD-ROM for transfer.
    c. Copies files from the CD-ROM to a suitable installation location.

11. A .CAB file contains
    a. Compressed Windows installation files.
    b. Security information and passwords.
    c. Application program data files.
12. Windows 95 applications are
    a. Not compatible with new Windows operating systems.
    b. Not compatible with old Windows 3.x operating systems.
    c. Not a factor as far as the operating system is concerned.
13. When a program is uninstalled, the application files are
    a. Deleted by the uninstall process.
    b. Moved to the Recycle Bin by the uninstall process.
    c. Moved into the TEMP directory.
14. To complete an installation, you may need to
    a. Uninstall the application.
    b. Reinstall the operating system.
    c. Reboot the computer.
15. To remove an application,
    a. Use Install/Uninstall from Add/Remove Programs.
    b. Simply delete its associated installation directory.
    c. Use the REMOVE command in DOS.

Completion

Fill in the blank or blanks with the best answers.

16. The _____ _____ on the Windows Setup menu show which components are installed.
17. Files are copied to a hard drive during a(n) _____ process.
18. _____ compatibility means Windows 3.x software can be installed on a Windows 95, 98, or NT system.
19. The system Registry contains information about each registered _____ _____.
20. Access to the Add/Remove Programs icon is found on the _____ _____.
21. A Setup _____ may guide an installation process.
22. .CAB files are stored on the Windows _____.

FAMILIARIZATION ACTIVITY

1. Determine the status of each component in the Windows Setup window. Which ones are completely installed, partially installed, or not installed at all?
2. Determine what application programs have been installed.
3. Visit the Microsoft Web site to examine support information and data.

QUESTIONS/ACTIVITIES

1. What are .CAB files?
2. What actions does a Setup Wizard program perform?
3. What actions are performed by an uninstall program?

REVIEW QUIZ

Under the supervision of your instructor

1. Determine which Windows applications are currently installed.
2. Locate the updates for the Windows operating system on the Microsoft Web site.
3. Discuss the general steps required to perform an application installation and deinstallation.

19

Installing New Hardware

INTRODUCTION

Joe Tekk had just returned from a troubleshooting assignment. Don, the senior technician, asked him how it went.

"Well, I've never encountered this problem before. The customer complained that her keyboard was not working properly. I thought it might be a bad connector, a stuck key, or dirt inside the keyboard, but everything looked fine."

Don's eyebrows shot up. "So, what was it?"

"I have no idea, Don. I was poking around inside Device Manager and I decided to delete the keyboard from the system. I rebooted, reinstalled the keyboard, rebooted again, and the problem went away."

Don thought about Joe's story. "I never would have tried that."

Joe laughed. "Me neither. It was the act of a desperate man."

PERFORMANCE OBJECTIVES

Upon completion of this exercise, you will be able to

1. Remove and install a piece of hardware.
2. Examine/adjust hardware settings.
3. Verify correct operation of the new hardware.

KEY TERMS

Driver   Virtual device driver
Real-mode device driver

BACKGROUND INFORMATION

In this exercise we will watch what happens as we install a new piece of hardware. If you use your computer a great deal, eventually you will feel a need for improvement. Adding a faster processor (on a new motherboard) or an additional hard drive, upgrading your video adapter to one that supports DirectX, or installing more DRAM all lead down the same path. The installation either goes well or something goes wrong that sets you back hours when the job should have taken 20 minutes.

STARTING OFF

Unit V has plenty of exercises that describe how different types of hardware are installed and what the requirements are. The purpose of this exercise is to examine the built-in hardware installation support Windows offers. To demonstrate, we will work through the installation of a modem. Physically, the modem must be plugged into its associated slot, secured, and properly connected to the telephone lines.

When Windows 95/98 comes up for the first time after the modem has been installed, it may find a new plug-and-play device and proceed with the installation automatically, in accordance with the plug-and-play guidelines. If the modem is not a plug-and-play device, it is necessary to run the Add New Hardware utility in Control Panel. The first window should look like Figure 19.1. Notice the Back and Next buttons. You are able to move back and forth in the installation process until you are satisfied with the results. Hardware installation in Windows NT will be covered at the end of this exercise.

DETECTING NEW HARDWARE

Figure 19.2 shows the next installation screen. Be careful when you let Windows 95/98 detect the new hardware by itself. Windows 95/98 is not a perfect operating system. It may get the model or make of your network card wrong, or not detect a plug-and-play device. Windows 95/98 may hang during the detection, forcing you to simply reboot without being

FIGURE 19.1   The first Add New Hardware dialog box

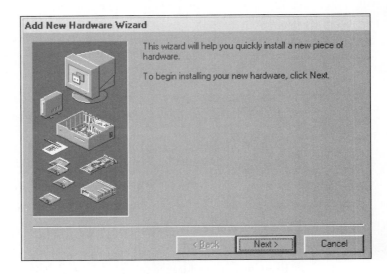

FIGURE 19.2   Add New Hardware Wizard menu

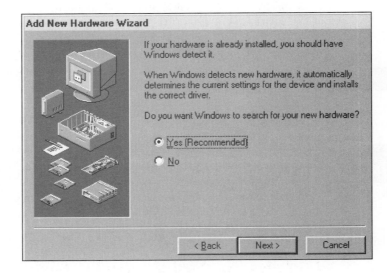

FIGURE 19.3   Add New Hardware Wizard informational message

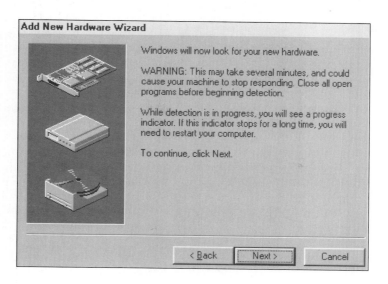

FIGURE 19.3   Add New Hardware Wizard informational message

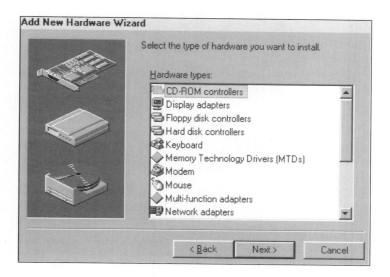

FIGURE 19.4   Choosing a hardware device to add

able to properly shut down first. The good thing is that Windows 95/98 is always on the watch for new hardware, so if it does not see the device the first time you reboot, it may the second time. Figure 19.3 is evidence that your installation may not go as smoothly as planned.

If you click the No button in the screen shown in Figure 19.2 and do not let Windows 95/98 detect the new hardware, you will have to navigate through a series of menus asking what type of hardware is being installed and other details. Figure 19.4 shows the menu you get if you click "No." Windows 95/98 has a wealth of hardware device types ready to go.

While the detection phase is going on, the display will look similar to Figure 19.5.

THE RESULTS

When Windows 95/98 finishes detecting the new hardware, the window shown in Figure 19.6 will appear. It is a good idea to click the Details button to see what Windows 95/98 actually found. Figure 19.7 indicates that a 14.4 Data FAX Modem was detected.

FINISHING UP

Clicking the Finish button completes the hardware installation and verifies that the modem was installed correctly. Figure 19.8 indicates that you are given one last chance to back out

FIGURE 19.5   The Detection progress meter

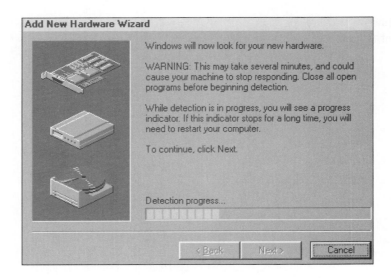

FIGURE 19.6   New hardware has been detected

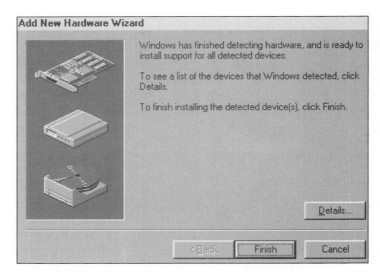

FIGURE 19.7   The new hardware details

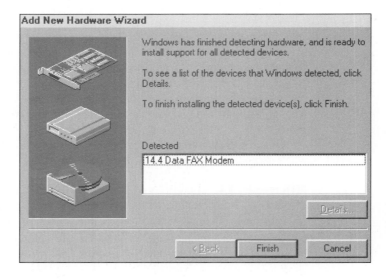

and not install the new hardware, or change the detected device. If you click the Change button, Windows 95/98 presents you with a menu of modems to choose from, the same menu you get if you skip the hardware detection phase by answering "No" in the first screen. Figure 19.9 shows the menu of modems.

FIGURE 19.8 Verifying the new hardware information

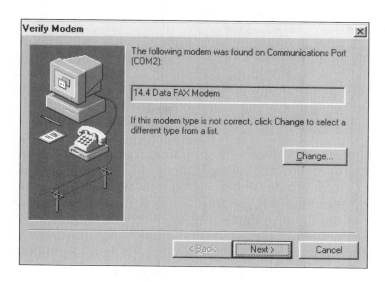

FIGURE 19.9 Choosing a modem

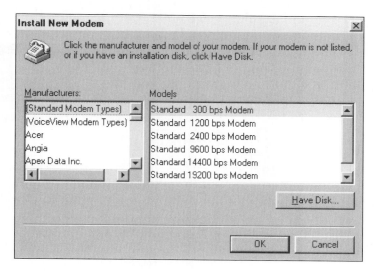

GENERAL COMMENTS

No matter what hardware is being installed, there are two types of software ***drivers*** available for use, typically ***real-mode device drivers*** and *protected-mode*, or ***virtual***, ***device drivers***. Since Windows likes to do as much work as possible in protected mode (the most powerful mode of operation in the Intel processors; see Exercises 50 and 53 for details), a large number of virtual device drivers (.VXD extension on Windows 95/98 machines) are provided for all types of hardware. If you have an older hardware device that requires a real-mode device driver, or your hardware is not supported by Windows (no .VXD file), you have no choice but to use a real-mode driver. In general, real-mode device drivers lower the performance of Windows, which must switch from protected mode to real mode to use the device driver (and back when finished). The mode-switching overhead could be a factor if the driver is used often (as in a modem).

In addition, there is bound to come a time when an interrupt or I/O location assigned to a new device conflicts with an existing assignment (for example, interrupt number 5 or port number 330 being used by two devices). Windows does a good job of catching these types of conflicts, as indicated in Figure 19.10. The Conflict information box indicates that the interrupt 10 setting for the hardware being configured conflicts with the network interface card. It should be possible to change the interrupt and avoid the conflict.

Some new devices require less effort to install than others. Adding another 64MB of DRAM is as easy as sliding the SIMM or DIMM into its socket and pushing it into place. BIOS will find the additional memory the next time the machine boots, as will Windows.

FIGURE 19.10   Interrupt 10 is in conflict

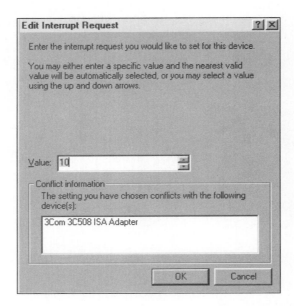

FIGURE 19.11   Windows NT CD-ROM contents

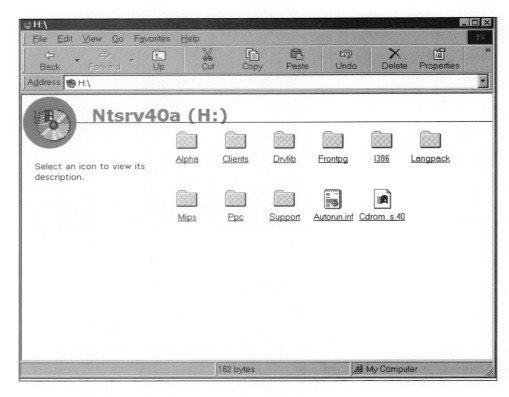

INSTALLING HARDWARE IN WINDOWS NT

Installing new hardware on a Windows NT computer is done much differently from the Windows 95/98 environment, which provides an Add New Hardware icon in the Control Panel. Device types and drivers that can be added to Windows NT are provided on the Windows NT CD-ROM.

To begin the installation process, it is necessary to open the Windows NT CD-ROM, the contents of which are shown in Figure 19.11. The Drvlib folder contains 12 folders that categorize the different types of hardware that can be added. These folders are shown in Figure 19.12. One of the most common types of devices to add to a Windows NT computer is a sound card, which is located in the Audio folder. The Windows NT CD-ROM provides drivers that support ESS and Sound Blaster hardware. Other Windows NT drivers are supplied

FIGURE 19.12   Drvlib folder

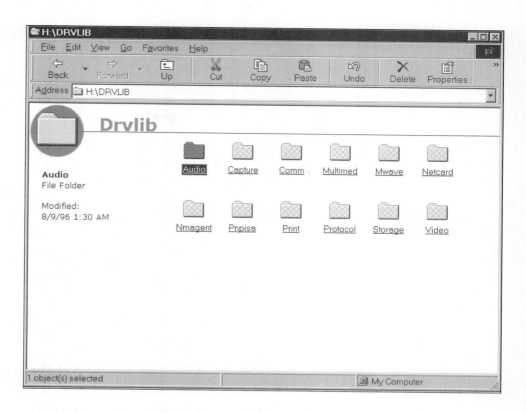

FIGURE 19.13   Sound Blaster
files

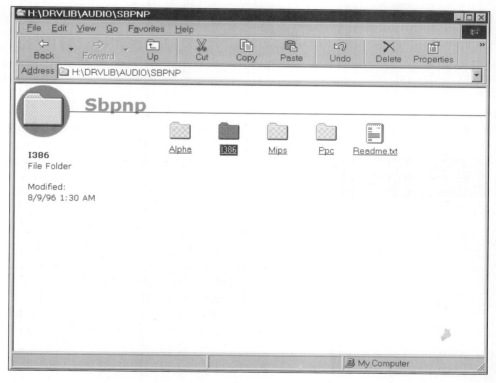

by the individual manufacturers using either a floppy disk, CD-ROM, or directly from their respective Web sites.

The Audio folder contains a folder called Sbpnp (Sound Blaster Plug-and-Play shown in Figure 19.13), which contains another set of folders that provide the necessary drivers for all of the Windows NT support hardware platforms. These platforms include the Alpha, Mips, Ppc, and Intel (and compatible) microprocessors. The Readme.txt file provides the specific details to install the device on any of the supported platforms. To install a Sound

243

FIGURE 19.14 Installing the new hardware

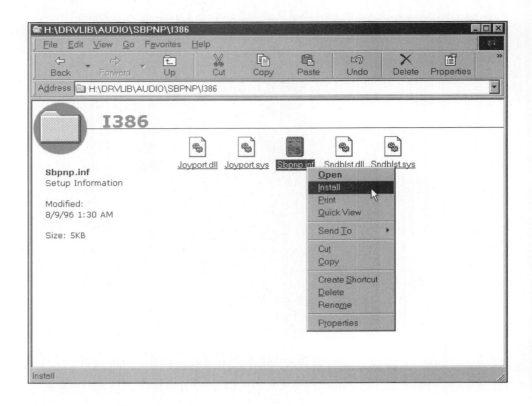

Blaster audio device on an Intel microprocessor platform, it is necessary to right-click on the Sbpnp.inf icon located in the I386 folder and then select the install option by left-clicking on Install as shown in Figure 19.14. The installation process copies all the required files to their proper locations and prepares for the driver files to be loaded after the system is rebooted. You are encouraged to explore each of the device folders found in the Drvlib folder.

TROUBLESHOOTING TECHNIQUES

A few things to keep in mind when installing new hardware:

- Be prepared to reboot a number of times.
- Do not be shocked if a plug-and-play card is not recognized or does not work when first installed.
- Avoid DOS (real-mode) device drivers if possible, using a virtual device driver (.VXD file) supplied by Windows.
- Sometimes it helps to delete a troublesome hardware device. The next reboot will reinstall it, hopefully solving the problem.
- Check the interrupt and I/O settings for conflicts.
- Windows contains many useful built-in troubleshooters, accessed via the Help menu. Follow the recommendations of the troubleshooter when trying to resolve conflicts.

SUMMARY

In this exercise we discovered that

- Windows may not detect new hardware properly.
- It is better to use a virtual device driver than a real-mode device driver.
- Hardware is added to a Windows NT computer differently than a Windows 95/98 machine.

This self-test is designed to help you check your understanding of the background information presented in this exercise.

True/False

Answer *true* or *false*.

1. Plug-and-play cards always work when first installed.
2. It is sometimes necessary to reboot when Windows is detecting hardware.
3. Windows contains a large database of hardware drivers.
4. It is good to use real-mode device drivers.
5. Windows cannot help at all when hardware conflicts arise.
6. Real-mode device drivers use the .VXD extension.
7. Windows NT has multiple hardware platforms.

Multiple Choice

Select the best answer.

8. Windows detects new hardware
    a. Every time a machine boots up.
    b. Only when you use a driver disk.
    c. Only during installation of the operating system.
9. When installing new hardware, the install wizard allows you to
    a. Click a single button that does everything.
    b. Always use a virtual device driver.
    c. Go back and change your mind.
10. If "No" is clicked on the menu in Figure 19.2, Windows will
    a. Abort the install process.
    b. Install a generic driver.
    c. Allow you to choose your hardware from a list.
11. After Windows detects new hardware it
    a. Verifies its operation.
    b. Runs a performance test to optimize the device.
    c. Disables all other hardware until the install is complete.
12. Virtual device drivers operate in
    a. Real mode.
    b. Imaginary mode.
    c. Protected mode.
13. Setup information for new hardware is found in
    a. .HDW files.
    b. .INF files.
    c. .VXD files.
14. To get new hardware to operate correctly, you may need to adjust its
    a. Interrupt settings.
    b. Input/output port settings.
    c. Both a and b.

Completion

Fill in the blank or blanks with the best answers.

15. VXD stands for _____ _____ _____.
16. You may need to _____ when detecting new hardware.
17. Two possible sources of conflict are _____ and I/O locations.
18. Real-mode device drivers require Windows to _____ back and forth from protected mode.
19. Real-mode device drivers _____ the performance of Windows.
20. Automatically detected hardware is called plug-and-_____.
21. The Drvlib folder contains hardware drivers for Windows _____.

1. If your lab machine has a modem, use Device Manager to select the modem and then click the Remove button to delete it from the system (a confirmation window will pop up). Power down and reboot. If your modem is a plug-and-play modem, Windows should detect it and begin installing it. Finish the installation and demonstrate that the modem works properly.
2. Repeat step 1 for any other piece of hardware your instructor deems acceptable.

1. If your familiarization activity used a plug-and-play card, how smooth was the installation? Were you really able to plug and play?
2. What is the problem with real-mode device drivers?

Under the supervision of your instructor

1. Remove and install a piece of hardware.
2. Examine/adjust hardware settings.
3. Verify correct operation of the new hardware.

20

Windows CE

INTRODUCTION

Joe Tekk was meeting with several sales representatives concerning the purchase of new digital telephony equipment. Joe was concerned about the specifications of the high-bandwidth leased line from the local phone provider.

"How much is the service agreement for the leased line?" he asked, hoping it was not prohibitively expensive.

Joyce Weylout, the lead sales representative, smiled and said, "Hold on a moment while I get the exact figures." She pulled a palm-size PC from her briefcase, tapped on its screen a few times with a pen like instrument, and smiled again. "Here they are."

Joe looked at the screen and saw the maintenance costs. He smiled too, for they were less than half of what he had budgeted for.

"I think we have an agreement, Joyce." He looked at her palm-size PC again.

"And by the way, where can I get one of those?"

PERFORMANCE OBJECTIVES

Upon completion of this exercise, you will be able to

1. Compare Windows CE with Windows 98/NT.
2. Discuss PPC and HPC devices.

KEY TERMS

Handheld PC (HPC)   Unicode character set
Palm-size PC (PPC)

BACKGROUND INFORMATION

Windows CE is a scaled-down version of Windows 98/NT designed for use on ***handheld PCs (HPCs)*** and ***palm-size PCs (PPCs)***. Figure 20.1 shows a PPC running Windows CE. A lot of power is packed into a PPC, which may utilize a 133 MHz CPU. Windows CE is used to provide many of the familiar controls and features found on more expensive, desktop personal computers.

In this exercise, we will examine the features of Windows CE and see how it compares with the Windows 98 and Windows NT operating systems.

**Figure 20.1   The Cassiopeia
PPC**

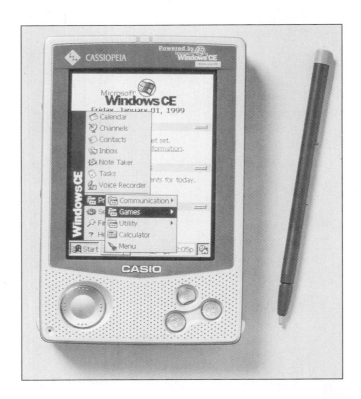

WINDOWS CE DEVICES

Windows CE devices fall into two catgories: PPCs and HPCs. PPCs fit in the palm of your hand. They do not contain a keyboard or mouse. Instead, a small stylus similar to a pen is used to gently tap the screen on the device to select an application, change a setting, or manipulate an object. A graphical keyboard can be used to enter text by tapping on the desired key symbols.

The display screen is small, typically 240 by 320 pixels, which is exactly one-fourth the size of a 640-by-480-pixel Windows 98/NT desktop.

HPCs are larger than PPCs (but still not as bulky as a laptop). They contain a miniature keyboard and a larger display (640 by 240 pixels). The stylus is also used on the HPC. Expansion is provided through a Type II PCMCIA slot (for additional memory or peripherals, such as a network connection).

HOW WINDOWS CE IS SIMILAR TO OTHER VERSIONS OF WINDOWS

Windows CE shares a few features with Windows 98/NT that give it the look and feel of the larger operating systems. For example, there is a Start button and a taskbar. Dialog boxes have the familiar Close buttons, there is support for sound, an internal clock keeps track of the time and date, and there are several built-in applications (Calculator, e-mail) for functionality. The core of Windows CE, like Windows 98/NT, is WIN32, with support for multitasking.

HOW WINDOWS CE IS DIFFERENT
FROM OTHER VERSIONS OF WINDOWS

Windows CE falls short of the power of Windows 98/NT when it comes to the area of memory. Unlike Windows 98/NT systems, Windows CE devices do not contain hard drives and utilize smaller memory systems, such as 4, 8, or 16MB. Applications must be designed to release memory to the Windows CE operating system when necessary, or run with reduced memory availability.

Although the WIN32 core is similar to Windows 98/NT, the functionality is reduced and applications compiled for Windows 98/NT need to be recompiled (and even rewritten) for use on Windows CE devices.

UNDER THE HOOD

Unlike Windows 98/NT systems, which utilize an Intel x86 or Pentium CPU or equivalent, Windows CE devices come with many different types of processors, such as x86, Ppc, Mips, and ARM. Applications written for Windows CE must correctly compile and execute on all processors.

COMMUNICATING WITH WINDOWS CE DEVICES

Windows CE devices can communicate in a variety of ways. A serial port is typically provided for connection directly to a PC or to an external modem. An infrared port is also provided, to allow easy transfer of data between two Windows CE devices.

Windows CE supports several important protocols, such as PPP and SLIP for serial connections; HTTP, FTP, and TCP/IP for use with the Internet; and IrDA over infrared. Programmers may use Winsock 1.1 functions in their networking code.

WINDOWS CE APPLICATIONS

Windows CE comes with many of the applications we are familiar with, such as Calculator, Solitaire, e-mail, and specialized versions of Word, Excel, and PowerPoint, plus additional applications for playing audio and video files, browsing the World Wide Web, and communication. Additional applications are found in the usual locations: on distribution CD-ROMs (to download from Windows 98/NT machines) and numerous Web sites.

THE SPECIAL PROGRAMMING REQUIREMENTS OF WINDOWS CE

The differences between Windows CE and Windows 98/NT require the programmer to follow certain guidelines. In general, the programmer must

- Use the Unicode character set.
- Manage memory properly.
- Adapt to specific devices.

The *Unicode character set* is an international standard for representing text characters from many different languages. Windows CE text is based on Unicode. Compilers for Windows CE provide the necessary Unicode elements.

Memory management is especially important, since Windows CE uses RAM for program memory and for simulated hard disk space. The user is even able to adjust the amount of memory allocated to each. So, an application must be constantly aware of available memory and also able to release some of its memory back to the operating system if requested.

An application should also be able to adapt to the Windows CE device it is running on. For example, the different screen sizes between PPCs and HPCs require adjustments to the way dialog boxes are displayed. One device may have a black-and-white screen, the other color. The memory capacities and even the processors also may be different. So, a programmer may need to support many different versions of the same application.

LOGO REQUIREMENTS

The "Powered by Windows CE" logo seen on Windows CE devices and related items is a guarantee that the product has met certain requirements deemed significant to the operation of Windows CE. For example, a programmer writing a new application for a Windows CE device must pay attention to the following:

- How the application installs itself.
- The design of the user interface (location and type of menu items, help features).
- Functionality (how application adapts to the hardware it is running on and takes advantage of services available).
- File handling issues.

Meeting the logo requirements provides the right to use the logo, participate in special advertising opportunities with Microsoft, and know that your application passes a rigid set of standards for an important new computing device.

WINDOWS CE RESOURCES

Here are several locations to check out for additional information on Windows CE software, devices, communications, and more:

- http://www.microsoft.com/windowsce/default.asp
- http://www.conitech.com/windows/ce.html
- http://www.socketcom.com/
- http://www.ceware.com/
- http://www.casio.com/hpc/

TROUBLESHOOTING TECHNIQUES

Even Windows CE devices come with Reset buttons. As powerful as they are for their size, Windows CE devices can suffer the same fate as the components on a motherboard. They can be damaged by static electricity or operate erratically when battery power is low. The Reset button is provided to restart the Windows CE device. Once reset, the operating system goes through its simplified boot procedure, configuring the touch screen (by having the user tap the stylus on a moving cursor), and setting the time, date, and personal information.

SUMMARY

In this exercise we discovered that

- Windows CE is a scaled-down operating system.
- Windows CE is designed for use on PPCs and HPCs.
- Windows CE has special programming requirements (memory management, low power).

SELF-TEST

This self-test is designed to help you check your understanding of the background information presented in this exercise.

True/False

Answer *true* or *false*.

1. Windows 98 applications run without modification on Windows CE.
2. Only Pentium processors are used in Windows CE devices.
3. PPP and SLIP are supported by Windows CE.
4. Windows CE does not support networking.
5. The "Powered by Windows CE" logo is available free from Microsoft.

Multiple Choice

Select the best answer.

6. Windows CE devices utilize
   a. A light-pen.
   b. A mouse.
   c. A stylus.
7. Windows CE provides the familiar
   a. Shutdown menu.
   b. Start button and taskbar.
   c. Login screen.
8. Networking functions are available through
   a. Winsock 1.1.
   b. Winsock 2.0.
   c. There are no network functions available.
9. Which is not a supported Windows CE processor?
   a. PowerPC.
   b. Mips.
   c. Z80.
10. Which is not a Windows CE protocol?
    a. TCP/IP.
    b. PPCPP.
    c. PPP.

Completion

Fill in the blank or blanks with the best answers.

11. PPCs and HPCs are two examples of Windows CE _____.
12. The IrDA protocol is used with the _____ port.
13. Windows CE has a _____ processing core similar to Windows 98/NT.
14. Windows CE uses the _____ character set.
15. Use of the "Powered by Windows CE" _____ has certain requirements that must be met.

FAMILIARIZATION ACTIVITY

PPCs and HPCs are sold in a variety of locations, from computer stores to general merchandise outlets. Find a PPC or HPC and examine its features. Take a tour through the Help screens. Talk to the salesperson about the device.

QUESTIONS/ACTIVITIES

1. Compare the cost of a black-and-white PPC with that of a color PPC.
2. Compare the price of a color PPC with a color HPC.
3. What is the current memory capacity of a PPC?
4. How fast is the processor on a new PPC?

REVIEW QUIZ

Under the supervision of your instructor

1. Compare Windows CE with Windows 98/NT.
2. Discuss PPC and HPC devices.

21

Other Network Operating Systems

INTRODUCTION

Joe Tekk visited his friend Marlene Hall, an educational planner for a high-technology consulting firm. Marlene was busy preparing color transparencies for a presentation.

"Hi, Joe," she said, handing him a thick pile of transparencies. "Look through these and tell me what you think."

Joe examined the transparencies with interest. Marlene had put together a detailed comparison of Windows NT and NetWare. "Who are these for?" he asked.

Marlene gave Joe a quick stare and then replied, "The president of our European division."

Two weeks later, Joe received a call from Marlene. The presentation had gone so well the president chose both operating systems, and hired Marlene to oversee their integration in the European office.

PERFORMANCE OBJECTIVES

Upon completion of this exercise, you will be able to

1. Compare the features of NetWare with Windows NT.
2. Discuss the file organization, protocols, and security available in NetWare.
3. Briefly describe the Linux operating system.

KEY TERMS

NetWare	Jukebox
In-place migration	High Capacity Storage System (HCSS)
Across-the-wire migration	IPX/SPX protocols
Network Directory Service (NDS)	Linux
Data migration	

BACKGROUND INFORMATION

In this exercise we will examine two additional network operating systems: NetWare and Linux. Let's start with a look at the features of NetWare.

NETWARE

The NetWare operating system originated in the early days of DOS, allowing users to share information, print documents on network printers, manage a set of users, and so on. One important difference between NetWare and Windows is in the area of application software. NetWare does not provide the 32-bit preemptive multitasking environment found in Windows. Applications written for Windows will not run on NetWare. They may, however, communicate over the network using NetWare's proprietary IPX/SPX protocol. The following sections describe many of the main features of NetWare.

INSTALLING/UPGRADING NETWARE

Unlike Windows 95, 98, and NT, the NetWare operating system runs on top of DOS (as Windows 3.x did). So before NetWare can be installed on a system, DOS must be up and running. NetWare 4.x and above provide a DOS environment as part of the installation process. Versions of NetWare older than 3.1x must be upgraded to 3.1x before they can be further upgraded to NetWare 4.x and above.

There are two ways to perform an upgrade: through *in-place migration* and through *across-the-wire migration*. In-place migration involves shutting down the NetWare server to perform the upgrade directly on the machine. Across-the-wire migration transfers all NetWare files from the current server to a new machine attached to the network. The new machine must already be running NetWare 4.x or above. This method allows the older 3.1x server to continue running during the upgrade.

NETWARE-WINDOWS TIMELINE

Table 21.1 shows the NetWare and Windows operating system releases during the last decade. Both operating systems have matured to provide significant network support, management, and productivity features.

NDS

The Network Directory Service (NDS) is the cornerstone of newer NetWare networks. Network administrators can manage all users and resources from one location. NDS allows users to access global resources regardless of their physical location, using a single login. The NDS database organizes information on each object in the network. These

TABLE 21.1   NetWare–Windows timeline

NetWare		Windows			
Version	Year	3.x/95/98/ME		NT/2000	
		Version	Year	Version	Year
3.x	1989	3.0	1990	NT3.1	1993
4.x	1993	3.1	1992	NT3.5	1994
5	1998	3.11	1993	NT3.51	1995
5.1	2000	95	1995	NT4.0	1996
		98	1998	NT5.0 beta	1997
		98 SE	1999	2000	2000
		ME	2000		

FIGURE 21.1 Typical tree structure

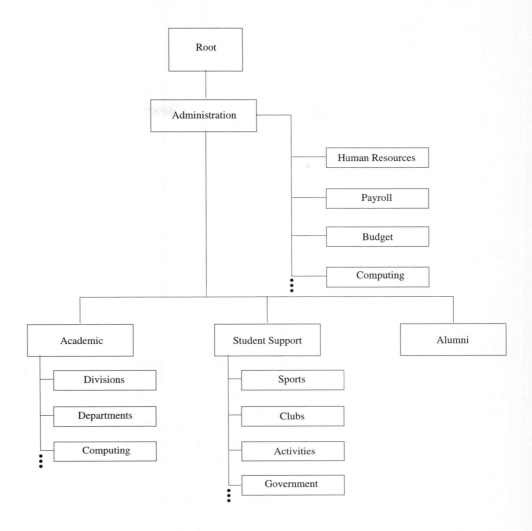

objects are users, groups, printers, and disk volumes. Typically these objects are organized into a hierarchical tree that matches the internal structure of an organization. Figure 21.1 shows a typical tree structure for a small two-year college. Each major area of the organization (Administration, Academic, Student Support, and Alumni) has its own unique requirements. The requirements are applied to all the users who are associated with each specific area. Over time, as the requirements of the organization change, elements in the hierarchical tree are added, modified, or removed very easily.

NetWare uses *data migration* to move data from one location to another to maintain effective use of available hard drive space. Large files are moved to a secondary storage system (such as a *jukebox*) and copied back (demigrated) to the hard drive when needed. Files that have been migrated still show up in directory listings.

Accurate timekeeping plays an important role in the operation of NDS. Multiple servers must agree on the network time so that file updates are performed in the correct sequence. NetWare uses several kinds of time servers to maintain an accurate Universal Coordinated Time (UTC). These servers are called reference, primary, secondary, and single-reference. Reference servers use a connection to an accurate time source (such as the U.S. Naval Observatory's Atomic Clock) to provide the network time. Primary and reference servers negotiate with each other to determine the network time. Secondary servers provide the time to NetWare clients. Single-reference time servers are designed for use on small networks in which one machine has total control over the network time.

NDS add-ons are also available for Windows NT and Unix environments, allowing those systems to fully participate in the NetWare environment.

FIGURE 21.2   NetWare
protocol suite

ISO/OSI Layer	Protocols		
7 = Application	NCP	SAP	RIP
6 = Presentation			
5 = Session			
4 = Transport	SPX		
3 = Network	IPX		
2 = Data Link	LSL 802.2 MLID		
1 = Physical	802.3 Ethernet 802.5 Token Ring		

HCSS

The High Capacity Storage System (HCSS) provided by NetWare allows for tremendously large volumes to be created that span up to 32 physical hard drives. When hard drives were quite small in comparison to the sizes available today, HCSS allowed for the creation of volumes up to 32GB in size. NetWare 5 introduced increased volume sizes up to 8 terabytes (still much larger than disks typically available today). In conjunction with a configuration of RAID (Redundant Array of Inexpensive Disks), data is protected even if one of the disks in the volume fails.

NETWORK PROTOCOLS

Figure 21.2 shows an outline of the major protocols used in the NetWare environment. Let's take a look at them and see what their roles are in the NetWare environment.

Internetwork Packet Exchange (IPX)

The Internetwork Packet Exchange (IPX) protocol is similar to UDP under TCP/IP, providing connectionless, unreliable network communication. IPX packets are used to carry higher-level protocols such as SPX, RIP, NCP, and SAP.

NetWare Core Protocol (NCP)

The NetWare Core Protocol (NCP) is the workhorse of NetWare, responsible for the majority of traffic on a NetWare network. Spanning three layers of the OSI protocol stack (Session, Presentation, and Application), NCP carries all file system traffic in addition to numerous other functions, including printing, name management, and establishing connection-oriented sessions between servers and workstations. NCP provides connection-oriented packet transmission allowing for positive acknowledgment of each packet. A more efficient implementation called NCPB (NetWare Core Protocol Burst) was added to reduce the amount of control traffic necessary when using NCP.

Sequenced Packet Exchange (SPX)

The Sequenced Packet Exchange (SPX) protocol is similar to TCP under TCP/IP, providing connection-oriented communication that is reliable (lost packets are retransmitted). SPX packets contain a sequence number that allow packets received out of order to be reassembled correctly. Flow control is used to synchronize both ends of the connection to achieve maximum throughput.

Service Advertising Protocol (SAP)

The Service Advertising Protocol (SAP) is a broadcast protocol that is used to maintain a database of servers and routers connected to the NetWare network.

Routing Information Protocol (RIP)

Like SAP, the Routing Information Protocol (RIP) is also a broadcast protocol. RIP is used by routers to exchange their routing tables. Multiple routers on the same network discover each other (during a RIP broadcast) and build entries for all networks that can be reached through each other. When conditions on the network change (a link goes down or is added), the change will be propagated from router to router (using RIP packets).

Multiple Link Interface Driver (MLID)

The Multiple Link Interface Driver (MLID) provides the interface between the network hardware and the network software. A specification called *ODI (Open Data-link Interface)* is supported by MLID. ODI allows NetWare clients to use multiple protocols over the same network interface card. The LSL (Link Support Layer) provides the ODI facilities.

Menus, Login Scripts, and Messaging

Several of the most important issues for the user involve the system menus, shared access to a common set of data, and electronic messaging capabilities. Access to the software located on each system is created using the menu generation program. Access to items in the menu is made available during the login process using a login script.

The login script contains a list of commands that are executed when each user logs in to the network. The commands are typically used to establish connections to network resources such as mapping of network drives. A login script is a property of a container, Profile, or User object. If a login script is defined for each of these objects, all associated login scripts will execute when a user logs in, allowing for a great deal of control over each user's environment.

Electronic messaging is provided for all users using information available through NDS. Each user's specific information is centrally maintained in the NDS database. The Message Handling Service (MHS) provides access to the X.400 standard implementation for e-mail. FirstMail client software is provided with Novell NetWare 4.1 and above, which is used to access the X.400 messaging services. Add-on products such as GroupWise offer more sophisticated support for electronic messaging. GroupWise also provides document management, calendaring, scheduling, task management, workflow, and imaging.

Security

The security features of the NetWare operating system offer the system and/or network administrator the ability to monitor all aspects of the system operation from a single location. There are two types of security: file system security and NDS security.

File System Security

Table 21.2 shows the various types of rights that may be assigned to a NetWare user. Note that similar rights are available under Windows NT.

Rights are inherited and/or modified via filters or masks that designate permissible operations.

TABLE 21.2   NetWare Rights

Rights Name	Rights Description
Access Control	May control the rights of other users to access files and directories
Create	May create new file or subdirectory
Erase	May delete existing files or directories
File Scan	May list the contents of a directory
Modify	May rename and change file attributes
Write	May write data into an existing file

NDS Security

In addition to encryption of login passwords, NDS security provides auditing features that allow one user to monitor events caused by other users (changes to the file system, resource utilization).

MANAGEMENT

Management of any network involves many different activities to ensure quality control. Some of these items include the following:

- Monitor network traffic to develop a baseline from which to make network-related decisions.
- Unusual activity monitoring, such as successive login failures, or file creation or file write errors.
- Disk resource utilization issues.
- Software and hardware installation and upgrades.
- Backup scheduling.
- Help desk support for problem resolution.
- Short-range and long-range planning.

Many other important items can be added to this list, depending on the organization. Some of these issues will be explored in the problems located at the end of this exercise.

PRINT SERVICES

A core component of NetWare consists of the services available for managing and using printers. Print jobs are first sent to a printer queue, where they are temporarily stored until the assigned printer is available. Printers may be attached to workstations, print servers, or even directly connected to the network.

NetWare 5 expands print services to allow for notification of print job completion or status, enhanced communication between printers and clients (printer features are shared), and support for multiple operating systems. An online database of printer drivers is also provided to assist with new printer installations.

NetWare Client Software

In addition to one or more NetWare servers, there will be many NetWare clients on the network taking advantage of the services provided. Windows users can install NetWare client software and have access to the power of NetWare while still being part of a Windows network environment.

Figure 21.3 shows the new items found in the Network Neighborhood window after the NetWare client software has been installed. NetWare's folders exist side by side with Windows 98's icon for the Raycast machine.

FIGURE 21.3 Network Neighborhood contents after NetWare installation

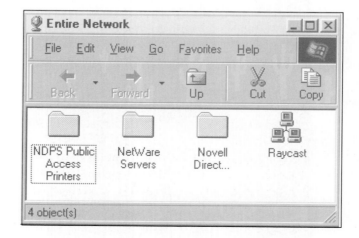

FIGURE 21.4   NetWare
controls

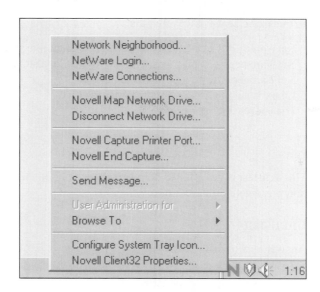

Figure 21.4 shows the context-sensitive menu that appears after right-clicking on the NetWare icon (N) in the system tray. Note all the different features accessible from the menu. Other NetWare properties can be examined and/or modified by selecting Novell NetWare Client properties in the Network window under Control Panel.

THE LINUX OPERATING SYSTEM

The Linux operating system is a free version of the popular Unix operating system used on many different workstations and servers. The Linux operating system was written by Linus Torvold, a Swedish computer science student, as a personal project that grew into a world-wide cooperative development effort. Complete with all of the bells and whistles of a proprietary Unix operating system, Linux provides a stable environment on which to run many of the popular applications.

Like Windows, Linux is a stand-alone network operating system that does not run on top of DOS. When Windows and Linux are installed on the same system, Linux allows access to the Windows drives. A significant feature of Linux is that the complete source code of the entire operating system is included on the CD-ROM. This allows for custom modifications to be made, if necessary, and modifications necessary to add new hardware devices as they become available.

Linux, like Unix, uses TCP/IP as the built-in network protocol, allowing any Linux-based computer the ability to be used as a server on the Internet. All the required elements such as Domain Name Services (DNS), Dynamic Host Configuration Protocol (DHCP), and the complete suite of server software for HTTP, FTP, Telnet, and so on, are also provided.

Because of its unique status as a fully functional, free-operating system, Linux is quickly gaining wide popularity. In addition, support for Linux is provided by the entire Linux community.

**TROUBLESHOOTING
TECHNIQUES**

Selecting a network operating system is based on many factors, including the features provided and the complexity of the installation and management procedures. Is the operating system centralized or distributed? Will the network be strictly TCP/IP or should multiple protocols be supported? What file system properties are desired?

Whatever the answers are, the end result is an operating system that will require specific troubleshooting methods to diagnose and repair problems.

SUMMARY

In this exercise we discovered that

- NetWare is not a full-fledged operating system like Windows.
- NetWare provides high-capacity distributed data storage and retrieval.
- NetWare uses its own protocol suite: IPX/SPX.
- Linux is an operating system similar to Unix.
- The source code for Linux is available.

SELF-TEST

This self-test is designed to help you check your understanding of the background information presented in this exercise.

True/False

Answer *true* or *false*.

1. NetWare is a distributed operating system.
2. NDS allows users global access to resources.
3. SPX is an unreliable protocol.
4. The source code for NetWare is provided on the installation CD-ROM.
5. Linux runs on top of DOS.

Multiple Choice

Select the best answer.

6. Two ways to upgrade a NetWare server are in-place migration and
    a. Out-of-place migration.
    b. Across-the-wire migration.
    c. Parallel migration.
7. SPX packets contain
    a. Sequence numbers for proper reassembly.
    b. Routing information.
    c. Service broadcast data.
8. Login passwords are
    a. Stored as plain text.
    b. Stored as encrypted text.
    c. Not stored.
9. Using a single login, each user
    a. Gains access to container files only.
    b. Gains access to all network resources.
    c. Gains access to local files only.
10. NetWare print services support
    a. Multiple operating systems.
    b. Notification of job completion.
    c. Both a and b.

Completion

Fill in the blank or blanks with the best answers.

11. Two types of security are file system security and _____ security.
12. Linux is based on _____.
13. The main protocol used for NetWare operations is _____.
14. RAID stands for _____ _____ _____ _____.
15. NetWare uses several types of _____ servers to synchronize network events.

NetWare

1. Try to measure the delay associated with demigrating a large file.
2. Watch the IPX network traffic with a protocol analyzer. What types of packets do you see?

Linux

1. Install Linux on a machine that already has an operating system on it. How are both operating systems accessible?
2. Search the Web for Linux download information. What is available?

1. Find a local business that uses NetWare. How many users are supported? Why did they choose NetWare? What version are they running?
2. Can you purchase NetWare in your local computer software store?

Under the supervision of your instructor

1. Compare the features of NetWare with Windows NT.
2. Discuss the file organization, protocols, and security available in NetWare.
3. Briefly describe the Linux operating system.

UNIT IV Computer Networks

22

What Is a Computer Network?

Joe Tekk was visiting his friend Julie Plume, an instructor at a local community college. Julie was interested in setting up a network in her classroom.

"Joe," she began, "I need to know a number of things. How much will it all cost? Where do I buy everything? Who can set it up for me?"

Joe laughed. "Hold on, Julie, one thing at a time. The cost depends on how many computers you want to network, the type of network used, and who you buy your equipment from. I have a number of networking catalogs you can look at, and you can also browse the Web for networking products."

Joe looked around the room. There were 14 computers, two laser printers, and a color scanner. "You could probably buy a 16-port Ethernet hub that would take care of this entire room. One network interface card for each PC, some UTP cable, and that's about it. Probably a few hundred dollars will do it. I could set it up with you some afternoon."

Julie had more questions. "Will I need to buy software?"

"Most of the stuff you'll want to do, such as network printing and sharing files, is already built into Windows. You may need to purchase special network versions of some of your software."

"Just one more question, Joe," Julie said. "How does it all work?"

PERFORMANCE OBJECTIVES

Upon completion of this exercise, you will be able to

1. Sketch and discuss the different types of network topologies and their advantages and disadvantages.
2. Sketch and explain examples of digital data encoding.
3. Discuss the OSI reference model.
4. Explain the basic operation of Ethernet and token-ring networks.

KEY TERMS

LAN (local area netwok)
WAN (wide area network)
Internet
Topology
Mesh network
Star network
Bus network

Ring network
Wireless
OSI reference model
Ethernet
Token-ring network
IEEE 802 Standards

A computer network is a collection of computers and devices connected so that they can share information. Such networks are called *local area networks* or *LANs* (networks in office buildings or on college campuses) and *wide area networks* or *WANs* (networks for very large geographical areas). Computer networks are becoming increasingly popular. With the *Internet* spanning the globe and the information superhighway (also called the National Information Infrastructure) providing many kinds of telecommunication services, the exchange of information among computer users is increasing every day. In this exercise we will examine the basic operation of a computer network, how it is connected, how it transmits information, and what is required to connect a computer to a network. This exercise lays the foundation for the remaining exercises in this unit.

COMPUTER NETWORK TOPOLOGY

Topology has to do with the way things are connected. The topology of a computer network is the way the individual computers or devices (called *nodes*) are connected. Figure 22.1 shows some common topologies.

Figure 22.1(a) illustrates a *fully connected network*. This kind of network is the most expensive to build, because every node must be connected to every other node in the network. The five-node network pictured requires 10 connections. A 20-node network would require 190 connections. The advantage of the fully connected network is that data need only traverse a single link to get from any node to any other node. This network is also called a *mesh* or *full-mesh network*.

Figure 22.1(b) shows the *star network*. Note that one node in the network is a centralized communications point. This makes the star connection inexpensive to build, since a minimum number of communication links are needed (always one less than the number of nodes). However, if the center node fails, the entire network shuts down. This does not happen in the fully connected network.

The *bus network* is shown in Figure 22.1(c). All nodes in the bus network are connected to the same communication link. One popular bus network is *Ethernet*, which we will be covering shortly. The communication link in an Ethernet network is often a coaxial cable connected to each node through a T-connector. The bus network is inexpensive to build, and it is easy to add a new node to the network just by tapping into the communication link. One thing to consider in the bus network is the maximum distance between two nodes, because this affects the time required to send data between the nodes at each end of the link.

FIGURE 22.1   Topologies for a five-node network

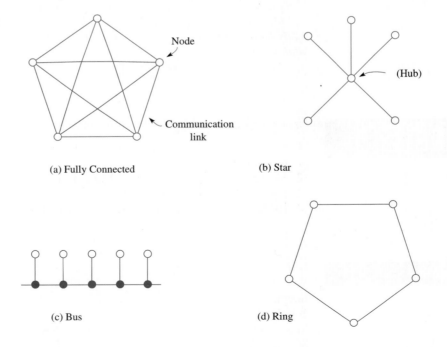

Node

Communication link

(a) Fully Connected

(Hub)

(b) Star

(c) Bus

(d) Ring

The last topology is the *ring network*, shown in Figure 22.1(d). This connection scheme puts the nodes into a circular communication path. Thus, as in the bus network, the maximum communication time depends on how many nodes there are in the network.

WIRED NETWORKS VERSUS WIRELESS NETWORKS

Pulling copper wire, or even fiber, throughout a building may be unsafe or prohibitively expensive. One solution involves the use of *wireless* networking equipment. In this topology, a base station connected to the network broadcasts data into the air in the form of a high-frequency RF signal (or through a line-of-sight infrared laser). Remote, or mobile, stations must stay within the range of the base station for reliable communication but are allowed to move about.

REPRESENTING DIGITAL DATA

The information exchanged between computers in a network is of necessity digital, the only form of data with which a computer can work. However, the actual way in which the digital data is represented varies. Figure 22.2 shows some of the more common methods used to represent digital data.

When an analog medium is used to transmit digital data (such as through the telephone system with a modem), the digital data may be represented by various forms of a *carrier-modulated* signal. Two forms of carrier modulation are amplitude modulation and frequency-shift keying. In amplitude modulation, the digital data controls the presence of a fixed-frequency carrier signal. In frequency-shift keying, the 0s and 1s are assigned two different frequencies, resulting in a shift in carrier frequency when the data changes from 0 to 1 or from 1 to 0. A third method is called *phase-shift keying*, in which the digital data controls the phase shift of the carrier signal.

When a digital medium is used to transmit digital data (between COM1 of two PCs, for example), some form of digital waveform is used to represent the data. A digital waveform is a waveform that contains only two different voltages. Inside the computer, these two voltages are usually 0 volts and 5 volts. Outside the computer, plus and minus 12 volts are often used for digital waveforms. Refer again to Figure 22.2. The nonreturn to zero (NRZ) technique simply uses a positive voltage to represent a 0 and a negative voltage to represent a 1. The signal *never* returns to zero.

Another popular method is Manchester encoding. In this technique, phase transitions are used to represent the digital data. A one-to-zero transition is used for 0s and a zero-to-one transition is used for 1s. Thus, each bit being transmitted causes a transition in the Manchester waveform. This is not the case for the NRZ waveform, which may have long

FIGURE 22.2   Methods of representing digital data

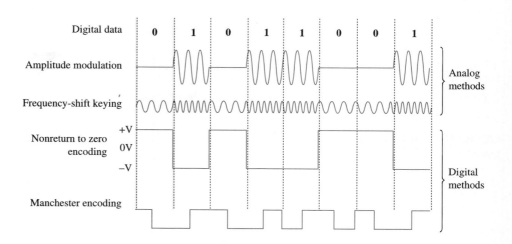

periods between transitions. The result is that Manchester encoding includes both data *and* a clock signal, which is helpful in extracting the original data in the receiver.

WORKING WITH DIGITAL DATA

What are some uses for digital data transmitted over a network? Let's examine a short list:

- Sharing files.
- Printing to a network printer.
- Loading a Web page.
- Sending e-mail.
- Listening to music (via streaming audio).
- Watching an MP3-encoded MPEG video.
- Making a phone call.
- Chatting.
- Making a purchase.
- Searching the Web.
- Playing a network game.

Many of these applications require large amounts of data to be exchanged. Some even require a secure connection. So, in addition to representing the digital data electronically (or physically), we must also represent it logically. Video files, for example, are compressed to reduce their storage requirements and downloading time. The information exchanged during a credit card purchase is typically encrypted to provide a measure of security. Compression and encryption are handled by software and also are supported by the use of communication protocols. Let's briefly examine the need for these protocols.

COMMUNICATION PROTOCOLS

Just throwing 1s and 0s onto a communication link is not enough to establish coherent communication between two nodes in a network. Both nodes must agree in advance on what the format of the information will look like. This format is called a ***protocol*** and is firmly defined. Figure 22.3 shows one of the accepted standards governing the use of protocols in computer networks. The ***OSI*** (Open Systems Interconnection) ***reference model*** defines seven layers required to establish reliable communication between two nodes. Different protocols are used between layers to handle such things as error recovery and information routing between nodes. A handy way to remember the names of each layer is contained in a simple statement: All Packets Should Take New Data Paths. The first letter of each word corresponds to the first letter of each OSI layer.

Not all of the seven layers are always used in a computer network. For example, Ethernet uses only the first two layers. The OSI reference model is really just a guide to establishing standards for network communications.

The Physical Layer

The Physical layer (layer 1) controls how the digital information is transmitted between nodes. In this layer, the encoding technique, the type of connector used, and the data rate,

FIGURE 22.3 OSI reference model

Layer	Function
7	Application
6	Presentation
5	Session
4	Transport
3	Network
2	Data Link
1	Physical

all of which are *physical* properties, are established. This layer is responsible for transmitting and receiving bits.

The Data-Link Layer

The Data-Link layer (layer 2) takes care of framing data, error detection, and it maintains flow control over the physical connection. The Data-Link layer consists of two sublayers: LLC (Logical Link Control) and MAC (Media Access Control).

The Network Layer

The Network layer (layer 3) is responsible for routing protocol-specific packets to their proper destination using logical IP addressing.

The Transport Layer

The Transport layer (layer 4) is the first layer that is not concerned with how the data actually gets from node to node. Instead, the Transport layer assumes that the physical data is error-free, and concentrates on providing correct communication between applications from a *logical* perspective. For example, the Transport layer guarantees that a large block of data transmitted in smaller chunks is reassembled in the proper order when received.

The Session Layer

The Session layer (layer 5) handles communication between processes running on two different nodes. For example, it allows two mail programs running on different nodes to establish a session to communicate with each other.

The Presentation Layer

The Presentation layer (layer 6) deals with matters such as text compression and encryption.

The Application Layer

The Application layer (layer 7) is where the actual user program executes and makes use of the lower layers. We will examine the operation of each layer in more detail in the remaining exercises.

ETHERNET

One of the most popular communication networks in use is *Ethernet*. Ethernet was developed jointly by Digital Equipment Corporation, Intel, and Xerox in 1980. Ethernet is referred to as a *baseband system*, which means that a single digital signal is transmitted. Contrast this with a *broadband system*, such as cable television, which uses multiple channels of data.

Ethernet transmits data at the rate of 10 million bits per second (which translates to 1.25 million bytes per second). This corresponds to a bit time of 100 ns. Manchester encoding is used for the digital data. New 100-Mbit and 1000-Mbit Ethernet is already being used.

Each device connected to the Ethernet must contain a *transceiver* that provides the electronic connection between the device and the coaxial cable commonly used to connect nodes. Figure 22.4 shows a typical Ethernet installation. The 11 devices on the Ethernet are grouped into three *segments*. Each segment consists of a coaxial cable with a *tap* for each device. It is very important to correctly terminate both ends of the coaxial cable in each segment; otherwise, signal reflections will distort the information on the network and result in poor communications. Segments are connected to each other through the use of *repeaters*, which allow two-way communication between segments.

Each Ethernet device has its own unique binary address. The Ethernet card in each device waits to see its own address on the coaxial cable before actually paying attention to the data being transmitted. Thus, when one device transmits data to another, every device listens. This is called *broadcasting*, much like the operation of a radio. However, Ethernet contains special hardware that detects when two or more devices attempt to transmit data at

FIGURE 22.4 Eleven-user
Ethernet LAN

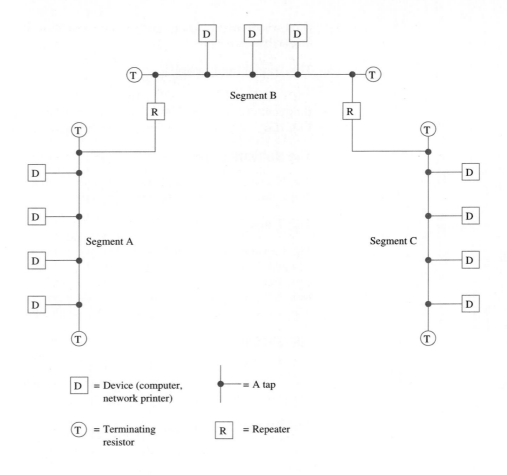

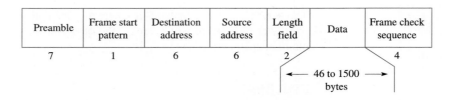

FIGURE 22.5 Ethernet frame
format

Preamble	Frame start pattern	Destination address	Source address	Length field	Data	Frame check sequence
7	1	6	6	2		4

46 to 1500 bytes

the same time (called a *collision*). When a collision occurs, all devices that are transmitting stop and wait a random period of time before transmitting the same data again. The random waiting period is designed to help reduce multiple collisions. This procedure represents a protocol called Carrier Sense Multiple Access with Collision Detection (CSMA/CD).

The format in which Ethernet transmits data is called a *frame*. Figure 22.5 details the individual components of the Ethernet frame. Recall that the Physical and Data-Link layers are responsible for handling data at this level. Note that the length of the data section is limited to a range of 46 to 1500 bytes, which means that frame lengths are also limited in range. Because of the 10-Mbit/second data rate and the format of the Ethernet frame, the lengths of the various segments making up an Ethernet LAN are limited (either 185 meters for 10base2 or 500 meters for 10base5). This guarantees that a collision can be detected no matter which two nodes on a segment are active.

TOKEN-RING LANs

Token-ring networks are not as popular as Ethernet but have their own advantages. The high collision rate of an Ethernet system with a lot of communication taking place is eliminated in a Token-ring setup.

The basic operation of a Token-ring network involves the use of a special token (just another binary pattern) that circulates between nodes in the ring. When a node receives the token, it simply transmits it to the next node if there is nothing else to transmit. But if a node has its own frame of data to transmit, it holds onto the token and transmits the frame instead. Token-ring frames are similar to Ethernet frames in that both contain source and destination addresses. Each node that receives the frame checks the frame's destination address with its own address. If they match, the node captures the frame data and then retransmits the frame to the next node. If the addresses do not match, the frame is simply retransmitted.

When the node that originated the frame receives its own frame again (a complete trip through the ring), it transmits the original token again. Thus, even with no data being transmitted between nodes, the token is still being circulated.

Unfortunately, only one node's frame can circulate at any one time. Other nodes waiting to send their own frames must wait until they receive the token, which tends to reduce the amount of data that can be transmitted over a period of time. However, this is a small price to pay for the elimination of collisions.

NETWORK OPERATING SYSTEMS

In addition to the communication protocols that enable reliable communication across a LAN or WAN, a computer network also requires software to control the communication protocols and provide all of the networking functions. Windows 95/98, Windows 2000, and Windows NT all contain built-in networking components, as do other network operating systems, such as Unix and Linux, Mac OS, and NetWare. Windows NT Server, in particular, is designed to manage large numbers of networked users through the services of a domain.

Windows provides a great deal of control over the operation of the network. Figure 22.6 shows the Network properties window. At a glance it is easy to see that the NetBEUI and TCP/IP protocols are installed and that the type of networking adapter is a Plug-and-Play ISA card.

FIGURE 22.6   Network properties window

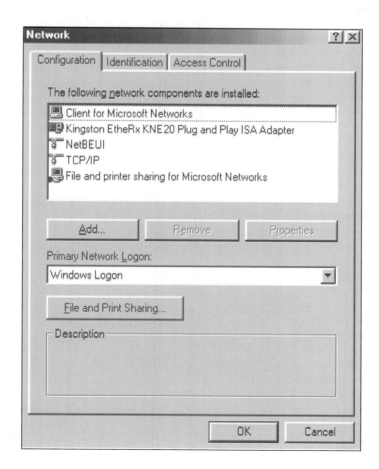

TABLE 22.1   IEEE 802 Standards

Standard	Purpose
802.1	Internetworking
802.2	Logical Link Control
802.3	Ethernet LAN (CSMA/CD)
802.4	Token-Bus LAN
802.5	Token-Ring LAN
802.6	Metropolitan Area Network (MAN)
802.7	Broadband Technical Advisory Group
802.8	Fiber-Optic Technical Advisory Group
802.9	Integrated Voice/Data Networks
802.10	Network Security
802.11	Wireless Networks
802.12	Demand Priority Access LAN (100 VG-AnyLAN)

IEEE 802 STANDARDS

The Institute of Electrical and Electronic Engineers (IEEE) has, over the years, established several committees dedicated to defining standards for computer networking. These standards are listed in Table 22.1. Any company entering the networking marketplace must manufacture networking hardware that complies with the published standards. For example, a new Ethernet network interface card must operate according to the standards presented in IEEE 802.2 and IEEE 802.3.

TROUBLESHOOTING TECHNIQUES

Troubleshooting a network problem can take many forms. Before the network is even installed, decisions must be made about it that will affect the way it is troubleshot in the future. For example, Ethernet and Token-ring networks use different data encoding schemes and connections, as well as different support software. Each has its own set of peculiar problems and solutions.

Troubleshooting a network may take you down a hardware path (bad crimps on the cable connectors causing intermittent errors), a software path (the machine does not have its network addresses set up correctly), or both. There may even be nothing wrong with the network, the failure coming from the application using the network. So, a good deal of trial and error may be required to determine the exact nature of the problem. In the remaining exercises, many of these troubleshooting scenarios will be discussed.

SUMMARY

In this exercise we discovered that

- Computer networks are connected using standard topologies.
- Two networking technologies are Ethernet and Token-ring.
- Communication protocols allow networked computers to communicate.
- The ISO reference model has seven network layers.
- Windows is a network operating system.
- IEEE 802 Standards exist that describe each networking technology.

SELF-TEST

This self-test is designed to help you check your understanding of the background information presented in this exercise.

True/False

Answer *true* or *false*.

1. The Internet is a computer network.
2. Both analog and digital media can be used to transmit digital data.
3. All seven layers of the OSI reference model are always used for communication in a network.
4. Ethernet uses collision detection to handle transmission errors.
5. The IEEE 802 Standards are only for Ethernet networks.
6. A domain is used to manage large groups of networked computers.

Multiple Choice

Select the best answer.

7. The term LAN stands for
   a. Logical access node.
   b. Local area network.
   c. Large access network.
8. Phase transitions for each bit are used in
   a. Amplitude modulation.
   b. Carrier modulation.
   c. Manchester encoding.
   d. NRZ encoding.
9. Ethernet transmits data in
   a. Continuous streams of 0s and 1s.
   b. Frames.
   c. Blocks of 256 bytes.
10. Which OSI network layer guarantees reliable data transmission?
    a. Physical
    b. Data-Link
    c. Network
11. Ethernet segments are connected together using
    a. Taps.
    b. Terminators.
    c. Repeaters.

Matching

Match a description of the topology property on the right with each item on the left.

12. Fully connected
13. Star
14. Bus
15. Ring

a. The whole network shuts down when the central node fails.
b. All nodes connect to the same communication link.
c. Uses a token to allow access to the network.
d. Most expensive to build.

Completion

Fill in the blank or blanks with the best answers.

16. The _____ of a network concerns how the nodes are connected.
17. A(n) _____ _____ network provides the fastest communication between any two nodes.
18. Using two different frequencies to represent digital data is called _____ _____.
19. The layer responsible for error detection and recovery is the _____ layer.
20. Collisions are eliminated in the _____ network.
21. _____ networking uses RF signals or infrared lasers.
22. The IEEE _____ Standards define the various network operations/technologies.

1. Visit the computer center of your school. Find out who the network administrator is and discuss the overall structure of the school's network with him or her.
2. Visit a local computer store and find out how much it would cost to set up a 16-user LAN.
3. Visit the Internet2 Web site (www.internet2.edu). What is the Internet2? What is QBone?

Under the supervision of your instructor

1. Sketch and discuss the different types of network topologies and their advantages and disadvantages.
2. Sketch and explain examples of digital data encoding.
3. Discuss the OSI reference model.
4. Explain the basic operation of Ethernet and Token-ring networks.

23

Network Topology

Joe Tekk met Don, the senior technician, at 6 A.M., outside the doors of a local high school.

"Are you ready, Joe?" Don asked. Joe had never accompanied Don on a site upgrade before.

"Sure, Don," Joe replied. "I'm looking forward to it."

For the next four hours, Joe crawled around on the floor, poked his head up into drop ceilings and underneath benches, and traced cables down long corridors, between floors, and down into the boiler room. When he finished, he was tired, bruised, and dirty.

"Well, Joe," Don said, "We've mapped the whole network out. Now we can begin the upgrade."

"Now?" Joe asked wearily.

"No," Don laughed, "later. We have to order the network components first. Go get some rest. You did a good job today."

PERFORMANCE OBJECTIVES

Upon completion of this exercise, you will be able to

1. Describe the difference between physical topology and logical topology.
2. Sketch the physical topologies of bus, star, ring, and fully connected networks.
3. Explain what is meant by network hierarchy.

KEY TERMS

Cloud
Physical topology
Logical topology
Virtual circuit
Virtual private network (VPN)
Hub
Concentrator
Collision

CSMA/CD
Segment
Repeater
Hierarchy
Network access point
Peering agreement
Point of presence (POP)

BACKGROUND INFORMATION

Topology concerns the structure of the connections between computers in a network. Figure 23.1 shows three computers (A, B, and C) and a network *cloud*, a graphic symbol used to describe a network without specifying the nature of the connections. The network cloud

FIGURE 23.1   Network cloud connecting three machines

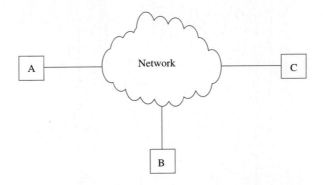

may comprise only the network found in a small laboratory, or it may represent a wide area network (WAN) such as the Internet.

The three computers in Figure 23.1 are connected in two different ways: physically and logically. Let's examine these two types of connections.

PHYSICAL TOPOLOGY VERSUS LOGICAL TOPOLOGY

Figure 23.2 shows the details of the connections inside the network cloud. Four intermediate network nodes (W, X, Y, and Z) are responsible for relaying data between each of the three machines A, B, and C. Five connections exist between the four intermediate nodes. This is the *physical topology* of the network. We will cover the details of each type of physical topology in the next few sections.

The *logical topology* has to do with the path a packet of data takes through the network. For example, from machine A to machine C there are three different paths. These paths are as follows:

1. Link 3
2. Link 1 to link 2
3. Link 1 to link 4 to link 5

Clearly, data sent on link 3 will get from machine A to machine C in the shortest time (assuming all links are identical in speed), while the third path (links 1, 4, and 5) takes the longest. Due to the nature of the network, we cannot guarantee that link 3 is the one that is always used. It may become too busy; its noise level may unexpectedly increase, making it unusable; or a tree might have fallen on the fiber carrying link 3's data. Thus, packets of data may take different routes through the network, arriving *out of order* at their destination. It is the job of the network software protocol to properly reassemble the packets into the correct sequence. Exercise 25 provides the details of the many network protocols in use.

FIGURE 23.2   Physical network topology

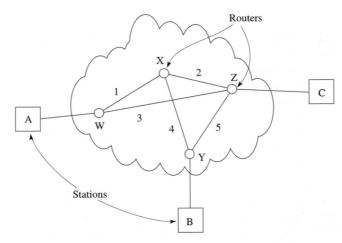

When a large amount of data must be sent between machines on a network, it is possible to set up a *virtual circuit* between the machines. A virtual circuit is a prearranged path through the network that *all* packets will travel for a particular session between machines. For example, for reasons based on the current state of the network, a virtual circuit is established between machines B and C through links 4 and 2. All packets exchanged between B and C will take links 4 and 2.

Another type of virtual connection is called a *virtual private network (VPN)*. A VPN uses public network connections (such as the Internet or the telephone system) to establish private communication by encrypting the data. The data travels in *tunnels*, logical connections between the nodes of the VPN. Virtual private networks are discussed in more detail in Exercise 27.

FULLY CONNECTED NETWORKS

Figure 23.3 shows four basic types of network connections. The fully connected network in Figure 23.3(a) is the most expensive to build, for each node has a link (communication channel) to every other node. Just adding one more node (for a total of six nodes) brings the number of links to 15. Seven fully connected nodes require 21 links. In general, the number of links (L) required in a fully connected network of N nodes is

$$L = \frac{N(N-1)}{2}$$

Table 23.1 shows the number of links for several values of N.

It is easy to see that the number of links required in a fully connected network quickly becomes unmanageable. Even so, fully connected networks provide quick communication between nodes, for there is a one-link path between every two nodes in the network. Even if a link goes down, the worst-case path only becomes two links long. So, fully connected networks are very reliable and somewhat secure, since many links have to fail for two nodes to lose contact.

STAR NETWORKS

Figure 23.3(b) shows a star network. All nodes connect to a central *hub* (also called a *concentrator*). For small networks, only a single hub is required. Four, five, eight, and even

FIGURE 23.3 Network topologies

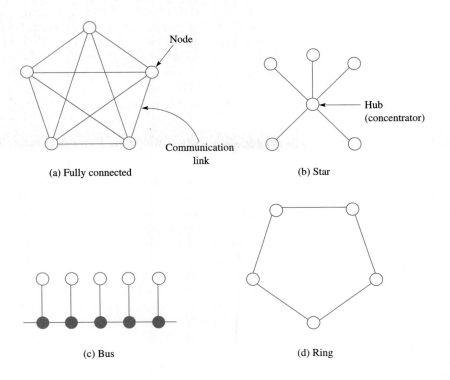

(a) Fully connected

(b) Star

(c) Bus

(d) Ring

TABLE 23.1   Number of links in a fully connected network

N	L
2	1
3	3
4	6
5	10
6	15
10	45
100	4950

16 or more connections are available on a single hub. Large star networks require multiple hubs, which increase the hardware and cabling costs. On the other hand, if a node on the network fails, the hub will isolate it so that the other nodes are not affected. Entire groups of nodes (machines) can be isolated at a time by disconnecting their hub. This helps narrow down the source of a network problem during troubleshooting.

BUS NETWORKS

A bus network uses a single common communication link that all nodes tap into. Figure 23.3(c) shows the bus connection; 10base2 and 10base5 Ethernet uses coaxial cable as the common connection. All nodes on the common bus compete with each other for possession, broadcasting their data when they detect the bus is idle. If two or more nodes transmit data at the same time, a *collision* occurs, requiring each node to stop and wait before retransmitting. This technique of sharing a common bus is known as *CSMA/CD* (Carrier Sense Multiple Access with Collision Detection) and is the basis of the Ethernet communication system.

Wiring a bus network is not too difficult. Suitable lengths of coaxial cable, properly terminated with BNC connectors on each end, are daisy-chained via T-connectors into one long *segment* of nodes. Each T-connector plugs into a network interface card. The problem with the daisy-chain bus connection is that bad crimps on the BNC connectors, poor connections in the T-connectors, or just an improperly terminated cable (no 50-ohm terminating resistor) can cause intermittent or excessive collisions; these problems can be difficult to find as well. A special piece of equipment called a *time domain reflectometer* (TDR) is used to send a pulse down the coaxial cable and determine where the fault is (by displaying a response curve for the cable).

In terms of convenience, the bus network is relatively easy to set up, with no significant hardware costs (no hubs are required). With 185 meters of cable possible in a segment (for 10base2 Ethernet), a large number of nodes can be wired together. Individual segments can be connected together with *repeaters* (more on this in Exercise 24).

RING NETWORKS

The last major network topology is the ring. As Figure 23.3(d) shows, each node in a ring is connected to exactly two other nodes. Data circulates in the ring, traveling through many intermediate nodes if necessary to get to its destination. Like the star connection, the number of links is the same as the number of nodes. The difference is that there is no central hub concentrating the nodes. Data sent between nodes will typically require paths of at least two links. If a link fails, the worst-case scenario requires a message to travel completely around the ring, through every link (except the one that failed). The increase in time required to relay messages around the bad link may be intolerable for some applications. The star network does not have this problem. If a link fails, only the node on that link is out of service.

Token-ring networks, although logically viewed as rings, are connected using central multistation access units (MAUs). The MAU provides a physical star connection.

HYBRID NETWORKS

A hybrid network combines the components of two or more network topologies. As Figure 23.4 indicates, two star networks are connected (with three additional nodes) via a bus. This is a common way to implement Ethernet, with coax running between classrooms or laboratories, and hubs in each room to form small subnetworks. Putting together a hybrid network takes careful planning, for there are various rules that dictate how the individual components may be connected and used. For example, when connecting Ethernet segments, a maximum of four repeaters may be used with five segments. Furthermore, if a 4-Mbit/second token-ring network is interfaced with a 10-Mbit/second Ethernet, there are performance issues that must be taken into consideration also (since any Ethernet traffic is slowed down to 4 Mbit/second on the token ring side). In addition, the overall organization of the hybrid network, from a logical viewpoint, must be planned out as well. This will become clearer in the next section.

NETWORK HIERARCHY

The machines networked together in Figure 23.4 are not organized into a **hierarchy** (a layered organization). Data transmitted by any machine is broadcast through both hubs. Everyone connected to the network sees the same data and competes with everyone else for bandwidth.

The same network is illustrated in Figure 23.5, except for a few hardware modifications. Both hubs have been replaced by 10-Mbit/second switches, which act like hubs except they

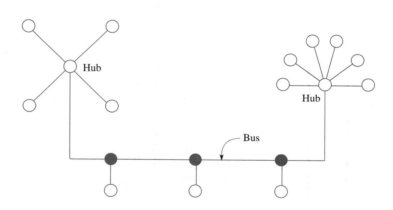

FIGURE 23.4   Hybrid network

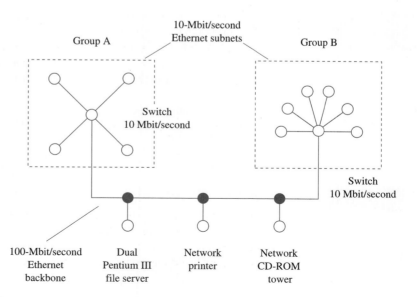

**FIGURE 23.5   Hybrid network
with hierarchy**

only forward data selectively. For example, any machine in group A can send data to any other machine in group A, through the switch, without the data being broadcast on the 100-Mbit/second backbone. The same is true for machines in group B. Their traffic is isolated from the important, high-speed 100-Mbit/second backbone.

Now, when a machine from group A requests service from the file server, which in turn accesses the CD-ROM tower, the hierarchy of the network allows the file server to communicate with the CD-ROM tower at a high speed, while sending data back to the machine in group A only when it needs to. Machines from both groups can communicate with each other over the backbone as well, without significantly interfering with the other backbone traffic (since all machines in each group operate at 10 Mbit/second). The switches enforce hierarchy by learning where data packets should be forwarded (based on their destination addresses). Since the backbone is the main communication link in the hybrid network, its 100-Mbit/second speed allows each network component to communicate at its best speed.

SUBNETS

In Exercise 25 you will cover the structure and use of *IP addresses*, 32-bit numbers used to locate and identify nodes on the Internet. For example, the IP of one computer in a classroom might be 192.203.131.137. Other computers on the same subnet may begin with 192.203.131 but have different numbers at the end, as in 192.203.131.130. This type of subnet is called a *class C subnet*.

NETWORK ACCESS POINTS

In the beginning of the computer networking age, an experimental network was created by connections between four major data processing facilities, located in Chicago, New York, San Francisco, and Washington, DC. Today, facilities all across the country called ***network access points (NAPs)*** provide access to national and global network traffic. Many companies have installed their own independent communication networks that connect to one or more NAPs, or even act as NAPs themselves. Figure 23.6 shows the Nap.Net national backbone, a fully connected DS3 (44.736 Mbits/second) ATM network.

Companies that connect to a NAP enter into ***peering agreements*** with each other that allow them to exchange traffic. When the traffic is Internet-based, the connection is called a ***POP***, or ***point-of-presence***.

TROUBLESHOOTING TECHNIQUES

It is a fact of life that we must worry about intentional harm being done to our network. In terms of security and reliability, we must concern ourselves with what is required to *partition* our network, breaking it up into at least two pieces that cannot communicate with each other. Figure 23.7 shows how bus, star, ring, and fully connected networks are partitioned. Note that the star network is completely partitioned (all nodes isolated) if the central hub fails.

When troubleshooting a network, knowledge of its topology, both physical and logical, is essential to proper partitioning, so that testing and repairing can proceed smoothly.

SUMMARY

In this exercise we discovered that

- Computer networks contain both physical and logical topology.
- Fully connected, star, ring, and bus topologies have their own advantages and disadvantages.
- A network may be organized into a hierarchy.

FIGURE 23.6 RWA Software national backbone

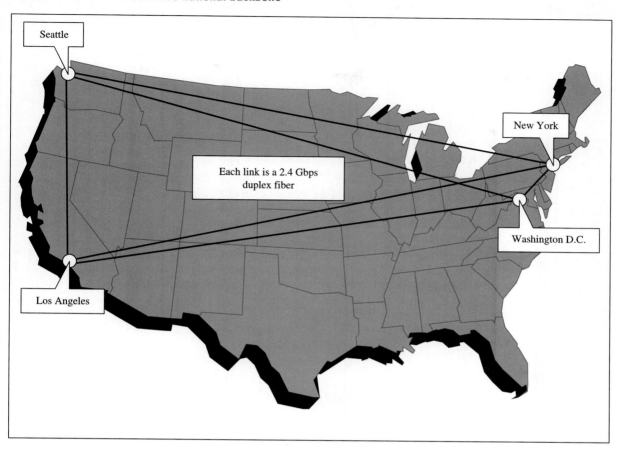

Seattle

New York

Each link is a 2.4 Gbps
duplex fiber

Washington D.C.

Los Angeles

FIGURE 23.7 Partitioning networks

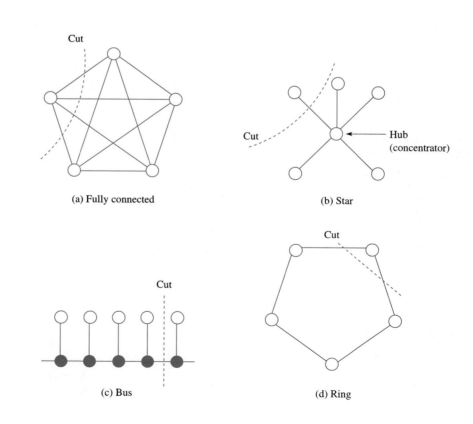

Cut

(a) Fully connected

Cut

Cut

Hub
(concentrator)

(b) Star

Cut

(c) Bus

Cut

(d) Ring

This self-test is designed to help you check your understanding of the background information presented in this exercise.

True/False

Answer *true* or *false*.

1. Packets never exit once they enter a network cloud.
2. Fully connected networks require more links than star networks.
3. 10base2 Ethernet allows for 2000-meter segments.
4. A collision is required to exchange data over an Ethernet cable.
5. A hub is used to enforce network hierarchy.
6. Data travels in tunnels on a VPN.
7. Packets may arrive out of order at their destination.

Multiple Choice

Select the best answer.

8. A fully connected network of 10 nodes requires
    a. 10 links.
    b. 45 links.
    c. 90 links.
9. Star networks
    a. Require hubs.
    b. Are limited to 16 nodes.
    c. Both a and b.
10. A portion of an Ethernet bus is called a(n)
    a. CSMA module.
    b. Etherpath.
    c. Segment.
11. Assuming any single link fails, the worst-case path through the ring network of Figure 23.3(d) is
    a. Three links.
    b. Four links.
    c. Five links.
12. What is required to partition a fully connected network of six nodes?
    a. Cut one link.
    b. Cut one link at each node.
    c. Cut all links at a single node.
13. Which network of 20 nodes is the most expensive to build?
    a. Bus.
    b. Star.
    c. Fully connected.
14. Which network is the hardest to partition?
    a. Bus.
    b. Star.
    c. Fully connected.

Matching

Match a description of the topology property on the right with each item on the left.

15. Bus
16. Star
17. Ring
18. Fully connected

a. All paths are one link long.
b. Each node has exactly two links.
c. All nodes share the same link.
d. All nodes share a central node.

Completion

Fill in the blank or blanks with the best answers.

19. A prearranged connection between computers is called a(n) _____ circuit.
20. Hubs are also called _____.
21. CSMA/CD stands for _____ _____ _____ _____ with _____ _____.
22. Breaking a link in a network may _____ it.
23. A network that uses encryption to send secure data over public communication links is called a _____ _____ network.
24. Companies connected to a NAP utilize _____ agreements.

FAMILIARIZATION ACTIVITY

1. Sketch all the different network configurations possible using six communication links.
2. Search the Web for FDDI (fiber distributed data interconnect). What type of network topology does FDDI use? What is FDDI used for?

QUESTIONS/ACTIVITIES

1. What are all the paths from machine B to machine C in the network of Figure 23.2?
2. Draw a fully connected eight-node network. How many links are there?
3. Repeat step 2 for a star network. Assume each hub has four connections available.
4. Find a laboratory or classroom that is networked. Make a diagram of the network, showing the various nodes and what they actually represent (computers, printers, etc.).

REVIEW QUIZ

Under the supervision of your instructor

1. Describe the difference between physical topology and logical topology.
2. Sketch the physical topologies of bus, star, ring, and fully connected networks.
3. Explain what is meant by network hierarchy.

24 Networking Hardware

INTRODUCTION

Don, the senior technician, had his hands full of UTP cable. He had carefully stripped off 1 inch of outer insulation and was gingerly holding all eight twisted-pair conductors in a neat row by pinching them between his thumb and forefinger. In his other hand he held a clear plastic RJ-45 crimp-on connector and was slowly pushing all eight wires into their thin grooves in the connector.

Joe walked up and slapped Don on the back in a friendly way. "What are you doing, Don?" he asked.

Don let out a long-suffering sigh as all eight wires popped out of the connector. "Starting over again, Joe," he replied.

PERFORMANCE OBJECTIVES

Upon completion of this exercise, you will be able to

1. List and describe the basic networking hardware components.
2. Explain the differences in 10base2 Ethernet and 10baseT Ethernet.
3. Compare the advantages of fiber over copper wire.

KEY TERMS

10base2, 10base5, 10baseT, 100baseT	NIC
Thinwire	Transceiver
Thickwire	Stackable
Backbone	Bridge
Network diameter	Switch
UTP	Store and forward
RJ-45	Router
Fiber	Cable modem
Mode	Cable tester

BACKGROUND INFORMATION

In this exercise we will examine many of the different hardware components involved in networking. You are encouraged to look inside your machine to view your network interface card, around your lab to locate hubs and trace cables, and around your campus (especially the computer center) to see what other exotic hardware you can find.

FIGURE 24.1   Coaxial cable construction

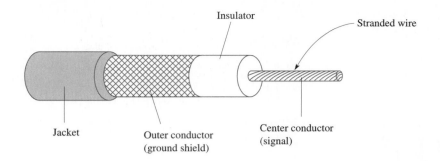

ETHERNET CABLING

We begin our hardware presentation with Ethernet cabling. Ethernet cables come in three main varieties. These are

1. RG-58 coaxial cable, used for 10base2 operation (also called ***thinwire***)
2. RG-11 coaxial cable, use for 10base5 operation (also called ***thickwire***)
3. Unshielded twisted pair (UTP), used for 10baseT and 100baseT operation

There are other, specialized cables, including fiber (10baseFL), that are used as well.

RG-58 cable is typically used for wiring laboratories and offices, or other small groups of computers. Figure 24.1 shows the construction of a coaxial cable.

The maximum length of a thinwire Ethernet segment is 185 meters (606 feet), which is due to the nature of the CSMA/CD method of operation, the cable attenuation, and the speed at which signals propagate inside the coax. The length is limited to guarantee that collisions are detected when machines that are far apart transmit at the same time. BNC connectors are used to terminate each end of the cable. Figure 24.2 shows several different cables and connectors, including BNC T-connectors (one containing a terminating resistor).

When many machines are connected to the same Ethernet segment, a daisy-chain approach is used, as shown in Figure 24.3(a). The BNC T-connector allows the network interface card (NIC) to tap into the coaxial cable and the coax to pass through the machine to the next machine. The last machines on each end of the cable (or simply the cable ends

FIGURE 24.2   (a) Assorted connectors and cables (photograph by John T. Butchko)

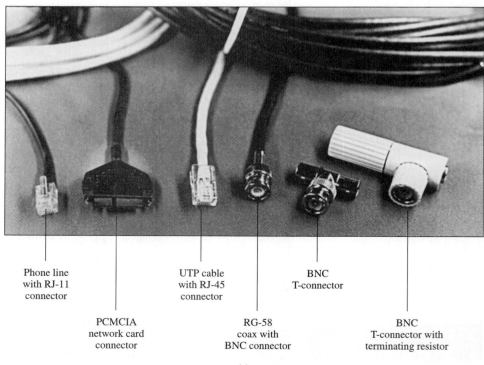

(a)

FIGURE 24.2 *(continued)*
(b) Fiber optic cable with ST connectors

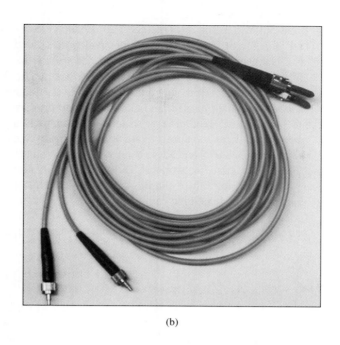

(b)

FIGURE 24.3   10base2 Ethernet wiring

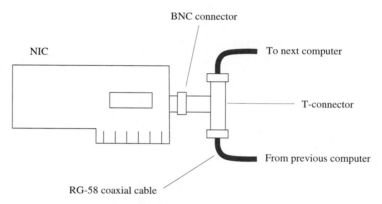

BNC connector

NIC

To next computer

T-connector

From previous computer

RG-58 coaxial cable

(a) Daisy-chain connection

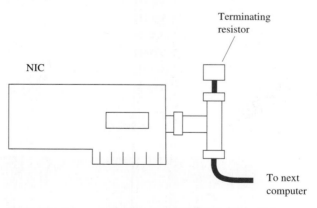

Terminating resistor

NIC

To next computer

(b) Terminating connection (required at each end of the cable)

themselves) must use a terminating resistor (50 ohms) to eliminate collision-causing reflections in the cable. This connection is illustrated in Figure 24.3(b).

RG-11 coaxial cable is used as a *backbone* cable, distributing Ethernet signals throughout a building, an office complex, or other large installation. RG-11 is thicker and more sturdy than RG-58 coax. Thickwire Ethernet segments may be up to 500 meters (1640 feet) long, with a maximum of five segments connected by repeaters. This gives a total distance of five times 500 meters, or 2500 meters. This is called the ***network diameter***. The network diameter is different for other cable types and signal speeds. RG-11 cable is typically orange, with black rings around the cable every 2.5 meters to allow taps into the cable. The taps, called *vampire taps*, are used by transceivers that transfer Ethernet data to and from the cable.

UTP cable, used with hubs and other 10/100baseT equipment, uses twisted pairs of wires to reduce noise and crosstalk and allow higher-speed data rates (100-Mbit/second category 5 UTP for Fast Ethernet). The twists tend to cause the small magnetic fields generated by currents in the wires to cancel, reducing noise on the signals. UTP cable length is limited to 100 meters (328 feet) and RJ-45 connectors are used for termination. The network diameter for UTP-based 10baseT networks is 500 meters. For 100baseT the diameter drops to 200 meters.

The structure of the 8-pin RJ-45 connector is shown in Figure 24.4. Its modular format is similar to the telephone companies' 6-pin RJ-11 connector.

Table 24.1 shows the wire color combinations used in UTP cabling. Note that only two pairs are required for 10baseT operation, one pair for transmit and the other for receive.

FIGURE 24.4   RJ-45 (10baseT) connector

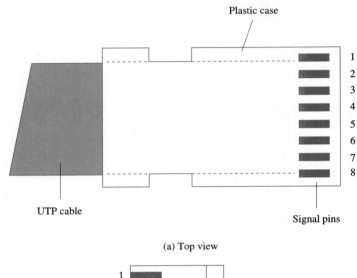

(a) Top view

(b) Side view

TABLE 24.1   RJ-45 pin assignments (568B standard)

Pin	Color	Function	Used for 10baseT
1	White/Orange	T2	✔
2	Orange/White	R2	✔
3	White/Green	T3	✔
4	Blue/White	R1	
5	White/Blue	T1	
6	Green/White	R3	✔
7	White/Brown	T4	
8	Brown/White	R4	

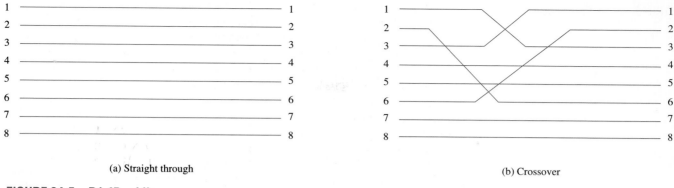

(a) Straight through

(b) Crossover

FIGURE 24.5   RJ-45 cabling

FIGURE 24.6   10baseT Ethernet wiring

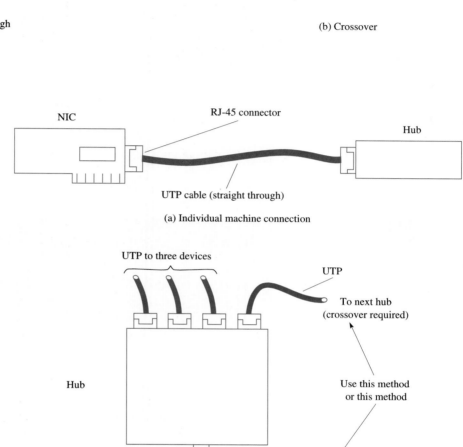

(a) Individual machine connection

(b) Connecting the hub

UTP cables are wired as straight-through or crossover cables. Figure 24.5 shows the wiring diagrams for each type of cable. Straight-through cables typically connect the computer's network interface card to a port on the hub. Crossover cables are used for NIC-to-NIC communication, and for hub-to-hub connections when no crossover port is available. Sample wiring configurations are shown in Figure 24.6.

Fiber optic cable relies on pulses of light to carry information. Figure 24.7(a) shows the basic construction of an optical *fiber*. Two types of plastic or glass with different physical properties are used (the inner *core* and the outer *cladding*) to allow a beam of light to reflect off the boundary between the core and the cladding. This is illustrated in Figure 24.7(b). Some fiber optic cables allow many different paths (or *modes*), others allow one single mode. These are called multimode and single-mode fibers, respectively. A popular multimode fiber has core/cladding dimensions of 62.5/125 nanometers.

FIGURE 24.7   Fiber optic cable

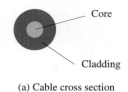

(a) Cable cross section

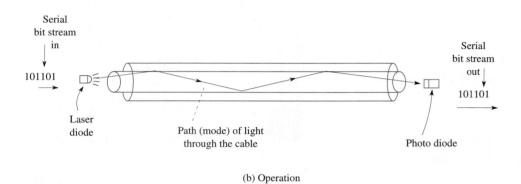

(b) Operation

TABLE 24.2   Comparing cabling systems

	10base5	10base2	10baseT	10baseFL
Cable Type	RG-11	RG-58	UTP	Fiber
Maximum Segment Length	500 m (1640 ft)	185 m (606 ft)	100 m (328 ft)	2000 m (6560 ft)
Max Nodes	100	30	2	2

High-speed laser diodes generate short bursts of light of a particular wavelength (typically 850 nanometers). These bursts travel down the fiber and affect a photodiode on the receiving end of the fiber. Since this light-based communication is one way, two fibers are used to make a two-way connection.

Fiber does not suffer from the problems found in copper wires, which are sensitive to electromagnetic interference, exhibit high signal loss, and are limited in bandwidth. The original 10baseFL fiber standard specifies a cable length of 2000 meters, significantly longer than the 10base5 segment length of 500 meters. In addition, fiber supports data rates in the gigabit range, providing the fastest communication method available for networking.

Table 24.2 compares each cabling system.

THE NIC

The network interface card (*NIC*) is the interface between the PC (or other networked device) and the physical network connection. In Ethernet systems, the NIC connects to a segment of coaxial or UTP cable (fiber NICs are available but not very common yet). As with any other type of adapter card, NICs come in ISA, PCMCIA, and PCI bus varieties. Figure 24.8 shows a typical Ethernet NIC. Since the NIC contains both BNC and RJ-45 connectors, it is called a *combo* card. The NE2000 Compatible stamp indicates that the NIC supports a widely accepted group of protocols.

The NIC in Figure 24.8 is an ISA adapter card. PCI networking cards are available in both non-bus-mastering and bus-mastering varieties. Bus-mastering means that the NIC

FIGURE 24.8   Network interface card (Ethernet) (photograph by John T. Butchko)

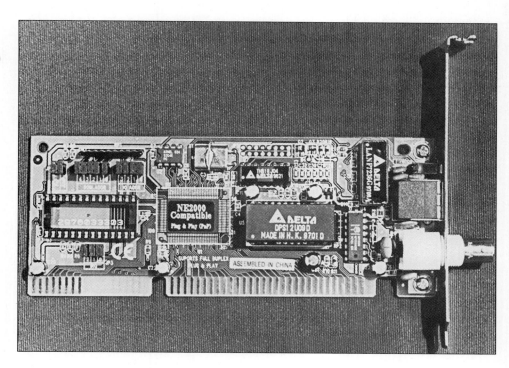

FIGURE 24.9   PCMCIA Ethernet card with cable (photograph by John T. Butchko)

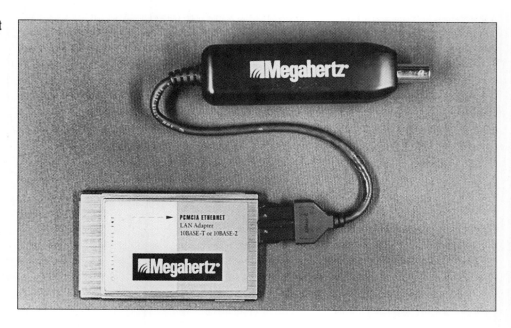

can take over the system bus and access memory directly. Figure 24.9 shows a PCMCIA Ethernet NIC and cable.

The NIC is responsible for operations that take place in the Physical layer of the OSI network model. It is only concerned with sending and receiving 0s and 1s, using the IEEE 802.3 Ethernet standard (or IEEE 802.5 token ring).

Windows 95/98 identifies the installed NIC in Network properties. Figure 24.10 shows the 3Com NIC entry. Note that the NetBEUI and TCP/IP protocols are *bound* to the 3Com adapter. To use a protocol with a NIC you must bind the protocol to the adapter card. This is typically done automatically when the protocol is added.

Double-clicking the 3Com 3C508 entry brings up its Properties window, which is shown in Figure 24.11. The indicated driver type is NDIS, Microsoft's network driver interface specification, which allows multiple protocols to use a single NIC. An ODI (open data-link

FIGURE 24.10   3Com 3C508 NIC entry

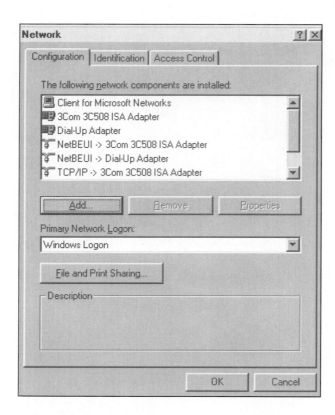

FIGURE 24.11   NIC Properties window

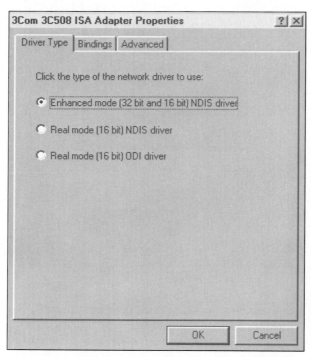

interface, developed by Novell) driver performs the same function for multiple protocol stacks used with the NetWare network operating system. Figure 24.12 shows the NDIS/ODI interface. Both are designed to *decouple* the protocols from the NIC. The protocols do not require any specific information about the NIC. They use the NDIS/ODI drivers to perform network operations, with the drivers responsible for their specific hardware.

It is important to mention that all NICs are manufactured with a unique 48-bit MAC address (for example, 00-60-97-2B-E6-0F). The first six digits (00-60-97) indicate the

**FIGURE 24.12   NDIS/ODI
interface**

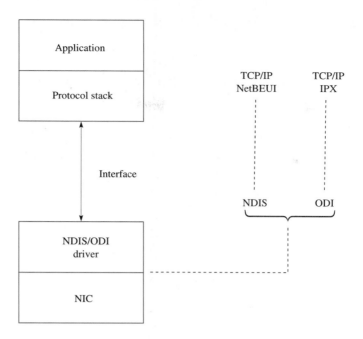

**FIGURE 24.13   Viewing the
NIC's MAC address**

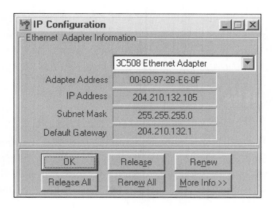

manufacturer (3Com). You can view your NIC's MAC address using the WINIPCFG utility (from the Run menu). Figure 24.13 shows the initial WINIPCFG screen.

TOKEN RING

The IEEE 802.5 standard describes the token-ring networking system. IBM developed the initial 4-Mbit/second standard in the mid-1980s, with 16-Mbit/second token ring also available.

Token-ring networks use a multistation access unit (MAU), which establishes a logical ring connection even though the physical connections to the MAU resemble a star. Figure 24.14 shows the basic operation of the MAU. Computers in the ring circulate a software *token*. The machine holding the token is allowed to transmit data to the next machine on the ring (even if the data is not meant for that machine). One machine (typically the first to boot and connect) is identified as the *active monitor* and keeps track of all token-ring operations. If the active monitor detects that a machine has gone down (or been shut off), the connection to that machine is bypassed. If the active monitor itself goes down, the other machines vote to elect a new active monitor. Thus, we see that token-ring networks are *self-healing*, unlike Ethernet, which is only capable of resolving collisions.

FIGURE 24.14 Token-ring network

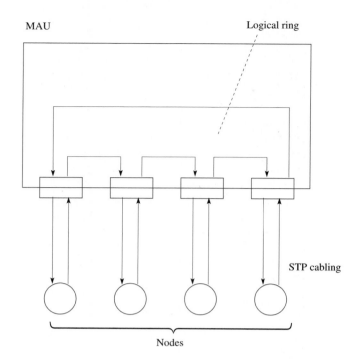

MAU

Logical ring

STP cabling

Nodes

Token-ring connections are made using STP (shielded twisted pair) cables. STP cable contains a metal shield around the twisted pairs that provides isolation from external crosstalk and noise. In general, do not substitute STP for UTP.

REPEATERS

A *repeater* connects two network segments and broadcasts packets between them. Since signal loss is a factor in the maximum length of a segment, a repeater is used to amplify the signal and extend the usable length. A common Ethernet rule is that no more than four repeaters may be used to join segments together. This is a physical limitation designed to keep collision detection working properly. Repeaters operate at layer 1 (Physical layer) of the OSI model.

TRANSCEIVERS

A **transceiver** converts from one media type to another. For example, a 10base2-to-fiber transceiver acts like a repeater, except it also interfaces 10base2 coaxial cable with a fiber optic cable. It is common to use more than one media type in an installation, so many different kinds of transceivers are available. Figure 24.15 shows two examples of Ethernet transceivers.

FIGURE 24.15 Transceivers

(a) Thinwire coax to fiber

(b) UTP to AUI

294

HUBS

Hubs, also called *concentrators*, expand one Ethernet connection into many. For example, a four-port hub connects up to four machines (or other network devices) via UTP cables. The hub provides a star connection for the four ports. Many hubs contain a single BNC connector as well to connect the hub to existing 10base2 network wiring. The hub can also be connected via one of its ports. One port is designed to operate in either straight-through or crossover mode, selected by a switch on the hub. Hubs that can connect in this fashion are called *stackable* hubs. Figure 24.16 shows an eight-port stackable Ethernet hub. Port 8 is switch selectable for straight through or crossover (cascade).

A hub is similar to a repeater, except it broadcasts data received by any port to all other ports on the hub. Most hubs contain a small amount of intelligence as well, examining received packets and checking them for integrity. If a bad packet arrives, or the hub determines that a port is unreliable, it will shut down the line until the error condition disappears.

Note that a hub also acts like a repeater. Because of its slight delay when processing a packet, the number of hubs that may be connected in series is also limited. Figure 24.17 shows how several hubs are used to connect five Ethernet segments, within the accepted limits. Since each UTP cable may be as long as 100 meters, the maximum distance between nodes is 500 meters (the network diameter).

BRIDGES/SWITCHES

When a network grows in size, it is often necessary to partition it into smaller groups of nodes to help isolate traffic and improve performance. One way to do this is to use a *bridge*, whose operation is indicated in Figure 24.18. The bridge keeps segment A traffic on the A

FIGURE 24.16   Ethernet hub (photographs by John T. Butchko)

(a) Front view

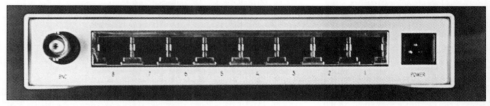

(b) Rear view

FIGURE 24.17   Connecting five segments with hubs

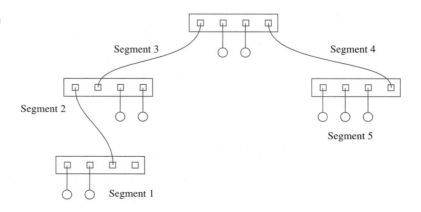

FIGURE 24.18   Operation of a bridge

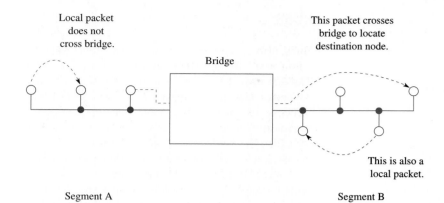

Local packet does not cross bridge.

This packet crosses bridge to locate destination node.

Bridge

This is also a local packet.

Segment A   Segment B

FIGURE 24.19   One configuration in an eight-port N-way switch

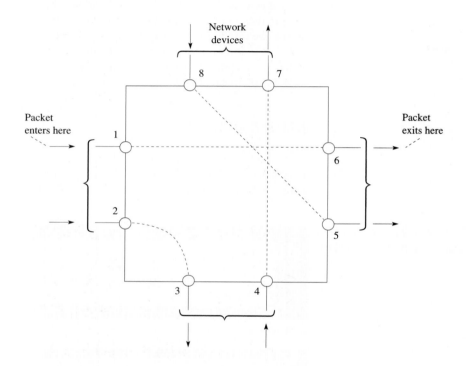

Network devices

Packet enters here

Packet exits here

side and segment B traffic on the B side. Packets from segment A that are meant for a node in segment B will cross the bridge (the bridge will permit the packet to cross). The same is true for packets going from B to A. The bridge learns which packets should cross as it is used.

A **switch** is similar to a bridge, with some important enhancements. First, a switch may have multiple ports, thus directing packets to several different segments, further partitioning and isolating network traffic in a way similar to a router. Figure 24.19 shows an eight-port N-way switch, which can route packets from any input to any output. Some or all of an incoming packet is examined to make the routing decision, depending on the switching method that is used. One common method is called **store and forward**, which stores the received packet before examining it to check for errors before retransmitting. Bad packets are not forwarded.

In addition, a switch typically has auto-sensing 10/100-Mbit/second ports and will adjust the speed of each port accordingly. Furthermore, a *managed* switch supports SNMP for further control over network traffic. Switches operate at layer 2 (Data Link) of the OSI model.

ROUTERS

A **router** is the basic building block of the Internet. Each router connects two or more networks together by providing an interface for each network to which it is connected. Figure 24.20(a) shows a router with an interface for an Ethernet network and a token-ring network. The router examines each packet of information to determine whether the packet must be translated from one network to another, performing a function similar to a bridge. Unlike a bridge, a router can connect networks that use different technologies, addressing methods, media types, frame formats, and speeds. Figures 24.20(b) and 24.20(c) show the top and rear views of the Cisco 1600 router. This router connects two Ethernet networks and has an expansion slot for a wide area network (WAN) connection.

A router is a special-purpose device designed to interconnect networks. For example, three different networks can be connected using two routers, as illustrated in Figure 24.21. If a computer in network A needs to send a packet of information to network C, both routers pass the packets from the source network to the destination network.

Routers maintain routing tables in their memories to store information about the physical connections on the network. The router examines each packet of data, checks the routing

FIGURE 24.20 (a) Router with two interfaces and (b) top view of Cisco 1600 router

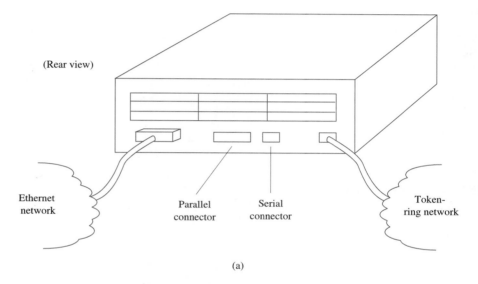

(a)

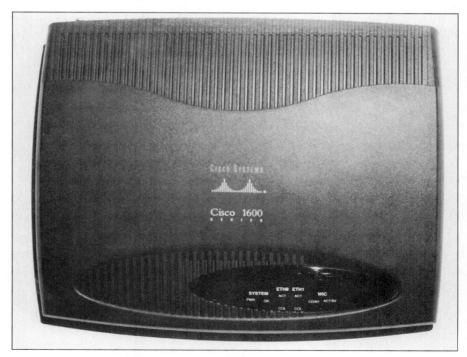

(b)

(continued on the next page)

297

FIGURE 24.20 *(continued)*
(c) Rear view of Cisco 1600 router

(c)

FIGURE 24.21   Two routers connecting three networks

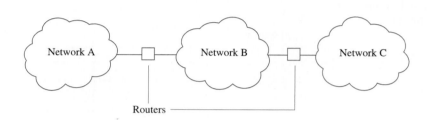

Network A — Network B — Network C

Routers

FIGURE 24.22   Packet routing

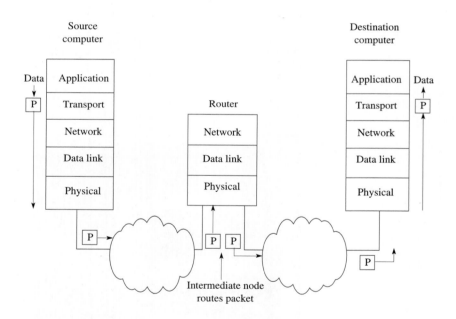

Source computer

Destination computer

Router

Intermediate node routes packet

table, and then forwards the packet if necessary. Every other router in the path (between a source and a destination network) performs a similar procedure, as illustrated in Figure 24.22. Note that a router does not maintain any state information about the packets; it simply moves them along the network. Routers operate at layer 3 (Network) of the OSI model.

CABLE MODEMS

A **cable modem** is a high-speed network device connected to a local cable television provider. The cable television company allocates a pair of channels (one for transmit, one for receive) on the cable system to transmit data. At the *head-end* of the network, located at the cable supplier offices, a traditional Internet service provider (ISP) service is established to service the network clients. The connection from the cable system ISP to the Internet uses traditional telecommunications devices, such as T1 or T3 lines.

The subscribers to the cable modem service use a splitter to create two cable wires. One wire is reconnected to the television and the second is connected to the new cable modem. This is illustrated in Figure 24.23. The cable modem itself requires just a few connections, as shown in Figure 24.24. After all the connections have been made, the power light on the front of the cable modem will be on. The other lights show the cable modem status. Both the cable and PC lights are on when the cable system ISP and the PC network card are set up properly. The test light is normally off, but comes on after a reset or when power is reapplied. Figure 24.25(a) shows the front panel display of a typical cable modem. Figure 24.25(b) shows the Motorola Wave modem.

There is no maintenance for the cable modem subscriber other than providing adequate ventilation and keeping the power applied to the cable modem at all times. The cable system ISP may update the internal software or run tests on them during off-peak hours.

SATELLITE NETWORK SYSTEM

The Hughes Corporation offers a unique solution to low-speed Internet connections. For a few hundred dollars, you can buy their DirecPC Internet satellite networking system. Figure 24.26 shows the basic operation. Internet data comes to your PC via satellite at 400 Kbps. Through your modem and ISP, data goes back to Hughes's network operations center (NOC), where it is uploaded to the Internet. This is an ideal situation for browsing, when you need to receive information fast (if a Web page contains many images) but only send information out (clicking on a new URL to load a new page) occasionally. A low-speed line for transmitted packets is acceptable, unless you are uploading large files to an FTP site or sending e-mail with large attachments. If there is no cable where you live or work (for a cable modem connection), DirecPC may be the answer for you.

FIGURE 24.23   Cable service connections

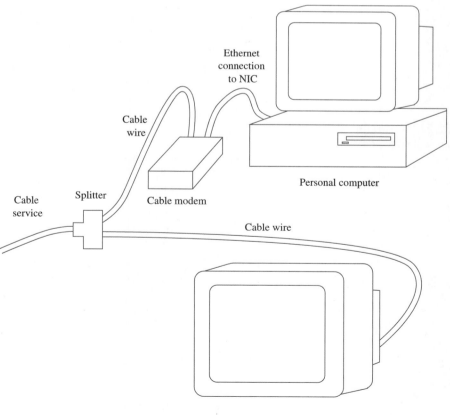

Ethernet connection to NIC

Cable wire

Personal computer

Cable service

Splitter

Cable modem

Cable wire

Television

EXOTIC HARDWARE AND SOFTWARE

We have only examined the basic types of networking hardware in this exercise. Many more exotic (and expensive) networking components are available. For example, instead of using multiple 16-port switches, a single industrial switch with 64 ports or more, including port management, may be used. IP addresses can be assigned to specific ports, ports can be activated/deactivated with software, and the port speed can be controlled.

For networks that must be distributed over a large geographic area (such as a college or industrial campus), line-of-sight infrared lasers can be used to link separate networks together. If fiber is used instead, fiber repeaters may be necessary to obtain the required distance for a link.

FIGURE 24.24   Cable modem connections

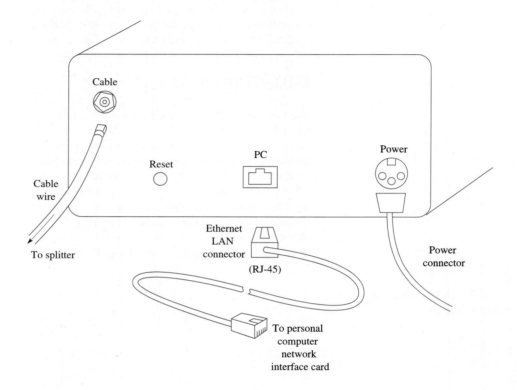

FIGURE 24.25   (a) Cable modem indicator lights

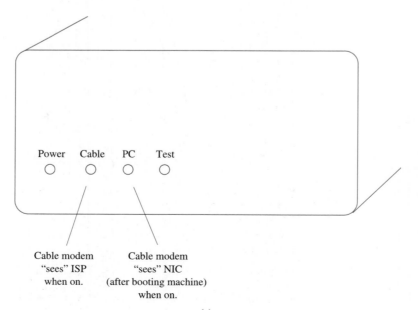

(a)

FIGURE 24.25 *(continued)*
(b) Motorola Wave modem

(b)

FIGURE 24.26 Satellite Internet

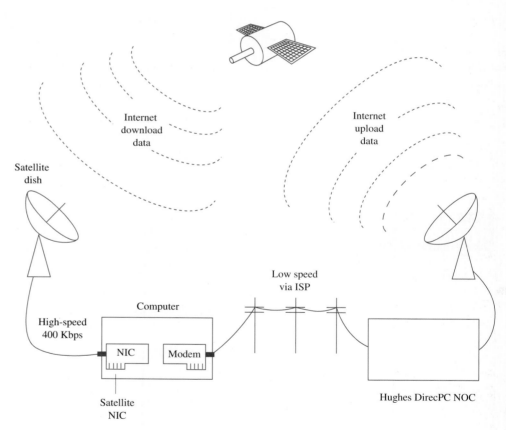

Computer users can now walk around with their laptops, relying on wireless Ethernet technology to maintain the connection with the mobile machine. Other, special-purpose devices called cable-modem routers act like switches but allow multiple machines to share a single Internet connection.

Network software has also evolved. Network management software will display a graphic diagram of your network, with pertinent information (IP/MAC addresses, link speed, activity, and status). One system allows the network technician to be paged when a problem occurs.

FIGURE 24.27 Electronic cable analyzer (photograph by John T. Butchko)

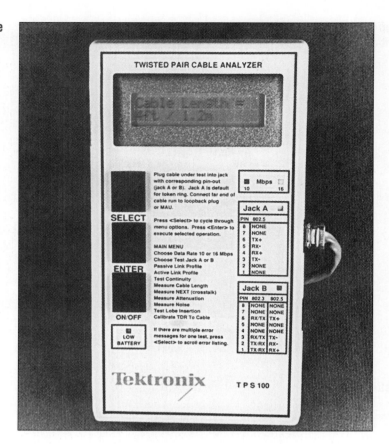

A tour through a local industry will introduce you to a nice variety of sophisticated networking components. Bring a notebook to write down the names and model numbers on the equipment so you can search the Web for more information about them later.

TROUBLESHOOTING TECHNIQUES

One of the most important phases of a network installation is making the cables required for all the nodes. It is much less expensive to buy BNC connectors and spools of coax and make custom-length cables than it is to buy finished cables. This may not be the case for UTP cable, which is harder to terminate (four or eight wires) due to the rigid requirements of the UTP specifications.

A valuable tool to have at your disposal when preparing or checking cables is a **cable tester**. Figure 24.27 shows an electronic cable tester, capable of performing these (and many other) tests on UTP cable:

- Passive and active profiles
- Continuity
- Cable length
- NEXT (near-end crosstalk)
- Attenuation
- Noise

Other, more sophisticated network test equipment, such as the Fluke LANmeter, capture and diagnose network packets of many different protocols, gather statistics (collisions, packets sent), perform standard network operations such as PING and TRACERT, and can transmit packets for troubleshooting purposes. The power of this type of network analyzer is well worth the cost.

In this exercise we discovered that

- Ethernet runs over coax, UTP, and fiber.
- Each cabling system has its own length requirements.
- The NIC operates at the Physical layer.
- MAC addresses can be viewed using WINIPCFG.
- Network hardware proceeds from repeaters and hubs, to switches, to routers.
- Switches learn where to forward packets.
- Routers connect networks together.

SELF-TEST

This self-test is designed to help you check your understanding of the background information presented in this exercise.

True/False

Answer *true* or *false*.

1. Only RG-58 coax is used in Ethernet systems.
2. Vampire taps are used with RG-11 cable.
3. All NICs have the same MAC address (for broadcasting).
4. Token-ring networks use an MAU.
5. Transceivers are only used with fiber and UTP.
6. The only type of fiber is single-mode fiber.
7. The network diameter is the same for all types of networks.

Multiple Choice

Select the best answer.

8. Hubs are also called
   a. Repeaters.
   b. Transceivers.
   c. Concentrators.
9. Hubs act like
   a. Repeaters.
   b. Transceivers.
   c. Routers.
10. A bridge between networks C and D
    a. Broadcasts all packets between C and D.
    b. Broadcasts selected packets between C and D.
    c. Broadcasts all packets from C to D and selected packets from D to C.
11. A router
    a. Connects two different networks.
    b. Ties all hubs together.
    c. Is not used for Internet connections.
12. Cable testers check
    a. Continuity.
    b. Crosstalk.
    c. Frequency response.
    d. a and b only.
13. Fiber optic cable consists of two parts: the core and the
    a. Cladding.
    b. Cloaking.
    c. Jacket.
14. To create a link between two network sites located 800 meters apart, use
    a. Seven hubs and eight 100 meters of UTP cable.
    b. Two 500-meter Thickwire segments and a repeater.
    c. A pair of infrared lasers.

Matching

Match each cable type on the right with each network system on the left.

15. 10base2
16. 10base5
17. 10baseT
18. 10baseFL
19. IEEE 802.5

a. Fiber
b. RG-11
c. RG-58
d. UTP
e. STP

Completion

Fill in the blank or blanks with the best answers.

20. Connecting to RG-11 coaxial cable requires a(n) _____ tap.
21. A NIC that contains both types of connectors is called a(n) _____ card.
22. Protocols must be _____ to a NIC before they can be used.
23. The number of Ethernet segments that may be connected via repeaters is _____.
24. One technique used by switches is store and _____.
25. The path a pulse of light takes through a fiber is called a(n) _____.
26. The maximum distance between two nodes on a network is the network _____.

FAMILIARIZATION ACTIVITY

1. Go to a local computer store and ask to speak with their network technician. If the technician has a service kit that he or she takes on installations or repair assignments, ask whether you may examine the contents. What kind of spare parts are involved? What test equipment and tools are required? Does the technician keep a log book?
2. Find out what types of networks are in use at your location. What devices (hubs, switches, routers) are used to connect them?

QUESTIONS/ACTIVITIES

1. Why do you think fiber has not totally replaced coax and UTP cable?
2. What is the difference between the following:
    a. A repeater and a hub.
    b. A hub and a switch.
    c. A switch and a router.
3. What makes a hub stackable?

REVIEW QUIZ

Under the supervision of your instructor

1. List and describe the basic networking hardware components.
2. Explain the differences in 10base2 Ethernet and 10baseT Ethernet.
3. Compare the advantages of fiber over copper wire.

25 Networking Protocols

INTRODUCTION

Joe Tekk walked by Don, the senior technician, muttering under his breath, "Arp, rarp. Arp, rarp."

Don looked at Joe with a quizzical expression. "What are you saying, Joe?" he asked.

Joe laughed. "Arp, rarp. They're abbreviations for two networking procotols. Arp stands for 'address resolution protocol.' Rarp stands for 'reverse address resolution protocol.' You know, arp, rarp. I like the sound of it."

Don just shook his head, bewildered. "You sound like my three-year-old grandson."

Joe was shocked. "Really? Does he know anything about networking?"

PERFORMANCE OBJECTIVES

Upon completion of this exercise, you will be able to

1. Explain the purpose of each layer of the ISO-OSI model.
2. Discuss the different protocols that make up the TCP/IP suite.
3. Describe the relationship between IP addresses and MAC addresses.

KEY TERMS

ISO-OSI network	MAC address
NetBEUI	ARP
IPX/SPX	RARP
TCP/IP	ICMP
RFC	FTP
IP datagram	SMTP
Best effort delivery	Telnet
TCP	SLIP
Connection	PPP
Port	SNMP
Socket	RIP
UDP	OSPF
DNS	NLSP
IP address	IGRP

Whenever you use a network to transfer data, an entire set of *protocols* is used to set up and maintain reliable data transfer between the two network stations. These protocols are used for many different purposes. Some report errors, some contain control information, others carry data meant for applications. One widely accepted standard that defines the layering of the network (and thus a division of protocols) is the ***ISO-OSI network model***, which is described in Table 25.1.

The ISO-OSI model breaks network communication into seven layers, each with its own responsibility for one part of the communication system. As indicated, the Physical layer is responsible only for sending and receiving digital data, nothing else. Any errors that show up in the data are handled by the next higher layer, Data Link. As we will see in this exercise, each layer has its own set of protocols.

THE NETBEUI PROTOCOL

The Network BIOS Extended User Interface (NetBEUI) protocol is the backbone of Windows for Workgroups, Windows 95/98, and Windows NT networking. File and printer sharing between these network operating systems is accomplished through the use of NetBEUI.

NetBEUI provides the means to gather information about the Network Neighborhood. Table 25.2 shows the advantages and disadvantages of the NetBEUI protocol. One of the main disadvantages with NetBEUI is that it is a nonroutable protocol. This means that a NetBEUI message cannot be routed across two different networks. It was designed to support small networks (up to 200 nodes) and becomes inefficient in larger installations.

THE IPX/SPX PROTOCOLS

The Internetworking Packet Exchange/Sequenced Packet Exchange (IPX/SPX) protocols were first developed by Novell for use with their NetWare network operating system. Windows supports IPX/SPX and thus connects easily with a NetWare network. Figure 25.1

TABLE 25.1 ISO-OSI network model

ISO-OSI Layer	Operation (purpose)
7 = Application	Use network services via an established protocol (TCP/IP, NetBEUI).
6 = Presentation	Format data for proper display and interpretation.
5 = Session	Establish, maintain, and teardown session between both networked computers.
4 = Transport	Break application data into network-sized packets.
3 = Network	Handle network addressing.
2 = Data Link	Flow control, reliable transfer of data.
1 = Physical	All hardware required to make the connection (NIC, cabling, etc.) and transmit/receive 0s and 1s.

TABLE 25.2 Advantages and disadvantages of NetBEUI

Advantages	Disadvantages
Easy to implement	Not routable
Good performance	Few support tools
Low memory requirements	Proprietary
Self-tuning efficiency	

FIGURE 25.1 NetWare
protocol suite

ISO-OSI Layer	Protocols		
7 = Application	NCP	SAP	RIP
6 = Presentation			
5 = Session			
4 = Transport	SPX		
3 = Network	IPX		
2 = Data Link	LSL 802.2 MLID		
1 = Physical	802.3 Ethernet 802.5 Token ring		

shows the relationship between the ISO-OSI model and the NetWare protocol suite. The different protocols in NetWare's model are defined as follows:

- NCP: NetWare Core Protocol
- SAP: Service Advertising Protocol
- RIP: Routing Information Protocol
- SPX: Sequenced Packet Exchange
- IPX: Internetworking Packet Exchange Protocol
- LSL: Link Support Layer
- MLID: Multiple Link Interface Driver

Although the suite of NetWare protocols is used in many different applications, a different protocol suite has gained much wider acceptance, largely due to its use on the World Wide Web. This is the TCP/IP protocol suite. Let's now examine its features.

THE TCP/IP PROTOCOL SUITE

The TCP/IP protocol is unquestionably one of the most popular networking protocols ever developed. TCP/IP has been used since the 1960s as a method to connect large mainframe computers together to share information among the research community and the Department of Defense. Now TCP/IP is used to support the largest computer network, the Internet. Most manufacturers now incorporate TCP/IP into their operating systems, allowing all types of computers to communicate with each other. Figure 25.2 shows the relationship between the ISO-OSI seven-layer networking model and the TCP/IP networking model. Since the protocols indicated in Figure 25.2 are so common, they will be explained in the following sections.

RFCs

The Network Information Center, or NIC, located at www.internic.net contains information about many different aspects of the Internet. One of the most important items stored at the Internic are the Request for Comments documents, or RFCs. These documents describe how each of the protocols contained in TCP/IP are implemented. Refer to Table 25.3 for a list of RFCs associated with some of the most popular TCP/IP protocols. You are encouraged to visit the Network Information Center to become familiar with all the information and services offered, such as the RFCs.

FIGURE 25.2   TCP/IP protocol suite

ISO-OSI Layer	TCP/IP Protocols	
7 = Application	Telnet	
6 = Presentation	FTP	SNMP
5 = Session	SMTP	DNS
4 = Transport	TCP	UDP
3 = Network	IP	
2 = Data Link	LLC 802.2 ------------------------------------ MAC	
1 = Physical	802.3 Ethernet 802.5 Token ring	

TABLE 25.3   Several important TCP/IP RFCs

Protocol	RFC	Name
Telnet	854	Remote Terminal Protocol
FTP	959	File Transfer Protocol
SMTP	821	Simple Mail Transfer Protocol
SNMP	1098	Simple Network Management Protocol
DNS	1034	Domain Name System
TCP	793	Transport Control Protocol
UDP	768	User Datagram Protocol
ARP	826	Address Resolution Protocol
RARP	903	Reverse Address Resolution Protocol
ICMP	792	Internet Control Message Protocol
BOOTP	951	Bootstrap Protocol
IP	791	Internet Protocol

IP

The Internet Protocol (IP) is the base layer of the TCP/IP protocol suite. All TCP/IP data is packaged in units called *IP datagrams*. All other TCP/IP protocols are encapsulated inside an IP datagram for delivery on the network. IP datagrams are eventually encapsulated inside a particular hardware frame, such as Ethernet or token ring. In general, the IP is considered to be unreliable because there is no guarantee the datagram will reach its destination. IP provides what is called *best effort delivery*. Usually, when an IP datagram runs into trouble on the network, it is simply discarded. An error contained in an ICMP message may or may not be returned to the sender. Figure 25.3 illustrates an example of how an ICMP message is encapsulated in an IP datagram. The upper layers of the IP such as TCP and UDP provide the required reliability.

IP datagrams are routed on the network using an IP address. The IP address consists of a 32-bit number. The IP address is divided into four sections, each containing 8 bits. Each section may take on a decimal value between 0 and 255. These four 8-bit groups are called

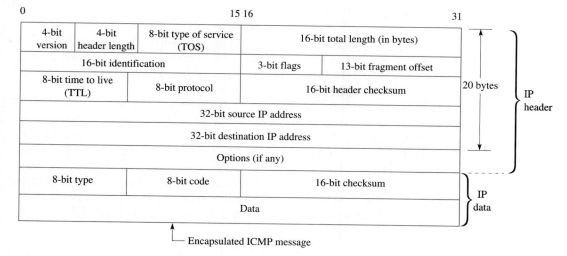

FIGURE 25.3   IP encapsulated message

octets and are separated by periods. This is called *dotted-decimal notation*. An example of dotted-decimal notation is shown in Figure 25.4 illustrating a class C Internet address.

The Internet addresses are classified into five address classes. These addresses are shown in Figure 25.5. Each address class consists of a network ID and a host ID.

The network ID in a class A address contains a maximum number of 128 possible networks and more than 16 million hosts. A class B address allocates more bits in the IP address to the network ID and fewer bits to the host ID, creating a possible 16,384 networks and 65,536 hosts. The class C address provides more than 2 million networks, each with a possible 256 hosts. Class D addresses are reserved for multicast data, and class E addresses are reserved.

Some of the IP addresses are reserved for special functions. For example, the class A address 127 is reserved for loopback testing, allowing a local method to test functionality of the TCP/IP software and applications. Other addresses are used as masks to identify each network type, such as 255.255.255.0 to identify a class C address.

FIGURE 25.4   Class C network IP

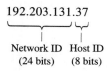

FIGURE 25.5   IP address classes

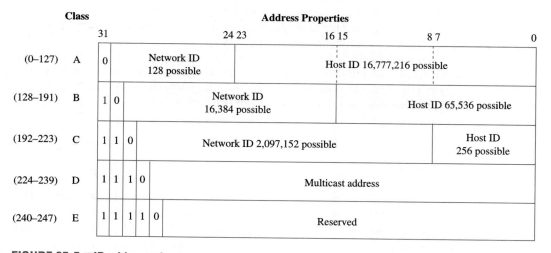

IP VERSION 6

One of the problems with the current version of the Internet is the lack of adequate addresses. Basically, there are no more available (through the Network Information Center). A new version of the Internet, sometimes called the next-generation Internet, has resolved many of the problems experienced by its predecessor.

For example, the address for the next-generation Internet is 128 bits as opposed to 32 bits. This additional address space is large enough to accommodate network growth for the foreseeable future. Addresses are grouped into three different categories: *unicast* (single computer), *multicast* (a set of computers with the same address), and *cluster* (a set of computers that share a common address prefix), routed to one computer closest to the sender.

Other changes include different header formats, new extension headers, and support for audio and video. Unlike IP version 4, IP version 6 does not specify all of the possible protocol features. This allows new features to be added without the need to update the protocol.

TCP

The Transmission Control Protocol (TCP) defines a standard way that two computers can communicate together over interconnected networks. Applications using TCP establish ***connections*** with each other through the use of predefined ***ports*** or ***sockets***. A TCP connection is reliable, with error checking, acknowledgment for received packets, and packet sequencing provided to guarantee the data arrives properly at its destination. Telnet and FTP are examples of TCP/IP applications that use TCP.

UDP

The User Datagram Protocol (UDP) is similar to TCP except it is *connectionless*. Data is transmitted with no acknowledgment of whether it is received. UDP is thus not as reliable as TCP. But some applications do not require the additional overhead of TCP handshaking, using UDP instead. For example, a multiplayer network game might use UDP because it is simple to implement, requires less overhead to manage than TCP, and because the game may not be severely affected if a few packets are lost now and then.

DNS

Every machine on a network must be uniquely identified. On the Internet, this identification takes the form of what is called an ***IP address***. The IP address consists of a 4-byte number, commonly represented in dotted-decimal notation. For example, 204.210.133.51 specifies the IP address of some machine on some network somewhere in the world. To make it easier to remember an address, a *host* name can be associated with an IP address. The Domain Name System (DNS) provides the means to convert from a host name to an IP address, and vice versa. Instead of using the IP address 204.210.133.51, we may instead enter "raycast.rwa.com" as the host name. Typically a DNS server application running on the local network has the responsibility of converting names to IP addresses.

ARP AND RARP

Before any packet can be transmitted, the address of its destination must be known. This address is called a ***MAC address***. MAC stands for media access control, and takes the form of a 48-bit binary number that uniquely identifies one machine from every other. Every network interface card manufactured responds to a preassigned MAC address.

The Address Resolution Protocol (ARP) is used in the Data-Link layer to obtain the MAC address for a given IP address. For example, an ARP request may say, "What is the MAC address for 204.210.133.51?" The ARP reply may be, "The MAC address is 00-60-97-2B-E6-0F."

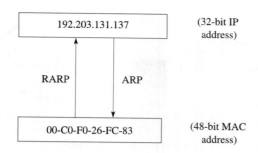

FIGURE 25.6   Using ARP and RARP

The Reverse Address Resolution Protocol (RARP) performs the opposite of ARP, providing the IP address for a specific MAC address. Figure 25.6 gives an example of both protocols at work.

ICMP

The Internet Control Message Protocol (ICMP) uses IP packets to provide updates on network error conditions. Some ICMP messages cause other messages to be sent. Table 25.4 shows a list of the different ICMP messages.

When Netscape gives you its "host unreachable" error message, an ICMP message is responsible.

FTP

The File Transfer Protocol (FTP) allows a user to log on to a remote computer and transfer files back and forth through simple commands. Many FTP sites allow you to log on as an *anonymous* user, an open account with limited privileges but still capable of file transfers. A typical FTP application might look like the one shown in Figure 25.7. Files may be transferred in either direction.

SMTP

The Simple Mail Transport Protocol (SMTP) is responsible for routing e-mail on the Internet using TCP and IP. The process usually requires connecting to a remote computer and transferring the e-mail message, but due to problems with the network or a remote computer, messages can be temporarily undeliverable. The e-mail server will try to deliver any messages by periodically trying to contact the remote destination. When the remote computer becomes available, the message is delivered using SMTP.

TELNET

A Telnet session allows a user to establish a terminal emulation connection on a remote computer. For example, an instructor may Telnet into his or her college's mainframe to do some work. Figure 25.8 shows how the Telnet connection is set up. Once the connection has been made, Telnet begins emulating the terminal selected in the Connection Dialog window. Figure 25.9 shows this mode of operation.

SLIP AND PPP

The Serial Line Interface Protocol (SLIP) and Point-to-Point Protocol (PPP) are used to transfer IP packets over serial connections, such as those provided by modems. Windows provides PPP service through its Dial-Up Networking software. SLIP, developed earlier than PPP, is limited to supporting only TCP/IP applications, whereas PPP supports TCP/IP,

TABLE 25.4   ICMP messages

Type	Code	Description	Query	Error
0	0	Echo reply	✔	
3		Destination unreachable		
	0	network unreachable		✔
	1	host unreachable		✔
	2	protocol unreachable		✔
	3	port unreachable		✔
	4	fragmentation needed but don't fragment bit set		✔
	5	source route failed		✔
	6	destination network unknown		✔
	7	destination host unknown		✔
	8	source host isolated (obsolete)		✔
	9	destination network administratively prohibited		✔
	10	destination host administratively prohibited		✔
	11	network unreachable for TOS		✔
	12	host unreachable for TOS		✔
	13	communication administratively prohibited by filtering		✔
	14	host precedence violation		✔
	15	precedence cutoff in effect		✔
4	0	Source quench (elementary flow control)		✔
5		Redirect		
	0	redirect for network		✔
	1	redirect for host		✔
	2	redirect for type-of-service and network		✔
	3	redirect for type-of-service and host		✔
8	0	Echo request	✔	
9	0	Router advertisement	✔	
10	0	Router solicitation	✔	
11		Time exceeded		
	0	time-to-live equals 0 during transit (Traceroute)		✔
	1	time-to-live equals 0 during reassembly		✔
12		Parameter problem		
	0	IP header bad (catchall error)		✔
	1	required option missing		✔
13	0	Timestamp request	✔	
14	0	Timestamp reply	✔	
15	0	Information request (obsolete)	✔	
16	0	Information reply (obsolete)	✔	
17	0	Address mask request	✔	
18	0	Address mask reply	✔	

IPX, and other protocols at the same time. These two protocols operate in the lower two network layers, as indicated by Figure 25.10. Note that PPP operates in the Physical *and* Data-Link layers, whereas SLIP only functions inside the Physical layer. This is further evidence that PPP has additional features.

FIGURE 25.7 FTP application

FIGURE 25.7 FTP application

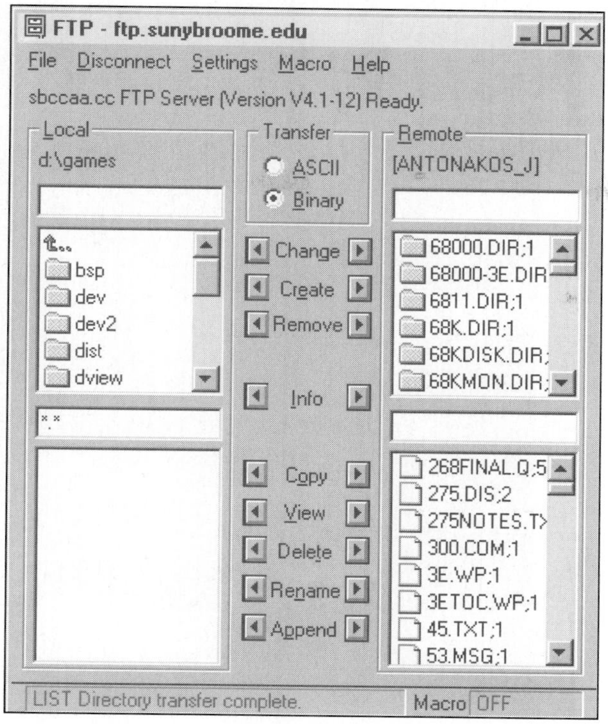

FIGURE 25.8 Establishing the Telnet connection

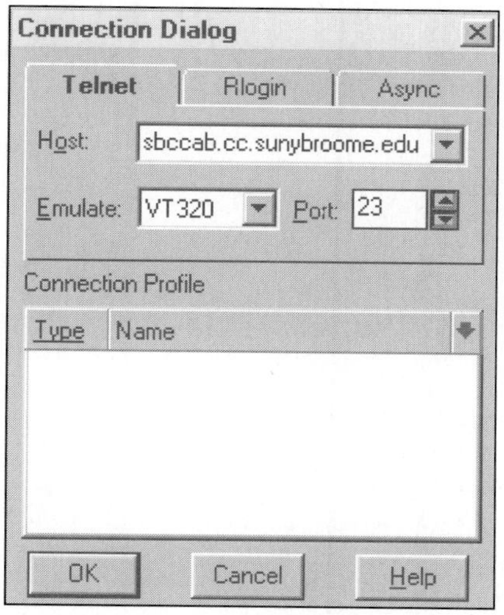

SNMP

Network managers responsible for monitoring and controlling the network hardware and software use the Simple Network Management Protocol (SNMP), which defines the format and meaning of messages exchanged by the manager and agents. The network manager (*manager*) uses SNMP to interrogate network devices (*agents*) such as routers, switches, and bridges in order to determine their status and also retrieve statistical information.

ROUTING PROTOCOLS

Routing protocols direct the flow of information within and between networks. Let's look at a small sample of the protocols in use today.

FIGURE 25.9 Sample Telnet session

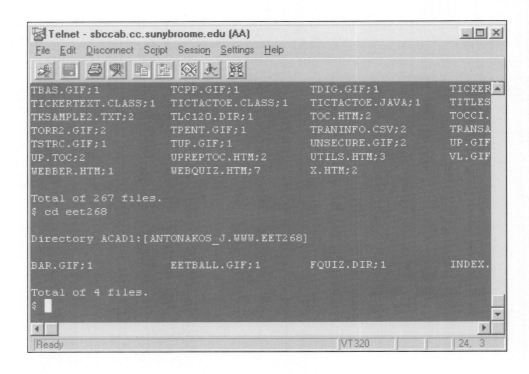

FIGURE 25.10 SLIP and PPP protocols

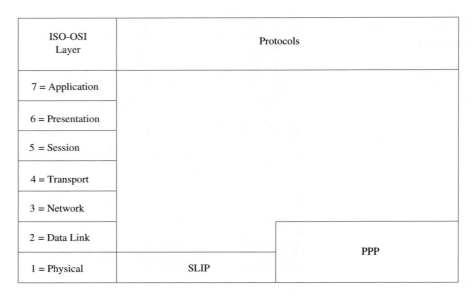

ISO-OSI Layer	Protocols	
7 = Application		
6 = Presentation		
5 = Session		
4 = Transport		
3 = Network		
2 = Data Link		PPP
1 = Physical	SLIP	

RIP

The Routing Information Protocol (RIP), used to route NetWare packets, relies on a method called *distance vector route discovery*, which determines the shortest number of hops to the destination by adding the cost of each subpath to the destination. To maintain accurate information, all routers using RIP periodically broadcast their routing tables to each other (typically once every 30 seconds). RIP is not a good routing protocol for large networks due to the large amount of traffic resulting from routing table broadcasts.

OSPF

Open Shortest Path First (OSPF) is a routing protocol used over TCP/IP that performs *link state routing*. This protocol has advantages over RIP's distance vector routing protocol, such as sharing routing tables less frequently (initially when the router powers up), performing *load balancing*, and constantly testing the state of each communication link (thus adjusting to network problems more quickly than RIP).

NLSP

Network Link Services Protocol (NLSP) is NetWare's link state routing protocol, capable of routing packets farther than RIP (127 hops versus RIP's 15 hops), load balancing, and fast response to changes in the network.

IGRP

The Interior Gateway Routing Protocol (IGRP) is similar to RIP, using distance vector route discovery and routing table broadcasting. Tables are broadcast every 90 seconds instead of every 30 seconds to reduce network traffic. This causes an increase in the time required for the network to recover from a lost route resulting from a broken connection.

IEEE 802 Standards

Several IEEE standards have been developed to support network implementations. Three of these standards are

- IEEE 802.2 Logical Link Control
- IEEE 802.3 Ethernet (CSMA/CD)
- IEEE 802.5 Token Ring

IEEE 802.2 is the initial protocol at work in the Data-Link layer. It interfaces with the Ethernet and token-ring protocols, which operate primarily in the Physical layer. Any company manufacturing network interface cards or other products must conform to the IEEE 802 standards.

The original Ethernet specification called for 10-Mbit/second capability. The demands of network users have resulted in a newer 100-Mbit/second Ethernet, with faster standards approaching. CSMA/CD (Carrier Sense Multiple Access with Collision Detection) is used to manage data transmission.

Token ring is available in speeds of 4 and 16 Mbit/second. This network protocol requires that a special token be passed from one node to another. Only the node holding the token can send data over the network.

PROTOCOL ANALYZERS

Protocol analyzers (or *sniffers*) are hardware or software devices that listen to the traffic on a network and capture various packets for examination. Hardware analyzers also double as cable testers.

Figure 25.11 shows a demo version of the LanExplorer protocol analyzer at work. LanExplorer displays an ongoing update of network traffic statistics; allows packets to be captured, disassembled, and saved; and can transmit packets to facilitate testing and troubleshooting.

TROUBLESHOOTING TECHNIQUES

Windows provides a very useful utility called WINIPCFG, which you can run from the Run menu. Figure 25.12 shows the display window with all details included. Notice the various IP addresses indicated. Because the network software used by the system receives an IP address on the fly, via DHCP (Dynamic Host Configuration Protocol), the IP address of the DHCP server (204.210.159.18) must be known by the system. DHCP is not used when your system has been allocated a fixed IP address by your network administrator.

If you have difficulty with your network connection, the information displayed by WINIPCFG will be valuable to the person who is troubleshooting the connection.

FIGURE 25.11 LanExplorer protocol analyzer window

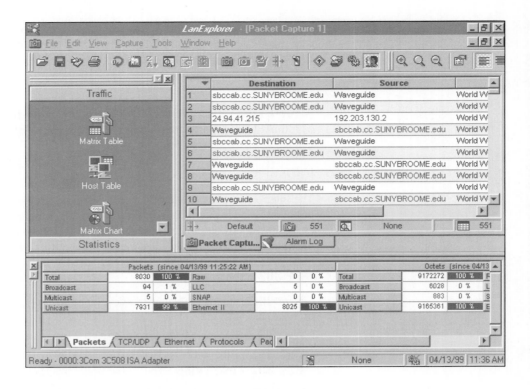

FIGURE 25.12 WINIPCFG display window

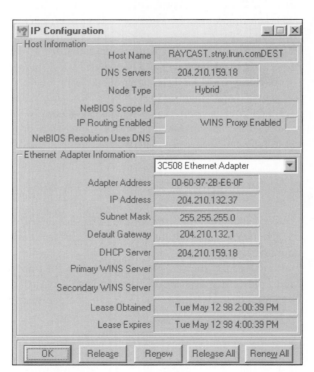

SUMMARY

In this exercise we discovered that

- NetBEUI is a nonroutable protocol for small networks.
- The IPX/SPX protocols are used with NetWare.
- TCP/IP protocols are described in RFCs.
- IP addresses are 32 bits.
- UDP is an unreliable, routable protocol.
- TCP is a reliable, routable protocol.

- DNS resolves host names into IP addresses.
- There are many routing protocols, such as RIP, OSPF, and IGRP.

This self-test is designed to help you check your understanding of the background information presented in this exercise.

True/False

Answer *true* or *false*.

1. The IPX/SPX protocols are the backbone for TCP/IP.
2. The ISO-OSI network model defines the nine network layers.
3. RFCs are documents that describe how each of the TCP/IP protocols are implemented.
4. IP datagrams are guaranteed to be delivered.
5. Loopback testing allows for a remote method to test TCP/IP and application functionality.
6. NetBEUI is a routable protocol.
7. PPP only supports TCP/IP connections.

Multiple Choice

Select the best answer.

8. TCP connections are
   a. Unreliable and must rely on the IP protocol to provide the necessary reliability.
   b. Created using predefined ports or sockets.
   c. Considered connectionless without the need to acknowledge the received data.
   d. Limited to class A addresses.
9. Dotted-decimal notation is used to
   a. Describe each of the five address classes.
   b. Identify all the networks and hosts on the Internet.
   c. Reserve addresses used for special purposes.
   d. All of the above.
   e. None of the above.
10. DNS servers provide
    a. The MAC address for every host address.
    b. The error-checking protocol for ICMP messages.
    c. A mapping from a host name to an IP address.
    d. A mechanism to transfer IP packets over serial lines.
11. PPP operates on
    a. All seven ISO-OSI networking layers.
    b. All nine ISO-OSI network layers.
    c. The Physical and Data-Link layers.
    d. Remote terminal emulation connections.
12. Routing protocols are designed to
    a. Direct the flow of information between networks.
    b. Interrogate network device agents to gather statistics.
    c. Broadcast traffic on all segments of a network.
    d. Troubleshoot TCP and UDP connection problems.
13. Which protocol offers unreliable, connectionless delivery?
    a. ARP
    b. TCP
    c. UDP
    d. ICMP
14. Which protocol allows anonymous users to transfer files?
    a. DNS
    b. FTP
    c. IGRP
    d. OSPF

Matching

Match a description of the items on the right with each item on the left.

15. IPX/SPX
16. NetBEUI
17. TCP/IP
18. PPP
19. FTP

  a. Dial-Up Networking protocol
  b. Connection-Oriented File Transfer protocol
  c. NCP, SAP, and RIP protocols
  d. Microsoft Windows network protocol
  e. UDP, TCP, and ICMP protocols

Completion

Fill in the blank or blanks with the best answers.

20. Most FTP sites allow for _____ login.
21. SMTP is used to exchange _____.
22. A(n) _____ describes each of the TCP/IP protocols.
23. IP addresses are commonly shown in _____ _____ _____.
24. The _____ protocol is designed for use on small networks and is not routable.
25. UDP datagrams offer _____ connections.
26. 210.38.154.27 is an example of a(n) _____ address.

FAMILIARIZATION ACTIVITY

1. Install LanExplorer from the companion CD. If possible, do the following:
   - Capture several packets and note the type of packets captured.
   - Watch the network statistics for several minutes.
   - Identify all the different protocols present on the network.
2. Visit the computer center at your school and ask the network technicians to share information about the IP classes in use, what network management protocols are used, and how many nodes are on the network.
3. Look up the term *broadcast storm* in a networking dictionary (or search the Web for it). What is a broadcast storm, what are its effects, and how can it be prevented?

QUESTIONS/ACTIVITIES

1. What is the purpose of a set of protocols?
2. Why are IP addresses used for routing instead of MAC addresses?

REVIEW QUIZ

Under the supervision of your instructor

1. Explain the purpose of each layer of the ISO-OSI model.
2. Discuss the different protocols that make up the TCP/IP suite.
3. Describe the relationship between IP addresses and MAC addresses.

26

Network Applications

INTRODUCTION

Joe Tekk answered the telephone in his office. It was Jeff Page, who sounded worried.
*   "Joe, PING my computer for me. I'm at 206.25.117.119."*
*   Joe entered the appropriate command. Several* Request timed out *error messages appeared.* "It doesn't look good, Jeff. PING doesn't see your machine."
*   Jeff groaned.* "Hold on, Joe."
*   A few minutes later Jeff said,* "Try it again."
*   Joe ran the PING utility again. This time he got the normal display statistics.*
*   "What did you do, Jeff?"* he asked, wanting to know what the problem was.*
*   "I don't know. I rebooted and everything works again."*

PERFORMANCE OBJECTIVES

Upon completion of this exercise, you will be able to

1. Discuss the role of PING and TRACERT.
2. Explain the numbers WINIPCFG displays.
3. Define DHCP and state its purpose.

KEY TERMS

PING
TRACERT

WINIPCFG
DHCP

BACKGROUND INFORMATION

In addition to the browser, e-mail, FTP, and Telnet applications previously discussed, there are a few additional network applications that deserve mention. Let's take a look at them and the network functions they perform.

PING

PING is a TCP/IP application that sends datagrams once every second in the hope of an echo response from the machine being PINGed. If the machine is connected and running a TCP/IP protocol stack, it should respond to the PING datagram with a datagram of its own.

PING displays the time of the return response in milliseconds or one of several error messages (such as "Request timed out" or "Destination host unreachable").

PING can be used to simply determine the IP address of a World Wide Web site if you know its host name. For example, the IP address for www.yahoo.com is 204.71.200.74, which can be found with PING by opening a DOS window, and entering

```
C> ping www.yahoo.com
```

You should see something similar to this:

```
Pinging www.yahoo.com [204.71.200.74] with 32 bytes of data:

Reply from 204.71.200.74: bytes=32 time=180ms TTL=245
Reply from 204.71.200.74: bytes=32 time=127ms TTL=245
Reply from 204.71.200.74: bytes=32 time=145ms TTL=245
Reply from 204.71.200.74: bytes=32 time=146ms TTL=245
```

So, PING performed DNS on the host name to find out the IP address, and then sent datagrams to Yahoo's host machine and displayed the responses. If you know the IP address you can enter it directly and PING will skip the DNS phase.

To get a list of PING's features, enter PING with no parameters. You should see something similar to this:

```
Usage: ping [-t] [-a] [-n count] [-l size] [-f] [-i TTL] [-v TOS]
            [-r count] [-s count] [[-j host-list] | [-k host-list]]
            [-w timeout] destination-list

Options:

    -t              Ping the specified host until interrupted.
    -a              Resolve addresses to hostnames.
    -n count        Number of echo requests to send.
    -l size         Send buffer size.
    -f              Set Don't Fragment flag in packet.
    -i TTL          Time To Live.
    -v TOS          Type Of Service.
    -r count        Record route for count hops.
    -s count        Timestamp for count hops.
    -j host-list    Loose source route along host-list.
    -k host-list    Strict source route along host-list.
    -w timeout      Timeout in milliseconds to wait for each reply.
```

You are encouraged to experiment with these parameters.

TRACERT

TRACERT (Trace Route) is a TCP/IP application that determines the path through the network to a destination entered by the user. For example, running

```
C> tracert www.yahoo.com
```

generates the following output:

```
Tracing route to www7.yahoo.com [204.71.200.72]
over a maximum of 30 hops:
  1    20 ms    19 ms    19 ms  bing100b.stny.lrun.com [204.210.132.1]
  2    10 ms    14 ms     9 ms  m2.stny.lrun.com [204.210.159.17]
  3    12 ms    24 ms    10 ms  ext_router.stny.lrun.com [204.210.155.18]
  4    46 ms    40 ms    44 ms  border3-serial4-0-6.Greensboro.mci.net [204.70.83.85]
  5    42 ms    53 ms    46 ms  core1-fddi-0.Greensboro.mci.net [204.70.80.17]
  6   109 ms   160 ms   122 ms  bordercore2.Bloomington.mci.net [166.48.176.1]
  7   123 ms   126 ms   113 ms  hssi1-0.br2.NUQ.globalcenter.net [166.48.177.254]
```

```
 8    125 ms    117 ms    115 ms   fe5-1.cr1.NUQ.globalcenter.net [206.251.1.33]
 9    114 ms    125 ms    113 ms   pos0-0.wr1.NUQ.globalcenter.net [206.251.0.122]
10    125 ms    124 ms    121 ms   pos1-0-OC12.wr1.SNV.globalcenter.net [206.251.0.74]
11    122 ms    139 ms    115 ms   pos5-0.cr1.SNV.globalcenter.net [206.251.0.105]
12    128 ms    129 ms    138 ms   www7.yahoo.com [204.71.200.72]

Trace complete.
```

The trace indicates that it took 12 *hops* to get to Yahoo. Every hop is a connection between two routers. Each router guides the test datagram from TRACERT one step closer to the destination. TRACERT specifically manipulates the TTL (time to live) parameter of the datagram, adding 1 to it each time it rebroadcasts the test datagram. Initially the TTL count is 1. This causes the very first router in the path to send back an ICMP time-exceeded message, which TRACERT uses to identify the router and display path information. When the TTL is increased to 2, the second router sends back the ICMP message, and so on, until the destination is reached (if it ever is).

It is fascinating to examine TRACERT's output. Notice that hop 5 contains a reference to *FDDI*, which means that the datagram spent some time traveling around a fiber distributed data interface network.

WINIPCFG

WINIPCFG (Windows IP Configuration) is a handy tool for examining all the numbers associated with your networking components. Figure 26.1 shows the initial WINIPCFG window, which displays some of the more important networking information. Figure 26.2 shows the results of clicking the "More Info >>" button. Note that there is an entry for DHCP Server. DHCP refers to dynamic host configuration protocol, a TCP/IP protocol that allows a DHCP client to request an IP address at boot time from a pool of free IP addresses (maintained by the DHCP server). The IP address is *leased* to the DHCP client for a limited time period (which can be renewed or extended). Every time the DHCP client boots, it may receive a different IP address. This in no way affects the way information is exchanged.

ECHO SERVERS AND CHAT SERVERS

Individuals teaching networking courses typically have on hand two applications that demonstrate how to send and receive messages over a network. An *echo server* simply sends any echo client messages back to the client unchanged. A *chat server* allows two or more users to send messages that are broadcast to all members of the group (with private messaging possible via software). These two applications are simple enough that they are used as beginning programming exercises in a course on network programming with *sockets*. A socket is a name for a TCP/IP connection, an application that "plugs" into the appropriate "socket."

FIGURE 26.1   Initial WINIPCFG window

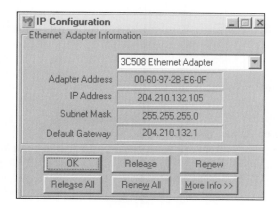

321

FIGURE 26.2   More WINIPCFG details

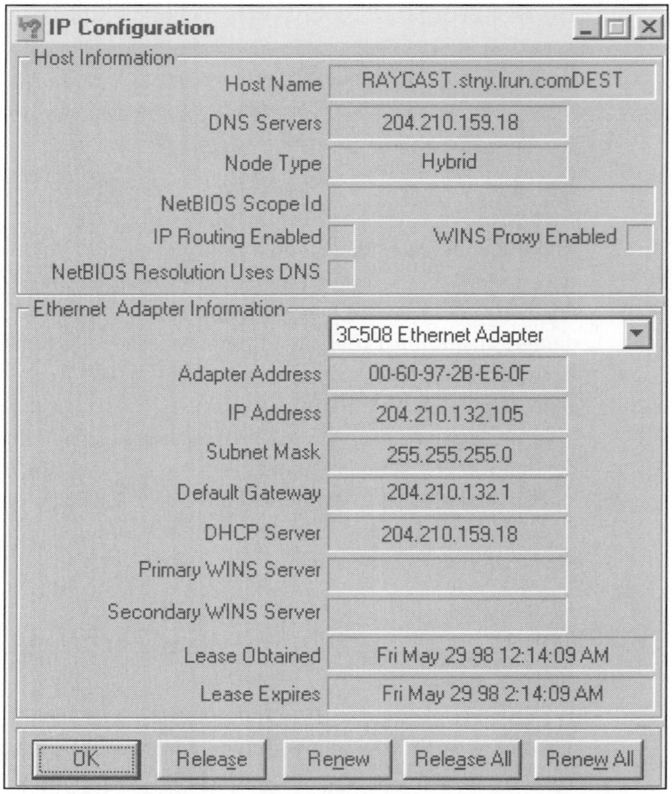

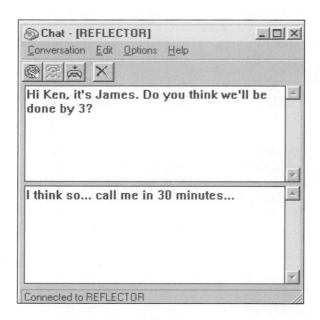

FIGURE 26.3   WINCHAT network utility

Windows contains an installable chat application called WINCHAT that allows one user to dial another by entering the machine name of the machine to chat with. Figure 26.3 shows a WINCHAT session in progress. The user James is communicating with Ken on the machine named REFLECTOR (NetBEUI must be operational for WINCHAT to work). Any machine on the Network Neighborhood that has a WINCHAT utility can be dialed by entering its name.

PCANYWHERE

pcANYWHERE is an excellent utility for connecting two machines over a network. One machine is set up as the host. The second machine remotely controls the host. The remote machine's pcANYWHERE window looks exactly like the desktop on the host. If you click on a program icon, the program will launch on the host. Essentially, you can do just about everything from the remote machine you could do on the host. pcANYWHERE is a TCP/IP application, so it can be used over large, routable networks (although a modem may also be used).

A screen shot of pcANYWHERE's control panel is shown in Figure 26.4. With the Remote Control button pushed, double-clicking on wave, acd, or ken will begin a network connection sequence (using TCP/IP) to the machine at the IP address indicated. Clicking wave, for example, brings up the initialization window shown in Figure 26.5.

If the connection is successful, the host desktop appears in the remote machine's pcANYWHERE window. Figure 26.6 illustrates this feature. Now you can do whatever you like with the host machine. Bear in mind that even over a high-speed network screen, updates might be sluggish due to the amount of data that must be transferred.

pcANYWHERE contains many powerful features, including built-in chat and file-transfer functions. Figure 26.7 shows the file-transfer screen, which is easy to use and navigate. The arrow at the top of the window indicates the direction of the transfer. Clicking the arrow changes its direction. Whatever files are highlighted in the file menus are transferred when the Send button is pressed. Entire directories can be sent with a single click.

Overall, pcANYWHERE adds a new dimension to remote computing.

FIGURE 26.4   pcANYWHERE control panel

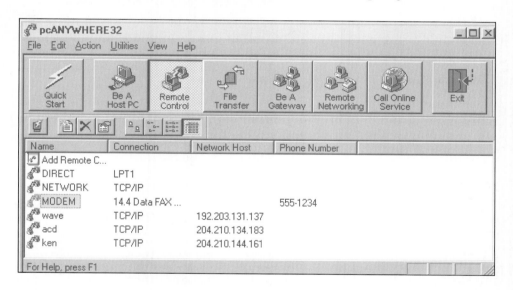

FIGURE 26.5   Connecting to the host machine

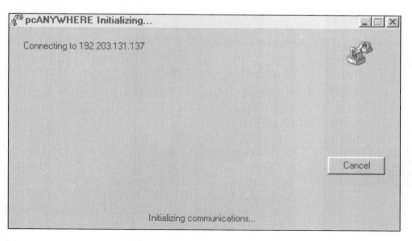

FIGURE 26.6   Host desktop

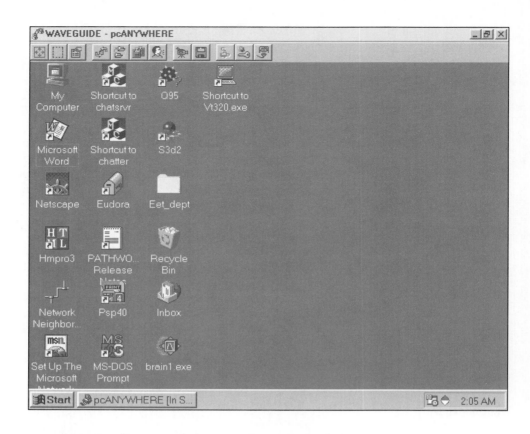

FIGURE 26.7   pcANYWHERE file-transfer screen

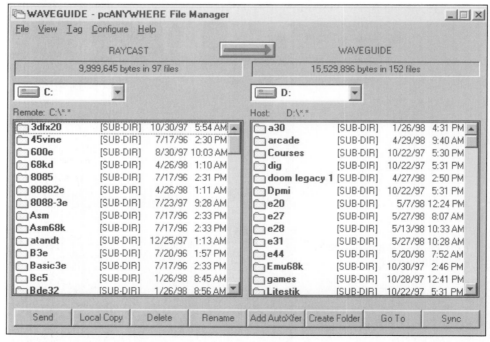

TROUBLESHOOTING TECHNIQUES

Although rebooting a machine is the last resort, in many cases troubleshooting a network problem may require several reboots. If you install a new protocol, change the name of your machine, or add a network printer, Windows recommends that you reboot so that the changes may take effect. You might as well follow Windows's advice and reboot. While the machine is doing so you have a few moments to yourself to review what is happening. Is the problem with the computer (protocol properties not set right), the network connection (external cabling, NIC interrupt or I/O settings, protocol binding problems), the machine on the other end, or the application software?

By the time the machine has rebooted you are probably waiting to try something else, a different theory, a new approach. Try it. Reboot if necessary after making adjustments, and remain patient. There is a reason the network is not working properly.

SUMMARY

In this exercise we discovered that

- PING, TRACERT, and WINIPCFG are network applications useful for examining network properties.
- WINIPCFG can be used to manage DHCP leases.

SELF-TEST

This self-test is designed to help you check your understanding of the background information presented in this exercise.

True/False

Answer *true* or *false*.

1. PING sends a sonar pulse onto the Internet.
2. Running TRACERT with an invalid URL will cause it to hang forever.
3. An echo server retransmits all received messages.
4. WINIPCFG only works for fixed IP addresses.
5. pcANYWHERE has built-in file-transfer capability.
6. PING determines the IP address of a destination.
7. WINIPCFG provides information about DHCP leases.

Multiple Choice

Select the best answer.

8. PING uses
   a. NetBEUI packets.
   b. TCP/IP packets.
   c. PING packets.
9. How many different class C networks are shown in the TRACERT output for Yahoo?
   a. 8
   b. 9
   c. 10
10. To find out the adapter address (MAC address), use
    a. PING.
    b. TRACERT.
    c. WINIPCFG.
11. Which utility is used to send text messages to another user?
    a. PING
    b. WINCHAT
    c. WINIPCFG
12. DHCP is used to
    a. Obtain a fixed IP address.
    b. Obtain a leased IP address.
    c. Carry PING requests.
13. The name for a TCP/IP connection that an application "plugs into" is called a(n)
    a. Address.
    b. I/O port.
    c. Socket.
14. To find the route to a destination, use
    a. PING.
    b. TRACERT.
    c. WINIPCFG.

Completion

Fill in the blank or blanks with the best answers.

15. WINCHAT uses the _____ protocol.
16. DHCP stands for _____ _____ _____ _____.
17. pcANYWHERE uses a remote computer to connect to a(n) _____ computer.
18. To determine an unknown IP address, use _____.
19. To determine the path to an IP address, use _____.
20. PING uses _____ to determine a host name's IP address.
21. IP addresses are _____ to DHCP clients.

1. PING the following addresses:
   - www.yahoo.com
   - www.intel.com
   - www.whitehouse.gov
   - www.nasa.gov
   - 192.203.131.137
2. Run TRACERT on the addresses in step 1. Comment on any interesting router names that show up.

1. How long is the leased IP address good for in Figure 26.2?
2. Why use WINCHAT when you can easily make a call on the telephone?
3. Search the Web for products similar to pcANYWHERE.

Under the supervision of your instructor

1. Discuss the role of PING and TRACERT.
2. Explain the numbers in WINIPCFG displays.
3. Define DHCP and state its purpose.

27

The Internet

It was 2:45 A.M. Joe Tekk was awake, sitting in his darkened living room. The only light in the room was coming from the monitor of his computer. Joe was exhausted, but didn't want to stop browsing the Web. He had stumbled onto a Web page containing links to computer graphics, game design, and protected-mode programming. For three hours, Joe had been going back and forth from one page to another, adding some links to his bookmarks and ignoring others. When he finally decided to quit, it was not because of lack of interest, but simply time to go to sleep.

"From now on, I'm only browsing for 30 minutes," Joe vowed to himself. But he knew he would have another late-night browsing session that would last much longer. It was too much fun having so much information available instantly.

PERFORMANCE OBJECTIVES

Upon completion of this exercise, you will be able to

1. Describe the basic organization of the Internet.
2. Explain the purpose of a browser and its relationship to HTML.
3. Discuss the usefulness of CGI and Java applications.
4. Identify the elements of a virtual private network.

KEY TERMS

Universal service
Hypertext Transport Protocol (HTTP)
Hypertext markup language
Universal resource locator (URL)

CGI
Tunneling
Instant messaging

BACKGROUND INFORMATION

The Internet started as a small network of computers connecting a few large mainframe computers. It has grown to become the largest computer network in the world, connecting virtually all types of computers. The Internet offers a method to achieve *universal service*, or a connection to virtually any computer, anywhere in the world, at any time. This concept is similar to the use of a telephone, which provides a voice connection anywhere at any time. The Internet provides a way to connect all types of computers together regardless of

FIGURE 27.1   Concept of
Internet connections

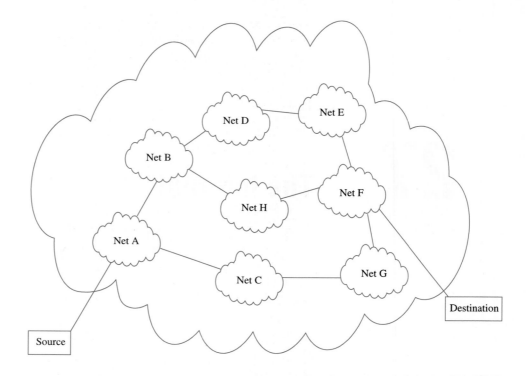

their manufacturer, size, and resources. The *one* requirement is a connection to the network. Figure 27.1 shows how several networks are connected together.

The type of connection to the Internet can take many different forms, such as a simple modem connection, a cable modem connection, a T1 line, a T3 line, or a frame relay connection.

THE ORGANIZATION OF THE INTERNET

The current version of the Internet (V4) is organized into several categories, as shown in Table 27.1. The name of an Internet host shows the category to which it is assigned. For example, the rwa.com domain is the name of a company, and the bcc.edu domain is an educational institution. Each domain is registered on the appropriate root server. For example, the domain rwa.com is known by the com root server. Then, within each domain, a locally administered Domain Name Server allows for each host to be configured.

WORLD WIDE WEB

The World Wide Web, or WWW as it is commonly referred to, is actually the **Hypertext Transport Protocol (HTTP)** in use on the Internet. The HTTP protocol allows for hypermedia

TABLE 27.1   Organization of
the Internet

Domain Type	Organization Type
edu	Educational institution
com	Commercial organization
gov	Government
mil	Military
net	Network providers and support
org	Other organizations not listed above
country code	A country code, for example, .us for United States, .ca for Canada, .jp for Japan

information to be exchanged, such as text, video, audio, animation, Java applets, images, and more. The ***hypertext markup language***, or ***HTML***, is used to determine how the hypermedia information is to be displayed on a WWW browser screen.

The WWW browser is used to navigate the Internet by selecting *links* on any WWW page or by specifying a ***Universal Resource Locator***, or ***URL***, to point to a specific *page* of information. There are many different WWW browsers. The two most popular are Microsoft Internet Explorer and Netscape Navigator, shown in Figures 27.2 and 27.3, respectively. Both of these browsers are available free over the Internet, and contain familiar pull-down menus and graphical toolbars to access the most commonly used functions such as forward, backward, stop, print, and reload.

FIGURE 27.2   Sample home page displayed using Internet Explorer

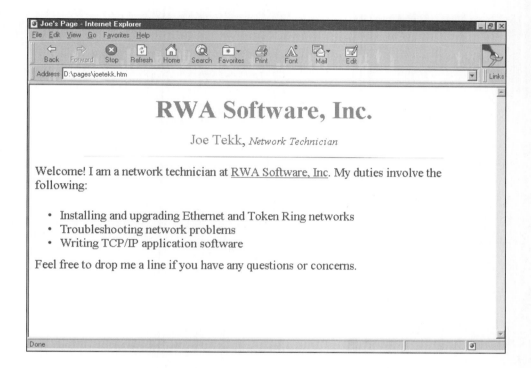

FIGURE 27.3   Sample home page displayed using Netscape Navigator

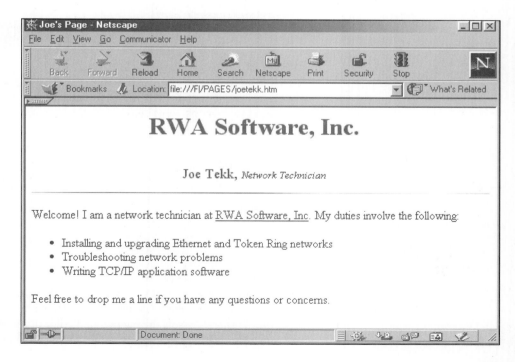

Note the differences in page layout between Figures 27.2 and 27.3. Individuals who design Web pages must take into account the different requirements of each browser so that the page looks acceptable in both browsers.

HTML

HTML stands for hypertext markup language. HTML is the core component of the information that composes a Web page. The HTML *source code* for a Web page has an overall syntax and structure that contains formatting commands (called *tags*) understood by a Web browser. Following is a sample HTML source. The actual Web page for this HTML code was shown in Figure 27.3.

```
<HTML>

<HEAD>
<TITLE>Joe's Page</TITLE>
</HEAD>

<BODY BGCOLOR="#FFFF80">

<P ALIGN="CENTER">
<B><FONT SIZE="+3" COLOR="#FF0000">RWA Software, Inc.</FONT></B>
</P>

<P ALIGN="CENTER">
<FONT SIZE="+1"><FONT COLOR="#008000">Joe Tekk</FONT>,
<FONT SIZE="-1"><I>Network Technician</I></FONT></FONT>
</P>

<P ALIGN="CENTER">
<IMG SRC="bar.gif" ALT="Color Bar">
</P>

<P ALIGN="LEFT">
Welcome! I am a network technician at
<A HREF="http://www.rwasoftware.com">RWA Software, Inc</A>. My duties
 involve the following:
</P>

<UL>
<LI>Installing and upgrading Ethernet and Token Ring networks</LI>
<LI>Troubleshooting network problems</LI>
<LI>Writing TCP/IP application software </LI>
</UL>

<P ALIGN="LEFT">
Feel free to drop me a line if you have any questions or
concerns.
</P>

</BODY>
</HTML>
```

The HTML source consists of many different tags that instruct the browser what to do when preparing the graphical Web page. Table 27.2 shows some of the more common tags. The main portion of the Web page is contained between the BODY tags. Note that BGCOLOR= "#FFFF80" sets the background color of the Web page. The six-digit hexadecimal number contains three pairs of values for the red, green, and blue color levels desired.

TABLE 27.2   Assorted HTML tags

Tag	Meaning
<P>	Begin paragraph
</P>	End paragraph
<B>	Bold
<I>	Italics
<IMG SRC...>	Image source
<UL>	Unordered list
<LI>	List item
<TABLE>	Table
<TR>	Table row
<TD>	Table data
<A>	Anchor

Pay attention to the tags used in the HTML source and what actually appears on the Web page in the browser (Figure 27.3). The browser ignores white space (multiple blanks between words or lines of text) when it processes the HTML source. For example, the anchor for the RWA Software link begins on its own line in the source, but the actual link for the anchor is displayed on the same line as the text that comes before and after it.

Many people use HTML editors, such as HoTMetaL or Front Page, to create and maintain their Web pages. Options to display the page in HTML format, or in WYSIWYG (what you see is what you get), are usually available, along with sample pages, image editing, and conversion tools that convert many different file types (such as a Word document) into HTML. Demo versions of these HTML editors, and others, can be downloaded from the Web. Figure 27.4 shows HoTMetaL's graphical page editor with Joe Tekk's page loaded.

WWW pages are classified into three categories: static, dynamic, and active. The easiest to make are static and involve only HTML code. The page content is determined by what is contained in the HTML code. Dynamic WWW pages contain a combination of HTML code and a "call" to a server using a Common Gateway Interface application, or a CGI

FIGURE 27.4   HoTMetaL PRO with sample page

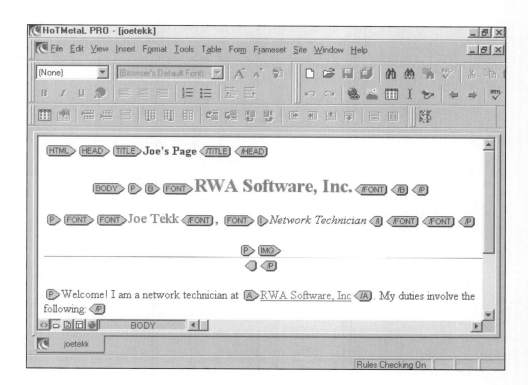

FIGURE 27.5   Web page with
FORM element

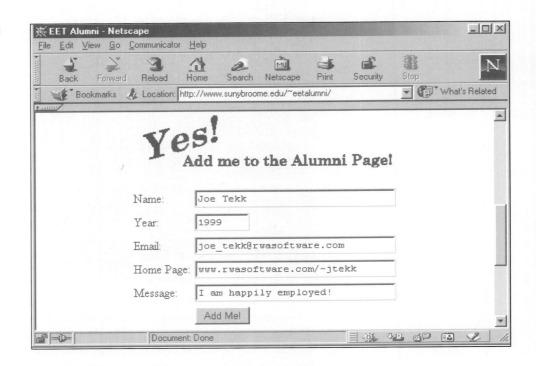

application. In this scenario, information supplied by the user into an HTML form is transferred back to a host computer for processing. The host computer then returns a dynamic customized WWW page. Active pages contain a combination of HTML code and applets. Therefore, the WWW page is not completely specified during the HTML coding process. Instead, using a Java applet, it is specified while being displayed by the WWW browser.

CGI

The Common Gateway Interface (CGI) is a software interface that allows a small amount of interactive processing to take place, with information provided on a Web page. For example, consider the Web page shown in Figure 27.5. The Web page contains a FORM element, which itself can contain many different types of inputs, such as text boxes, radio buttons, lists with scroll bars, and other types of buttons and elements. The user browsing the page enters his or her information and then clicks the Add Me! button. This begins the following chain of events:

1. The form data entered by the user is placed into a message.
2. The browser POSTs the form data (sends message to CGI server).
3. The CGI server application processes the form data.
4. The CGI server application sends the results back to the CGI client (Netscape or Internet Explorer).

Let's take a closer look. The HTML code for the alumni page form in Figure 27.5 looks like this:

```
<FORM ACTION="/htbin/cgi-mailto/eetalumni" METHOD="POST">
<P ALIGN="CENTER"><IMG SRC="yes.gif"></P>
<CENTER>
<TABLE WIDTH="50%" ALIGN="CENTER">
<TR><TD>Name:</TD>
<TD><INPUT TYPE="TEXT" NAME="name" SIZE="32"></TD></TR>
<TR><TD>Year:</TD>
<TD><INPUT TYPE="TEXT" NAME="year" SIZE="8"></TD></TR>
<TR><TD>Email:</TD>
```

```
<TD><INPUT TYPE="TEXT" NAME="email" SIZE="32"></TD></TR>
<TR><TD>Home Page:</TD>
<TD><INPUT TYPE="TEXT" NAME="home" SIZE="32"></TD></TR>
<TR><TD>Message:</TD>
<TD><INPUT TYPE="TEXT" NAME="msg" SIZE="32"></TD></TR>
<TR><TD></TD>
<TD><INPUT TYPE="SUBMIT" VALUE="Add Me!"></TD></TR>
</TABLE>
</CENTER>
</FORM>
```

The first line of the form element specifies POST as the method used to send the form data out for processing. The CGI application that will receive the POSTed form data is the cgi-mailto program in the *htbin* directory. Specifically, cgi-mailto processes the form data and sends an e-mail message to the *eetalumni* account. The e-mail message looks like this:

```
From:  SBCCVA::WWWSERVER
To:    eetalumni
CC:
Subj:

REMOTE_ADDRESS: 204.210.159.19
name: Joe Tekk
year: 1998
email: joe_tekk@rwa.software.com
home: www.rwasoftware.com/~jtekk
msg: I am happily employed!
```

Note that the identifiers (name, year, email, home, and msg) match the names used to identify the text input elements in the form.

Instead of e-mailing the form data, another CGI application might create a Web page on the fly containing custom information based on the form data submitted. CGI applications are written in C/C++, Visual BASIC, Java, Perl, and many other languages. The Web is full of sample forms and CGI applications available for download and inclusion in your own Web pages.

JAVA

The Java programming language is the method used to create active WWW pages using Java applets. An active WWW page is specified by the Java applet when the WWW page is displayed rather than during the HTML coding process. A Java applet is actually a program transferred from an Internet host to the WWW browser. The WWW browser executes the Java applet code on a Java virtual machine (which is built into the WWW browser). The Java language can be characterized by the following nonexhaustive list:

- General purpose
- High level
- Object oriented
- Dynamic
- Concurrent

Java consists of a programming language, a run-time environment, and a class library. The Java programming language resembles C++ and can be used to create conventional computer applications or applets. Only an applet is used to create an active WWW page. The run-time environment provides the facilities to execute an application or applet. The class library contains prewritten code that can simply be included in the application or applet. Table 27.3 shows the Java class library functional areas.

TABLE 27.3 Java class library
categories

Class	Description
Graphics	Abstract window tool kit (AWT)
Network I/O	Socket level connnections
File I/O	Local and remote file access
Event capture	User actions (mouse, keyboard, etc.)
Run-time system calls	Access to built-in functions
Exception handling	Method to handle any type of error condition
Server interaction	Built-in code to interact with a server

The following Java program is used to switch from one image to a second image (and back) whenever the mouse moves over the Java applet window. Furthermore, a mouse click while the mouse is over the applet window causes a new page to load.

```java
import java.awt.Graphics;
import java.awt.Image;
import java.awt.Color;
import java.awt.Event;
import java.net.URL;
import java.net.MalformedURLException;

public class myswitch extends java.applet.Applet implements Runnable
{
    Image swoffpic;
    Image swonpic;
    Image currentimg;
    Thread runner;

public void start()
{
    if (runner == null)
    {
        runner = new Thread(this);
        runner.start();
    }
}

public void stop() {
    if (runner != null)
    {
        runner.stop();
        runner = null;
    }
}

public void run()
{
    swoffpic = getImage(getCodeBase(), "swoff.gif");
    swonpic = getImage(getCodeBase(), "swon.gif");
    currentimg = swoffpic;
    setBackground(Color.red);
    repaint();
}

public void paint(Graphics g)
```

```
    {
        g.drawImage(currentimg, 8, 8, this);
    }

    public boolean mouseEnter(Event evt, int x, int y)
    {
        currentimg = swonpic;
        repaint();
        return(true);
    }

    public boolean mouseExit(Event evt, int x, int y)
    {
        currentimg = swoffpic;
        repaint();
        return(true);
    }

    public boolean mouseDown(Event evt, int x, int y)
    {
        URL destURL = null;
        String url = "http://www.sunybroome.edu/~eet_dept";

        try
        {
            destURL = new URL(url);
        }
        catch(MalformedURLException e)
        {
            System.out.println("Bad destination URL: " + destURL);
        }
        if (destURL != null)
            getAppletContext().showDocument(destURL);
        return(true);
    }

}
```

Programming in Java, like any other language, requires practice and skill. With its popularity still increasing, now would be a good time to experiment with Java yourself by downloading the free Java compiler and writing some applets.

VIRTUAL PRIVATE NETWORKS

A virtual private network (VPN) allows for remote private LANs to communicate securely through an untrusted public network such as the Internet. This is shown in Figure 27.6. This is in contrast to the traditional approach in which a large corporation or organization used private or leased lines to communicate between different sites in order to provide privacy of data. Using a VPN, only authorized members of the network are allowed access to the data. A VPN uses an IP tunneling protocol and security services that are transparent to the private network users.

Using a VPN, a private LAN connected to the Internet can be connected to other LANs using a combination of tunneling, encryption, and authentication. *Tunneling* means that data that is transferred through the public network is in an encapsulated form. This is illustrated in Figure 27.7. All of the data, including the addresses of the sender and destination, is enclosed within a packet. Although tunneling is sufficient to create a VPN, it does not ensure complete data security.

FIGURE 27.6   RWA Software VPN (physical view)

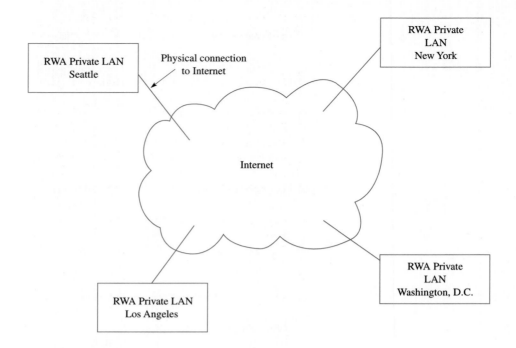

FIGURE 27.7   RWA Software VPN (logical view)

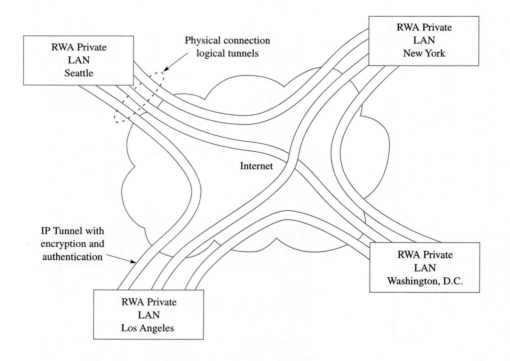

Complete security is accomplished when the data communication is also encrypted and authenticated. Packets that are protected by tunneling, encryption, and authentication (certified by an agreed-on certification authority such as verisign.com) offer the highest level of security. The IP Security (IPSec) standards provide a security protocol for tunneling as well as for data privacy, integrity, and authentication, creating a truly secure VPN.

IPSec is a set of protocols developed by the Internet Engineering Task Force that adds additional security solutions to TCP/IP networking. IPSec currently supports several encryption algorithms such as DES, 3DES, and Public Key Encryption and is designed to incorporate new algorithms as they are created. IPSec offers a solution to data privacy, integrity, and authentication that is network independent, application independent, and supports all IP services (e.g., HTTP, FTP, SNMP, etc.). Note that protocols such as PPTP

(Point-to-Point Tunneling Protocol) and L2TP (Layer 2 Tunneling Protocol) used to create a VPN support only tunneling, whereas IPSec includes tunneling, encryption, and authentication.

INSTANT MESSAGING

One of the latest tools used to communicate on the Internet is ***instant messaging***. Many different companies provide software to enable users to send electronic messages. Unlike electronic mail, instant messaging is an application that provides the capability for a user to send and receive messages instantly. A few of the most popular instant messaging applications are AOL's Instant Messenger and Microsoft's NetMeeting. Both of these programs allow for a user to send or receive instant messages.

There are many additional choices available for instant messaging. You are encouraged to search for, download, and test other instant messaging applications.

RELATED SITES

Following are a number of service, reference, and technology-based sites that may be of interest:

- www.prenhall.com   Engineering and technology textbooks
- www.yahoo.com   Search engine
- www.internic.net   Internet authority
- www.intel.com   Intel Corporation
- www.microsoft.com   Microsoft Corporation
- www.sunybroome.edu/~mansfield_k   Author's home page
- www.sunybroome.edu/~antonakos_j   Author's home page
- www.netscape.com   Netscape Corporation

The Internet is full of information about every aspect of the Web page development process. Many people put a tremendous amount of information on their own Web pages. You are encouraged to learn more about Web pages and Web programming.

TROUBLESHOOTING TECHNIQUES

The Internet and the World Wide Web are not the same thing. The Internet is a physical collection of networked computers. The World Wide Web is a logical collection of information contained on many of the computers comprising the Internet. To download a file from a Web page, the two computers (client machine running a browser and server machine hosting the Web page) must exchange the file data, along with other control information. If the download speed is slow, what could be the cause? A short list identifies many suspects:

- Noise in the communication channel forces retransmission of many packets.
- The path through the Internet introduces delay.
- The server is sending data at a limited rate.
- The Internet service provider has limited bandwidth.

So, before buying a new modem or upgrading your network, determine where the bottleneck is. The Internet gets more popular every day. New home pages are added, additional files are placed on FTP sites for downloading, news and entertainment services are coming online and broadcasting digitally, and more and more machines are being connected. The 10- and 100-Mbit Ethernet technology is already hard-pressed to keep up with the Internet traffic. Gigabit networking is coming, but it will only provide a short respite from the ever-increasing demands of global information exchange.

SUMMARY

In this exercise we discovered that

- The Internet offers a method to obtain universal service.
- The Internet is organized into domains.

- HTML is the source code of a Web page. Web pages are transferred using HTTP and displayed using browsers.
- Interactivity is added to a Web page using CGI and Java.
- A virtual private network uses tunneling of encrypted messages.

This self-test is designed to help you check your understanding of the background information presented in this exercise.

True/False

Answer *true* or *false*.

1. The hypertext markup language is used to encode GIF images.
2. CGI stands for Common Gateway Interchange.
3. HTML contains formatting commands called tags.
4. The main portion of a Web page is contained between the HEADER tags.
5. Java is a Web browser produced by Microsoft Corporation.
6. A VPN is a type of Web page.
7. IPSec is a set of protocols that add security to TCP/IP networking.

Multiple Choice

Select the best answer.

8. CGI applications can be written using
   a. Perl, Java, C, C++, and Visual BASIC.
   b. Only the Javascript language.
   c. An HTML editor.
9. The three different categories of Web pages are
   a. Large, medium, and small.
   b. Active, passive, and neutral.
   c. Static, dynamic, and active.
10. When the network is slow,
    a. Turn off all power to the computer and perform a reset.
    b. Try to determine where the bottleneck is located.
    c. Immediately upgrade to the newest, most expensive hardware available.
11. CGI applications use FORMs to
    a. Receive the input required for processing.
    b. Post the data to an e-mail application.
    c. Send information to the browser display.
12. The concept of universal service and the Internet involves
    a. Being able to connect to a universal router on the Internet.
    b. Being able to exchange information between computers at any time or place.
    c. Allowing all users to access the universal Internet database.
13. One method of sending data from a FORM is called
    a. FORM-SEND.
    b. GET.
    c. POST.
14. <P ALIGN="LEFT"> is an example of an
    a. HTML tag.
    b. HTTP tag.
    c. HTTP URL.

Completion

Fill in the blank or blanks with the best answers.

15. The same _____ code is displayed differently using different Internet browsers.
16. A CGI application provides the ability to create _____ _____ on the fly.

17. Java is used to create _____ Web pages.
18. _____ information includes text, video, audio, Java applets, and images.
19. The _____ protocol is used to exchange hypermedia information.
20. VPNs use IP _____ protocols.
21. _____ messaging allows you to send a message to another computer instantly.

FAMILIARIZATION
ACTIVITY

WWW

1. Examine each of the pull-down menu items in Netscape Navigator and Internet Explorer.
2. Read the online help to learn about browser features.
3. Identify similarities between the two browsers discussed in this exercise.
4. Identify differences between the two browsers discussed in this exercise.

CGI

1. Search the Web to locate information about Perl.
2. Locate a source for Perl, available free of charge.
3. Download Perl.
4. Install Perl.
5. Run some of the sample Perl scripts.

Java

1. Search the Web to locate information about the Java language.
2. Locate the source of a Java compiler available free of charge.
3. Download the Java compiler.
4. Install the Java compiler.
5. Compile some of the sample Java applets.
6. Execute a sample Java applet.

QUESTIONS/ACTIVITIES

1. Determine how to clear the browser's cache memory.
2. Determine the current allocation settings for the browser cache.
3. Search the Web to locate some useful resources related to active Web page development.

REVIEW QUIZ

Under the supervision of your instructor

1. Describe the basic organization of the Internet.
2. Explain the purpose of a browser and its relationship to HTML.
3. Discuss the usefulness of CGI applications.
4. Identify the elements of a virtual private network.

28 Electronic Mail

INTRODUCTION

Joe Tekk checked his e-mail for the fifth time in 10 minutes. He was getting impatient, having gotten used to the quick turnaround at his favorite software download site.

He checked it for a sixth time and was excited to see the e-mail he had been waiting for, an account verification response. The e-mail came with a small attachment containing a custom client application used with the software download site. Joe installed the client software and connected to the site to verify its operation.

Then Joe scanned several recent photographs and e-mailed the images to his sister, sent copies of a new network application he wrote to several friends, and finally handled the nine e-mail messages waiting in RWA Software's help desk mailbox.

Satisfied, Joe went to lunch. When he returned 45 minutes later, he had 18 additional e-mails to go through.

PERFORMANCE OBJECTIVES

Upon completion of this exercise, you will be able to

1. Describe the features of e-mail communication software.
2. Configure an electronic mail client.
3. Send and receive electronic mail.
4. Organize electronic mail messages into folders.

KEY TERMS

Electronic mail (e-mail)	MIME
Mailbox	POP3
Header	SMTP

BACKGROUND INFORMATION

Communication tools are at the heart of the personal computer revolution. In this exercise, we will explore *electronic mail* (commonly referred to as *e-mail*), one of the most common communication tools available. This exercise will cover the basic features of electronic mail, how to configure client software, how to send and receive electronic mail, and how to organize e-mail messages on a computer that is connected to the Internet.

WHAT IS E-MAIL?

In the early days of computer networking, a simple electronic mail program was used to exchange text messages between two people. Since then, electronic mail has evolved into personal communication tool that can be used to

- Send a message to several recipients.
- Send a message that contains text, graphics, and even multimedia audio and video files.
- Send a message that a computer program will respond to, such as a mailing list program or mail exploder.

Electronic mail combines the speed of electronic communication with features similar to the postal mail service. The major difference between the postal mail service and e-mail is that a computer can transmit a message across a computer network almost instantly.

When using electronic mail, several common features are available to the computer user. For example, it is possible for every user to

- Compose an e-mail message.
- Send an e-mail message.
- Receive notification that an e-mail message has arrived.
- Read an e-mail message.
- Forward a copy of an e-mail message.
- Reply to an e-mail message.

Let's begin our examination of e-mail by looking at how e-mail actually works on the Internet.

HOW E-MAIL WORKS

In order for a computer to use e-mail, it is necessary to install an e-mail client. Electronic mail uses the client-server method to allow for mail to be exchanged. Client computers exchange messages with a server that is ultimately responsible for delivering the e-mail messages to the destination.

On the server computer each user is assigned a specific mailbox. Each electronic *mailbox* or e-mail address has a unique address. It is divided into two parts: a mailbox name and a computer host name, which are separated using an "at" sign, @, such as

```
mailbox_name@computer_name
```

Together, both of these components provide for a unique e-mail address.

The mailbox portion of the address is often made from a user's name. The host name part of the address is chosen by a network administrator. For example, Joe Tekk has the e-mail address joetekk@stny.rr.com. From the example, this indicates that joetekk is the mailbox name and stny.rr.com is the computer name. Notice that Joe Tekk's e-mail address ends in .com. The .com indicates that stny.rr is a commercial organization. You will observe that the last three characters of an e-mail address will normally end with a limited number of domain name categories. These categories are shown in Table 28.1.

E-mail messages are actually exchanged using the client-server environment illustrated in Figure 28.1. Note that both of computers in Figure 28.1 are called e-mail servers. When the mail message is exchanged, the mail transfer program on the sending computer temporarily becomes a client and connects to the mail transfer program running as a server on the receiving computer. In this way, depending on whether mail is being sent or received, the mail transfer program acts as a client or a server.

FORMAT OF E-MAIL MESSAGES

The format of an e-mail message exchanged between the servers is quite simple. Each message consists of ASCII text that is separated into two parts. A blank line is used as the

TABLE 28.1 Common domain names

Domain Name	Assigned Group
com	A company or commercial organization
edu	An educational institution
gov	A government organization
mil	A military organization
net	Network service provider
org	Other organizations
country code	A country code, such as .us for United States or .jp for Japan

FIGURE 28.1 How e-mail is exchanged between servers

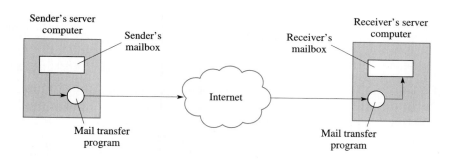

TABLE 28.2 Typical e-mail header keywords

Header Keyword	Description
To	The mail recipient's e-mail address
From	The sender's e-mail address
Cc	List of carbon copy addresses
Bcc	List of blind carbon copy addresses
Date	The date when the message was sent
Subject	The subject of the message
Reply-to	The address to which a reply should be sent

separator between the parts. The first part of the message is called a **header**. A header consists of a keyword followed by a colon and additional information. Some of the most common header keywords are shown in Table 28.2.

The second part of the message is called the *body* and contains the actual text of the message. Note that because it is possible to attach many different file types to an e-mail message, a special scheme called Multipurpose Internet Mail Extensions (MIME) was developed. This provides a way for binary programs, graphical images, or other types of files to be attached to an e-mail message. The sending computer encodes the message into text for transport and the receiving computer decodes the message back into the original form.

E-MAIL CLIENT SOFTWARE

One of the most popular client software e-mail client programs is Microsoft Outlook Express. It is installed as a part of the Windows operating system. There are usually several different ways to access the Outlook Express program. For example, there may be an icon on the desktop that may be double-clicked, a small Outlook Express icon may be placed on the taskbar, or it may be a program that can be selected from the Windows start menu. In any case, after the Outlook Express program is started, the computer user is presented with a screen display similar to Figure 28.2.

FIGURE 28.2 Microsoft Outlook Express displaying a message

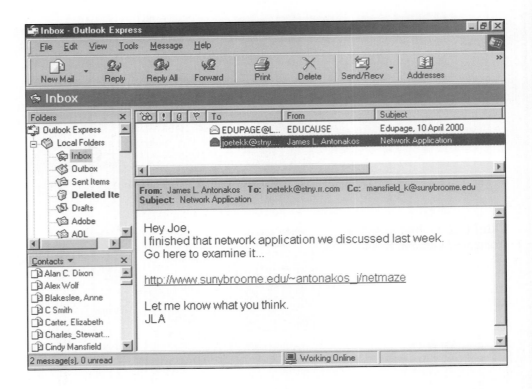

FIGURE 28.3 Outlook Express general e-mail properties

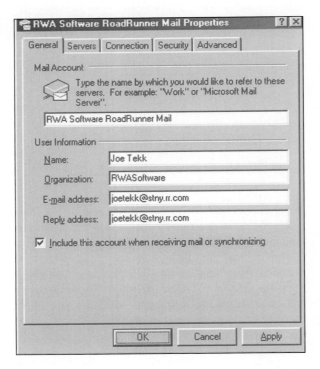

In order to use the Outlook Express program, it must be configured properly. This configuration consists of providing user information such as the user name, organization, e-mail address, and reply address. These items are located on the General E-mail Properties tab as shown in Figure 28.3. It is also necessary to identify the server computer to which the client will connect to send and receive mail. This information is found on the Servers tab of the Mail Properties window shown in Figure 28.4. There is a server associated with both incoming and outgoing mail.

FIGURE 28.4   Outlook Express Mail Servers

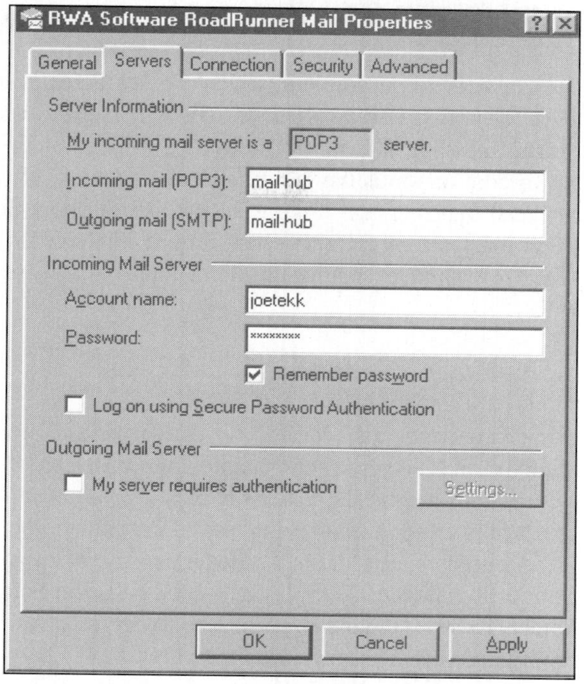

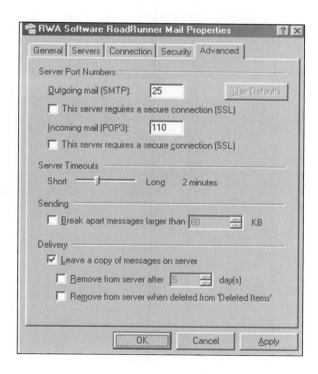

FIGURE 28.5   Outlook Express Advanced Mail Properties

Incoming mail uses POP3, the Post Office Protocol, whereas the outgoing mail server uses SMTP, the Simple Mail Transport Protocol. Notice that this is the screen where the user enters an Incoming Mail Server account name and password. As an added convenience, it is possible for Outlook Express to save or remember the password for future use.

Sometimes it is necessary to change some of the mail parameters. For example, it may be necessary to change the server timeout value of the mail program or change the setting that determines if a copy of a mail message is to be left on the server computer after it has been transferred to the client. As you can see from Figure 28.5, there are several different settings that can be modified. It is always a good idea to leave the settings alone unless there is a good reason to change them.

SENDING AN E-MAIL MESSAGE

Let's consider an example in which Joe Tekk creates the e-mail message shown in Figure 28.6. Joe uses the e-mail client to send a message to windy@alpha.com.

The message is sent by Joe to the e-mail server at stny.rr.com. The mail server at stny.rr.com forwards the message to the e-mail server at alpha.com where the user Windy can read that message. Figure 28.7 illustrates how the e-mail message is sent using the Microsoft Outlook Express client program. Notice that SMTP is used to transfer the message everywhere except for the client connection at the destination, which uses POP3.

RECEIVING AN E-MAIL MESSAGE

E-mail messages are received by the server and stored in the Inbox inside of a user's mailbox until it is read. For example, Figure 28.8 shows a message from the Java Developer Connection mailing list. After the message has been read, it can be deleted or saved. If a message is saved, it is normally moved to a folder other than the Inbox. This allows for mail to be stored in user-defined categories. To create a new folder for the Java Developer Connection message, simply right-click on Local Folders in the folder list and select New Folder. To move the message into the folder, drag it from the Inbox message list to the appropriate folder. This provides for an easy way to keep track of all related messages.

Note that Outlook Express provides the capability to store as many messages as necessary (as long as there is enough disk space available), although it is a good idea to keep the mailbox clean.

FIGURE 28.6   Creating a new e-mail message

FIGURE 28.7   Sending and receiving e-mail

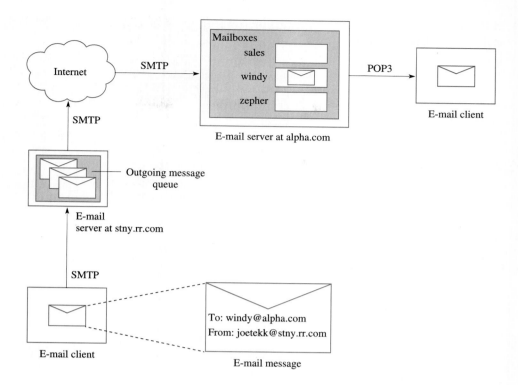

FIGURE 28.8   Reading a message in the Outlook Express Inbox

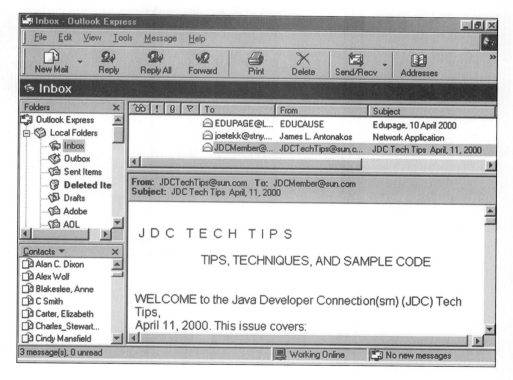

E-MAIL ERROR MESSAGES

There are several reasons why an error message may be generated when trying to send e-mail. Two of the most common errors stem from the user incorrectly specifying either the mailbox name or the computer name. In either case, a message will be sent back to the sender indicating

FIGURE 28.9 E-mail message indicating an invalid recipient

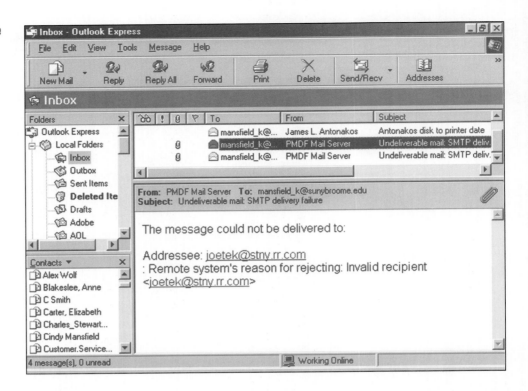

FIGURE 28.10 E-mail message indicating an invalid host/domain name

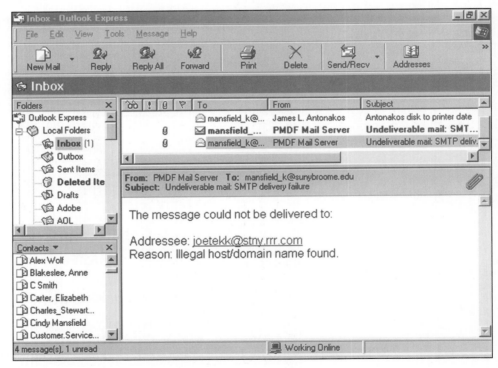

what type of error has occurred. Figure 28.9 illustrates an error with the mailbox portion of the address, whereas Figure 28.10 indicates a problem with the computer portion. Other problems with the mail will have specific messages that may help to resolve the problem.

ACCESS TO E-MAIL USING THE WWW

Some e-mail servers allow access to the mail system using a World Wide Web browser. The browser acts the same as an e-mail client that allows a user to send and receive e-mail

FIGURE 28.11 Accessing the e-mail server using the Web

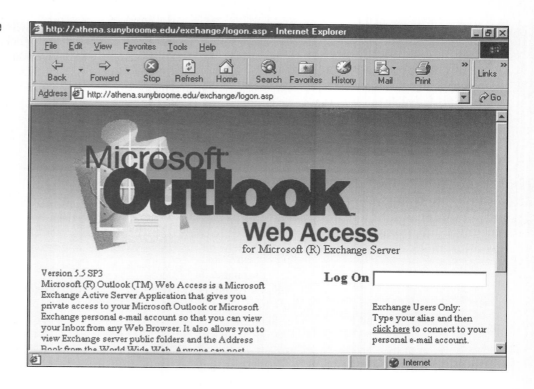

messages. Figure 28.11 shows the Microsoft Outlook Web Access screen that uses the Microsoft Exchange Server. Note that a user name and password are required in order to access any mail files.

TROUBLESHOOTING TECHNIQUES

One reason it might be helpful to know a few basic POP3 commands has to do with a real-world situation in which several e-mail messages were queued up behind an e-mail with a very large (over 4MB) attachment. Unfortunately, a network router problem creating frequent packet losses prevented the e-mail with the attachment from being properly transferred to the recipient's e-mail client. To get at the queued-up e-mail messages, the user used Telnet to connect to the POP3 server and deleted the e-mail message containing the large attachment. This allowed the remaining messages to transfer to the e-mail client. Since the messages were small, they transferred quickly, with only a slight delay introduced by the router problem.

SUMMARY

In this exercise we discovered that

- Electronic mail is sent using SMTP to a mailbox at a destination host computer.
- Electronic mail is read from a mailbox using POP3.
- E-mail messages contain a header and a body (and possibly MIME attachments).

SELF-TEST

This self-test is designed to help you check your understanding of the background information presented in this exercise.

True/False

Answer *true* or *false*.

1. An e-mail address consists of three parts.
2. The header and body of an e-mail message are separated by a blank line.

3. MIME provides a way to send binary attachments to an e-mail message.
4. Microsoft Outlook Express is an e-mail server program.
5. E-mail messages can be read using a World Wide Web browser.
6. POP3 is used to send e-mail.
7. SMTP is used to read e-mail.

Multiple Choice

Select the best answer.

8. In order to keep all related e-mail messages together
   a. Create a message bucket.
   b. Create a message folder.
   c. Keep all messages in the Outbox.
9. The mailbox portion of an e-mail address is typically
   a. A server.
   b. A client.
   c. A user name.
10. Outlook Express is a(n) _____ e-mail program.
    a. Client.
    b. Server.
    c. Both a and b.
11. E-mail messages are delivered on the Internet using the
    a. POP3 protocol.
    b. SMTP protocol.
    c. Both a and b.
12. To create a folder, it is necessary to right-click on the
    a. Inbox folder.
    b. Outbox folder.
    c. Local Folders folder.
13. In the e-mail address packets@nicard.com, the mailbox name is
    a. packets.
    b. nicard.
    c. nicard.com.
14. A program may be _____ to an e-mail message.
    a. Attached.
    b. Connected.
    c. MIME'd.

Completion

Fill in the blank or blanks with the best answers.

15. MIME stands for _____ _____ _____ _____.
16. The Telnet application can be used to connect to a(n) _____ server.
17. The _____ header keyword is used to send a carbon copy of an e-mail message.
18. After reading an e-mail message, it may be saved or _____.
19. The e-mail _____ is ultimately responsible for delivering e-mail messages to their destination.
20. Outgoing and incoming messages may be held in a(n) _____ with other messages.
21. Outlook Express is an e-mail _____.

FAMILIARIZATION ACTIVITY

1. Create and send an e-mail message to an invalid mailbox. What type of error message is generated?
2. Create and send an e-mail message to an invalid computer. What type of error message is generated?
3. Identify the name of the server computer for your e-mail address.

1. Why is it necessary to organize e-mail messages into folders?
2. Search the Web to locate information on POP3 commands. Make a detailed list.
3. Search the Web for free e-mail client programs. Are there any limitations or restrictions on using the software?

Under the supervision of your instructor

1. Describe the features of e-mail communication software.
2. Configure an electronic mail client.
3. Send and receive electronic mail.
4. Organize electronic mail messages into folders.

29 Network Design and Troubleshooting Scenarios

INTRODUCTION

Joe Tekk and Don, the senior technician, were meeting with their boss, Bill Bestman, the president of RWA Software, Inc.

"Bill," Don was saying, "you need to upgrade the office network."

"Why, Don?" President Bestman asked.

Don looked at Joe, who had begged him weeks ago to do the president's presentation.

"I'm going to let Joe explain the reasons, Bill."

Joe took over the meeting, expertly explaining why RWA Software needed to upgrade its existing network right away. Joe had researched the network hardware, looked up prices, and even estimated the number of man-hours required to complete the upgrade. Using his upgrade plan, no users would experience network downtime, since he planned on performing the upgrades in the evening, one department at a time.

When Joe finished the president said, "It sure sounds expensive, Joe. Do we really need to do this?"

"Yes, Bill, we do," Joe replied confidently. "There are local high schools that have better networks than we do."

President Bestman glanced at Don and then back to Joe. "Make it a showcase, you two, the best network in the area. I'm tripling your estimate, Joe. Buy whatever you need."

PERFORMANCE OBJECTIVES

Upon completion of this exercise, you will be able to

1. Discuss several considerations that must be made when networking computers together.
2. Estimate the hardware components needed for a specific network.
3. Discuss some initial steps to take when toubleshooting a network.

KEY TERMS

Network Neighborhood	Baseband
Broadcast storm	Broadband

BACKGROUND INFORMATION

In this exercise we will take a look at several different network scenarios, each more complex than the last. The goal is to provide you with ideas to begin designing your own network. Several network troubleshooting examples will also be examined.

NETWORKING TWO COMPUTERS

Connecting just two computers (in a dorm room or basement office) can be done several ways using

- Direct cable connection
- Network interface cards
- Modems

These methods are illustrated in Figure 29.1. The least expensive route is the ***direct cable connection***. The connection may be through a serial cable or a parallel cable. Go to Start, Programs, Accessories, and finally Communications to initiate a connection. Figure 29.2 shows the initial Direct Cable Connection window. Note that the computer is set up as a Guest. Left-clicking the Change tab allows you to switch between Guest mode and Host

FIGURE 29.1   Connecting two computers

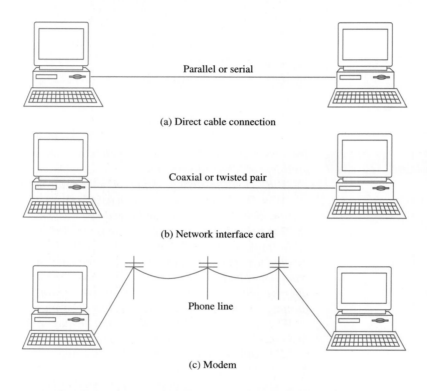

(a) Direct cable connection

(b) Network interface card

(c) Modem

FIGURE 29.2   Direct Cable Connection window

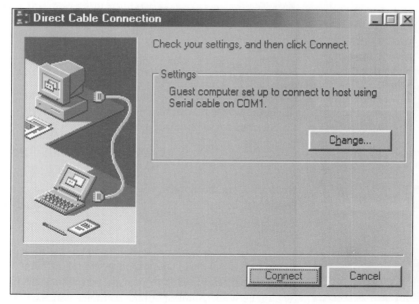

mode. The Host computer provides the resources the Guest computer wants to access over the connection. Use a parallel cable to get the fastest data transfer speed.

Network interface cards do not cost as much as modems, and no hub is required to connect only two computers together, just a crossover cable (for 10/100baseT) or length of coax (10base2).

Modems, by far the slowest connection, are useful when the two computers are separated by a large distance.

NETWORKING A SMALL LAB

The last few years have seen a tremendous increase in the number of schools connected to the Internet. On college campuses, many departments are networking laboratories to share resources, save money on equipment, and provide Internet access to their students and faculty.

What is required to network a small laboratory? Figure 29.3 shows an overhead diagram of the laboratory indicating the positions of computers, printers, and other devices.

Altogether a total of 11 machines are to be networked. With Ethernet as the desired technology, two possibilities exist:

1. Use one or more hubs or switches (10/100baseT).
2. Use coax (10base2).

Using hubs is more expensive, but may have advantages over using coax. For example, buying a 16-port hub will leave five ports free for future expansion. Using a four-port hub and an eight-port hub will leave one port available (each hub connected via its 10base2 port), and allow the network to be partitioned if necessary. Switches may also be used if it is necessary to establish a network hierarchy or guarantee bandwidth.

Using coaxial cable saves on the hardware cost, since we need only buy a spool of cable, a box of BNC connectors, and some T-connectors. More time is required to install this type of network, unless preterminated lengths of coax are purchased.

In addition to the hardware, network software must also be configured. Windows machines with built-in networking support will automatically communicate over the network via NetBEUI (file sharing, Network Neighborhood). It may be necessary to also assign each machine an IP address (static or dynamic). Use any class C address in the range 192.168.xxx.xxx if the organization does not have its own assigned address.

FIGURE 29.3   A small laboratory

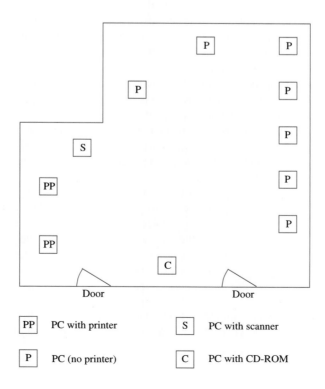

| PP | PC with printer | S | PC with scanner |
| P | PC (no printer) | C | PC with CD-ROM |

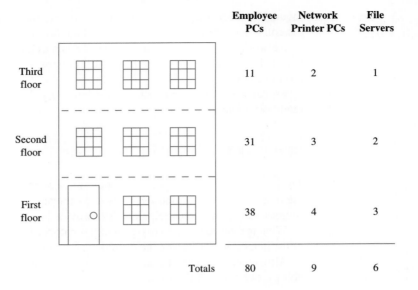

FIGURE 29.4 A small business

	Employee PCs	Network Printer PCs	File Servers
Third floor	11	2	1
Second floor	31	3	2
First floor	38	4	3
Totals	80	9	6

NETWORKING A SMALL BUSINESS

For the sake of this discussion, let's consider a small business with 80 employees. These employees are spread out over several floors of a small office building. In addition to one PC per employee, there are 15 additional PCs in various locations, mostly performing special duties as file servers or network print servers. Figure 29.4 shows the distribution of machines throughout the office building.

A network of this size (both the number of machines and the physical size of the office building) almost requires a hybrid network, with hubs or switches used to group bunches of PCs together, and coax, UTP, or fiber optic cable used to connect the hubs or switches. If hubs are used, whenever anyone sends a job to one of the network printers or requests a file from a file server, everyone (all 95 machines) must contend with the traffic. If switches are used, a network hierarchy can be established that isolates switched groups of users to their own local network printer so only members within the group contend with the printer traffic, not the entire network. File servers should be isolated in the same way. Using switches instead of hubs also allows the network to be repartitioned at a later time, to tweak performance or add new machines. A combination of switches and hubs also may be used, with switches isolating each floor of the building and hubs connecting all the users on a single floor.

It would be a mistake to wire the entire office building using only coax. First, the size of the building may prevent the use of a single length of coax, requiring different network segments connected by repeaters. Second, 95 *pairs* of crimps will be necessary to link all the machines together in a daisy chain. This situation is a disaster waiting to happen when just one of the 190 crimps goes bad. In addition, a coax-only network would provide fewer than 106,000 bits/second for each machine, a significantly smaller bandwidth than a switch would provide.

Figure 29.5(a) and (b) show two switch-based topologies that could be used to guarantee 10/100 Mbits/second to each machine. Using the star topology allows entire groups of machines to be disconnected for troubleshooting purposes.

If the nature of the business is heavily data dependent (multimedia presentations, streaming audio/video), it may be necessary to connect each floor via fiber and utilize Fast or Gigabit Ethernet technology. This can be done using a fiber switch or a fiber ring topology. The high-speed fiber backbone between floors guarantees that floor-to-floor bandwidth is available for all applications. Figure 29.5(c) shows how duplex fiber is used to daisy chain the fiber-10/100baseT switches. A star topology with a central fiber-only switch would also be acceptable, although at the cost of an extra switch.

FIGURE 29.5 Sample network topologies for office building

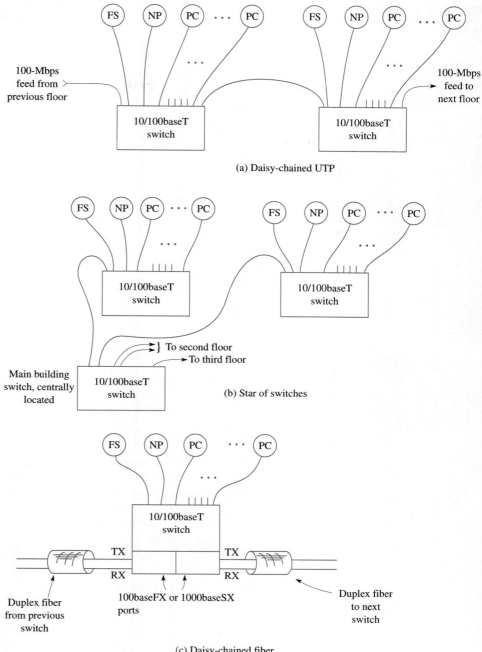

(a) Daisy-chained UTP

(b) Star of switches

(c) Daisy-chained fiber

Key: FS: File server
NP: Network printer
PC: Machine

NETWORKING A COLLEGE CAMPUS

A typical community college may employ several hundred faculty and staff and host several thousand students. Computers for student use are grouped into laboratories, with several laboratories in each building on campus. Our sample college in Figure 29.6 has a total of 14 laboratories, each one containing 16 machines and a network printer (stand-alone, no PC required). The number of labs in each building is circled in the figure.

FIGURE 29.6 A college campus

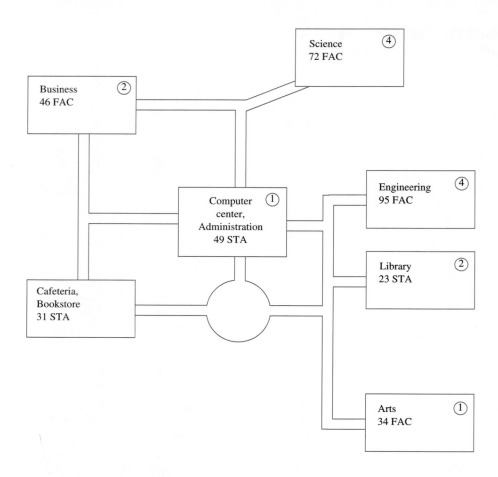

In addition to the lab computers, there are 350 faculty and staff wired to the network. Their numbers are indicated by the FAC and STA terms.

Each building connects to a central communications rack in the Computer Center. A pair of fibers (duplex cable) runs from each building to the Computer Center, where they all plug into a 100-Mbit/second fiber switch. Fiber was used instead of coax or UTP because of environmental concerns, as the college's geographical area is prone to thunderstorms. Fiber transceivers in each building convert between fiber and the 10base5 coaxial backbone cable used to distribute the network. Each floor has its own switch to isolate traffic. Figure 29.7 shows the layout of a typical campus building.

In the Computer Center, two mainframes connect to the central communications rack switch. One mainframe is for administrative use, the other for faculty/student use. The switch provides the necessary hierarchy separating the mainframes and their associated users. In addition, a router connected to the switch performs gateway duties, connecting the college to the Internet through a leased T1 (1.54-Mbit/second) line.

Because of the large number of machines (close to 240 in labs plus 350 faculty/staff), the college uses three class C subnets. With a number of IP addresses from each subnet reserved by the Computer Center, there are still over 700 IP addresses available for campus use. Taking away the lab and employee computers leaves 110 IP addresses, all available for future expansion.

Unfortunately, the use of 10base5 coaxial cable for the backbone in each building limits the network speed to 10 Mbits/second. Although the 10base5 technology was originally chosen for its 500-meter segment length (in order to connect all floors of a building), the bandwidth is inadequate for the campus traffic. Faculty and staff complaints about the "slow network" have prompted the administration to spend some money on a network upgrade. The proposed upgrade plan is as follows:

1. Replace the fiber to 10base5 transceiver with a fiber to 100baseT switch. Feed each floor with its own 100baseT cable.
2. Replace all hubs with 10/100baseT switches.
3. Install new 10/100baseT NICs in selected machines.

FIGURE 29.7   Network structure of a typical campus building

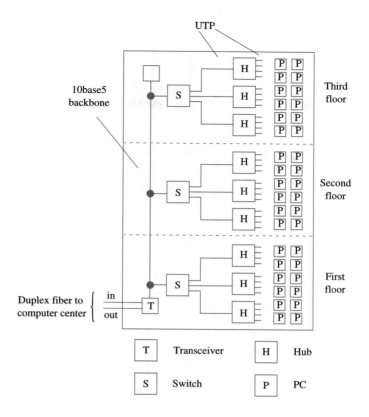

The administration did not want to spend the money to run fiber to each floor. Even so, their upgrade plan gets many users up to 100 Mbits/second.

A counterproposal made by a committee of faculty, staff, and students came in with a cost 23% higher than the administration proposal but provided for fiber to each floor, a gigabit backbone, and an additional T1 line.

The administration accepted the committee's proposal.

TROUBLESHOOTING TECHNIQUES

This section will present a number of troubleshooting tips and also a few case studies involving actual networking problems and their solutions. Bear in mind that troubleshooting network problems requires time, patience, and logical thinking. The tips presented here offer a place to begin troubleshooting a network problem.

CHECKING THE HARDWARE

Right-clicking on the My Computer icon and selecting Properties will bring up the System Properties window. Left-click the Device Manager tab to examine the list of installed hardware. If a small exclamation mark is present on the Network adapters device, there is a problem with the NIC or its driver. It is not uncommon for an interrupt conflict to arise when a sound card wants to use the same interrupt as the network adapter. Sometimes you can successfully change the interrupt of one of the devices. You may also be able to find and install a newer driver for the network card (on the Web).

USING TEST EQUIPMENT

For really nasty hardware problems, such as intermittent connections, it may be necessary to use sophisticated test equipment, such as a cable tester (UTP), time domain reflectometer (TDR) for coax, optical TDR (for fiber), or a network analyzer.

WHAT'S MY IP?

It is a good idea to run WINIPCFG and view the network information (addresses, mask, lease details, etc.). Figure 29.8(a) shows a WINIPCFG display for a machine that was not able to communicate with its network. Notice that the DNS Servers and Default Gateway fields are blank and that the DHCP Server address is 255.255.255.255. Compare this display with the one shown in Figure 29.8(b). In addition to different addresses in each field, the Lease Expires field is now filled in, and the Host Name has changed (indicating that the machine WAVEGUIDE is now part of the stny.rr.com domain).

CHECK THE NETWORK NEIGHBORHOOD

If your machine is properly networked you should be able to open up Network Neighborhood and see hosts sharing resources on the network. If all you get is a flashlight waving back and forth, there is a problem with the NetBEUI protocol or some other low-level network component.

CAN YOU PING?

Just being able to PING another network host (or not being able to) is valuable information. By successfully PINGing the host, we have proof that the network hardware and software are operating correctly. A sample PING report looks like this:

```
C> ping www.sunybroome.edu

Pinging sbccab.cc.sunybroome.edu [192.203.130.2] with 32 bytes of data:

Reply from 192.203.130.2: bytes=32 time=66ms TTL=245
Reply from 192.203.130.2: bytes=32 time=82ms TTL=245
```

FIGURE 29.8(a)   WINIPCFG display indicating invalid network information

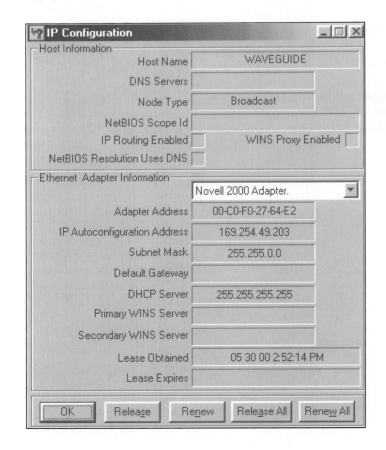

FIGURE 29.8(b) (continued)
WINIPCFG display of valid
network information

```
Reply from 192.203.130.2: bytes=32 time=92ms TTL=245
Reply from 192.203.130.2: bytes=32 time=122ms TTL=245

Ping statistics for 192.203.130.2:
    Packets: Sent = 4, Received = 4, Lost = 0 (0% loss),
Approximate round trip times in milli-seconds:
    Minimum = 66ms, Maximum =  122ms, Average =  90ms
```

If you can PING a host using its IP address but not its domain name, there could be a problem with your DNS server.

CASE STUDY #1: CANNOT BROWSE THE NETWORK NEIGHBORHOOD

One student in a networking lab was distressed because his machine was the only one that could not view the Network Neighborhood. The instructor spent a good deal of time checking properties before finally rebooting the machine and asking the student to try again. The instructor watched in surprise as the student cancelled the logon process to get to the desktop quicker. He then explained to the student that network privileges are only given to users who log on to Windows.

CASE STUDY #2: RJ-45 CONNECTOR PARTIALLY INSERTED

One user was confused when her machine would not connect to the network, even though the light on her hub was lit and the machine had been working on the network the previous evening. After spending a good deal of time on the phone with a technician and checking all hardware and software properties, she was about to give up. She began recalling her steps, thinking about everything she had done or seen since the computer last worked. She remembered that she pulled the minitower case to the front of her computer desk to adjust the volume control on her sound card. She carefully pulled the computer across the desk again, and saw that the RJ-45 connector was partially pulled out of its socket on the NIC. When she pushed on it she heard a snap as it clicked into place. After rebooting, her machine worked fine again.

CASE STUDY #3: MISSING TERMINATOR

After rearranging the old networking hardware in a closet, a technician discovered that the entire 10base2 network was not functioning. Not a single machine worked on the network. Careful inspection of the rearranged equipment revealed that the technician forgot to reconnect the terminating resistor on the end of the coax.

CASE STUDY #4: WRONG UTP CABLE TYPE

Of the eight machines connected to a hub in a new networking lab, only seven can connect to the network. When the UTP cables are swapped between the bad machine and a good one, the problem moves to the good machine and the bad machine is able to connect to the network. When the two cables are swapped into different ports on the hub the problem also moves. After examining the ends of the original cable, it was discovered that the cable was a crossover cable, not a straight-through cable that normally connects a NIC to a hub or switch port.

CASE STUDY #5: CANNOT MAP A NETWORK DRIVE

One user tried everything to try to map a network drive (no password required) on the mainframe at his business. He changed NICs, reinstalled all the networking protocols, and finally reinstalled the operating system. Finally, with nothing else to try, he changed from using a static IP address to obtaining one via DHCP. After rebooting, he was able to map the network drive. Discussing the problem and its accidental solution with everyone he knew always brought up the same response: "What does a static IP address have to do with the problem?" The answer is that it should not make a difference. No satisfactory explanation for the solution exists. We just feel lucky to have solved it. It is now something else to try if a similar problem shows up in the future.

CASE STUDY #6: BROADCAST STORM

One business suffered frequent *broadcast storms*, a flooding of its network with so many packets that its switches were forced to drop packets to try to maintain their buffers. Eventually, by capturing network traffic with a protocol analyzer and examining it, the network technicians found a network packet used as a trigger for the broadcast storm, a message sent to a broadcast address on the business's subnet. Someone was PINGing their broadcast address!

After more digging, the technicians learned that a software engineer at the company was experimenting with a network application he downloaded from a hacker Web site. The application, whether intentionally or not, was responsible for the broadcast storms.

CASE STUDY #7: DHCP NOT WORKING

On a particular college campus a DHCP server is used to dynamically assign IP addresses to machines when they boot. One day the entire system stopped working, and no one could obtain an IP address. After isolating the building causing the DHCP server (in the Computer Center) to fail, the campus network technician went from lab to lab in the affected building. He finally found the problem: a Linux machine set up by a student as the DHCP server for a project and accidentally connected to the college network. Once the machine was disconnected, the normal DHCP service resumed.

BASEBAND VERSUS BROADBAND

All the networks discussed in this exercise were *baseband* communication networks. A baseband network has a single information carrier that is modulated with the digital network data. *Broadband* communication networks, like the television cable system, use many different carriers and thus support multiple channels of data. Broadband communication systems require more expensive hardware than baseband systems and are typically used for high-bandwidth applications, such as broadcast video and FM audio.

SUMMARY

In this exercise we discovered that

- Both network hardware and network software must be working to use a network.
- Network troubleshooting requires checking the hardware and the software.
- There are baseband and broadband communication networks.

SELF-TEST

This self-test is designed to help you check your understanding of the background information presented in this exercise.

True/False

Answer *true* or *false*.

1. Only the direct cable connection can be used to connect two computers.
2. Coax should always be used when networking small laboratories.
3. Coax should always be used when networking small businesses.
4. Regarding Figure 29.7, machine-to-machine traffic on the second floor is also broadcast to the fiber.
5. Regarding Figure 29.7, machine-to-machine traffic from the second floor to the third floor is also broadcast to the fiber.
6. Interrupt conflicts do not prevent a NIC from operating correctly.
7. You must log on to Windows to have access to network services.

Multiple Choice

Select the best answer.

8. Modems are useful for
   a. PCs that are in the same room.
   b. PCs that have old 386 processors.
   c. PCs that are separated by large distances.
9. How many four-port 10baseT hubs are needed to network the machines in the laboratory of Figure 29.3? Each hub also contains a 10base2 port.
   a. 2 hubs.
   b. 3 hubs.
   c. 4 hubs.
10. Assuming only 16-port hubs (with 10base2 ports) are used, how many are required for the small business shown in Figure 29.4?
    a. 4 hubs.
    b. 5 hubs.
    c. 6 hubs.
11. How many network connections are required for all the labs in Figure 29.6?
    a. 224.
    b. 238.
    c. 240.

12. To connect to the Internet you must use a
    a. Hub.
    b. Switch.
    c. Router.
13. Direct cable connection has two modes of operation, Guest and
    a. Client.
    b. Host.
    c. Server.
14. A network burdened by excessive packets is experiencing a
    a. Broadcast flood.
    b. Broadcast storm.
    c. Packet flood.

Completion

Fill in the blank or blanks with the best answers.

15. The direct cable connection is the _____ expensive way to connect two computers.
16. Modems provide the _____ speed connection between two computers.
17. Switches are used to establish a network _____.
18. Bad _____ are a problem in large coaxial (10base2) networks.
19. Fiber is used in the college campus scenario because of _____ conditions.
20. The IP address of a machine is displayed by the _____ utility.
21. Normally, a NIC connects to a port on a hub using a(n) _____ UTP cable.

FAMILIARIZATION ACTIVITY

1. Using a networking hardware catalog (or the Web), look up the prices on each of the following networking components:
   • Cables
   • Network interface cards (ISA vs. PCI, 10 Mbit vs. 100 Mbit)
   • Repeaters and transceivers
   • Hubs
   • Switches
   • Routers
2. Using the prices from step 1, determine the cost of a network containing 10 PCs with 100-Mbit PCI network interface cards, a four-port hub and an eight-port hub, and all necessary UTP cabling.
3. How does the cost change if one 16-port hub is used?
4. How does the cost change if switches are used?
5. Log on to a computer that is networked and run WINIPCFG. Copy down all addresses and other information.
6. On the same computer, open up Network Neighborhood. How many machines are shown? What are their names?

QUESTIONS/ACTIVITIES

Analyze the laboratory diagram in Figure 29.9. Determine how to network the computers. Assume that four-port and eight-port hubs (with 10base2 ports) are available for use, if necessary. Explain why you would use 10base2 over 10/100baseT, or vice versa, how you considered the costs, future expansion of the lab, the amount of network traffic, and any other parameter that was significant.

FIGURE 29.9   Lab setup for networking activity

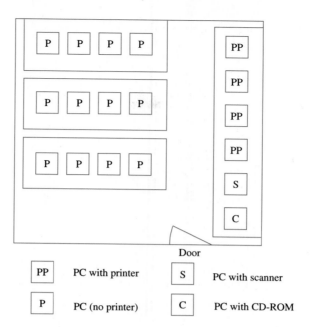

| PP | PC with printer | | S | PC with scanner |
| P | PC (no printer) | | C | PC with CD-ROM |

REVIEW QUIZ

Under the supervision of your instructor

1. Discuss several considerations that must be made when networking computers together.
2. Estimate the hardware components needed for a specific network.
3. Discuss some initial steps to take when troubleshooting a network.

30 Windows NT Domains

INTRODUCTION

Joe Tekk was very excited. He was finally given the opportunity to set up the new Windows NT domain for RWA Software, Inc. Joe thought that a Windows NT network was necessary because it was becoming harder and harder to maintain the workgroup that was originally installed several years ago.

Joe told Don, the senior technician, "RWA has grown so much since I started working here. The NT Server operating system is going make it so much easier to administer the network."

Don replied, "If you say so, Joe. I'll leave the network administration up to you." He continued, "Joe, please let me know what you are planning before we make any big changes. We don't want to make any avoidable mistakes."

Joe responded, "I'm glad you mentioned that, Don. I laid out the plan on paper so everyone can understand how the new network will operate. I have also set up a timetable to get everyone up and running."

Don smiled and said, "Great job, Joe. Keep it up!"

Joe spent most of his spare time reading about Windows NT. When he finally received his copy of the Windows NT Server CD-ROM, he could not wait to get started.

PERFORMANCE OBJECTIVES

Upon completion of this exercise, you will be able to

1. Describe the benefits of creating a Windows NT domain.
2. Explain some different types of Windows NT domains.
3. Discuss the different types of clients able to join a Windows NT domain.

KEY TERMS

Primary domain controller
Backup domain controller

Trust relationship
Security accounts manager (SAM) database

BACKGROUND INFORMATION

Any group of personal computers can be joined together to form either a workgroup or a domain. In a workgroup, each computer is managed independently but may share some of its resources with the other members of the network, such as printers, disks, or a scanner. Unfortunately, as the number of computers in the workgroup grows, it becomes more and more difficult to manage the network. This is exactly the situation in which a Windows NT

367

TABLE 30.1 Comparing a workgroup and a domain

Workgroup	Domain
Small networks	Large networks
Peer-to-peer	Client–server
No central server	Central server
Low cost	Higher cost
Decentralized	Centralized

domain can be used. A *domain* offers a centralized mechanism to relieve much of the administrative burden commonly experienced in a workgroup. A domain requires at least one computer running the Windows NT Server operating system. Table 30.1 illustrates the characteristics of a workgroup and a domain.

DOMAINS

Each Windows NT domain can be configured independently or as a group in which all computers are members of the same domain. Figure 30.1 shows two independent domains. Each domain consists of at least one Windows NT primary domain controller (PDC) and any number of backup domain controllers (BDC). One shared directory database is used to store user account information and security settings for the entire domain.

A BDC can be promoted to a PDC in the event the current PDC on the network becomes unavailable for any reason. A promotion can be initiated manually, causing the current PDC to be demoted to a backup. Figure 30.2 shows a domain containing two Windows NT Server computers. One computer is the PDC and the other computer is the BDC.

Windows NT can administer the following types of domains:

- Windows NT Server domains
- Windows NT Server and LAN Manager 2.x domains
- LAN Manager 2.x domains

A LAN Manager 2.x domain is a previous version of Microsoft networking software used by older MS-DOS and Windows computers.

The different types of activities that can be performed on a domain include the following:

- Create a new domain
- Modify an existing domain
- Join a domain
- Add a computer to a domain
- Remove a computer from a domain
- Synchronize files in a domain
- Promote a BDC to a PDC
- Establish trust relationships

When a system is set up as a PDC, the new domain name is required in order to proceed through the Windows NT installation process. This domain name is required by all other computer users who want to join the domain. Note that each domain can contain only *one*

FIGURE 30.1 Independent Windows NT domains

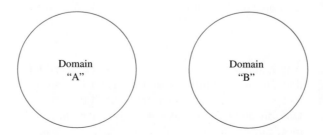

FIGURE 30.2 Domain "A"
configuration

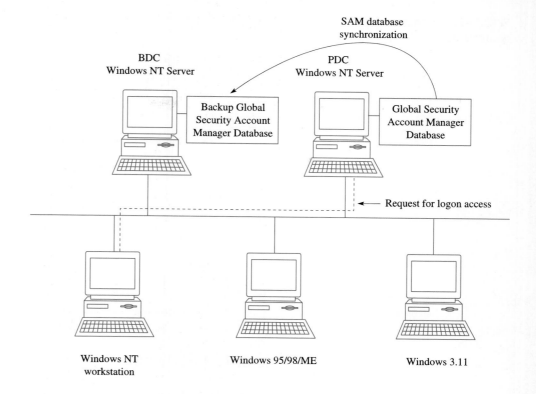

primary domain controller. All other Windows NT Server computers can be designated as backups or ones that do not participate in the domain control process at all.

A computer can be configured to join a domain during the Windows NT installation process, using the Network icon in the system Control Panel, or by using the Server Manager tool. A computer can be removed using the Network icon in the system Control Panel or the Server Manager tool.

Synchronizing a domain involves exchanging information between a primary domain controller and any secondary or backup domain controllers as shown in Figure 30.2. The synchronization interval for a Windows NT computer is five minutes. This means account information entered on the primary domain controller takes only five minutes to be exchanged with all secondary computers. This synchronization is performed automatically by Windows NT.

Domains can also be set up to offer *trust relationships*. A trust relationship involves either providing or receiving services from an external domain, as shown in Figure 30.3. A trust relationship can permit users in one domain to use the resources of another domain. A trust relationship can be a one-way trust or a two-way trust, offering the ability to handle many types of requirements.

A one-way trust relationship as shown in Figure 30.3(a) identifies domain "B" as a trusted source for domain "A." A two-way trust, shown in Figure 30.3(b), involves two separate domains sharing their resources with each other. Each domain considers the other to be a trusted source. Extreme caution must be exercised when setting up trust relationships. If the trusted domain is really untrustworthy, valuable information can be lost using the "trusted" accounts.

DOMAIN CLIENTS

A Windows NT domain can support many different types of clients, such as

- Windows NT Servers
- Windows NT workstations
- Windows 95/98 clients
- Windows 3.11 clients
- Windows 3.1 clients
- MS-DOS clients
- OS/2 workstations

369

FIGURE 30.3   Domain trust
relationships

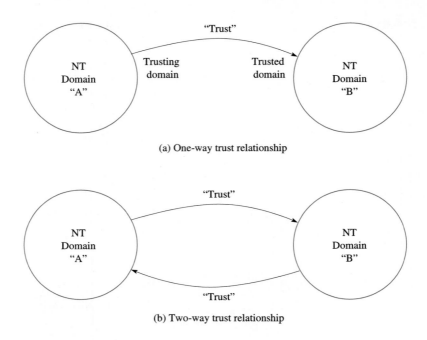

(a) One-way trust relationship

(b) Two-way trust relationship

LOGGING ONTO A NETWORK

When a computer is configured to run in a network, each user must be authorized before
access to the computer can be granted. Figure 30.4 shows a typical Windows 95/98 logon
screen. Each user must supply a valid user name and a valid password in order to gain access
to the computer and any network resources. In a *workgroup* setting, all password information
is stored locally on each computer in PWL files. The PWL files are named using the
following format: the first eight letters of the user name entered in the logon screen followed
by the .PWL file extension. The PWL files contain account and password information stored
in encrypted form. These files are typically stored in the Windows directory. Figure 30.5
shows the concept of a workgroup where each computer is administered independently.

In a *domain* setting, a centralized computer running Windows NT is contacted to verify
the user name and password. If the information provided to the server is valid, access is
granted to the local machine. If either the user name or password is invalid, access to the
local computer is denied. As you might think, this method offers tremendously more flexibility as far as the administration is concerned. This concept is illustrated in Figure 30.6.

RUNNING A NETWORK SERVER

Running a network server involves installing the Windows NT operating system and then
configuring it to run as a primary or secondary domain controller during the installation
process. After a PDC is created during the installation, the domain exists on the network.

FIGURE 30.4   Windows 95/98
logon screen

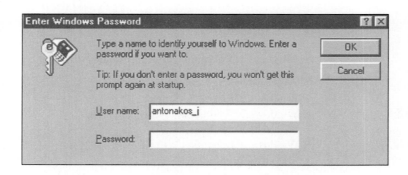

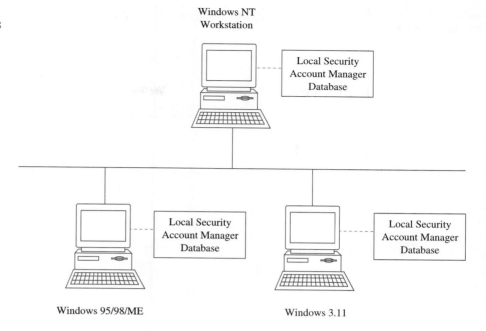

FIGURE 30.6 Domain concept

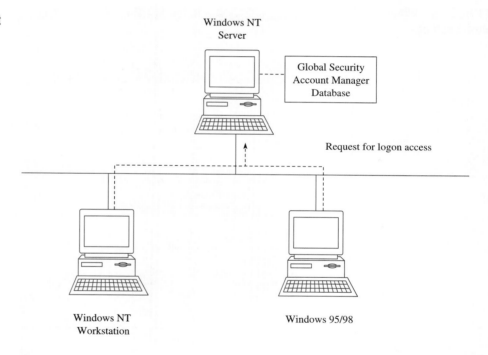

Windows NT computers can then join the domain by changing the Member of Domain as shown in Figure 30.7. Windows computers join the domain by changing individual settings on each computer. Figure 30.8 shows the Primary Network Logon, selecting the Client for Microsoft Networks option. Then, by selecting the properties for Client for Microsoft Networks, the specific domain can be identified as illustrated in Figure 30.9. After making these changes, a system reset is necessary to make the changes active.

Network server computers are also assigned the task of running more applications to manage both the server and network. For example, a Windows NT Server may be used to add fault tolerance to disks using a Redundant Array of Inexpensive Disks (RAID) technology. A server may also run the WWW server application, Windows Internet Naming System (WINS), Dynamic Host Configuration Protocol (DHCP), and Remote Access Server (RAS). These services are usually required 24 hours a day, seven days per week.

**FIGURE 30.7 Configuring a
Windows NT Server**

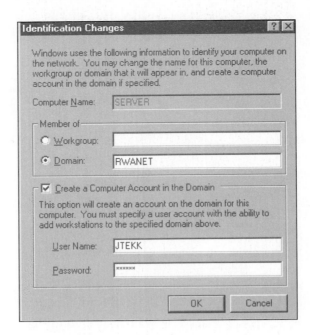

**FIGURE 30.8   Windows 95/98
Network settings**

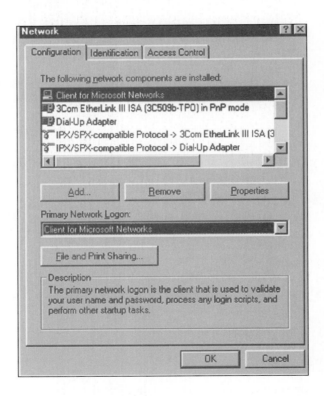

Windows NT Server computers are designed to handle the computing workload for entire organizations, corporations, or any other type of enterprise. In these cases, many servers (including a PDC and several BDCs) are made available to guarantee the availability of any required services.

USER PROFILES

In a domain, the primary domain controller maintains all user profiles. This allows for centralized control of the *security accounts manager (SAM) database*. Two programs are

FIGURE 30.9 Configuring Windows 95/98 to log on to a domain

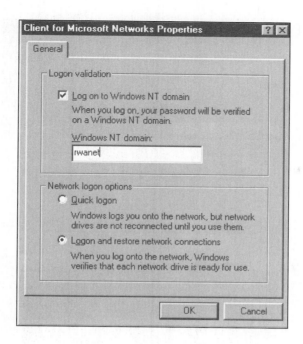

FIGURE 30.10 New User dialog box

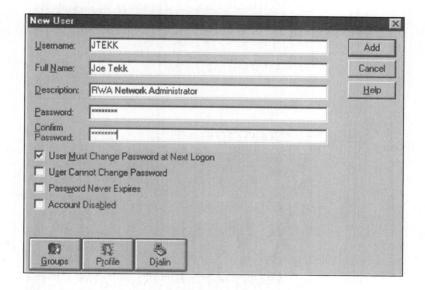

provided to update the SAM database. One of the programs is used in a stand-alone (no domain) environment and the other is for use where a domain is specified. Otherwise the programs operate in the same way. Let's examine what is involved when setting up a user account as illustrated in Figure 30.10. Information must be specified about each user account including user name, full name, a description of the account, and the password setting. The check boxes are used to further modify the account, such as requiring a password change during the first logon, restricting changing the password, extending the life of a password, and, lastly, disabling the account.

The three buttons at the bottom of the New User window (Figure 30.10) allow for each new account to be added to different *groups* as shown in Figure 30.11. It is a good idea to grant access to groups on an individual basis as certain privileges are granted by simply belonging to the group, such as administrator.

The User Environment Profile screen specifies the path to an individual profile and any required logon script name. Additionally, the home directory may be specified as shown in Figure 30.12.

FIGURE 30.11   Group
Memberships selection screen

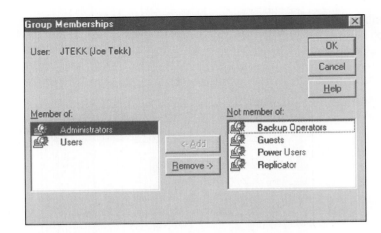

FIGURE 30.12   User
Environment Profile screen

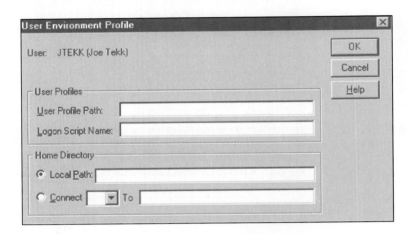

FIGURE 30.13   Dialin
Information settings

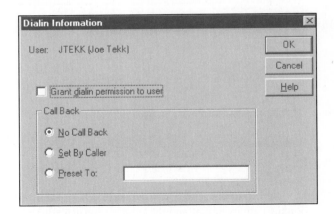

Lastly, the Dialin Information window determines if a Windows NT account has access to Dial-Up Networking. The Call Back option may also be configured to require the computer to call the user back. This is an additional security feature that may be implemented if necessary. Figure 30.13 shows these settings.

SECURITY

Windows NT is a C2 compliant operating system, when it is configured properly as defined by the National Computer Security Center (NCSC). C2 compliance involves properly configuring Windows NT to use the built-in safeguards. An application tool supplied with the operating system (C2CONFIG.EXE) examines the operating system setting against a recommended setting. Any exceptions are noted.

FIGURE 30.14 System events display

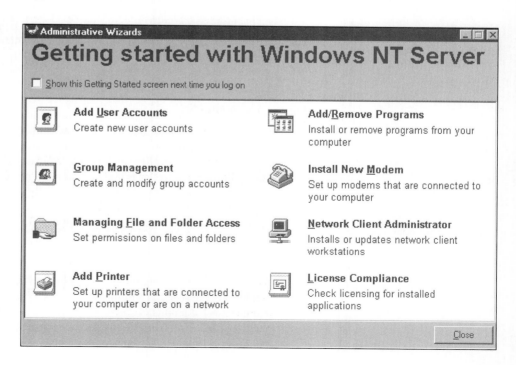

	Date	Time	Source	Category	Event	User	Computer
	Event Viewer - System Log on \\SERVER						
●	5/5/98	11:26:08 PM	Service Control Mar	None	7000	N/A	SERVER
❶	5/5/98	11:24:50 PM	BROWSER	None	8033	N/A	SERVER
①	5/5/98	10:55:30 PM	Srv	None	2013	N/A	SERVER
①	5/5/98	10:55:30 PM	Srv	None	2013	N/A	SERVER
①	5/5/98	10:55:30 PM	Srv	None	2013	N/A	SERVER
●	5/5/98	10:50:29 PM	Service Control Mar	None	7001	N/A	SERVER
●	5/5/98	10:50:29 PM	Service Control Mar	None	7001	N/A	SERVER
●	5/5/98	10:50:27 PM	Service Control Mar	None	7000	N/A	SERVER
●	5/5/98	10:50:27 PM	Service Control Mar	None	7000	N/A	SERVER
❶	5/5/98	10:50:14 PM	EventLog	None	6005	N/A	SERVER
●	5/5/98	10:50:27 PM	Service Control Mar	None	7000	N/A	SERVER
❶	5/5/98	3:12:37 PM	BROWSER	None	8033	N/A	SERVER
①	5/5/98	3:10:18 PM	Srv	None	2013	N/A	SERVER
①	5/5/98	3:10:18 PM	Srv	None	2013	N/A	SERVER
①	5/5/98	3:10:18 PM	Srv	None	2013	N/A	SERVER
●	5/5/98	3:05:07 PM	Service Control Mar	None	7001	N/A	SERVER
●	5/5/98	3:05:07 PM	Service Control Mar	None	7001	N/A	SERVER
●	5/5/98	3:05:05 PM	Service Control Mar	None	7000	N/A	SERVER
●	5/5/98	3:05:05 PM	Service Control Mar	None	7000	N/A	SERVER
❶	5/5/98	3:04:51 PM	EventLog	None	6005	N/A	SERVER
●	5/5/98	3:05:05 PM	Service Control Mar	None	7000	N/A	SERVER
❶	5/5/98	3:03:46 PM	BROWSER	None	8033	N/A	SERVER
●	5/5/98	3:01:55 PM	Service Control Mar	None	7001	N/A	SERVER

Windows NT provides security-logging features designed to track all types of system activities, such as logon attempts, file transfers, Telnet sessions, and many more. Typically the system administrator will determine which types of events are logged by the system. Figure 30.14 shows the system log. The icons along the left margin are color coded to draw attention to more serious events. Event logs should be reviewed daily.

TROUBLESHOOTING TECHNIQUES

A networked computer environment (especially when using Windows NT) can become somewhat complex, requiring the system or network administrator to have many technical skills. Fortunately, Windows NT also provides many resources designed to tackle most networking tasks. For example, the Administrative Tools menu contains the Administrative Wizards option shown in Figure 30.15. Most of these wizards perform the activities that are necessary to get a network up and running.

FIGURE 30.15 Administrative Wizards display

Getting started with Windows NT Server

☐ Show this Getting Started screen next time you log on

Add User Accounts
Create new user accounts

Add/Remove Programs
Install or remove programs from your computer

Group Management
Create and modify group accounts

Install New Modem
Set up modems that are connected to your computer

Managing File and Folder Access
Set permissions on files and folders

Network Client Administrator
Installs or updates network client workstations

Add Printer
Set up printers that are connected to your computer or are on a network

License Compliance
Check licensing for installed applications

Close

FIGURE 30.16   Windows NT
Help display

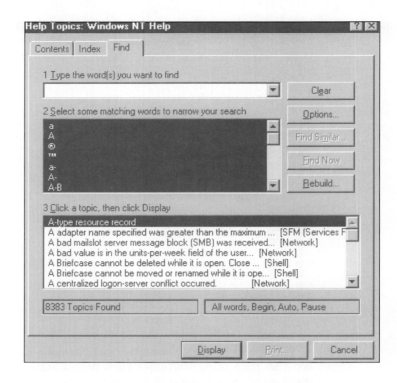

It is also a good idea to examine the online help system to get additional information, which may simplify any task. Figure 30.16 shows a Help screen that contains a total of 8383 topics, many of which contain information about networking.

SUMMARY

In this exercise we discovered that

- Windows NT can operate as a PDC, BDC, or neither.
- A PDC administers a domain.
- Trust relationships exist between domains.
- User profiles are maintained in the SAM database.

SELF-TEST

This self-test is designed to help you check your understanding of the background information presented in this exercise.

True/False

Answer *true* or *false*.

1. A workgroup uses a centralized server to administer the network.
2. Each Windows NT domain can be configured independently.
3. A primary domain controller can be demoted to a backup domain controller.
4. A backup domain controller is updated every 10 minutes.
5. Windows 95 computers function only marginally in a Windows NT domain.
6. A single PDC can operate several domains.
7. Windows NT has no security features built into it.

Multiple Choice

Select the best answer.

8. A Windows NT server can administer
   a. Windows NT domains.
   b. Windows NT and TCP/IP domains.
   c. Windows NT and LAN Manager domains.

9. Windows computers are added to a Windows NT domain by
   a. Double-clicking on the Windows NT computer in the Network Neighborhood.
   b. Modifying the properties of the TCP/IP network settings.
   c. Modifying the properties of the Client for Microsoft Network settings.
10. A Windows NT Server can be a
    a. Parent domain controller and a child domain controller.
    b. Secondary domain controller and a backup domain controller.
    c. Primary domain controller and a secondary domain controller.
11. A trusted domain
    a. Contains only one primary domain controller and no secondary controller.
    b. Is granted special access to all Windows computers in the trusted domain.
    c. Permits users in one domain to use the resources of another domain.
12. Running a network server involves
    a. Installing and configuring a Windows NT workstation computer.
    b. Installing and configuring a Windows NT Server computer.
    c. Connecting Windows 95 computers to Windows NT workstation computers.
13. If the PDC fails,
    a. The domain shuts down.
    b. All files revert to read-only status.
    c. A BDC takes over.
14. Network servers may run the following:
    a. WWW server, WINS, DHCP server.
    b. RAID.
    c. Both a and b.

Matching

Match a description of the networking topic on the right with each item on the left.

15. Windows NT client
16. Windows NT server
17. Domain types
18. Trust relationships
19. Domain activity

a. Windows NT and LAN Manager
b. One-way and two-way
c. PDC, BDC, none
d. DOS, OS/2, Windows 95/98
e. Establish trust relationships

Completion

Fill in the blank or blanks with the best answers.

20. A backup domain controller is _____ to a primary domain controller.
21. A large number of computers cannot be managed effectively in a(n) _____ setting.
22. Computers administered centrally are part of a(n) _____.
23. Each domain must contain _____ primary domain controller.
24. A Windows NT Server may be either a(n) _____, _____, or not involved in the domain controller process.
25. The Tools menu contains the Administrative _____ to help manage the NT network.
26. Exchanging information between a PDC and its BDCs is called _____.

FAMILIARIZATION ACTIVITY

1. Configure a Windows NT computer to function as a primary domain controller.
2. Configure a Windows NT computer to function as a backup domain controller.
3. Change the name of the domain.
4. Add some client computers to the Windows NT domain.
5. Examine networking topics using the online help system.
6. Run the Administrative Wizard applications to become familiar with account and network information.

1. Under what circumstances can a Windows NT workstation computer become a primary domain controller? A Windows NT Server computer?
2. When should a Windows NT domain be used instead of a workgroup?
3. What is necessary for an operating system to become a network client?
4. What is C2 security?
5. Where can additional C2 information be found?

Under the supervision of your instructor

1. Describe the benefits of creating a Windows NT domain.
2. Explain some different types of Windows NT domains.
3. Discuss the different types of clients able to join a Windows NT domain.

31

An Introduction to Telecommunications

INTRODUCTION

Joe Tekk stood in the equipment room of a local Internet service provider. He watched silently while the ISP manager, Dave Guza, described all the hardware and software.

"We service 800 local individuals and businesses from this room," Dave explained. "The three servers on the floor are for e-mail, Web pages, and news. We have two T1 connections that constitute our main Internet connection, with a dedicated 56K baud backup for emergencies."

While Dave was speaking he moved to the back of several tall racks of equipment. "These are our modem banks. They service our 750 dial-up access lines."

Joe watched as the modem lights blinked on and off. All the modems looked busy. "Is it this busy all the time?" he asked.

Dave smiled at Joe. "No, it is usually busier."

PERFORMANCE OBJECTIVES

Upon completion of this exercise, you will be able to

1. Describe the different telecommunication technologies.
2. Discuss reasons for choosing one technology over another.

KEY TERMS

TDM	Virtual channel
T1 carrier	ISDN
Circuit switching	Basic rate interface
Packet switching	SONET
Frame relay	FDDI
ATM	DWDM
Cell relay	DFWMAC

BACKGROUND INFORMATION

The world of telecommunications is getting both larger and smaller at the same time. From a hardware standpoint, more equipment is being installed every day, connecting more and more people, businesses, and organizations.

At the same time, the pervasiveness of the World Wide Web has made it easy to communicate with someone practically anywhere on the planet. The world does not seem as large as it once did.

In this exercise we will examine the many different telecommunication technologies available and see how they take part in our everyday communication.

TDM

Time-division multiplexing, or TDM, is a technique used by telephone companies to combine multiple digitized voice channels over a single wire. Telephone conversations are digitized into 8-bit PCM (pulse code modulation) samples and sampled 8000 times per second. This gives 64,000 bps for a single conversation. Now, using a multiplexer, if we rapidly switch from one channel to another, it is possible to transmit the 8-bit samples for 24 different conversations over a single wire. All that is required is a fast bit rate on the single wire. Figure 31.1 shows a timing diagram for the TDM scheme on a basic carrier called a **T1 carrier**. The T1 provides 1.544 Mbps multiplexed data for twenty-four 64,000 bps channels. The 8 bits for each channel (192 bits total) plus a framing bit (a total of 193 bits) are transmitted 8000 times per second.

Table 31.1 shows the different levels of T-carrier service available.

CIRCUIT SWITCHING

In the early days of the telephone system, large rotary switches were used to switch communication lines and make the necessary connections to allow end-to-end communication. The switches completed a circuit, hence the name *circuit switching*.

Eventually these slow, mechanical switches were replaced with fast, electronic switches. Also called an *interconnection network*, a switch is used to direct a signal to a specific output (such as the telephone you are calling).

Figure 31.2 shows one way to switch a set of eight signals. This type of switch is called a *crossbar switch*. Connections between input and output signals are made by closing switches at specific intersections within the 8-by-8 grid of switches. Only one switch is turned on in any row or column (unless we are broadcasting). Since each intersection contains a switch that may be open or closed, one control bit is required to represent the position of each switch. The pattern for the first row of switches in Figure 31.2 is 01000000. The pattern for the second row is 00000100. A total of 64 control bits is required.

FIGURE 31.1   Time-division multiplexing

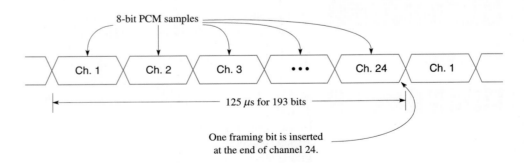

TABLE 31.1   T-carrier services

Level	Number of voice channels	Data Rate (Mbps)
1	24	1.544
2	96	6.312
3	672	44.736
4	4032	274.176

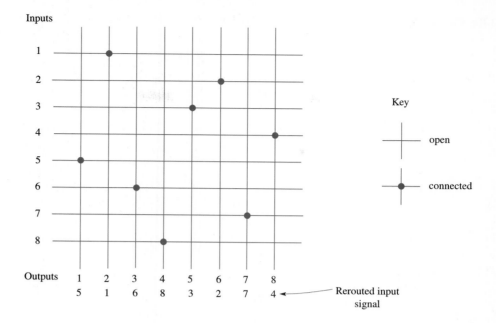

A nice feature of the crossbar switch is that any mapping between input and output is possible.

If the cost of 64 switches is too much for your communication budget, a different type of switch can be used to switch eight signals, but with less than half the number of switches. Called a *multistage switch*, it relies on several stages of smaller switches connected in complex ways. Figure 31.3 shows a sample three-stage switch capable of switching eight signals. Each smaller switch can be configured as a straight-through or crossover switch, with a single control bit specifying the mode. Now, with only 12 smaller switches, the control information has shrunk from 64 bits in the crossbar switch to only 12 bits. The number of switches is 24 (one switch for straight-through, one switch for crossover, times 12), which is less than half of the 64 required in the crossbar switch.

FIGURE 31.3 Eight-signal multistage switch

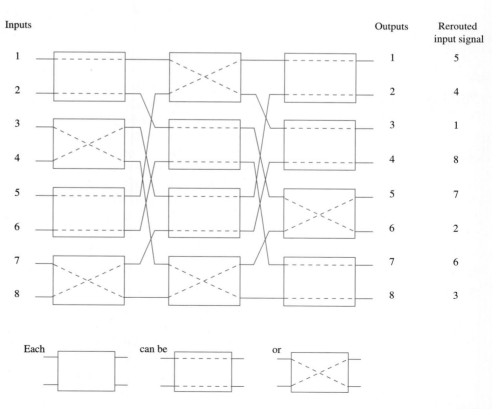

FIGURE 31.4   Sample WAN

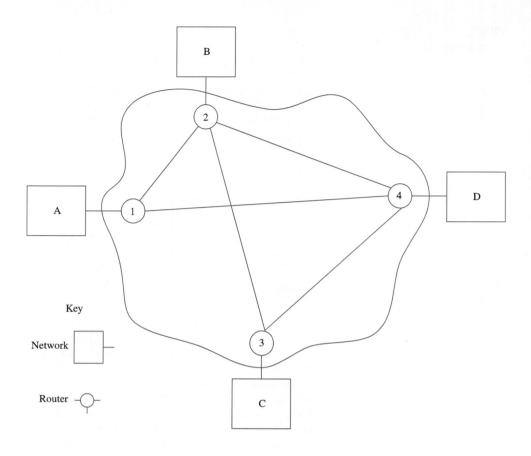

The price we pay for the simplified hardware in the multistage switch is a smaller number of switching possibilities. For example, in Figure 31.3, is it possible to set up the 12 smaller switches so that the output maps to 87654321? The answer is no, indicating that the multistage switch may block some signals from getting to the correct output. This problem is usually temporary, since memory buffers are typically used to store data that cannot be transmitted right away.

PACKET SWITCHING

Figure 31.4 shows a simple WAN connecting four networks (A, B, C, and D). Suppose that a machine on network A wants to send a large chunk of data to a machine on network D. Using packet switching, the large chunk of data is broken down into smaller blocks and transmitted as a series of packets.

Due to the nature of traffic on shared networks, some packets may go directly from router 1 to router 4 (one hop), while others may go from router 1 to router 2, then to router 4 (two hops). A three-hop route is also possible. Thus, it is likely that packets arrive at network D out of order. This is a characteristic of *packet switching*. Packets can be reassembled in the correct order by including a sequence number within the packet. Even so, this characteristic makes packet switching unsuitable for digitized phone conversations, which, unlike an e-mail message, cannot wait for gaps to be filled in at some unknown later time. These features provide a means for choosing between circuit switching and packet switching.

FRAME RELAY

Packet switching was designed during a time when digital communication channels were not very reliable. To compensate for errors in a channel, a handshaking arrangement of send-and-acknowledge packets was used to guarantee reliable data transfers. This error protocol added time-consuming overhead to the packet switching network, with transmitting stations constantly waiting for acknowledgments before continuing.

Frame relay takes advantage of the improvement in communication technology (fiber links, for example, have a very low error rate compared with copper links) and relies on fewer acknowledgments during a transfer. Only the receiving station need send an acknowledgment.

With fewer acknowledgments and a lower error rate, frame relay provides a significant improvement in communications technology.

ATM

Asynchronous transfer mode (ATM), also called *cell relay*, uses fixed-size cells of data and supports voice, data, or video at either 155.52 Mbps or 622.08 Mbps. Cells are 53 bytes each, with 5 bytes reserved for a header and the remaining 48 for data, as indicated in Figure 31.5. The reason for using fixed-size cells is to simplify routing decisions at intermediate nodes in the ATM system.

ATM uses communication connections called *virtual channel* connections. A virtual channel is set up between the end-to-end stations on the network and fixed-size cells are sent back and forth. Decisions concerning routing are resolved using information supplied in the ATM header, which is shown in Figure 31.6. Notice the entries for virtual path and virtual channel identifiers.

FIGURE 31.5   An ATM cell

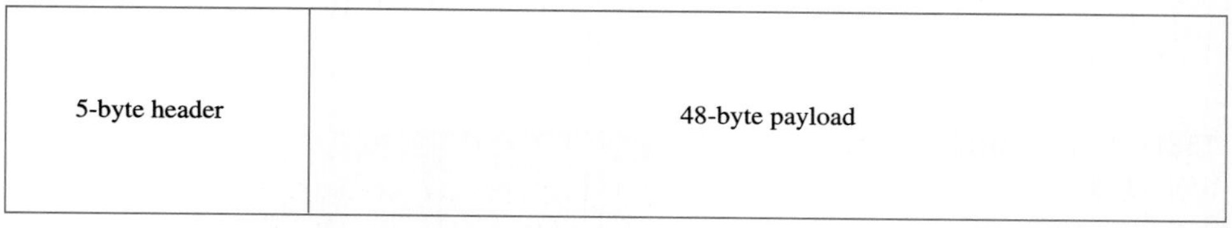

5-byte header	48-byte payload

FIGURE 31.6   ATM header fields

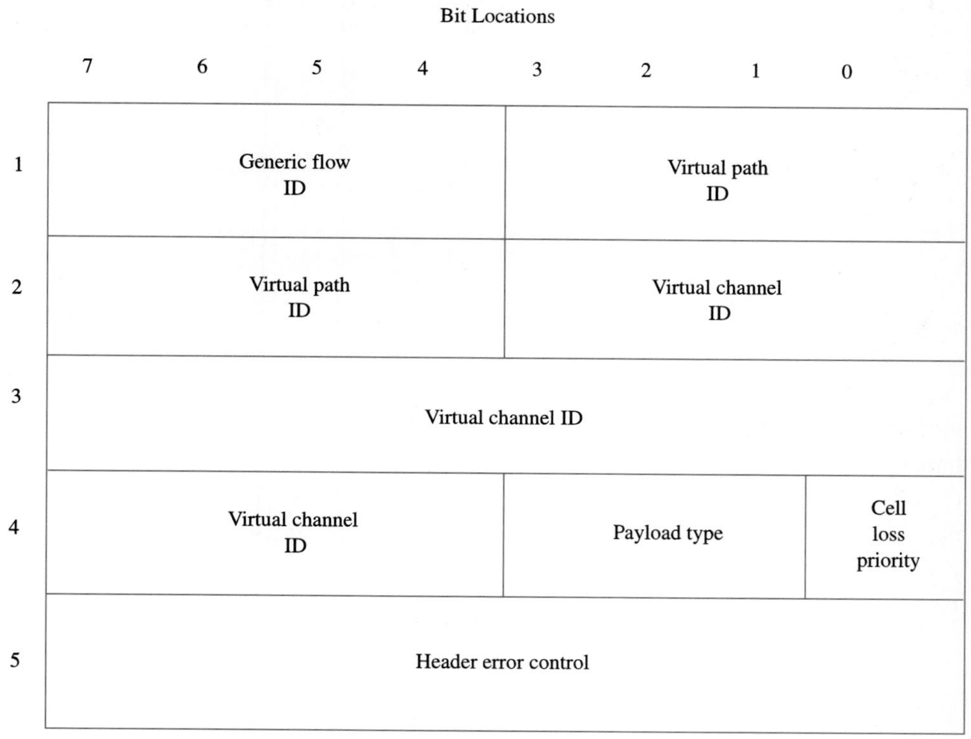

Bit Locations

	7	6	5	4	3	2	1	0
1	Generic flow ID				Virtual path ID			
2	Virtual path ID				Virtual channel ID			
3	Virtual channel ID							
4	Virtual channel ID			Payload type			Cell loss priority	
5	Header error control							

ISDN

The simplest Integrated Services Data Network (ISDN) connection is called a ***basic rate interface*** and consists of two 64 Kbps B channels (for data) and one 16 Kbps D channel (for signaling). The design of ISDN supports circuit-switching, packet-switching, and frame operation. When ISDN is carried over a T1 line (1.544 Mbps), twenty-three 64 Kbps B channels and one 64 Kbps D channel are possible.

SONET

The Synchronous Optical Network (SONET) technology was designed to take advantage of the high speed of a fiber connection between networks. As Table 31.2 indicates, the lowest-speed SONET signal level (STS-1) runs at 51.84 Mbps. That is equivalent to more than nine hundred 56-Kbps modems running simultaneously (minus a few for overhead). STS-48 has 48 times the bandwidth of STS-1, so you can imagine how many telephone calls can be carried over a single fiber link.

There are additional benefits to using fiber: It is not susceptible to electrical noise, it can be run farther distances than copper wire before requiring a repeater to extend the signal, and it is easier to repair.

Figure 31.7 shows the format of a SONET STS-1 frame. A total of 810 bytes are transmitted in a 125-microsecond time slot. Several bytes from each row of the frame are used for control/status information, such as several 64 Kbps user channels; 192 Kbps and 576 Kbps control, maintenance, and status channels; and several additional signaling items.

TABLE 31.2   SONET signal hierarchy

SONET Level	Data Rate (Mbps)
STS*-1	51.84
STS-3	155.52
STS-9	466.56
STS-12	622.08
STS-18	933.12
STS-24	1244.16
STS-36	1866.24
STS-48	2488.32

*STS (Synchronous Transport Signal)

FIGURE 31.7   SONET STS-1 frame format

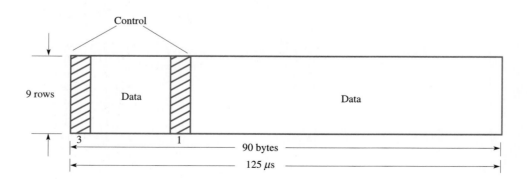

384

FDDI

The Fiber Distributed Data Intrface (FDDI) was developed to provide 100 Mbps connections between LANs over a wide geographical area. Dual fiber rings are used, with the second ring serving as a backup for the first ring, called the *primary* ring. A token-passing scheme similar to token-ring technology is used to allow access to the ring.

The use of fiber allows longer distances between stations (or LANs). The FDDI physical layer allows for up to 100 fiber repeaters in the ring, with a spacing of 2 kilometers between repeaters. Thus, the size of the FDDI ring covers a perimeter of 200 kilometers (more than 124 miles). This is an attractive technology for long-distance communication.

DENSE WAVE DIVISION MULTIPLEXING (DWDM)

The demands of network users and their applications has pushed even the speed limits of the available fiber-based SONET and FDDI technologies. To provide relief, technology was developed to allow *multiple* channels of light to coexist on the same fiber (by carefully varying the wavelength of the optical carrier, hence the terms "dense" and "division"). Thus, instead of a single OC-48 optical stream (2.4 G bps), there may be as many as 40 different OC-48 streams, giving a total of 100 G bps bandwidth. Even more than 40 optical channels will be possible in the future, with 128 channels already being discussed.

Additional benefits of DWDM are

- Easy mixing of different optical carriers
- Longer fiber segments (800 kilometers or more)
- Good for long-haul, point-to-point connections

With fiber optics being the fastest communication medium currently available, DWDM provides one way to keep up with the ever-increasing demand for bandwidth.

MOBILE COMMUNICATION

Almost by definition, mobile communication implies the use of wireless technologies. The traditional cellular technologies are quickly migrating from analog to digital signals that offer additional features and significantly enhanced security benefits. Older geosynchronous satellite communication systems are being replaced by low earth orbit satellite communication systems that can provide wireless coverage for the entire planet.

Wireless technology is based on the concept of having transmitters and receivers. The transmission of wireless signals falls into two categories: omnidirectional and directional. Omnidirectional signals propagate from the transmitter in all directions similar to the transmitter used for an AM or FM radio station. A directional signal is focused at the receiver. Using a combination of these two types of signals, many different applications of the technology are possible. Some of these applications are shown in Table 31.3.

To accompany the new wireless technologies, the IEEE 802.11 specifications provide a software framework on which to build. A new protocol, DFWMAC (Distributed Foundation

TABLE 31.3 Wireless technologies applications

Wireless Technology	Application
Digital Cellular	Voice, Data
Wireless LAN	Voice, Data, Video
Personal Communication System	Voice, Data, Video, Fax, Global Positioning

Wireless MAC), was created to work in the MAC layer of the OSI network model. A modified version of Ethernet called CSMA/CA (Collision Sense Multiple Access with Collision Avoidance) is used to transmit data in the network.

TROUBLESHOOTING TECHNIQUES

The sophistication of the wide variety of telecommunication equipment requires expertise that is typically beyond that obtained in an ordinary electronics or engineering technology program. Fully developed telecommunication degree programs are now available that train the student in all aspects of the field, using state-of-the-art equipment, such as Optical Time Domain Reflectometers, network analyzers, and digital sampling oscilloscopes. Becoming a telecommunication technician or engineer would be a challenging and rewarding pursuit.

SUMMARY

In this exercise we discovered that

- Telecommunications covers a broad area of electronic communications.
- Technologies include TDM, circuit and packet switching, frame and cell relay, ISDN, SONET, and FDDI.
- DWDM is pushing the bandwidth limits of fiber technology.
- Mobile communications are becoming standardized.

SELF-TEST

This self-test is designed to help you check your understanding of the background information presented in this exercise.

True/False

Answer *true* or *false*.

1. Using TDM, multiple data channels are sent over a single wire.
2. A multistage switch may block a signal from getting to its destination.
3. Frame relay uses fewer acknowledgment packets than packet switching.
4. FDDI is used for networks in small geographical areas.
5. Wireless communication requires transmitters and receivers.

Multiple Choice

Select the best answer.

6. Voice conversations are sampled using
   a. AM.
   b. FM.
   c. PCM.
7. A crossbar switch allows
   a. Limited switching between input and output.
   b. Full switching between input and output.
   c. Fixed-paths only between input and output.
8. In packet switching, packets may arrive
   a. Out of order.
   b. Not at all.
   c. Both a and b.
9. ATM cells are
   a. Fixed in length at 53 bytes.
   b. Fixed in length at 64 bytes.
   c. Variable in length.

10. How large can an FDDI ring be?
    a. 2 kilometers.
    b. 100 kilometers.
    c. 200 kilometers.

Completion

Fill in the blank or blanks with the best answers.

11. Twenty-four 64,000 bps channels are available in a(n) _____ carrier.
12. The bit pattern needed to connect input signal 3 to output signal 7 in the crossbar switch is _____.
13. When all switches are configured as straight-through in the multistage switch, the output signal sequence is _____.
14. When all switches are configured as crossover in the multistage switch, the output signal sequence is _____.
15. ATM is also called _____ relay.

FAMILIARIZATION ACTIVITY

1. Visit a local Internet service provider. Ask for a tour of their facility. What kind of telecommunication equipment do they have? How many individuals do they serve? What kind of service do they provide?
2. Search the Web for telecommunication equipment. Compare prices and features among similar equipment.

QUESTIONS/ACTIVITIES

1. How much does your local telephone provider charge for a leased 56K baud line? How much for a T1 connection?
2. What types of telecommunication technologies are in use at your educational site?

REVIEW QUIZ

Under the supervision of your instructor

1. Describe the different telecommunication technologies.
2. Discuss reasons for choosing one technology over another.

UNIT V Microcomputer Hardware

32 Computer Environments

Joe Tekk walked off the elevator on the 11th floor of a high-rise office building. He was on a service call to one of RWA Software's clients, a small business with a network of 20 computers.

The office workers reported frequent erratic behavior by their machines. As soon as Joe walked into the office, he was concerned, because he saw a deep shag rug carpeting the floor. Joe got a shock when he grabbed the doorknob when entering the office and another shock when he touched a table as he sat down to check a computer. He checked the fans on several mini-towers; they were clogged with dust. One keyboard looked as if a drink had been spilled on it.

Joe spoke with Debbie Grant, the office manager. "Even though there are many conditions in your office environment that need to be corrected, I still do not see the cause of the problems you are experiencing. I'm going to go check some of the other offices on your floor."

Forty-five minutes later Joe returned. "I think I have good news, Debbie. After talking to your neighbors, who are having similar problems, I spoke with the building superintendent. He told me that the elevator was repaired recently, just over a week ago. Isn't that when your problems started?"

Debbie agreed. "Yes, they were here last Tuesday. They worked on the elevator for hours, because it got stuck between floors."

Joe continued. "Apparently they rewired the elevator's control panel, and tapped into the circuit that feeds your office and the other three on this side of the building. I think the electrical noise from the elevator is causing your problems. Let's try adding surge protectors to your equipment to see if that helps."

PERFORMANCE OBJECTIVES

Upon completion of this exercise, you will be able to

1. Design an environmental checklist for an assigned computer workstation.
2. Perform a computer environmental check of an assigned computer workstation.

KEY TERMS

Physical environment	Ambient temperature
Electrical environment	Internal temperature
Heat	Power cycling

Thermal shock	Electrical noise
Dust	Uninterruptible power source (UPS)
Corrosion	Electrostatic discharge (ESD)
Galvanic corrosion	Environmental checklist
Magnetic field	

Before going further with the hardware aspects of personal computers, this is a good time to present some of the important requirements for a computer operating environment.

The major causes of the environmental problems that can occur with personal computers are

- High temperatures
- Dust
- Corrosion
- Magnetic fields
- Electrical noise
- Electrical power variations

These causes can be classified into two major environment areas: the *physical environment* and the *electrical environment*.

PHYSICAL ENVIRONMENT

Heat

Figure 32.1 shows the relationship between the *ambient temperature* and the *internal temperature*. The ambient temperature is the temperature of the surrounding air.

When a computer system has been turned off for some time, its internal temperature is the same as its ambient temperature. However, once a computer has been on for 15 minutes or more, its internal temperature is much higher than its ambient temperature. Depending on the type of computer, the temperature difference between ambient and internal can be as much as 40°F.

The recommended ambient temperature for a PC is

Turned off:	60 to 90°F
Turned on:	50 to 110°F

Excessive internal heat in a computer can cause many serious problems to occur. Some of these problems are illustrated in Figure 32.2.

In many cases, excessive internal heat is one of the major causes of computer failure.

FIGURE 32.1 Ambient and internal temperature

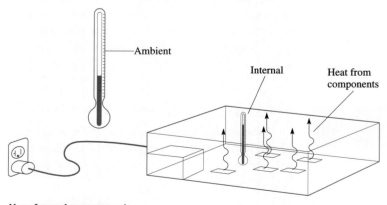

Note: Internal temperature is higher than ambient temperature.

FIGURE 32.2   Typical problems caused by excessive internal heat

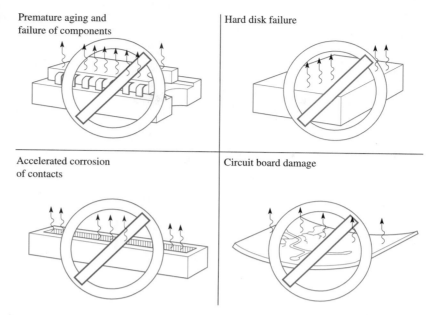

Premature aging and failure of components

Hard disk failure

Accelerated corrosion of contacts

Circuit board damage

You can do much to prevent the problems caused by excessive heat. For example, you can recommend that air conditioning be used when the computer is to be operated in a small construction office that does not normally have air conditioning. Some of the sources of excessive heat are listed in Table 32.1 and illustrated in Figure 32.3.

Power Cycling—Thermal Shock

When a computer or any other electrical system is turned on, it undergoes a rapid increase in temperature. This increase in temperature comes about because of the normal power loss produced by electrical circuits. This power loss results in the production of heat.

Thermal shock occurs when a rapid increase or decrease in temperature causes an undue mechanical stress on the electrical components that make up the computer circuits. Figure 32.4 illustrates the concept of thermal shock.

As shown in Figure 32.4, thermal shock is brought about by the mechanical expansion and contraction caused by changes in temperature. The greater the change in temperature, the greater the thermal shock. The amount and speed of contraction or expansion also are determined by the type of material. Every time you turn the computer on, it undergoes some degree of thermal shock. Consider, for example, coming into a cold office on a Monday morning and turning on your computer. In this case, the thermal shock may be quite large, because the computer undergoes a large increase in internal temperature. Table 32.2 lists some of the undesirable effects that can be brought about by thermal shock.

Rapid changes in system temperature can also cause problems with hard disk drives. As

TABLE 32.1   Sources of excessive heat

Problem	Solution
High ambient temperature	Use air conditioning in the computer room or area.
Direct sunlight	Remove the computer from contact with the rays of the sun. This may be accomplished by moving the computer or by using drapes or other sun-blocking material.
Fan outlet and equipment vents blocked	Do not place the rear of the computer directly against a wall or other object. Do not cover vent holes with any material that would block the flow of air.

FIGURE 32.3   Sources of undesirable heat

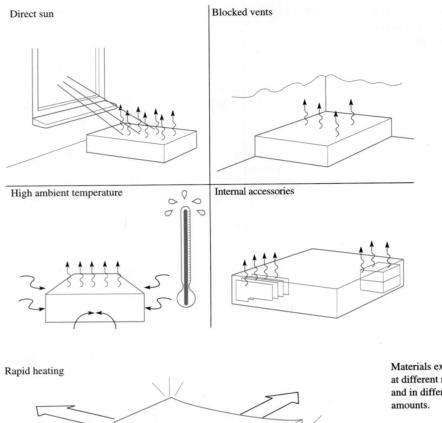

Direct sun

Blocked vents

High ambient temperature

Internal accessories

FIGURE 32.4   Concept of thermal shock

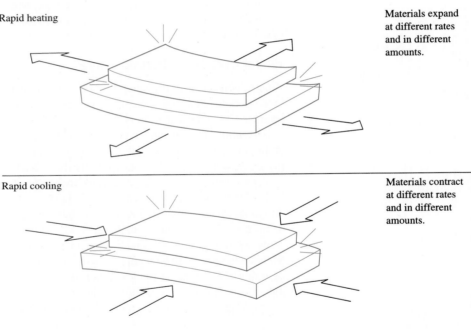

Rapid heating

Materials expand at different rates and in different amounts.

Rapid cooling

Materials contract at different rates and in different amounts.

TABLE 32.2   Some undesirable effects of thermal shock

Action	Results
Chip creep	Rapid expansion or contraction of IC leads can cause them to gradually lift out of the chip sockets.
Circuit board foil separation	Because the material of the circuit board is different from the copper foil used to connect the circuits, the foil may separate and actually form hairline cracks, resulting in intermittent system operation.
Broken solder joints	Again, because of the differences in solder material and the material containing the solder, a broken solder joint can result from rapid temperature change in the system. This can lead to unreliable system operation.

FIGURE 32.5   How thermal shock affects computer components

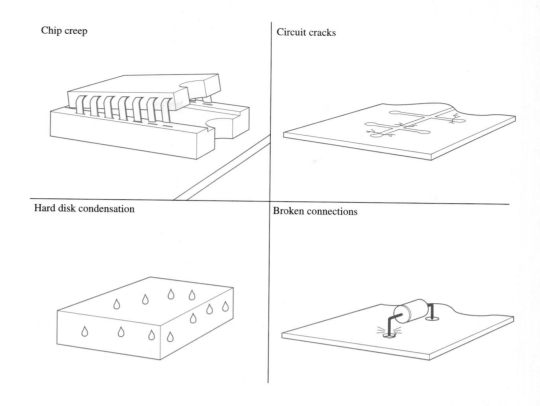

Chip creep

Circuit cracks

Hard disk condensation

Broken connections

TABLE 32.3   Methods of preventing rapid changes in internal temperature

Method	Comment
Never turn the computer off.	Some computer users never turn off a computer. Doing this eliminates thermal shock. The belief here is that any long-term wear from constantly keeping the system on will be less than problems caused by thermal shock.
Warm the room and wait for the computer to reach room temperature.	Some users do not leave their computers on over the weekend or overnight. At the same time, during the winter, automated thermostats reduce the room temperature to around 50°F. When starting up in the morning, heat up the room first, allowing the computer to heat up slowly before turning it on.
Leave the computer on during the workday.	The computer should not be constantly turned on and off during the workday, even if there are long periods in the day when it is not being used.
Do not reboot by using the ON/OFF switch.	It is considered poor practice to reboot using the ON/OFF switch. If for any reason you need to reboot the system, use the Ctrl-Alt-Delete key combination (or shutdown) or a RESET button if your computer has one.

an example, if a hard drive is shipped during the winter and is then placed in a computer system (before having a chance to warm up), moisture can condense on the hard disks, rendering the drive useless. Some of the undesirable effects of thermal shock are illustrated in Figure 32.5.

There are several ways of preventing the harmful effects of rapid changes in temperature. These are listed in Table 32.3.

There is another point to consider about rebooting the computer by turning it off and then on again. Every time you turn your computer on, the power supply, hard disk drive motor, and other electrical components undergo surges of electrical current. Although a power surge lasts for only a brief period of time, it can wipe out a component or power supply. Recall that when a lightbulb burns out, it usually does so just after you turn it on.

Effects of Dust

Dust can be a silent killer of your computer and its disk drives. This means that any computer system should be housed in as clean an environment as possible. There are some steps that can be taken to reduce the long-term effects of dust. One is not to burn materials (such as tobacco products) around a computer, and the other is to cover the computer with a dust cover when it is not in use.

Note: Never operate a computer with a dust cover on it because this will cause the computer to overheat.

One of the adverse effects of dust is that, with time, it builds up a thick coating over the internal chips and other components inside your computer. This thick layer of dust can act as a heat insulator, resulting in a higher-than-normal internal temperature in your components. Another problem with dust is that even the smallest dust particle can permanently damage a disk. Table 32.4 lists some of the actions that can be taken to reduce the bad effects of dust, and Figure 32.6 shows some of the methods used to remove dust from a computer.

Effects of Corrosion

Corrosion inside a computer is the process of metal pin connectors, wires, interface cards, and chip pins undergoing chemical changes. These chemical changes can gradually eat away critical parts of the circuit, causing improper operation of the system. The main sources of corrosion are chemicals in the atmosphere and contact between computer parts and human hands.

If your computer is in an area that is subject to atmospheric pollution such as smog, you may need to use high-quality air filters on air-intake ducts leading into the area where the computers are kept. If filters are used, it is important to change them regularly in order to keep them working at their peak performance.

The sulfates found in tobacco smoke have a devastating corrosive effect on computer circuits. The burning of tobacco products should never be allowed in a closed area with a computer. Circulating air does not help here, because the increased circulation could actually cause more of the corrosive by-products of tobacco smoke to get into the computer.

Never touch metal computer parts directly with your bare hands. Doing so will leave a thin coat of your natural skin oil on the parts. This oil residue could then promote *galvanic corrosion*. Galvanic corrosion creates the effect of a small battery, causing a tiny current to flow between the metals. This process gradually eats away the metal, causing computer failure.

TABLE 32.4   Methods of reducing the effects of dust

Method	Effect
Use dust covers.	Slows down the accumulation of dust inside the computer system.
Close windows.	Reduces some of the sources of dust.
Prohibit indoor fires.	Not allowing the burning of tobacco or other products indoors greatly reduces the amount of dust in the computer.
Keep foods away from work area.	Certain foods, especially the "snack" variety, can produce small particles that find their way into the computer system.
Use a cleaning schedule.	Removes dust that has already accumulated on and in the computer system.

FIGURE 32.6 Methods used for removing computer dust

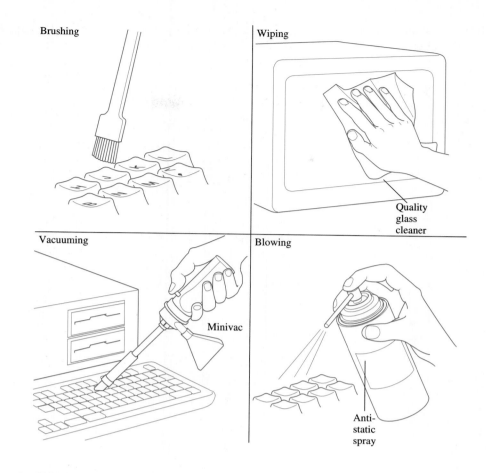

Brushing

Wiping

Quality glass cleaner

Vacuuming

Minivac

Blowing

Anti-static spray

ELECTRICAL ENVIRONMENT

Magnetic Fields

The best way to reduce the effects of *magnetic fields* is to be aware of where they are found and then remove the sources. Figure 32.7 shows some of the most common sources of magnetic fields that may be found around a computer station.

Electrical Noise

Electrical noise can be generally classified as any undesirable and sometimes unpredictable random changes caused by sources of electricity. There are several types of electrical noise, as characterized in Table 32.5.

Figure 32.8 illustrates various kinds of electrical noise and their relationship to the computer. All the types of electrical noise listed in Table 32.5 can have adverse effects on your computer or on other equipment such as television and radio receivers. In terms of your computer, electrical noise can cause the system to lock up, produce inconsistent results, damage information on disks, cause "garbage" to appear on the screen, or produce some other "mysterious" type of improper operation.

The effects of electrical interference are not always consistent and not always easy to detect. The best rule when working with them is to reduce the possibility of any of them happening. Figure 32.9 illustrates some of the causes of electrical interference.

Reducing Electrical Noise

Because of the electromagnetic radiation (EMR) that can be caused by computers, the FCC (Federal Communications Commission) has established specifications for the legal amount

FIGURE 32.7   Common sources
of magnetic fields

Mechanical ringing phones

Magnetic clips

Speakers/headsets

Scissors/screwdrivers

TABLE 32.5   Types of electrical noise

Type	Definition
EMR	Electromagnetic radiation is a general category of electrical noise that affects computers and other electronic equipment. It can be radiated or otherwise transmitted through space or conducted along wires or other conductors.
EMI	Electromagnetic interference is a more specific category of electrical noise that occurs in the frequency range of 1 Hz to about 10 KHz.
Transient EMI	Transient electromagnetic interference is a short-term undesirable electrical response that may appear when equipment is first turned on or off. This includes power line transients and electrical discharges from lightning.
Internal EMI	Internal electromagnetic interference is caused by the internal circuitry of the computer, such as faulty chips or improperly installed wires on the motherboard or other system components.
RFI	Radio-frequency interference is a more specific category of electrical noise that occurs at a frequency of 10 KHz and above.
Conducted RFI	Radio-frequency interference is sometimes conducted along the power line from the PC out into the power system.
Radiated RFI	Radiated radio-frequency interference is sometimes transmitted from your computer into the surrounding space.
ESD	Electrostatic discharge is like the static electricity that builds up in your body when you walk across a carpet on a dry day and is released when you touch a grounded metal object.

FIGURE 32.8 Various kinds of electrical noise

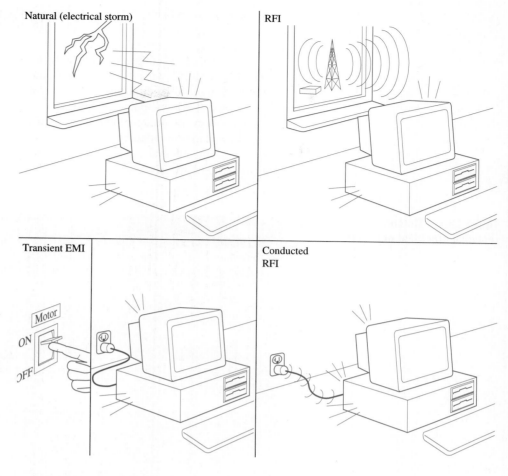

Natural (electrical storm)

RFI

Transient EMI

Motor
ON
OFF

Conducted
RFI

FIGURE 32.9 Causes of electrical interference

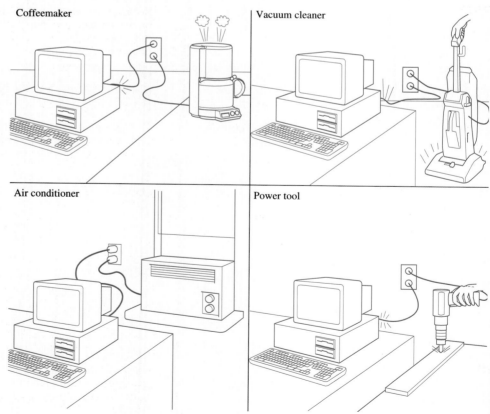

Coffeemaker

Vacuum cleaner

Air conditioner

Power tool

of radiated noise allowed to be emitted from a personal computer. The FCC has created two categories for personal computers: Class A and Class B. The Class A category is for industrial computing devices in commercial and business use and not usually sold to the general public. Class B consists of consumer computing devices. The PC is considered a Class B computing device. All PCs and compatibles must meet FCC EMR standards. Therefore, all PCs that are legally sold in the United States for personal or industrial use must be approved by the FCC. Each computer is required to have a statement affixed to it stating that it meets or exceeds the minimum requirements of the FCC.

Some of the methods of reducing the effects of electrical noise are listed in Table 32.6. If you check out a computer system using this table as a guide, you will have reduced or eliminated the majority of the causes of electrical noise.

TABLE 32.6 Methods of reducing electrical noise effects

Method	Effect
Keep computer parts inside the metal enclosure.	The switching power supply is a big source of EMI. The computer's metal case helps shield against EMR and is required in order to meet minimum FCC standards.
Cover openings with metal inserts.	Do not leave openings on your computer case. This includes the openings in the rear of the system where the peripheral cards are to be inserted. Covering these openings helps reduce the effects of EMR.
Use metal honeycomb screens over cooling vents.	This practice also helps reduce the effects of EMR.
Use shielded cables.	Using shielded cables greatly reduces the effects of EMR. The shield on the connecting cable helps prevent radiation from reaching the cable.
Ground cables.	Make sure that interconnnecting cables, such as those between your computer and other devices such as the printer, have their ground wires attached as required by the manufacturer. Doing this helps reduce or eliminate a major source of electrical interference.
Use fiber optics where and when available.	One of the best methods of reducing electrical interference is to use properly installed fiber optics, which do not radiate any undesirable EMR.
Turn off cordless telephones.	Cordless telephones are actually tiny transmitters of EMR. They should be removed from the vicinity of any PC.
Avoid any high-speed digital circuits or equipment.	Digital equipment is also capable of transmitting EMR that may interfere with the operation of your computer. This includes digital thermostats, certain types of burglar alarms, and industrial controllers.
Remove any type of radio transmitters.	Radio or televison transmitting equipment is, by its nature, a major source of EMR. These systems should be removed from the vicinity of any PC.
Turn off certain types of communication receivers.	Many communication receivers use what is called the "superheterodyne" principle in their operation. These include standard AM and FM radios. These receivers, which contain a circuit called a local oscillator, can emit EMR if not properly shielded. These devices should be removed from the vicinity of a computer.
Avoid electrical machinery.	Electrical machinery, such as electrical motors, air conditioners, compressors, and heaters, is a potential source of electrical interference. Again, the computer should not be in the immediate vicinity of this type of equipment.

Power Source Problems

Probably one of the most important environmental concerns for your computer is a reliable and "clean" source of electrical power. Computers are more sensitive to power line variations than most other electrical equipment. Office or room lights can tolerate wide variations in the power source with very little adverse effect on operation. This is not the case with a computer. The major types of power line problems and their usual causes are listed in Table 32.7 and illustrated in Figure 32.10.

TABLE 32.7   Power line problems and their causes

Type of Problem	Cause(s)
Brownouts	Lowered output voltage from the wall outlet caused by an over-demand for electrical power. This overdemand can be caused by anything from a whole region using more power for air conditioning on a hot day to someone plugging in a coffeemaker on the same outlet as your computer.
Blackouts	A total loss of electrical power from the wall outlet. Blackouts are caused by a variety of conditions, such as power lines knocked down by natural or human-made disasters.
Power transients	Large and potentially dangerous voltage spikes appearing on the power line. These can be caused by lightning strikes or by the turning on or off of industrial machinery in the same building as your computer.

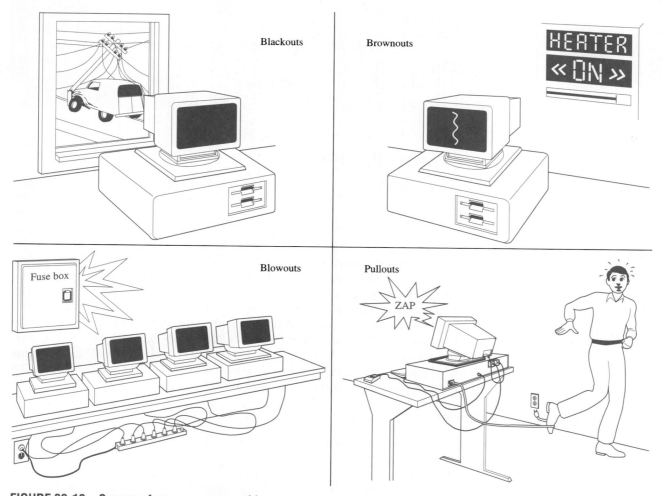

FIGURE 32.10   Causes of power source problems

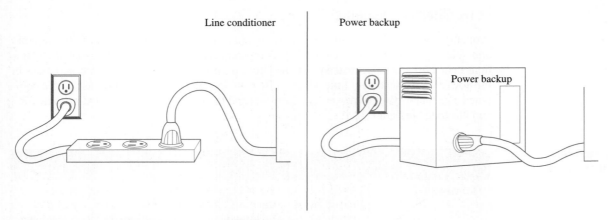

FIGURE 32.11 Methods of reducing power problems

TABLE 32.8 Types of power-protection schemes

Method	Description
Power isolators	Good at filtering out high-frequency voltage spikes. Not good at reducing the effects of a slowly changing condition, such as a brownout.
Power regulators	Good at helping maintain a constant line voltage for line voltage variations of not much more than 10%. Not effective against power spikes and brownouts.
Power filters	Good at removing electrical noise on the power line such as EMI and RFI. Does not stop voltage spikes or help with brownouts.
Uninterruptible power source (UPS)	A power source that keeps your computer on when the power is out or there is a brownout.

Reducing Power Problems

If you determine that power line problems are common in the area where your computer will be kept, there are some methods you can use to prevent such problems. One is conditioning the power line used by the computer, and the other is to have a power supply backup system. Both methods are illustrated in Figure 32.11. Various kinds of power-protection schemes are listed in Table 32.8.

An *uninterruptible power source (UPS)* can also provide protection against voltage spikes and electrical noise on the power line. It will maintain electrical power to your computer for a limited time when it loses its own source of power. The length of time the UPS should maintain power depends on the conditions of the area where the computer is kept. If power failures are frequent, a UPS that will supply emergency power for several hours should be considered. If power failures are less frequent, a less expensive UPS that supplies power for a few minutes may be all that is necessary. The few minutes could be used to shut down or to store data to a disk when the power has failed. Figure 32.12 shows a small UPS capable of maintaining power for five minutes at full load.

Electrostatic Discharge

Surprisingly, people and office furniture can accumulate large charges of static electricity. These charges can amount to several thousand volts. If this kind of voltage is discharged through your computer, it can severely damage some of the sensitive integrated circuits. Such discharge is referred to as *electrostatic discharge (ESD)*. Several methods of preventing or reducing the effects of electrostatic discharge are listed in Table 32.9.

FIGURE 32.12   Uninterruptible power source

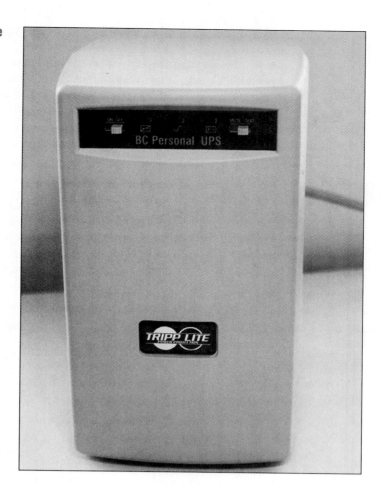

TABLE 32.9   Methods of reducing or eliminating the effects of ESD

Method	Discussion
Touch a conductive tabletop.	Touch the conductive table before touching anything else when first sitting down at the computer.
Use antistatic spray.	Antistatic sprays for rugs and computer equipment help reduce and control the effects of ESD.
Consider using a room humidifier.	Be cautious here. Many humidifiers help reduce the effects of static electricity but may produce undersirable corrosive side effects from the residue they leave on computers. You may have to run such a humidifier with distilled water. This could be an expensive solution.
Use antistatic floor mats.	Placing antistatic floor mats under chairs is a popular practice in offices.
Use antistatic cleaning solutions on floors.	Floors are usually cleaned with antistatic solutions in computer-manufacturing areas. One application usually lasts for several months.

TROUBLESHOOTING TECHNIQUES

Determining the cause of an environmental problem requires you to behave like a detective. You need to gather all the facts. When did the equipment stop working? Was there a major change in procedure recently?

Examine the equipment carefully. Are the connectors clean? Are they snug in their sockets? Is there adequate ventilation around the equipment? Have any antistatic precautions been taken?

403

Observe the individuals operating the equipment. Are they doing anything that could contribute to the problems being experienced?

Look for obvious signs that equipment has been mistreated. For example, an employee may have drilled a hole in the plaster to hang a new picture on the wall beside his computer. A coating of plaster dust on his computer would show he did not cover it when drilling the hole. There is no telling what type of physical or other damage can result from careless activities.

SUMMARY

In this exercise we discovered that

- There are physical environment problems and electrical environment problems.
- Heat, thermal shock, dust, and corrosion are physical environment problems.
- Magnetic fields, electrical noise, and electrostatic discharge are electrical environment problems.
- An uninterruptible power source can sustain power for a short period after a power loss.

SELF-TEST

This self-test is designed to help you check your understanding of the background information presented in this exercise.

True/False

Answer *true* or *false*.

1. The temperature inside a computer that has been powered up is always higher than the ambient temperature.
2. The speed at which the operating temperature of a computer changes does not affect the operation of the computer.
3. Having the computer in contact with direct sunlight is not a problem as long as the room is kept cool.
4. Physical environment and electrical environment are the same thing.
5. Chip creep is due to rapid expansion and contraction of IC leads.

Multiple Choice

Select the best answer.

6. An operating computer should not be placed directly against a backing wall because
   a. Of static electricity.
   b. The exhaust fan may become blocked.
   c. The wall may catch on fire.
   d. None of the above.
7. A mechanical stress on the computer components brought about by a rapid change in temperature is called
   a. Temperature runaway.
   b. Thermal runaway.
   c. Thermal shock.
   d. Thermal stress.
8. Undesirable electrical disturbances that cause improper computer operations may be caused by
   a. Electrical machinery.
   b. Cordless telephones.
   c. Digital thermostats.
   d. All of the above.
9. Touching a metal part with your bare hands can lead to
   a. Ionic corrosion.
   b. Galvanic corrosion.
   c. Sporadic corrosion.

10. EMR standards are published by the
    a. ABC.
    b. NSA.
    c. FCC.

Matching

Match a description of electrical noise on the right with each stated type on the left.

11. RFI   a. Occurs in the frequency range of 1 Hz to 10 KHz.
12. EMI   b. General category of electrical noise.
13. EMR   c. Occurs in the frequency range of 10 KHz and above.
                                         d. None of the above.

Completion

Fill in the blanks with the best answers.

14. A power _____ is good at filtering out high-frequency voltage spikes.
15. A(n) _____ can maintain a constant power source to your computer during an electrical outage.
16. The effect of corrosion is to cause a(n) _____ change of the material inside the computer.
17. ESD stands for _____ static discharge.
18. Thermal _____ can result when a system undergoes a rapid change in temperature.

FAMILIARIZATION ACTIVITY

This section contains an *environmental checklist*. This checklist is a guide to help you check out possible physical and electrical environmental hazards at a computer station. Your instructor may assign you and your lab group to an actual computer area located outside of your normal lab. This activity may be done during your normal lab time or as an outside assignment.

Keep in mind that doing an environmental check of a computer station is not always an exact activity with absolute yes or no answers. Much of what you will be doing will depend on your judgment. What is important is that you and those using the computer system are aware of what environmental factors may cause the computer to malfunction.

Environmental Checklist

Item to Check	Problem			Comments/Action
	Y	N	?	
Ambient temperature				
Direct sun				
Blocking of fan and vents				
Power cycling				
Dust covers needed				
Windows closed				
Indoor smoke				
Food/drinks				
Potential magnetic fields				
EMI				
RFI				
ESD				
Power source				
Corrosion				

To use the environmental checklist, look at the "Item to Check." Then, in the "Problem" column, indicate if this item is a potential problem: Y means yes, there is a potential problem; N means no, there is not a potential problem; and ? means you cannot tell from what you are observing. The "Comments/Action" column is provided so you can record any necessary corrective action(s) or to make comments on further studies that need to be made before you can decide if there is a particular problem.

You may not have room on the checklist to write all your comments or recommended actions, so use a separate sheet of paper as needed.

After completing this checklist, turn it in to your instructor in order to complete performance objective 2 for this exercise.

QUESTIONS/ACTIVITIES

Answer the following questions regarding the results of your environmental check in the familiarization activity for this exercise.

1. What major environmental hazards, if any, did you discover? What did you recommend as a solution?
2. What areas were you not sure of? Why?
3. Did you recommend that the computer(s) always be left on? If this was not done, what was the reason?
4. Did you discover any sources of magnetic fields? Explain.
5. When was the last time the computers were cleaned? Is there a regular cleaning program?
6. State the preventive steps that you found were used to reduce or eliminate the effects of ESD.

REVIEW QUIZ

Under the supervision of your instructor

1. Design an environmental checklist for an assigned computer workstation.
2. Perform a computer environmental check of an assigned computer workstation.

33

System Teardown and Assembly

Joe Tekk was pushing a cart of computer components down the hallway at a local community college. He was guest lecturing in a computer hardware laboratory. The purpose of the lab was to completely build a computer from individual parts, install Windows NT, and connect to the school's network. The cart contained all the components of a multimedia PC, from power supply to CD-ROM to adapter cards.

Joe was told the lab was a three-hour lab. He looked at all the components on the cart. He hoped there was enough time.

It took the students just over two hours to do everything. They spent the third hour surfing the Web.

PERFORMANCE OBJECTIVES

Upon completion of this exercise, you will be able to remove the cover of a personal computer and identify the following major sections:

1. Power supply
2. Floppy disk drive(s)
3. Hard disk drive(s)
4. CD-ROM drive
5. Motherboard
6. Expansion cards

KEY TERMS

Disassembly procedure   Assembly procedure

BACKGROUND INFORMATION

Up to this point in the exercises, we have concentrated on specific topics related to various skills needed for computer operation. In this exercise, you are given an overview of the complete computer system. Here you discover how to take the computer system apart, identify the major sections, and explain the purpose of each of these sections.

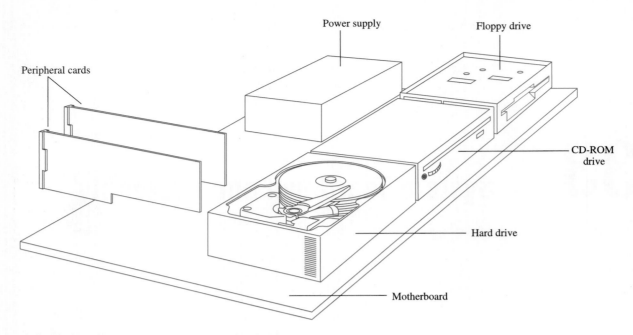

FIGURE 33.1   Major sections of a personal computer

TABLE 33.1   Purpose of each major computer section

Section	Purpose
Power supply	Converts the 120-V AC electricity from the line cord to DC voltages that are needed by the computer system.
Floppy drive(s)	Allows information to be stored and read from removable floppy disks.
Hard drive	Allows information to be stored and read from nonremovable hard disks.
CD-ROM drive	For reading and writing CDs.
Motherboard	Holds the microprocessor and memory circuitry and provides expansion connectors.
Peripheral cards	Allow accessory features and interconnect the computer to input and output devices such as drives, printers, monitors, and other external devices.

SYSTEM OVERVIEW

Figure 33.1 shows the major sections of a personal computer, and Table 33.1 explains the purpose of each of these major sections.

DISASSEMBLY PROCEDURES

The process of physically disassembling and reassembling the personal computer is not difficult. It is, however, a process you should experience before going further into other specific sections of the computer. You will find that the physical arrangement of the major components (such as the power supply) is similar in almost all computer systems, even those from different manufacturers.

When taking a computer apart, it pays to be organized. Keep these points in mind as you proceed with your disassembly:

- For safety, boot the machine and enter the BIOS setup program. Record all important BIOS settings in case they are lost during the disassembly procedure.
- Turn power off and unplug the computer (both ends of the power cord).
- Disconnect all attached devices, such as keyboard, mouse, monitor, and network or modem cable.
- Remove the case.
- Draw a diagram of the motherboard. Be accurate in regard to the position and number of expansion connectors, which adapter cards are plugged in, and what cables are plugged into the motherboard. Draw the diagram large so that you can make notes on it.
- Keep all screws and loose parts in a container. If different types of screws are removed, place them in separate containers and make notes to yourself about where they came from.
- Examine each off-board connector before and after it is removed. Look for pin-1 orientation marks (red or blue stripe on floppy drive, hard drive, and CD-ROM cables) or other labeling that helps distinguish or identify the connector. Many of the small two-conductor twisted pairs that go to the panel buttons and indicators have writing on them that identifies their function (Turbo LED, Reset switch, Speaker) as well as color-coded wiring.
- If it is necessary to remove the hard drive cage to get the motherboard out, be sure to disconnect the power and signal cables to the drive before removing the cage. For multiple-drive systems, note the location of each drive so that they are returned to their original locations.
- When disconnecting the floppy drive cable, note the relationship between the red or blue stripe on the ribbon cable and the drive connector. It is important to know where pin 1 is. In a two-drive system, watch for the twist in the cable to identify drive A:.
- When pulling ribbon cable connectors (such as the 40-pin IDE connector) out of their sockets, do not yank them out by pulling on the ribbon cable itself. Instead, grab the sides of the connector and gently rock it back and forth as you pull it out. It is not difficult to bend a connector pin when removing a connector. If you do bend a pin by accident, use long-nose pliers to carefully straighten it out. If the connector socket has a label (primary IDE, secondary IDE, COM1, and so on), read it and verify that the cable in fact goes to the indicated hardware. If the connector is not labeled, write your own label on your motherboard diagram.
- Note the shape of the keyed side of each power connector as it is removed. The important thing to remember is that the two sets of black conductors should always be in the middle of the 10 power conductors.
- When removing the motherboard, look for several screws or plastic standoffs that are used to hold it in place. Do not set the motherboard down or handle it without enclosing it in an antistatic bag.

ASSEMBLY PROCEDURES

Paying attention to details will save time and effort when assembling your computer. Here are some tips you may find useful:

- Connect the power cable to each component (hard drive, CD-ROM drive, floppy drive, motherboard). Nothing is more embarrassing than forgetting to connect the power cable.
- Do not force any connections. If a cable is not seating properly, examine the cable and your diagram to help determine the problem.
- Put the hard drives back in their cages, but do not tighten the screws until the connectors have been seated.
- Check the primary and secondary IDE connectors to be sure that all 40 pins have been seated properly. It is very easy to make an improper connection and leave pins unconnected.
- Use the proper connector on the floppy drive cable to reconnect the floppy drive(s). This is especially important in a single-drive system.

- Secure each expansion card with the retaining screw.
- If the fan was removed from the processor during teardown, make sure it is reattached to the processor. Forgetting to do so may result in erratic behavior that is difficult to diagnose unless you happen to notice that the processor is really hot or that the fan is missing.
- Reconnect the monitor, keyboard, mouse, and network or modem cable.
- Save the off-board front panel connections (such as the Turbo LED) for last as these are not essential to proper operation. If the machine has been reassembled correctly it will boot normally without them—you just will not see the hard drive light flash or be able to reset the machine or hear the speaker. The advantage of saving them until last is if the machine fails to boot, it cannot be attributed to these connections.
- If necessary, restore the BIOS settings recorded during teardown.
- After the computer has booted successfully, finish the assembly by replacing the case.

OTHER CONSIDERATIONS

If a computer is being assembled from scratch, and not from a previously torn down machine, there are probably a million reasons why it may not work when you turn it on. All of your skills as a troubleshooter are required to determine the cause of the problem. Windows may provide some assistance by telling you which interrupts or I/O devices are assigned improperly or that you may need to reinstall drivers for a particular device. It may be necessary to boot Windows into safe-mode (press F8 during boot to get a start-up menu) to investigate tough problems.

If the motherboard comes without memory, consult the motherboard manual for the type of memory used and how to install it. Typically, when memory is added or removed from the motherboard, the BIOS power-on self-test sees the change and may require action. To install a SIMM, place it into the SIMM connector at an angle. Push the SIMM slowly forward from each end until it is straight up in the socket. Small keying holds on each end of the SIMM should snap into place if the SIMM is seated properly. A similar procedure is used for DIMMs.

If the motherboard comes without a processor, locate the appropriate Socket 5, Socket 7, or Slot 1 processor and install it. Jumpers on the motherboard are used to select the clock speed of the processor. As before, attach a cooling fan to the processor. *Note:* Pentium II and Pentium III processors, which come in the Slot 1 style, contain their own onboard cooling fan.

TROUBLESHOOTING TECHNIQUES

Whenever components from different manufacturers are mixed together during construction of a new computer, there will inevitably be times when two pieces do not fit well with each other. It may be as simple as a missing hole for a second screw on the hard drive cage. It may be a serious problem, such as the power supply covering the RAM sockets or the motherboard mounting holes not aligning with the insulated supports on the chassis.

No matter what the situation, try to find an acceptable solution. You may have to be creative, such as making a custom mounting bracket. You may need to exchange one part for another. In the end, your efforts will be rewarded with a working system and the experience of having done it yourself.

SUMMARY

In this exercise we discovered that

- Care and observation should be used when disassembling and assembling a computer system.
- It is important to record all BIOS settings, pin-1 identifiers on cables and connectors, and what connections are made to the motherboard.

This self-test is designed to help you check your understanding of the background information presented in this exercise.

True/False

Answer *true* or *false*.

1. All computers have three major sections: the main unit, the monitor, and the printer.
2. The purpose of the power supply is to supply power to the computer when there is a blackout.
3. A peripheral card is used to connect the printer to the main part of the computer.
4. It is not necessary to record the BIOS settings before disassembling a computer.
5. Jumpers on the motherboard are used to select the CPU clock speed.

Multiple Choice

Select the best answer.

6. The section of the computer that holds and electrically interconnects the major sections of the computer system is called the
   a. Power supply.
   b. Big board.
   c. Motherboard.
   d. Peripheral card.
7. One difference between a hard drive and a floppy drive is that in a hard drive, you
   a. Cannot remove the disks.
   b. Can only read information, not write it.
   c. Can only write information, not read it.
   d. None of the above.
8. When unplugging a ribbon cable
   a. Pull as hard as you can on the cable until it comes out.
   b. Pry it off with a screwdriver.
   c. Gently rock the connector back and forth as you pull up from each side.
   d. None of the above.
9. Which processors have onboard cooling fans?
   a. Pentium.
   b. Pentium II and III.
   c. None.
10. During assembly and disassembly, the power supply should be
    a. Left on to preserve BIOS settings.
    b. Turned off and unplugged for safety.
    c. Opened and inspected.

Matching

Match one wiring cable on the right with each item on the left.

11. Floppy drive cable
12. Hard drive cable
13. Speaker wires

    a. 40-pin connector, no twist in cable.
    b. 34-pin connector, cable has a twist.
    c. Two-conductor twisted pair.
    d. None of the above.

Completion

Fill in the blank or blanks with the best answers.

14. The _____ conductors on the motherboard power supply connectors should be in the middle of the 10 conductors.
15. Erratic behavior in the PC may be the result of a missing _____ on the processor.
16. Before disassembling a computer, you should always unplug the _____ _____ to ensure that no electrical power can get to the computer.

17. Motherboards should be placed in _____ bags.
18. A red or blue stripe on a ribbon cable indicates pin _____.

In this activity you will actually remove the cover of your assigned computer and identify the following:

- Power supply
- Motherboard
- Peripheral cards
- Floppy disk drive(s)
- Hard disk drive(s)
- CD-ROM drive

Make sure you refer to the section of the background material that has the closest relationship to the computer with which you are working. You should have a large, flat, smooth surface upon which to work. A container such as a small, clean cup should be available for storing small pieces of hardware (such as screws). You may want to place a clean towel or other soft material over your workspace. Always remember to take your time and work carefully with the computer. Be sure to follow all the proper safety procedures.

1. Using the proper procedure for disassembling your assigned computer, remove its cover and identify the major sections for your lab partner(s).
2. Using the proper procedure for assembling your assigned computer, replace its cover.

1. If your computer system was different from any of those discussed in this section, state how it was different.
2. Why do you think it is recommended that you unplug both ends of the power cord before disassembling the computer?
3. Name the major sections of the computer and state the purpose of each.

Under the supervision of your instructor, remove the cover of a personal computer and identify the following major sections:

1. Power supply
2. Floppy disk drive(s)
3. Hard disk drive(s)
4. CD-ROM drive
5. Motherboard
6. Expansion cards

34

Power Supplies

INTRODUCTION

Joe Tekk was examining a computer that had suddenly stopped working. Nothing at all happened when the power was applied. Even the fan on the power supply stayed off. This bothered Joe. He took the cover off and gave the entire computer a careful visual inspection. When he looked at the display adapter card, he saw that the card was pulled slightly out of its socket, with its connector pins at a slight angle to the motherboard connector. The mounting-bracket screw that should have held the card in place was missing.

Joe reseated the card and put a screw in the card's mounting bracket. He plugged the power cord into its socket and turned the computer on. It booted normally.

Later, Joe discovered that the video monitor for the machine had been changed recently, and there had been difficulty getting the VGA connector disconnected from the display adapter card. Joe reasoned that, with no screw to hold it in place, the card had been yanked out of position during the struggle with the VGA connector.

PERFORMANCE OBJECTIVES

Upon completion of this exercise, you will be able to

1. Remove and replace the computer power supply, using proper safety procedures.
2. Explain why different computers require different power supplies.

KEY TERMS

Power supply
Switching power supply
Current (amperes)

Voltage (volts)
Power (watts)

BACKGROUND INFORMATION

All personal computers get their electrical energy from 120V AC wall outlets. Alternating current is used by the power company because it is more economical to transmit electricity this way. However, the circuits used by your computer must have a steady low voltage called a DC voltage.

It is the job of the computer *power supply* to convert the 120V AC to a low and steady DC voltage that can be used by the tiny circuits inside the computer. Low-voltage DC is used for computers because it is easier to control. The function of the power supply is illustrated in Figure 34.1(a). Figure 34.1(b) shows a typical PC power supply (rated at 235W).

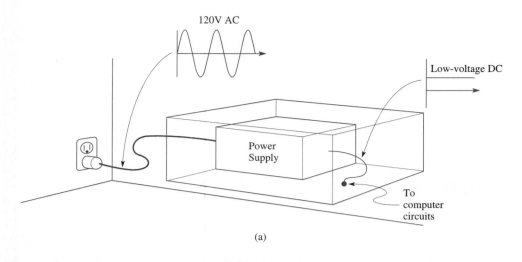

(a)

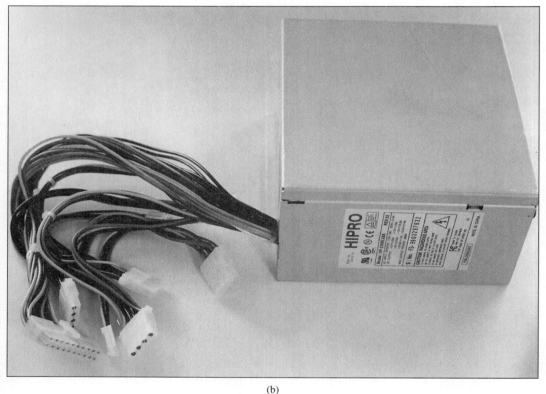

(b)

FIGURE 34.1   (a) Function of the power supply and (b) typical PC power supply

POWER SUPPLY CHARACTERISTICS

The type of power supply used in personal computers is called a ***switching power supply***. A switching power supply converts the 60-Hz power line frequency into a much higher 20,000-Hz frequency. This higher frequency allows the power supply to use much smaller and more economical filtering circuits and transformers. The higher frequency used in the power supply is switched ON or OFF at a very rapid rate, according to the requirements of the system. This switching is used because it is a very efficient system for regulating the electrical energy required by the computer system, resulting in less heat loss from the power supply.

FIGURE 34.2 Power supply types

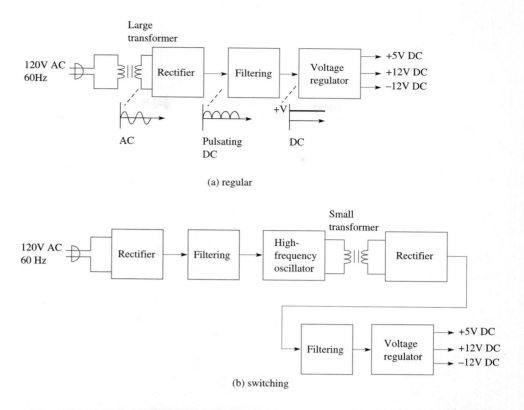

(a) regular

(b) switching

TABLE 34.1 Electrical characteristics of power supplies

Electrical Unit	Meaning
Current	The amount of movement of electrical charge. Current is measured in amps (A). The letter symbol for current is I.
Voltage	The amount of electrical potential that can cause a current flow. Voltage is measured in volts (V). The letter symbol for voltage is E.
Power	The amount of electrical energy. Power is measured in watts (W) and is calculated by $P = IE$, where P is the power in watts, I is the current in amps, and E is the voltage in volts.

Figure 34.2 shows the differences between a regular power supply and a switching power supply. In Figure 34.2(a), the regular power supply uses a rectifier to convert AC voltage into pulsating DC, a filter circuit to convert pulsating DC into DC, and a voltage regulator to adjust the output voltage to a desired level.

The switching power supply in Figure 34.2(b) uses the same three circuits as the regular power supply. However, several additional circuits are used to rectify the AC line voltage and use it to power a high-frequency oscillator.

Table 34.1 lists the three electrical characteristics of power supplies.

The power used by your computer has several different output voltages available to service the requirements of various internal sections. Figure 34.3 shows a typical power supply and some of the standard output voltages along with their typical maximum current ratings.

The power supply inside your computer must supply power to the following sections:

- Motherboard
- Floppy disk drives
- Hard disk drive
- Peripheral cards

These requirements are illustrated in Figure 34.4.

FIGURE 34.3 **Typical output voltages**

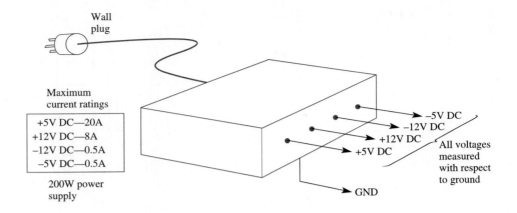

Wall plug

Maximum current ratings

| +5V DC—20A |
| +12V DC—8A |
| −12V DC—0.5A |
| −5V DC—0.5A |

200W power supply

−5V DC
−12V DC
+12V DC
+5V DC

All voltages measured with respect to ground

GND

FIGURE 34.4 **Circuits using the power supply**

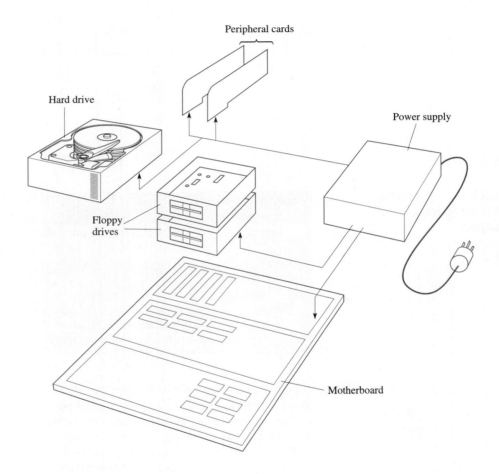

Peripheral cards

Hard drive

Power supply

Floppy drives

Motherboard

Different circuits inside your computer use different voltages. For example, peripheral cards use +5V. In order to conserve power, many processors now operate on 3.3V instead of 5V. Newer ATX power supplies provide this voltage. It is important that these voltages remain steady. The amount of current delivered by each voltage from the power supply will depend on the number of circuits and disk drives the power supply must service. The more of these circuits and drives, the more current the power unit must supply and the greater will be the power demand, as illustrated graphically in Figure 34.5. This is the reason why the same power supply is not used in all computers.

As you can see from Figure 34.5, the more features that are added to the inside of a computer (including external disk drives), the greater will be the power requirements of the power supply. If the power requirements of the computer exceed those of the power supply, the power supply will shut down and the system will not operate. The solution to this is to

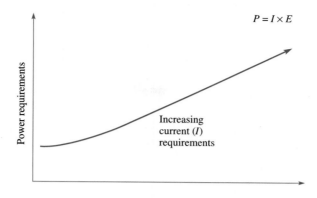

FIGURE 34.5 Relationship of system expansion and power requirements

$$P = I \times E$$

Power requirements (vertical axis)

Increasing current (*I*) requirements

Number of circuits using power supply

FIGURE 34.6 Various voltage connectors from the power supply

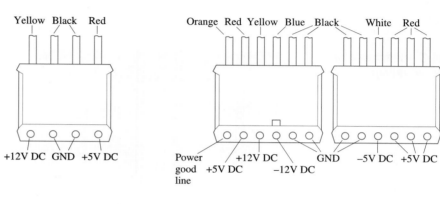

Yellow Black Red

+12V DC GND +5V DC

(a) Disk drive connector

Orange Red Yellow Blue Black White Red

Power good line +5V DC +12V DC −12V DC GND −5V DC +5V DC

(b) Motherboard connectors

3.3V 3.3V COM 5V COM 5V COM PW-OK 5VSB 12V

| 1 | 2 | 3 | 4 | 5 | 6 | 7 | 8 | 9 | 10 |
| 11 | 12 | 13 | 14 | 15 | 16 | 17 | 18 | 19 | 20 |

3.3V −12V COM PS-ON COM COM COM −5V 5V 5V

(c) ATX power connector

replace the existing power supply with a power supply of a higher power (wattage) rating. Always use the following rule when replacing power supplies:

Use the same voltage values and a power rating that is the same or larger.

Figure 34.6 shows the various connectors used by some of the most common computers. Since these are *switching* power supplies, never operate the power supply without its being plugged into the motherboard and with at least one disk drive connected. This means that you should never remove a computer power supply, place it on your test bench, plug it in, and expect to measure its output voltages. Doing this will not only give you incorrect readings, it could also damage the power supply. This is illustrated in Figure 34.7.

Table 34.2 lists the different power supply capabilities for some of the most common types of computers.

Table 34.3 lists the voltage outputs and compares maximum current outputs of a 63.5W power supply with that of a 200W power supply.

FIGURE 34.7 Possible damage to a computer power supply

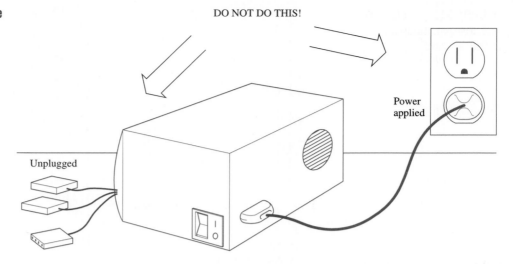

DO NOT DO THIS!

Power applied

Unplugged

TABLE 34.2 Common power supply power ratings

Power Rating	Computer Type
63.5W	PC
130W	XT
192W	AT
225W	Mini-Tower
300W	ATX

TABLE 34.3 Comparison of maximum current outputs

63.5W Unit			200W Unit		
Voltage	Maximum Current	Power	Voltage	Maximum Current	Power
+5V	7A	35W	+5V	20A	100W
+12V	2A	24W	+12V	8A	96W
−5V	0.3A	1.5W	−5V	0.5A	2.5W
−12V	0.25A	3W	−12V	0.5A	6W
Total Wattage		63.5W	Total Wattage		204.5W

As Table 34.3 shows, the power rating of a power supply indicates the maximum amount of current each of the output voltages can deliver. The power rating is also the sum of the individual maximum power outputs of each of the voltages. Note that the 200W power supply is actually a 204.5W unit. It may be interesting to note that using a computer with a 200W power supply for six hours per day costs about 16¢ per day (assuming the electric company charges 13¢ per kilowatt-hour).

TROUBLESHOOTING TECHNIQUES

Keep the following tips in mind when working with power supplies:

• A power supply that suddenly quits working has done so for a specific reason. Try to recall everything that happened to the computer since the last time the power supply worked. For example, it is very easy to forget that you bumped the machine by accident when it was off, or that you just installed additional motherboard cache and managed to plug the ICs in backward.

- Erratic behavior of your computer could be a sign of a failing or overloaded power supply. There may be a significant amount of ripple voltage on the DC outputs, much larger than that tolerated by the digital circuitry. You may want to examine the outputs (with power on) with an oscilloscope, or a DMM, to determine whether the voltage level is constant and within acceptable limits.
- Troubleshooting a failed power supply down to the bad component or components requires great skill, knowledge of power supply theory (rectification, filtering, current limiting, and so on), a suitable set of test equipment, and plenty of time. It may be a lot easier to simply purchase a new power supply with a higher power rating.
- If a bad system component (on the motherboard or in any of the adapter cards) caused the power supply to fail, a new supply may fail also. Always check the system components when replacing a power supply. A good visual may help identify the problem (a burned trace or pad, an out-of-alignment connector). Adapter cards may need to be removed and reinstalled one by one to find the card that is at fault.

SUMMARY

In this exercise we discovered that

- The personal computer uses a switching power supply.
- Switching power supplies contain more circuitry than regular power supplies.
- AC is converted into DC by a rectifier and filtering circuit.
- Power in watts equals voltage in volts times current in amperes.

SELF-TEST

This self-test is designed to help you check your understanding of the background information presented in this exercise.

True/False

Answer *true* or *false*.

1. The circuits inside a computer use steady, low-voltage DC.
2. It is the job of the computer power supply to convert the 120V AC from the wall plug to the low DC voltages used by the computer.
3. Power supplies used by computers are called *switching* power supplies and as such cannot safely be operated if they are disconnected from the circuit.
4. A switching power supply is more complex than a regular power supply.
5. A rectifier converts AC into pulsating DC.

Multiple Choice

Select the best answer.

6. The use of a switching power supply results in
   a. Less heat loss.
   b. More economical components.
   c. Greater power dissipation.
   d. Both a and b.
7. Electrical power is equal to
   a. Current times voltage.
   b. Current plus voltage.
   c. Current divided by voltage.
   d. Voltage divided by current.
8. The more circuits the power supply must power,
   a. The more voltage will be required.
   b. The more current will be required.
   c. The more power will be required.
   d. Both b and c.

9. Pulsating DC is turned into DC by a
   a. Rectifier.
   b. Transformer.
   c. Filtering circuit.
10. How much power is required when using 10A from a 5V supply?
    a. 2W.
    b. 15W.
    c. 50W.

Matching

Match a definition on the right with each electrical unit on the left.

11. Power   a. Electrical potential.
12. Voltage   b. Electrical energy.
13. Current   c. Movement of electrical charge.
                                         d. None of the above.

Completion

Fill in the blank or blanks with the best answers.

14. Five volts with a maximum current of 20A will produce a maximum power of _____W.
15. The power rating of a power supply indicates the maximum amount of _____ each of its output voltages can deliver.
16. Before removing the power supply, make sure you have completely disconnected the _____ _____.
17. A voltage _____ converts DC into the desired output voltage.
18. A switching power supply contains a high _____ oscillator.

FAMILIARIZATION ACTIVITY

Be sure to use the proper safety precautions when removing your power supply. You will need to remove the cover of the computer unit assigned to your lab group. You need to remove the cover only once for this exercise; then each lab partner can remove and insert the power supply.

1. Using the proper safety procedures, remove the power supply from the computer assigned to your lab group.
2. Reinstall the power supply in your computer.

QUESTIONS/ACTIVITIES

1. What was the power rating of the power supply you used in this exercise? How did you determine this?
2. Explain why the power requirements of a power supply increase when an extra disk drive is added to the computer system.
3. What should you look for when replacing the power supply of a computer?
4. Is it acceptable to completely remove a computer power supply from the computer, set it on the workbench, and plug it in so that you can measure its output voltages? Explain.
5. State, using diagrams where necessary, the purpose of a computer power supply.

REVIEW QUIZ

Under the supervision of your instructor

1. Remove and replace the computer power supply, using proper safety procedures.
2. Explain why different computers require different power supplies.

35

Floppy Disk Drives

INTRODUCTION

Joe Tekk unlocked one of several filing cabinets in the storage room at RWA Software. It was filled with 5.25-inch disks, old software archived years ago in case of fire. Joe laughed to himself. It had been a long time since he had seen a 5.25-inch disk.

The 3.5-inch disk is so pervasive, he thought. Just today he had seen the 3.5-inch floppy used three different ways. There was a miniature floppy drive on his laptop. His friend just bought a digital CCD camera that saves images as .JPG files on a 3.5-inch floppy (formatting it if necessary).

Joe thought about the third use of the 3.5-inch floppy. One manager at RWA Software carries one in his shirt pocket, to every meeting. He uses the floppy to bring PowerPoint presentations to the meetings where he will be speaking.

Joe realized that the 5.25-inch floppy never had a chance. It was simply too big.

PERFORMANCE OBJECTIVES

Upon completion of this exercise, you will be able to

1. Remove and replace a 3.5-inch floppy disk drive and boot the system.
2. Check the hardware settings of a floppy drive.

KEY TERMS

Floppy disk drive
Zip drive

Jaz drive
SuperDisk

BACKGROUND INFORMATION

OVERVIEW

A *floppy disk drive* is a device that enables a computer to read and write information on a floppy disk.

Floppy disk drives (FDDs) are located at the front panel of the computer. The most common is the 3.5-inch drive (the old 5.25-inch drive is almost obsolete). These two disk drives are illustrated in Figure 35.1.

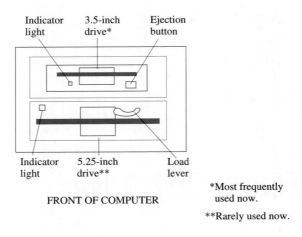

FIGURE 35.1   Typical floppy disk drives

Indicator light   3.5-inch drive*   Ejection button

Indicator light   5.25-inch drive**   Load lever

FRONT OF COMPUTER

*Most frequently used now.

**Rarely used now.

HOW A FLOPPY DISK DRIVE WORKS

Figure 35.2(a) is a simplified drawing of an FDD with its major components. Figure 35.2(b) shows a typical FDD unit. Table 35.1 summarizes the purpose of each major component of the FDD.

OPERATING SEQUENCE

The operating sequence of a typical 3.5-inch drive is as follows. Pushing the floppy disk into the drive causes the disk to be properly seated. The initial start-up for the drive consists of determining where track 0 is located. This is usually accomplished by a mechanical device, which is activated once the drive head is over track 0. When information is read, the stepping motor moves the read/write heads to their proper location. When information is written, the disk's write-protect status is checked and then new information is added to the disk.

FIGURE 35.2   (a) Major components of a floppy disk drive

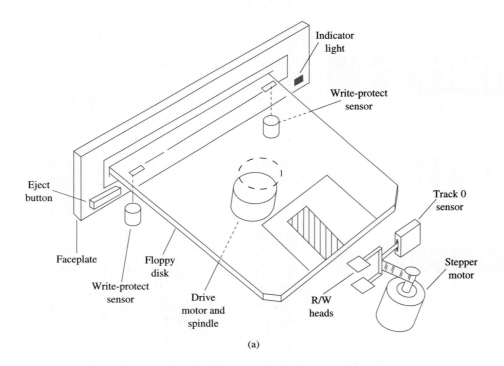

Indicator light

Write-protect sensor

Track 0 sensor

Stepper motor

R/W heads

Drive motor and spindle

Write-protect sensor

Floppy disk

Faceplate

Eject button

(a)

FIGURE 35.2 *(continued)*
(b) typical FDD unit

(b)

TABLE 35.1 Main parts of an FDD

Part	Purpose
Eject button	Used to eject a disk from the drive.
Write-protect sensor	Checks the condition of the floppy disk's write-protect system.
Read/write heads	Read and write information magnetically on the floppy disk. The heads move together, each working from its own side of the disk.
Track 0 sensor	Indicates when the read/write head is over track 0 of the floppy disk.
Drive motor	Spins the floppy disk inside the FDD.
Stepper motor	Moves the read/write head to different positions on the floppy disk.
Indicator light	Indicates if the disk drive is active.

DISK DRIVE SUPPORT SYSTEM

For proper operation, each part of the FDD support system must function properly.

1. *OS.* The operating system must be compatible with the media on the floppy disks.
2. *Floppy disk.* The disk itself must contain accurately recorded information in the proper format.
3. *Disk drive controller.* The drive controller conditions the signals between the motherboard and the FDD. Originally, a controller card that plugged into the motherboard was used to control the floppy drive. Most motherboards now have the floppy controllers built in.
4. *Disk drive electronic assembly.* This assembly consists of circuit boards that control the logical operations of the FDD. They act as an electrical interface between the disk drive controller and the electromechanical parts of the FDD.
5. *Disk drive mechanical assembly.* This assembly ensures proper alignment of the disk and read/write heads for reading and writing information.
6. *System power supply.* The power supply provides electrical power for all parts of the FDD, including the motors.

7. *Interconnecting cable.* Ribbon cable is used to transfer signals between the disk drive electrical assembly and the disk drive controller card.
8. *Power cable.* The DC power cable supplies electrical power to all parts of the FDD.

Let's take a closer look at many of these important components.

The OS

The operating system plays an important role in the operation of the floppy drive. Beginning with the system BIOS, all drive parameters must be known by the operating system so that data can be properly exchanged. Windows contains many applications designed specifically for disk drive operations. For example, right-clicking the drive A: icon in the My Computer window produces a Properties window similar to that shown in Figure 35.3. A pie chart is used to graphically illustrate used/free space on the drive. The volume label can be changed by entering a new one in the text box.

The Floppy Disk

There are several built-in Windows tools available for working with floppy disks. These tools are contained in the Tools submenu, as indicated in Figure 35.4.

The disk can be scanned for errors (using ScanDisk), backed up, or *defragmented.* A disk that has had many files created and deleted on it eventually becomes fragmented, with the files broken up into groups of sectors and scattered all over the disk (but still logically connected through the use of the FAT). This fragmentation increases the amount of time required to read or write entire files to the disk. By defragmenting the disk, all the files are reorganized and are stored in consecutive groups of sectors at the beginning of the storage space on the disk.

A fourth tool is included that allows you to *compress* the data on your floppy, increasing the amount of free space available. The Compression submenu shown in Figure 35.5 indicates what will be gained by compressing the current disk.

The Disk Drive Controller

The disk drive controller used to be an individual chip on a controller card. Now, the controller is just one part of a multifunction peripheral IC designed to operate the floppy and hard drives, the printer, and the serial ports. As always, I/O ports and interrupts are used to

FIGURE 35.3   Drive A: Properties window

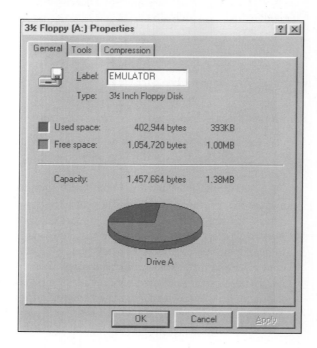

FIGURE 35.4 **Floppy disk tools**

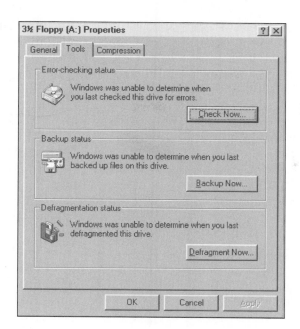

FIGURE 35.5 **Compression submenu**

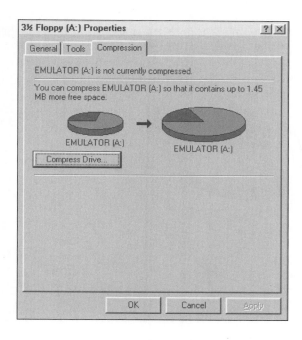

control the floppy disk drive. The settings used by Windows 95/98 can be examined/changed by using the Device Manager submenu of System Properties. The settings used by Windows NT can be examined from the Windows NT Diagnostics menu. Figure 35.6 shows the hardware configuration of a typical A: drive. These settings can be changed if necessary.

The Drive Cable

A 34-conductor ribbon cable is used to connect one or two disk drives to the controller. A twist in the cable between the two drive connectors reverses the signals on pins 9 through 16 at each connector. This twist differentiates the two drive connectors, forcing them to be used specifically for drive A: or drive B:. Figure 35.7 shows the cable details.

The meanings of the signals on the drive cable are shown in Table 35.2. Note that the signals affected by the twist in the cable are the select and enable signals for each drive.

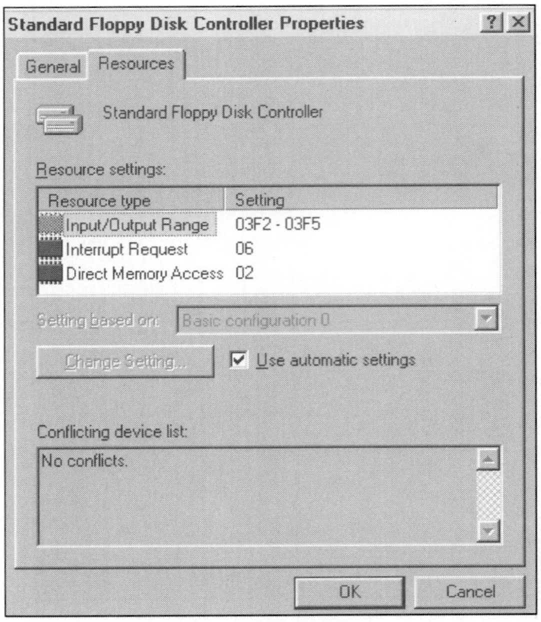

FIGURE 35.6 (a) Hardware settings for floppy drive and (b) Windows NT Floppy Properties

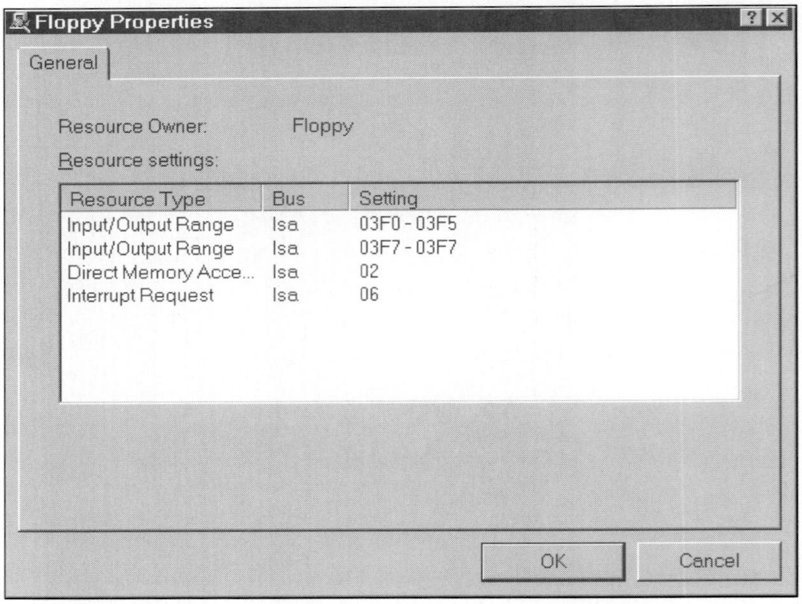

(a)

(b)

The Power Cable

Like any other peripheral, the disk drive needs power to operate the drive and stepper motors, the read/write amplifiers and logic, and the other drive electronics. Figure 35.8 shows the pinouts of the standard power connectors used on the floppy drive. The connectors are keyed so that they only plug in one way.

Zip Drives

A device similar to the floppy drive is the *Zip drive*, manufactured by Iomega. Zip drives connect to the printer or USB ports, have removable 100-MB cartridges, and boast a data transfer rate of 60MB/minute (using an SCSI connection). The 100-MB disks spin at

FIGURE 35.7 Floppy drive cable

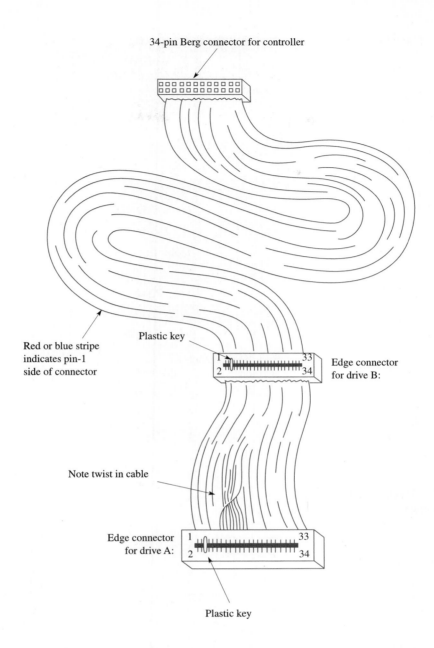

34-pin Berg connector for controller

Red or blue stripe indicates pin-1 side of connector

Plastic key

1   33
2   34

Edge connector for drive B:

Note twist in cable

Edge connector for drive A:

1   33
2   34

Plastic key

2941 RPM, have an average access time of 29 milliseconds, and are relatively inexpensive. Newer 250-MB cartridges are also available.

The software driver for the Zip drive uses the signal assignments shown in Figure 35.9 to control the Zip drive through the printer port. Using the printer port to control the Zip drive allows you to easily exchange data between two computers.

Jaz Drives

Similar to the Zip drive, the *Jaz drive* uses a 1-GB removable cartridge that spins at 5400 RPM, has an average seek time around 10 milliseconds, and has a sustained data transfer rate of more than 6MB/second. A Jaz drive operates similarly to a hard drive, except the drive media is removable.

The 120-MB SuperDisk

The *SuperDisk* is a new type of floppy drive with a 120-MB capacity. SuperDisk drives can read/write both 120-MB SuperDisk floppies and 1.44/2.88-MB 3.5-inch disks.

TABLE 35.2 Floppy drive cable signals

Conductor (Pin)	Signal
1–33 odd	Ground
2	Unused
4	Unused
6	Unused
8	Index
10	Motor Enable A
12	Drive Select B
14	Drive Select A
16	Motor Enable B
18	Stepper Motor Direction
20	Step Pulse
22	Write Data
24	Write Enable
26	Track 0
28	Write Protect
30	Read Data
32	Select Head 1
34	Ground

FIGURE 35.8 Floppy drive power connectors

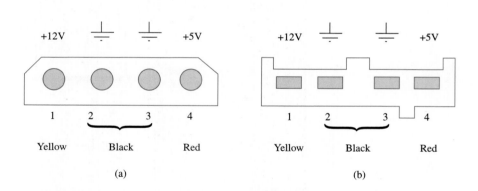

(a)

(b)

FIGURE 35.9 Zip drive parallel interface

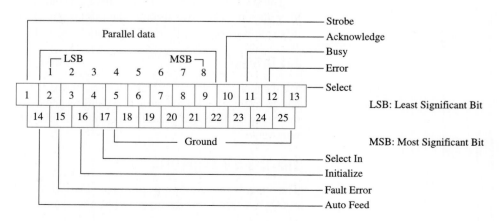

LSB: Least Significant Bit

MSB: Most Significant Bit

Working with Floppies

A few tips to keep in mind when working with floppy drives:

- Sometimes disks that are formatted on one system cannot be read by another. This may be due to differences in the read/write head alignments between both drives.
- If a floppy gives unexpected read errors, try ejecting the floppy and reinserting it.
- Run ScanDisk or some other suitable disk tool (such as Norton Utilities) to check a troublesome floppy.
- Always beware of disks given to you by someone else. Before you begin using them, scan them for viruses. This is especially important in an educational setting, where students and instructors often exchange disks as a normal part of class or lab.

TROUBLESHOOTING TECHNIQUES

SYSTEM OVERVIEW

Probable causes of what appears to be an FDD failure may be in one of the areas shown in Figure 35.10. This figure illustrates the relationship of the FDD to the entire computer system. At one end is the software on the disk; at the other extreme is the power cord connection to the electrical outlet. Every part of this system must be functioning properly for the disk drive to do its part. What is important here is to ensure that what appears to be a disk drive problem is not actually a problem caused by one of these other areas.

TROUBLESHOOTING LOGIC

The first step in troubleshooting the disk drive is to classify the problem as occurring in one of the areas shown in Figure 35.10. Once the area at fault is determined, corrective action may be taken. Figure 35.11 is a troubleshooting diagram for determining which of these areas may be at fault.

FIGURE 35.10   System relationship to floppy disk drive

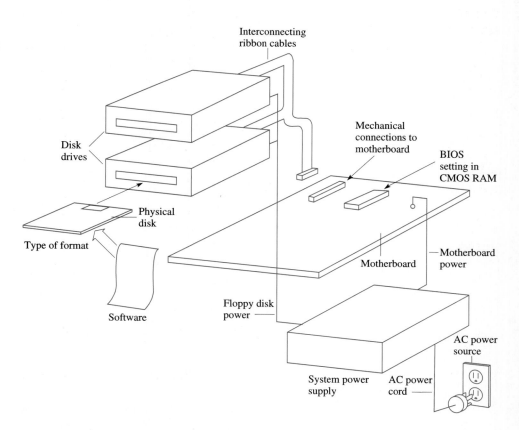

FIGURE 35.11 FDD troubleshooting chart

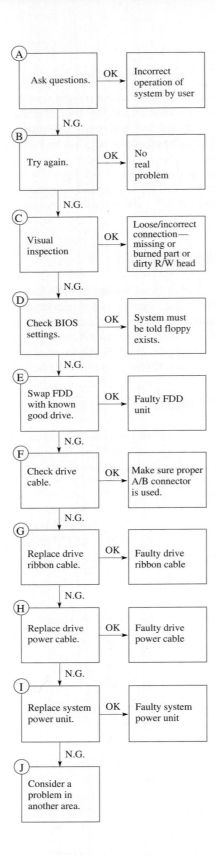

TROUBLESHOOTING STEPS

1. *Ask questions.* With all computer servicing, inquire about the history of the system. Was it recently modified? Was there an attempt to make any changes or repairs? Did the user buy the system used? For example, if the user recently installed the disk drive, it

may be that certain settings need to be set differently. In this case, you will need to refer to the documentation for the system as well as the newly installed FDD. Asking questions may help you quickly spot the problem, saving you time and money.

2. *Try again.* Sometimes reseating the floppy in the drive fixes any read errors encountered. You could also try using the floppy in a different drive.

3. *Visual inspection.* A good visual inspection may reveal a burned component, an improperly seated cable, a dirty read/write head, or other mechanical evidence of the problem. Remember, a good visual inspection is an important part of any troubleshooting process.

4. *Check BIOS settings.* The system must be aware that a floppy drive exists. This is accomplished by running the BIOS setup program at boot time.

5. *Swap FDD with known good drive.* There are some precautions you need to take when doing this. It may be that the A: drive has a terminating resistor. You need to refer to documentation for the disk drive in question.

6. *Check drive cable.* In a one-drive system, make sure the connector with the twist is connected to drive A:. In a two-drive system, make sure both connectors are in the correct drives.

7. *Replace drive ribbon cable.* Here, another good visual inspection may be needed. Make sure you refer to the servicing manual to ensure that the cable was correctly installed in the first place. Replacing the cable with a known good one will help determine if the original cable or its connectors are at fault.

8. *Replace drive power cable.* This may be difficult on some systems because the power cable may be permanently attached to the power supply. If this is the case, measure the voltages with a volt meter.

9. *Replace system power unit.* Doing this will help eliminate the problem of a power unit supplying the correct voltages when the power demands on it are small, but failing, as a result of the increased power demand, when a disk drive motor is activated (such as when the FDD is attempting to read a disk). Make a note of the power rating of the power supply; if it is below 100W, substitute another known good power unit with a higher power rating.

10. *Consider a problem in another area.* If all the preceding steps fail to locate the problem, the problem is in another area of the unit. The most likely area in this case is the motherboard.

SUMMARY

In this exercise we discovered that

- A floppy disk drive contains many electrical and mechanical parts, such as stepper motor, track 0 sensor, read/write heads, and spindle.
- The floppy disk has its own controller.
- The operating system contains drivers for the floppy disk drive and other software (such as ScanDisk) for maintaining the drive.

SELF-TEST

This self-test is designed to help you check your understanding of the background information presented in this exercise.

True/False

Answer *true* or *false.*

1. The most common type of floppy disk drive is the 3.5-inch size.
2. The purpose of a floppy disk drive is to read and write information from and to a floppy disk.
3. On a properly operating floppy disk drive, the indicator light is on only when the disk is in the writing mode.
4. The write-protect system protects information on the floppy disk from being written over.

5. The purpose of the eject button is to make sure the floppy disk is properly seated in its jacket.

6. The floppy disk does not require a controller.

7. The track 0 sensor determines when track 0 is full of data.

Multiple Choice

Select the best answer.

8. On a floppy disk drive, information is placed on the disk by
   a. A single read/write head.
   b. Two heads, one for reading and the other for writing.
   c. Two heads, one on each side of the disk, each of which reads or writes on its side.

9. The floppy disk is turned inside its jacket by the operation of the
   a. Drive motor.
   b. Stepper motor.
   c. Spindle motor.
   d. None of the above.

10. The purpose of the twist in the ribbon cable is to
   a. Swap data in/out signals.
   b. Swap drive A/B select signals.
   c. Invert the drive data.

11. The purpose of the stepper motor is to
   a. Step the floppy disk correctly in its jacket.
   b. Turn the floppy disk in its jacket.
   c. Ensure that the floppy disk is correctly aligned.
   d. Position the read/write heads along the floppy disk.

12. The position of the floppy disk can be determined by the
   a. Track 0 sensor.
   b. Indicator light sensor.
   c. Drive assembly.
   d. Index sensor.

13. The 120-MB SuperDisk can read
   a. 1.44-MB floppies.
   b. 120-MB SuperDisk floppies.
   c. Both a and b.

14. Disk fragmentation _____ the time required to read/write files.
   a. Decreases
   b. Increases
   c. Has no effect on

Matching

Match one or more conditions on the right with each item on the left.

15. Writing information
16. Reading information
17. Inserting the disk
18. Drive light on
19. Initial start-up

a. Disk is properly seated.
b. Write-protect notch is checked.
c. Stepper motor is active.
d. Heads are moved to track 0.
e. None of the above.

Completion

Fill in the blank or blanks with the best answers.

20. The floppy disk drive consists of two major sections, the _____ assembly and the electrical assembly.

21. Signals are conditioned between the motherboard and the disk drive by the _____ _____ controller.

22. A(n) _____ _____ is used to transfer information between the system and the disk drive.

23. The power cable connects the system _____ _____ and the floppy disk drive.
24. The dual-floppy drive cable has a(n) _____ between the two drive connectors.
25. The floppy disk drive ribbon cable contains _____ conductors.
26. Two large-storage drives similar to the floppy disk drive are the Zip drive and the _____ drive.

Open-Ended

27. Describe the types of disk drives used by your system.
28. How many disk drives does your system have?
29. Who manufactures your disk drives? How did you find this out?
30. Determine if the price of one Zip disk is a bargain compared to the price of 83 floppies.
31. Is it necessary to have any software in your system for your floppy disk drives to work properly? Explain.

FAMILIARIZATION ACTIVITY

1. Survey the machines in your laboratory. How many have one floppy? How many have two? Do any of the machines have 5.25-inch drives?
2. Remove the floppy disk drive from your system, using the appropriate tools and safety precautions.
3. In the following space, describe the steps you used to remove the drive.

4. In the following space, make a sketch of the drive you removed, showing the following: all retaining hardware, all cable connections, any areas in which you experienced difficulty in removing the drive, and any other notes you may want to use for future reference.

5. Replace the floppy disk drive.
6. In the following space, describe the steps you used to replace the drive.

7. Have your instructor check your work before proceeding to the next step.

Instructor OK: _____

8. Using the proper assembly procedure, reassemble the rest of your unit; if your unit was a functional one, demonstrate to your instructor that it is now working properly.

Instructor OK: _____

9. Check the hardware settings for your floppy disk drive using BIOS setup and Device Manager.

QUESTIONS/ACTIVITIES

1. Describe, in your own words and using illustrative diagrams, the method of removing and installing the floppy disk drive on your system. Do this as if the description you are writing were to be used by another service technician for the purpose of removing and installing a floppy disk drive on the same kind of system.
2. Using a recent computer parts catalog, find the list cost of a floppy disk drive that would be an exact replacement for the one you worked with in this exercise.
3. Contact at least one local computer repair shop and find out what they would charge to replace a floppy disk drive in a computer of the type you used in this exercise.
4. If you could make a major improvement in floppy disk drives, what would it be?
5. Your instructor may require that you troubleshoot a floppy disk drive type problem in the lab. If this is a requirement, state what the problem was and how you found it. If you did not find the problem, state what you could do differently in the future to find and correct a similar problem.

REVIEW QUIZ

Under the supervision of your instructor

1. Remove and replace a 3.5-inch floppy disk drive and boot the system.
2. Check the hardware settings of a floppy drive.

36

The Motherboard Microprocessor and Coprocessor

INTRODUCTION

The phone rang on Joe Tekk's desk. It was Joe's friend, a 13-year-old boy Joe met at a computer show. "Joe, it's Stephen. Are you busy?"

"No," Joe said, smiling. "What's up?" He was always happy to hear what Stephen was up to.

"Can you check the upgrade path for my 100-MHz Pentium for me? I need MMX technology for some course work at school."

Joe grabbed a computer catalog off a tall stack of catalogs piled near his desk. He leafed through it and found the microprocessor section. There was a table showing the various ways you could upgrade your microprocessor. "You can get a 166-MHz processor with MMX technology for around $100. How's that?"

Stephen was not satisfied. "That's not fast enough."

Joe reexamined the table in the catalog. "Do you have Socket 7 on your motherboard?"

Stephen said, "Hold on," and was silent for only an instant before saying, "Yes! Socket 7!" happily into the phone.

Joe was glad. "Great! You can get a 200-MHz version now for $125."

Stephen barely finished saying, "Order one, I'll pay you later," before the line went dead.

PERFORMANCE OBJECTIVES

Upon completion of this exercise, you will be able to

Use a standard personal computer motherboard to explain the location, purpose, bus size, and speed of the

- Microprocessor
- Math coprocessor

KEY TERMS

Motherboard
System board
Planar
Microprocessor
CPU
Bus

Memory location
SIMD
Speculative execution
Superpipelined
Coprocessor
Math chip

DEFINITION OF THE MOTHERBOARD

The main system board of the computer is commonly referred to as the *motherboard*. Several typical motherboards are shown in Figure 36.1. Sometimes the motherboard is referred to as the *system board*, or the *planar*.

FIGURE 36.1   (a) Typical PC motherboard (photo by John T. Butchko) and (b) AMD K6-2 motherboard

(a)

(b)

FIGURE 36.1 *(continued)*
(c) Pentium II motherboard

(c)

CONTENTS OF THE MOTHERBOARD

The motherboard holds and electrically interconnects all the major components of a PC. The motherboard contains the following:

- The microprocessor
- The math coprocessor (only on older 386 motherboards)
- BIOS ROM
- RAM (Dynamic RAM, or DRAM, as well as level-2 cache)
- The expansion slots
- Connectors for IDE drives, floppies, and COM ports

Table 36.1 lists these major parts and gives a brief overview of the purpose of each part.

TABLE 36.1 Purposes of major motherboard parts

Part	Purpose
Microprocessor	Interprets the instructions for the computer and performs the required process for each of these instructions.
Math coprocessor	Used to take over arithmetic functions from the microprocessor.
BIOS ROM	Read-only memory. Memory programmed at the factory that cannot be changed or altered by the user.
RAM	Read/write memory. Memory used to store computer programs and interact with them.
Expansion slots	Connectors used for the purpose of interconnecting adapter cards to the motherboard.
Connectors	Integrated controller on motherboard provides signals for IDE and floppy drives, the printer, and the COM ports.

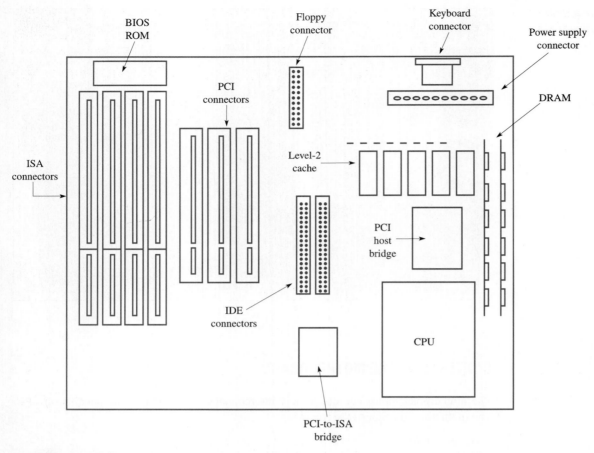

FIGURE 36.2   Motherboard layout

Figure 36.2 shows a typical motherboard layout and the locations of the major motherboard parts.

In this exercise, you will have the opportunity to learn more details about the microprocessor and the coprocessor. In the following exercises, you will learn about the other areas of the motherboard.

THE MICROPROCESSOR

You can think of the ***microprocessor*** in a computer as the central processing unit (CPU), or the "brain," so to speak, of the computer. The microprocessor sets the stage for everything else in the computer system. Several major features distinguish one microprocessor from another. These features are listed in Table 36.2.

You can think of a ***bus*** as nothing more than a group of wires all dedicated to a specific task. For example, all microprocessors have the following buses:

Data bus   Group of wires for handling data. This determines the data path size.
Address bus   Group of wires for getting and placing data in different locations. This helps determine the maximum memory that can be used by the microprocessor.
Control bus   Group of wires for exercising different controls over the microprocessor.
Power bus   Group of wires for supplying electrical power to the microprocessor.

Figure 36.3 shows the bus structure of a typical microprocessor.

Since all the data that goes in and out of a microprocessor is in the form of 1s and 0s, the more wires used in the data bus, the more information the microprocessor can handle at one

TABLE 36.2 Microprocessor features

Feature	Description
Bus structure	The number of connectors used for specific tasks
Word size	The largest number that can be used by the microprocessor in one operation
Data path size	The largest number that can be copied to or from the microprocessor in one operation
Maximum memory	The largest amount of memory that can be used by the microprocessor
Level-1 cache	Internal, high-speed memory used by the microprocessor to speed up memory accesses
Level-2 cache	External, high-speed memory located between CPU and memory
Speed	The number of operations that can be performed per unit time
Code efficiency	The number of steps required for the microprocessor to perform its processes

FIGURE 36.3 Typical microprocessor bus structure

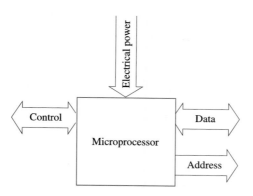

FIGURE 36.4 Relationship between data and address

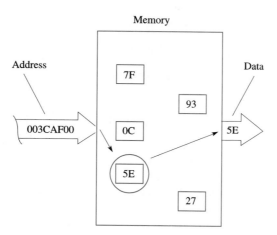

time. For example, some microprocessors have eight lines (wires or pins) in their data buses, others have 16, and some have 32 or 64.

The number of lines used for the address bus determines how many different places the microprocessor can use for getting and placing data. The *places* that the microprocessor uses for getting and placing data are referred to as ***memory locations***. The relationship between the data and the address is shown in Figure 36.4.

TABLE 36.3 Types of Intel microprocessors used in the PC

Microprocessor	Lines		Maximum Clock Speed	Addressable Memory
	Data	Address		
8088	8	20	8 MHz	1MB
8086	16	20	8 MHz	1MB
80286	16	24	20 MHz	16MB
80386SX	16	24	20 MHz	16MB
80386	32	32	33 MHz	4GB
80486	32	32	66 MHz	4GB
Pentium	64	32	233+ MHz	4GB
Pentium Pro	64	36	200+ MHz	64GB
Pentium II	64	36	400+ MHz	64GB
Pentium III	64	36	1000 MHz	64GB
Pentium 4	64	36	1500 MHz	64GB

The greater the number of lines used in the address bus of a microprocessor, the greater the number of memory locations the microprocessor can use. Table 36.3 lists the common microprocessors used in the PC. All of these microprocessors are manufactured by Intel Inc.

Note from Table 36.3 that the greater the number of address lines, the more memory the microprocessor is capable of addressing. In the table, 1MB = 1,048,576 memory locations, and 4GB = 4,294,967,296 memory locations.

Figure 36.5 shows some of the packaging used for microprocessors.

Compatible CPUs

A number of companies manufacture processors that compete with Intel for use in PC motherboards and other applications. Two of these companies are AMD and Cyrix. Table 36.4 shows sets of compatible CPUs.

Having more than one processor to choose from allows you to examine pricing, chip features, and other factors of importance when making a decision.

About the 80x86 Architecture

The advanced nature of the Pentium microprocessor requires us to think differently about the nature of computing. The Pentium architecture contains exotic techniques such as branch prediction, pipelining, and superscalar processing to pave the way for improved performance. Let's take a quick look at some other improvements from Intel:

- Intel has added MMX technology to its line of Pentium processors (Pentium, Pentium Pro, and Pentium II/III). A total of 57 new instructions enhance the processors' ability to manipulate audio, graphic, and video data. Intel accomplished this major architectural addition by *reusing* the 80-bit floating-point registers in the FPU. Using a method called *SIMD* (single instruction multiple data), one MMX instruction is capable of operating on 64 bits of data stored in an FPU register. AMD has its own multimedia instructions, called 3DNow!, that extend the capabilities of MMX.
- The Pentium Pro processor (and also Pentium II/III) use a technique called *speculative execution*. In this technique, multiple instructions are fetched and executed, possibly out of order, in order to keep the pipeline busy. The results of each instruction are speculative until the processor determines that they are needed (based on the result of branch instructions and other program variables). Overall, a high level of parallelism is maintained.

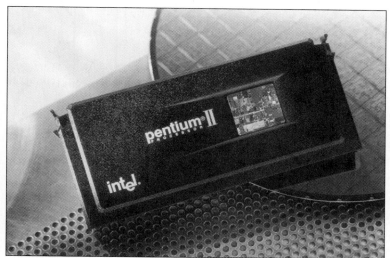

FIGURE 36.5   Microprocessor packaging (courtesy Intel Corporation)

TABLE 36.4   Comparing CPUs

Intel	AMD	Cyrix
Pentium	K5	5x86*
Pentium II	K6–2	6x86MX and MII
Pentium III	K6–III	VIA Cyrix III

*Pentium performance, pin compatible with the 80486.

TABLE 36.5 Matching math coprocessors

Microprocessor	Math Coprocessor
8086	8087
8088	8087
80286	80287
80386	80387
80386SX	80387SX
80486DX	Built-in coprocessor enabled
80486SX	Coprocessor disabled
Pentium, Pentium Pro, and Pentium II/III/4	Built-in coprocessor enabled always

- First used in the Pentium Pro, a bus technology called Dual Independent Bus architecture uses two data buses to transfer data between the processor and main memory (including the level-2 cache). One bus is for main memory, the second is for the level-2 cache. The buses may be used independently or in parallel, significantly improving the bus performance over that of a single-bus machine.
- The five-stage Pentium pipeline was redesigned for the Pentium Pro into a *superpipelined* 14-stage pipeline. By adding more stages, less logic can be used in each stage, which allows the pipeline to be clocked at a higher speed. Although there are drawbacks to superpipelining, such as bigger branch penalties during an incorrect prediction, its benefits are well worth the price.

Spend some time on the Web reading material about these changes, and others. It will be time well invested.

THE COPROCESSOR

Each Intel microprocessor released before the 80486 has a companion to help it do arithmetic calculations. This companion is called a *coprocessor*. For most software, the coprocessor is optional. However, some programs (such as CAD, computer-aided design, programs) have so many math calculations to perform that they need the assistance of the math coprocessor; the main microprocessor simply cannot keep up with the math demand.

These *math chips*, as they are sometimes called, are capable of performing mathematical calculations 10 to 100 times faster than their companion microprocessors and with a higher degree of accuracy. This doesn't mean that if your system is without a coprocessor it can't do math; it simply means that your microprocessor will be handling all the math along with everything else, such as displaying graphics and reading the keyboard.

Table 36.5 lists the math coprocessors that go with various microprocessors. Note that the 80486 and higher processors have built-in coprocessors.

For a math coprocessor chip to be used by software, the software must be specifically designed to look for the chip and use it if it is there. Some spreadsheet programs look for the presence of this chip and use the microprocessor for math if the coprocessor is not present. If the coprocessor is present, the software uses it instead. Some programs, such as word-processing programs, have no use for the math functions of the coprocessor and do not use the coprocessor at all. Therefore, the fact that a system has a coprocessor doesn't necessarily mean that the coprocessor will improve the overall system performance. Improvement will take place only if the software is specifically designed to use the coprocessor and there are many complex math functions involved in the program.

TROUBLESHOOTING TECHNIQUES

Windows identifies the processor it is running on. Use System Properties in Control Panel to check the processor type, as shown in Figure 36.6. Notice that Windows 95 has identified a Pentium as the CPU. It is interesting to note that other Pentium-compatible CPUs, such as a Cyrix 5x86, are not recognized by Windows as Pentiums and are reported in System Properties as 80486 CPUs. This kind of information is not easily found in existing documentation. It is good to talk directly to the manufacturer of the motherboard to determine how your performance might be affected.

FIGURE 36.6 **Processor identification, (a) Windows 95 System Properties, (b) Windows 98 System Properties**

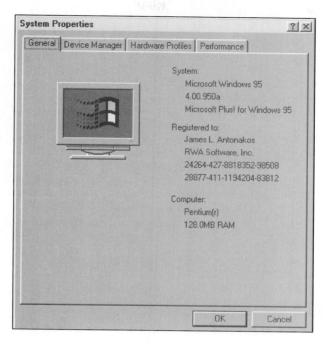

(a)

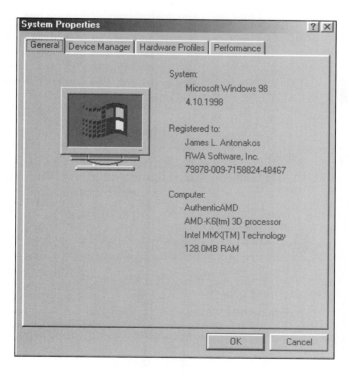

(b)

(continued on the next page)

FIGURE 36.6 *(continued)*
(c) Windows NT System Properties

(c)

SUMMARY

SUMMARY

In this exercise we discovered that

- The microprocessor is one component of many found on the motherboard.
- Microprocessors are categorized by bus width, the number of address lines, clock speed, and the features of its machine code.
- A bus carries multiple bits of information.
- Intel, AMD, and Cyrix manufacture competing processors.

SELF-TEST

This self-test is designed to help you check your understanding of the background information presented in this exercise.

True/False

Answer *true* or *false*.

1. The motherboard is sometimes referred to as the system board.
2. All the major components in the computer are interconnected through the motherboard.
3. ROM is computer memory that is programmed at the factory and cannot be changed directly by the user.
4. A dedicated math processor is called a coprocessor.
5. Internal processor cache is called level-1 cache.

Multiple Choice

Select the best answer.

6. You can think of the microprocessor in the computer as the
    a. CPU.
    b. "Brains of the computer."
    c. Main controlling part of the computer.
    d. All of the above.

7. The largest number that can be used by the computer in one operation is determined by the
   a. Amount of available memory.
   b. Speed of the computer.
   c. Word size.
   d. None of the above.
8. In a computer, a group of wires dedicated to a specific task is called the
   a. Bus.
   b. Track.
   c. Data path.
   d. Through path.
9. How many bytes of memory can be accessed using 32 address lines?
   a. 1MB.
   b. 16MB.
   c. 4096MB.
10. SIMD stands for
    a. Single instruction multiple data.
    b. Simple datapath.
    c. Simple instruction multimedia data.

Matching

Match a characteristic on the right with each microprocessor on the left.

11. 80386	a. 33-MHz clock speed.
12. 80486	b. 4GB of addressable memory.
13. Pentium	c. 64GB of addressable memory.
14. Pentium II	d. None of the above.

Completion

Fill in the blank or blanks with the best answers.

15. A(n) _____ is a chip used to help the microprocessor perform mathematical computations.
16. The video technology added to the Pentium architecture uses the abbreviation _____.
17. Whether or not a math coprocessor chip is actually used is determined by the system _____.
18. The _____ and _____ processors have built-in coprocessors.
19. The multimedia technology developed by AMD for their processors is called _____.
20. The Intel _____ microprocessor was the first to address 4GB of memory.

FAMILIARIZATION ACTIVITY

Use the proper procedure to open the case of your microcomputer. With your lab partner(s), locate the microprocessor and the coprocessor (if there is no coprocessor, locate the slot where the coprocessor would be located).

From the markings on the microprocessor, use the information presented in the background information section to determine the

1. Speed of the microprocessor.
2. Number of data lines.
3. Maximum amount of addressable memory.
4. Type of numeric coprocessor to use.

Questions

1. What are the specifications of the microprocessor used in your computer?
2. In your own words, state the purpose of the microprocessor.
3. Explain the difference in use of the microprocessor data bus and the address bus.
4. In your own words, state the purpose of a math coprocessor.
5. Explain when a math coprocessor is needed. What determines whether a math coprocessor actually will be used once it is installed?

Activities

1. Make a sketch of the motherboard used in your computer. Indicate the locations of the microprocessor and coprocessor.
2. Make a sketch of the microprocessor used in your system. Be sure to indicate all the information printed on the microprocessor chip.

Under the supervision of your instructor

Use a standard personal computer motherboard to explain the location, purpose, bus size, and speed of the

- Microprocessor
- Math coprocessor

37   The Motherboard Memory

INTRODUCTION

The phone rang on Joe Tekk's desk. It was Joe's friend Ken Koder. Ken was a software engineer for a local aerospace firm.

"Joe, I'm at the computer show. You know, the one that comes to the arena every two months."

"Sure, that's a great show," Joe said. Joe liked going to the shows. They always had good prices for brand-new equipment and components.

Ken continued. "They are selling 32MB EDO DRAM for $10. Do you need any?"

Joe thought a moment. He knew there were two open memory slots on his motherboard in his machine at home. He could pull the two 16MB RAMs from the other two slots and get four new 32MB RAMs, giving him 128MB at home. His office machine had two open slots as well. "Sure, Ken. Get me six if you can."

"Just six? I'm getting eight! That will give me 256MB of RAM for simulations. Anyway, I'll bring your memory over later today."

Joe thought that Ken could probably easily use 512MB for simulations. He himself was happy with 128MB just for Windows 98.

PERFORMANCE OBJECTIVES

Upon completion of this exercise, you will be able to

1. Demonstrate
   a. Methods of increasing the memory usage of a personal computer.
   b. How to determine the organization of the computer's memory.
2. From a motherboard selected by your instructor, determine
   a. The locations of the RAM and ROM.
   b. The size of the DRAM.
   c. The speed of the DRAM.

KEY TERMS

Magneticware
Primary storage
Secondary storage
Mass storage

Read-only memory (ROM)
Read/write memory
Random access memory (RAM)
Volatile memory

Flash EPROM
Parity checking
Parity bit
Single inline memory module (SIMM)
Dual inline memory module (DIMM)
Synchronous DRAM (SDRAM)
EDO DRAM
Dual-ported RAM
Level-2 cache
Nanosecond

Base memory
Memory map
Upper memory
Conventional memory
Extended memory
Expanded memory
Virtual 8086 mode
Virtual memory
Flat addressing

COMPUTER MEMORY

Computer memory consists of any device capable of copying a pattern of 1s and 0s that represents some meaningful information to the computer. Computer memory can be contained in *hardware*, such as in chips, or in **magneticware**, such as floppy and hard disks (or other magnetic material such as magnetic tape). Computer memory is not limited to just these two major areas. For example, a laser disk uses light to read large amounts of information into the computer; this too is a form of computer memory. For the purpose of discussion here, computer memory will be divided into two major areas: hardware memory and magneticware memory.

The hardware memory of a computer is referred to as ***primary storage***. The magneticware of a computer is referred to as ***secondary storage***, or ***mass storage***. Here are some facts about each.

Primary Storage

- Immediately accessible to the computer.
- Any part of the memory may be immediately accessed.
- Short-term storage.
- Limited capacity.

Secondary Storage

- Holds very large amounts of information.
- Not immediately accessible.
- May be sequentially accessed.
- To be used, must be transferred to primary storage.
- Long-term storage.

In this exercise, you will see how primary and secondary computer memories are used (see Figure 37.1) and how they can work with each other to produce an almost unlimited amount of computer memory. First, let's learn about primary storage.

FIGURE 37.1 Two major areas of computer memory

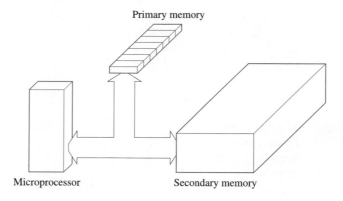

Primary memory

Microprocessor

Secondary memory

448

PRIMARY STORAGE—RAM AND ROM

There are two basic kinds of primary storage: one kind that the computer can quickly store information in and retrieve information from and another kind that the computer can only receive information from. Figure 37.2 shows the two basic kinds of primary storage memory.

The kind of memory that the computer can get information (read) from but cannot store information (write) to is called *read-only memory (ROM)*. The advantage of having ROM is that it can contain programs that the computer needs when it is first turned on; these programs (called the Basic Input/Output System, or BIOS) are needed by the computer so it knows what to do each time it turns on (such as reading the disk and starting the booting process). Obviously, these programs should not be able to be changed by the computer user, because doing so could jeopardize the operation of the system. Therefore, ROM consists of chips that are programmed at the factory. The programs in these chips are permanent and stay that way even when the computer is turned off; they are there when the computer is turned on again.

The kind of memory that the computer can write to as well as read from is called *read/write memory*. The acronym for read/write memory is RWM, which is hard to say. Because of this, read/write memory is called *RAM*, which stands for *random access memory*. Both ROM and read/write memory are randomly accessible, meaning that the computer can get information from any location without first going through other memory locations. However, read/write memory is traditionally referred to as RAM.

Unlike ROM, RAM loses anything that is stored in it when the power is turned off. Because the information in RAM is not permanent, it is referred to as *volatile memory*. Figure 37.3 shows this difference.

FIGURE 37.2   Two basic kinds of primary storage memory

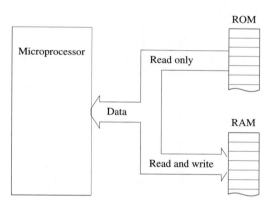

FIGURE 37.3   ROM and RAM

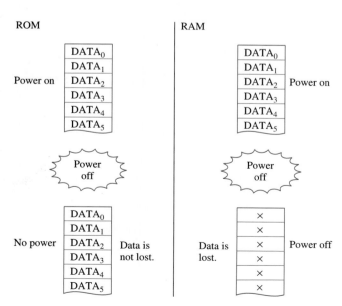

The system ROM chip for the PC contains two main programs, the Power-On Self-Test (POST) and the Basic Input/Output System (BIOS). The programs in the ROM chip set the personality of the computer. As a matter of fact, how compatible a computer is can be determined primarily by the programs in these ROM chips. The ROM chips have changed over time as systems have been improved and upgraded. There have been, for example, more than 20 changes in the ROM BIOS programs by IBM for its different PCs.

You may need to update your old BIOS to use new hardware in your system (large IDE hard drives, for example). To upgrade your BIOS ROM, you may (1) replace the ROM with a new one or (2) run a special upgrade program (typically available for download off the Web) that makes changes to a *flash EPROM*, an EPROM that can be electrically reprogrammed.

BITS, BYTES, AND WORDS

Recall that a *bit* is a single binary digit. It has only two possible conditions: ON and OFF. Everything in your computer is stored and computed with ONs and OFFs. The bits inside your computer are arranged in such a way as to work in units. The most basic unit, or group, of bits is called a *byte*. A byte consists of 8 bits. Mathematically, 8 bits have 256 unique ON and OFF combinations. You can figure this out with your pocket calculator—just calculate 2^8, which is 2 multiplied by itself eight times. A *word* is 16 bits, or 2 bytes. When 4 bytes are taken together, such as in 32-bit microprocessors, they are called a *double word*. These different arrangements are shown in Figure 37.4.

In PCs a method called *parity checking* is used to help detect errors. There are times when, in the process of working with computer bits, a bit within a byte may accidentally change from ON to OFF or from OFF to ON. To check for such an error, parity checking uses an extra bit called the *parity bit*. IBM and most compatibles use what is called *even parity* to check their bits. Even parity means that there will always be an even number of ONs for each byte, including the parity bit. Even parity checking is illustrated in Figure 37.5.

FIGURE 37.4 Arrangement of computer data

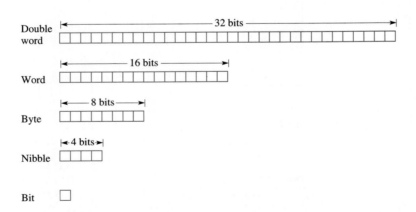

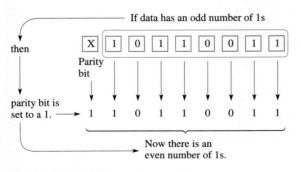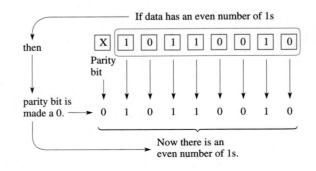

FIGURE 37.5 Even parity checking

SIMM

The *single inline memory module*, or **SIMM**, is another way of physically organizing memory. It is a small "boardlet" with several memory chips soldered to it. This boardlet is inserted into a system slot. Figure 37.6 shows a 16MB SIMM.

SIMMs came about in an attempt to solve two problems. The first problem was "chip creep." Recall that chip creep occurs when a chip works its way out of a socket as a result of thermal expansion and contraction. The old solution to this problem—soldering memory chips into the board—was not a good solution, because it made them harder to replace. So the SIMM was created. The only problem with the SIMM is that if only 1 bit in any of its chips goes bad, the whole SIMM must be replaced. This is more expensive than replacing only one chip. SIMMs come in 256K, 1MB, 4MB, and 16MB sizes. A similar type of memory module, called a SIPP, contains metal pins that allow the SIPP to be soldered directly onto the motherboard.

Regarding parity bits in a SIMM, a 32-bit SIMM is nonparity, and a 36-bit SIMM stores one parity bit for each byte of data. Pentium processors incorporate parity in their address and data buses.

DIMM

The *dual inline memory module (DIMM)* was created to fill the need of Pentium-class processors containing 64-bit data buses. A DIMM is like having two SIMMs side by side, and they come in 168-pin packages (more than twice that of a 72-pin SIMM). Ordinarily, SIMMs must be added in pairs on a Pentium motherboard to get the 64-bit bus width required by the Pentium. Figure 37.6 shows a 168-pin DIMM containing 128MB of SDRAM.

SDRAM

Synchronous DRAM (SDRAM) is very fast (up to 133-MHz operation) and is designed to synchronize with the system clock to provide high-speed data transfers.

EDO DRAM

Extended Data Out DRAM (EDO DRAM) is used with bus speeds at or below 66 MHz and is capable of starting a new access while the previous one is being completed. This ties in nicely with the bus architecture of the Pentium, which is capable of back-to-back pipelined bus cycles. Burst EDO (BEDO RAM) contains pipelining hardware to support pipelined burst transfers.

FIGURE 37.6   Assorted memories

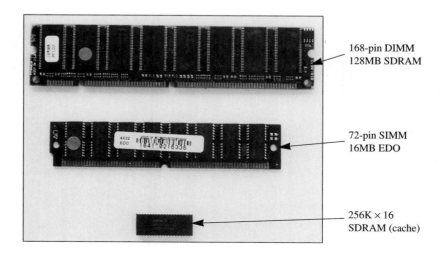

168-pin DIMM
128MB SDRAM

72-pin SIMM
16MB EDO

256K × 16
SDRAM (cache)

VRAM

Video RAM (VRAM) is a special dual-ported RAM that allows two accesses at the same time. In a display adapter, the video electronics needs access to the VRAM (to display the Windows desktop, for example) and so does the processor (to open a new window on the desktop). This type of memory is typically local to the display adapter card.

LEVEL-2 CACHE

Cache is a special high-speed memory capable of providing data within one clock cycle and is typically 10 times faster than regular DRAM. Although the processor itself contains a small amount of internal cache (8K for instructions and 8K for data in the original Pentium), you can add additional *level-2 cache* on the motherboard, between the CPU and main memory, as indicated in Figure 37.7. Level-2 cache adds 64K to 2MB of external cache to complement the small internal cache of the processor. The basic operation of the cache is to speed up the average access time by storing copies of frequently accessed data. A 512-K level-2 cache is shown in Figure 37.6. Newer processors (such as the Pentium II and above) use a separate bus to access the level-2 cache. In this case the cache is referred to as *backside* cache.

CHIP SPEED

When replacing a bad memory module, you must pay attention to the *speed* requirements of that module. If you do not, the replacement will not work. You can use memory chips that have a higher speed or the same speed as the replacement, but not a lower speed. The faster the memory, the more it costs. Adding faster memory to your system may not improve overall performance at all because the speed of your computer is determined by the system clock, among other things.

Memory chip speed is measured in ***nanoseconds***. A nanosecond is 0.000000001 second. To check the speed rating of a RAM chip, look at the coding on the top of the chip. Typical DRAM speeds are 60 ns, 10 ns, and 8 ns.

HOW MEMORY IS ORGANIZED

It is important that you have an understanding of the organization of memory in the computer. The 8088 and 8086 microprocessors are able to address up to 1MB of memory. Since

FIGURE 37.7　Using cache in a memory system

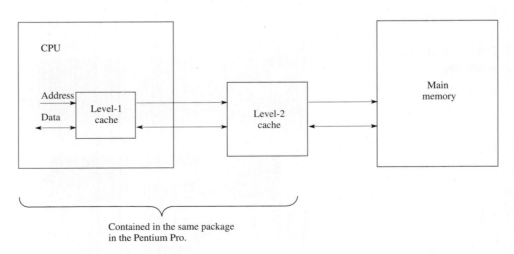

some of the first PCs used the 8088 and 8086 microprocessors, when the 80286, 80386, 80486, and Pentium microprocessors were introduced, they were made *downward compatible* with their predecessors. This meant that software that worked on an older PC would still work on the newer systems.

In order to keep downward compatibility, the newer microprocessors (80286, 80386, 80486, and Pentiums) come with two modes of operation (there is one other mode that will be presented later in this exercise). One mode is called the *real mode*, in which the microprocessors behave like their earlier models (and are limited to 1MB of addressable memory). The other mode, called the *protected mode*, allows the microprocessors to use the newer power designed into them (such as addressing up to 16MB for the 80286 and 4GB for the 80386, 80486, and Pentiums).

All PCs have what is called a *base memory*, which is the 1MB of memory that is addressable by the 8088, 8086, and newer microprocessors running in real mode. A common way of viewing the organization of memory is through the use of a *memory map*. A memory map is simply a way of graphically showing what is located at different addresses in memory. Figure 37.8 shows the memory map of the PC in real mode.

As you can see from the memory map, several areas of memory are designated for particular functions; not all the 1MB of memory is available for your programs. As a matter of fact, only 640KB can be used by DOS-operated systems. Table 37.1 lists the definitions of the various memory sections.

The memory above the conventional 640K of memory is referred to as *upper memory*.

FIGURE 37.8 Memory map of PC in real mode

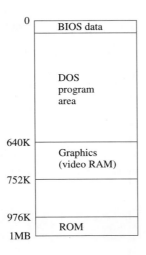

TABLE 37.1 Purpose of allocated memory

Assignment	Definition
Base memory	This refers to the amount of memory actually installed in the conventional memory area.
Conventional or user memory	This is the 640K of memory that is usable by DOS-based programs.
Video or graphics memory	This area of memory (128K) is reserved for storing text and graphics material for display on the monitor. As you will see in Exercise 44, the amount of this space actually used by the system depends on the requirements of the video monitor.
Motherboard ROM	This is space reserved for the use of the ROM chips on the motherboard.

FIGURE 37.9 Relationships among conventional, extended, and expanded memory

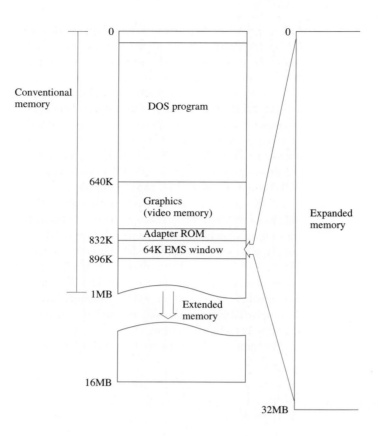

HOW MEMORY IS USED

There are three ways memory can be allocated: as ***conventional memory***, ***extended memory***, or ***expanded memory***. Figure 37.9 shows the relationships among the three types of memories.

Table 37.2 explains the uses of the three different memory allocation methods. It is important to note that DOS and systems that use DOS are limited to 1MB of addressable space. The

TABLE 37.2 Memory allocation methods

Memory Type	Comments
Conventional	1. Memory between 0K and 1MB, with 640K usable and 340K reserved. 2. Completely usable by DOS-based systems. 3. Uses real mode of 8086, 8088, 80286, 80386, 80486, and Pentium.
Extended	1. Uses protected mode of 80286 (up to 16MB), 80386, 80486, and Pentium (up to 4GB). 2. Cannot be used by DOS-based systems (which are limited to 1MB of memory). 3. Can all be accessed by the IBM OS/2 operating system. 4. Can be used by a virtual disk in DOS systems. 5. Is the type of memory to use with the 80386, 80486, and Pentium microprocessors.
Expanded	1. Uses "bank-switching" techniques. 2. Requires special hardware and software. 3. Is not a continuous memory but consists of chunks of memory that can be switched in and out of conventional memory. 4. Sometimes referred to as EMS (expanded memory specification) memory.

454

reason for this is that DOS is made for microprocessors running in the real mode. This memory limit exists because computer and program designers thought that 1MB of memory would be all that anyone would ever need on a PC for years to come. Since 340K of the 1MB of addressable DOS memory is reserved by the system, DOS really has only 640K left for user programs. Thus, DOS is said to have a 640K limit. However, as you will see, there are ways of allowing DOS to store data in an addressable memory location that is beyond this DOS limit.

BREAKING THE DOS BARRIER

The 80386, 80486, and Pentium microprocessors have a third mode of operation, called the *virtual 8086 mode*. This mode allows a single 80386, 80486, or Pentium microprocessor to divide its memory up into many "virtual" computers. The beauty of this is that each of these virtual computers can run its own program in total isolation from other programs running in the other virtual computers. This means that more than one DOS program can be run in the same computer at the same time.

You can use the DOS VDISK command to create a virtual disk in extended memory. The command put into the CONFIG.SYS file is

For IBM DOS:	DEVICE = VDISK.SYS[size][sectors][files]/E
For MS-DOS:	DEVICE = RAMDRIVE.SYS[size][sectors][files]/E

Both of these commands will create a virtual disk in extended memory: this is the purpose of the /E modifier.

In DOS 5.0, you can force DOS to use as much of high memory as possible. These commands are also put into the CONFIG.SYS file:

$$\text{DEVICE} = \text{C:\DOS\HIMEM.SYS}$$
$$\text{DOS} \quad = \text{HIGH}$$
$$\text{DEVICE} = \text{C:\DOS\EMM386.EXE}$$

The first line tells DOS to install its high-memory manager. The second line actually loads part of DOS into this region of memory. The third line tells DOS to load the EMM386 Expanded Memory Manager to make the expanded memory available. DOS 5.0 has a feature that releases even more of the 640K of user memory—another command placed in the CONFIG.SYS file:

$$\text{DOS} = \text{UMB}$$

or

$$\text{DOS} = \text{HIGH, UMB}$$

Once the UMB (upper-memory support) feature is enabled, you can use the same CONFIG.SYS file to load a device driver (such as a mouse) into high memory:

$$\text{DEVICEHIGH} = \text{[path name and name of the device]}$$

If you have a TSR (terminate and stay-resident program) that you want to have loaded into high memory (such as Borland's SideKick), DOS 5.0 gives you another option. Include the following in your AUTOEXEC.BAT file:

$$\text{LOADHIGH [program name]}$$

Another command available in DOS 4.01 and later is

$$\text{MEM /PROGRAM or MEM /P}$$

This command will output the memory allocation for the entire system. Figure 37.10 shows a typical printout. The memory locations are given in hexadecimal values. You can use your pocket calculator to convert the hex values to decimal numbers.

VIRTUAL MEMORY

Another method of extending memory is through the use of *virtual memory*. Virtual memory is memory that is not made up of real, physical memory chips. Virtual memory is memory made

FIGURE 37.10   Printout of
MEM /P

```
Address      Name         Size        Type

000000                    000400      Interrupt Vector
000400                    000100      ROM Communication Area
000500                    000200      DOS Communication Area

000700       IO           0021C0      System Program

0028C0       MSDOS        008E20      System Program

00B6E0       IO           00AEA0      System Data
             ANSI         001180       DEVICE=
             CCDRIVER     002AB0       DEVICE=
             MOUSE        003540       DEVICE=
                          000380       FILES=
                          000100       FCBS=
                          0029A0       BUFFERS=
                          0001C0       LASTDRIVE=
                          000CD0       STACKS=
016590       COMMAND      000070      Data
016610       MSDOS        000040      -- Free --
016660       FASTOPEN     002780      Program
018DF0       COMMAND      001640      Program
01A440       COMMAND      000100      Environment
01A550       APPEND       001E20      Program
01C380       GRAPHICS     0014A0      Program
01D830       COMMAND      000040      Data
01D880       SHELLB       000E80      Program
01E710       COMMAND      001640      Program
01FD60       COMMAND      0000A0      Environment
01FE10       MEM          000070      Environment
01FE90       MEM          012F00      Program
032DA0       MSDOS        06D250      -- Free --


  655360 bytes total memory
  655360 bytes available
  524640 largest executable program size

  393216 bytes total extended memory
  393216 bytes available extended memory
```

FIGURE 37.11   Virtual memory

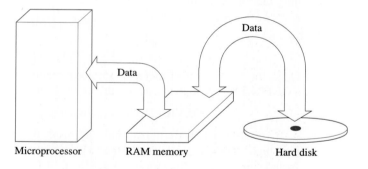

Microprocessor   RAM memory   Hard disk

up of mass storage devices such as disks. In the use of virtual memory, the computer senses when its usable real memory is used up, stores what it deems necessary onto a disk (usually the hard disk), and then uses what it needs of the freed-up real memory. If it again needs the data it stored on the disk, it simply frees up some more real memory (by placing its contents on the disk) and then reads what it needs from the disk back into the freed-up memory. The concept of virtual memory is illustrated in Figure 37.11. Windows 95/98 uses demand-paging virtual memory to manage memory, a technique supported by features of protected mode.

MEMORY USAGE IN WINDOWS

It is not difficult to determine why an operating system performs better if it has 32MB of RAM available, rather than only 8MB. With 32MB of RAM, the operating system will be able to support more simultaneous processes without having to use the hard drive for virtual memory backup. Additional memory will also be available for the graphical user interface (multiple overlapped windows open at the same time).

It is no secret that the Windows 3.x architecture did not use memory efficiently, typically requiring at least 8MB or 16MB to get a reasonable amount of performance on a 386 or 486 CPU. Windows 95/98 also performs much better when given a large amount of RAM to work with. A minimum of 16MB or 32MB is recommended.

What does Windows 95/98 use memory for? Conventional memory, the first 640K of RAM, still plays an important role supporting real-mode device drivers and DOS applications. For instance, if you want DOSKEY installed as part of your DOS environment under Windows 95, place its command line in AUTOEXEC.BAT as you normally would.

Upper memory, the next 360K of the first 1MB of RAM, can be used to place DOS and other memory-resident applications above the 640K limit, freeing more RAM for DOS applications.

Extended memory, everything above the first 1MB of RAM (4096MB total), is where Windows 95/98 runs most applications, using an addressing scheme called *flat addressing*. The flat addressing model uses 32-bit addresses to access any location in physical memory, without the need to worry about the segmented memory scheme normally used.

Windows 95/98 provides the Resource Meter (in the System Tools folder under Accessories) to monitor system resources. As indicated in Figure 37.12, the amount of resources available is shown graphically. The display is updated as resources are used and freed up.

Another useful tool is the System Monitor in Windows 95/98 and the Performance Monitor in Windows NT (shown in Figure 37.13), which display a running tally of resource usage over a period of time. The display format is selectable (bar, line, or numeric charts), as are the colors and type of information displayed.

FIGURE 37.12   Resource Meter display

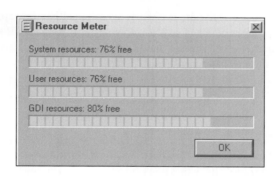

FIGURE 37.13   (a) Windows 95/98 System Monitor display

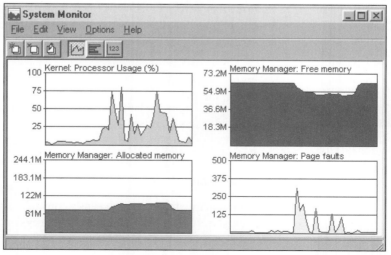

(a)

(continued on the next page)

FIGURE 37.13 *(continued)*
(b) Windows NT Performance
Monitor display

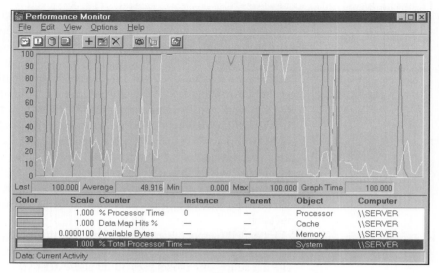

(b)

TROUBLESHOOTING TECHNIQUES

One of the simplest ways to determine whether your Windows system has enough RAM to handle its workload is to watch the hard drive light. No or little activity, except when opening or closing an application, is a good sign.

If the hard drive activates sporadically, doing a little work every now and then, the system is borderline. If the activity increases when additional applications are opened, there is a definite lack of RAM.

Frustrated with hard drive activity when only a few applications were open, one user increased the amount of RAM in his system from 32MB to 128MB (taking advantage of a drop in memory prices at that time). Now, even with a taskbar full of applications, the hard drive remains inactive.

SUMMARY

In this exercise we discovered that

- RAM and ROM are primary storage memories.
- RAM forgets its data when power is off; ROM does not.
- Data is organized as bits, bytes, words, and even larger quantities.
- There are many types of memory, such as SIMM, DIMM, cache, and SDRAM.
- A memory map describes how memory is allocated in a system.
- The illusion of a very large memory is provided by virtual memory.

SELF-TEST

This self-test is designed to help you check your understanding of the background information presented in this exercise.

True/False

Answer *true* or *false*.

1. All computer memory simply keeps a copy of a pattern of 1s and 0s.
2. The only useful computer memory is memory contained in computer chips.
3. When power is turned off, all computer memory is lost.

4. EDO DRAM works the same way as cache.
5. Flash EPROM acts like ROM but allows its contents to be modified.
6. RAM and ROM are secondary storage.

Multiple Choice

Select the best answer.

7. RAM is
   a. Random access memory.
   b. Read/write memory.
   c. Volatile memory.
   d. All of the above.
8. A byte is
   a. Larger than a bit.
   b. Smaller than a word.
   c. Equal to 8 bits.
   d. All of the above.
9. Base memory is
   a. All the memory that can be addressed by the microprocessor.
   b. The first 1MB of memory.
   c. The actual memory installed in the system.
   d. None of the above.
10. SDRAM stands for
    a. Static DRAM.
    b. Synchronous DRAM.
    c. Sideways DRAM.
11. How many bytes are carried by a 64-bit data bus?
    a. 8
    b. 32
    c. 64
12. How many bits wide is a SIMM?
    a. 8
    b. 32
    c. 64

Matching

Match a term on the right with each definition on the left.

13. Memory between 0K and 1MB
14. Memory that uses the real mode for the 80386
15. Chunks of extra memory that can be switched in and out of conventional memory

a. Extended memory.
b. Conventional memory
c. Expanded memory.
d. None of the above.

Completion

Fill in the blanks with the best answers.

16. _____ checking is a method used by the computer to check for errors.
17. When active memory interacts with the _____ disk, this is referred to as using _____ memory.
18. In practical applications, only _____ of memory is available for DOS systems.
19. External cache is also called _____ cache.
20. A memory wide enough for the Pentium's 64-bit data bus is the _____.
21. RAM that allows two accesses at the same time is called dual-_____.
22. A memory _____ shows what is located at different addresses in memory.

The first part of this familiarization activity gives you the opportunity to have your computer system use as much of its memory as possible. The second half of this activity has you examine the location and type of RAM used by the system.

1. Make sure you have DOS 5.0 or higher in your system.
2. Use the following DOS command to check how memory is used in your computer. If you have a printer available, redirect the output to the printer:

   ```
   A> MEM /PROGRAM
   ```

3. From step 2, determine whether your system has *extended* or *expanded* memory and how much. If you have problems with this, check with your instructor.
4. In this step, you force DOS to use as much higher memory as possible. Using a text editor, add the following DOS commands to your CONFIG.SYS file:

   > DEVICE = [drive][path]HIMEM.SYS
   > DOS   = HIGH
   > DEVICE = [drive][path]EMM386.EXE

5. Now reboot your system. Again, run the DOS memory program:

   ```
   A> MEM /PROGRAM
   ```

   What differences do you now see in the memory allocation of your system?
6. If your system has a device driver, such as a mouse, you can use DOS to try to force the driver into high memory. Modify your CONFIG.SYS file to add the following:

   > DOS = HIGH, UMB
   > DEVICEHIGH = [path and name of device]

7. Again, reboot your system. Using the DOS memory program

   ```
   A> MEM /PROGRAM
   ```

   determine what difference there is now in the allocation of your system memory.
8. If your system has a TSR program, attempt to force as much of it as possible into high memory. Add the following to your AUTOEXEC.BAT file:

   > LOADHIGH [program name]

9. Again, reboot your system. Using the DOS memory program

   ```
   A> MEM /PROGRAM
   ```

   determine what difference there is now in the allocation of your system memory.
10. In order to use as much extended memory as possible, install a virtual disk in high memory. To do so, make sure the program VDISK.SYS (for IBM DOS) or RAM-DRIVE.SYS (for MS-DOS) is on your DOS disk. Then add the following command, as appropriate to CONFIG.SYS:

    > (IBM DOS)   DEVICE = VDISK.SYS/E
    > (MS-DOS)   DEVICE = RAMDRIVE.SYS/E

11. Again, reboot the system and verify that the virtual disk has been installed. Use the DOS memory program

    ```
    A> MEM /PROGRAM
    ```

    Determine what difference there is now in the allocation of your system memory.
12. Using the proper procedures, remove the case of your computer and determine the following:
    a. The locations of the RAM and ROM.
    b. The type of RAM (how much memory in each RAM chip).
    c. The speed of the RAM.
13. Check the memory statistics of your machine using the System Monitor. Watch it for 10 minutes while you open and close applications. What do you find?

1. Does your system have any extended or expanded memory? How did you determine this?
2. Which step in the familiarization activity freed up the most user memory?
3. State how you could determine whether the virtual disk you installed was placed in high memory.
4. What is the size of the DRAM used in your system? How did you determine this?
5. What is the speed of the DRAM used in your system? How did you determine this?

Under the supervision of your instructor

1. Demonstrate
   a. Methods of increasing the memory usage of a personal computer.
   b. How to determine the organization of the computer's memory.
2. From a motherboard selected by your instructor, determine
   a. The locations of the RAM and ROM.
   b. The size of the DRAM.
   c. The speed of the DRAM.

38

Motherboard Expansion Slots

INTRODUCTION

Joe Tekk was thumbing through a PC hardware catalog. One ad for a motherboard contained a description that read "Expansion: 4 PCI, 3 ISA, AGP."

Joe wondered what AGP stood for; it was a new term to him. He brought up Netscape and used Yahoo to search for "AGP." There were several hits. Joe found out that AGP stands for Accelerated Graphics Port, a high-speed hardware interface that allows 3-D graphics cards to use PC memory efficiently, for better multimedia performance.

Joe found other ads for AGP-equipped motherboards, all of which required the Pentium II/III processor. Joe laughed to himself. "Just what I needed, one more reason to replace my old motherboard."

PERFORMANCE OBJECTIVES

Upon completion of this exercise, you will be able to

1. Identify the type of system, bus structure, and expansion slots for each of at least three computer motherboards and/or expansion cards.
2. Briefly describe the differences among the various bus architectures.

KEY TERMS

Expansion slot
ISA expansion slot
Direct memory access
Nonmaskable interrupt
AT expansion slot
Micro-channel architecture (MCA)
EISA expansion slot
Video coprocessor

Matched memory extension
Local bus
PCI expansion slot
PCMCIA expansion slot
Hot swapping
AGP expansion slot
Universal serial bus

BACKGROUND INFORMATION

OVERVIEW

Expansion slots serve a very important function in personal computers. They allow you to plug in electronic cards to expand and enhance the operation of your computer. The

FIGURE 38.1 **Expansion slots**

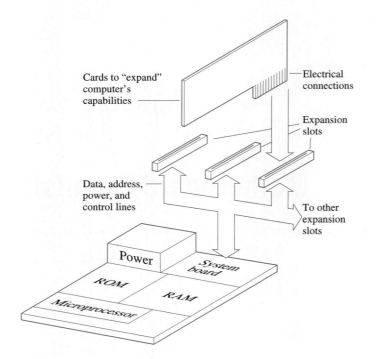

concept of expansion slots is simple; however, in practical terms, there are many things to consider. Figure 38.1 is a simple illustration of the function of expansion slots.

An expansion slot must be able to communicate with the computer. This communication usually includes access to the microprocessor. In achieving this access, the expansion bus must not interfere with the normal operation of the microprocessor. This means that the expansion bus must have access not only to the address and data lines used by the microprocessor but also to special control signals.

You need to be familiar with the functions of the various types of expansion buses that provide the means to add new features to your computer. One of the many things you will be doing for your customers is to help them make decisions about what kinds of added features they want to have for their computers. Most of these added features, such as extra memory, additional types of monitor displays, extra or different disk drives, telephone communications, and other enhancements, are added in part by electrical cards that fit into expansion slots on the computer (peripheral cards). However, as you will soon see, not all expansion slots are the same. It is important that you know their differences.

MAKEUP OF AN EXPANSION SLOT

Expansion slots have more similarities than differences. Table 38.1 lists the purposes of the different lines that are connected to the expansion slots. It should be noted that not all expansion slots use every one of these lines. The terminology used in this table does, however, apply to all expansion slots that use any of these lines.

ISA Expansion Slots

Figure 38.2 shows the ISA (Industry Standard Architecture) expansion slot. True PC compatibles also use the same kind of expansion slot. Pin assignments are shown in Figure 38.3.

The major features of the ISA expansion slots are listed in Table 38.2. The features are described in terms of a *bus*. Recall that a bus is nothing more than a group of conductors treated as a unit; as a *data bus*, it is a group of conductors used to carry data. In terms of an expansion slot, a bus can be thought of as a group of connectors that is connected to the bus on the motherboard.

TABLE 38.1 Expansion slot terminology

Connections	Purpose
Power lines	Power lines supply the voltages that may be needed by the various expansion cards. The power lines are +5V DC, −5V DC, +12V DC, −12V DC, and ground.
Data lines	Data lines are used to transfer programming information between the expansion card and the computer. One of the major differences among expansion slots in different computers is the number of data lines available.
Address lines	Address lines are used to select different memory locations. Another major difference among expansion slots is the number of address lines available.
Interrupt request lines	Interrupt request lines are used for hardware signals. These signals come from various devices, including the expansion card itself. Interrupt request signals are used to get the attention of the microprocessor. This is done so that the expansion card can temporarily use the services of the microprocessor.
DMA lines	DMA stands for *direct memory access*. DMA lines are control lines that provide direct access to memory (without having to go through the microprocessor, which tends to slow things down). DMA lines are also used to indicate when memory access is temporarily unavailable because it is being used by some other part of the system. DMA lines are used to indicate that direct memory access is being requested (called a DMA request line) and to acknowledge that request (called a DMA acknowledge).
NMI line	NMI stands for *nonmaskable interrupt*. This line is so called because it cannot be "masked," or switched off, by software. It is primarily used when a parity check error occurs in the system.
Memory-read, memory-write lines	The memory-read and memory-write lines are used to indicate that memory is either being written to or read from.
I/O read, I/O write lines	The I/O read and write lines are used to indicate that an input or output device (such as a disk drive) is to be written to or read from.
Special lines	Another one of the major differences among expansion slots in different types of computers is the number (and the types) of specialized lines used. For example, some systems offer an *audio channel line* for the purpose of carrying a sound signal.

FIGURE 38.2 ISA expansion slot and pin numbering

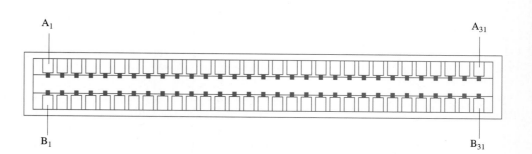

A₁ A₃₁ B₁ B₃₁

FIGURE 38.3   ISA expansion slot pin assignments

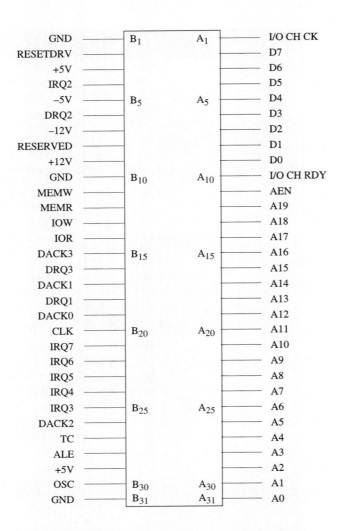

GND — B1	A1 — I/O CH CK
RESETDRV —	— D7
+5V —	— D6
IRQ2 —	— D5
−5V — B5	A5 — D4
DRQ2 —	— D3
−12V —	— D2
RESERVED —	— D1
+12V —	— D0
GND — B10	A10 — I/O CH RDY
MEMW —	— AEN
MEMR —	— A19
IOW —	— A18
IOR —	— A17
DACK3 — B15	A15 — A16
DRQ3 —	— A15
DACK1 —	— A14
DRQ1 —	— A13
DACK0 —	— A12
CLK — B20	A20 — A11
IRQ7 —	— A10
IRQ6 —	— A9
IRQ5 —	— A8
IRQ4 —	— A7
IRQ3 — B25	A25 — A6
DACK2 —	— A5
TC —	— A4
ALE —	— A3
+5V —	— A2
OSC — B30	A30 — A1
GND — B31	A31 — A0

TABLE 38.2   Major features of ISA expansion slots

Type of Bus	Comments
Total pins	62 separate connectors
Data bus	8 data lines
Address bus	20 address lines (1MB addressable memory)

FIGURE 38.4   Typical ISA expansion card

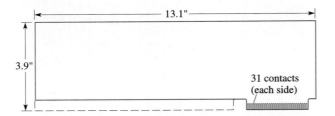

13.1"

3.9"

31 contacts (each side)

Figure 38.4 shows the design of an expansion card used in a PC. Note that there are two major types of PC expansion cards: one type goes straight back from the connector and the other has a skirt that dips back down to the board level. This distinction becomes important in the design of expansion slots used in other types of computers to accommodate PC expansion cards.

AT (16-Bit ISA) Expansion Slots

Figure 38.5 shows the AT expansion slots. These expansion slots are designed to accommodate the older 8-bit ISA expansion cards as well as the newer AT expansion cards.

The reason for the two different types of slots is to accommodate the PC expansion card containing the skirt, which comes down to the board level. Note from Figure 38.5 that the expansion slots are divided into two sections. The first section has the 62 pins that are electrically identical to the 62 pins of the 8-bit ISA expansion slots. The second section contains an additional 36 pins. This gives a total of 98 electrical connections. The additional 36 pins are used to handle the additional requirements of the 80286 microprocessor used by the AT systems. Table 38.3 gives the major features of AT expansion slots.

Figure 38.6 shows a typical AT expansion card. Note that the AT expansion card has two separate rows of connectors. The second, smaller row is designed to accommodate the additional requirements of the 80286 microprocessor (as well as the 16-bit 80386SX).

FIGURE 38.5   Typical AT expansion slots

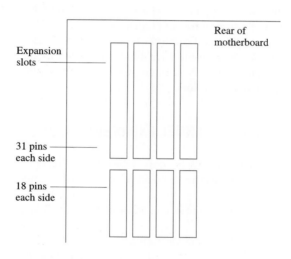

TABLE 38.3   Major features of AT expansion slots

Type of Bus	Comments
Total pins	98 separate connectors divided into two sections: one 62-pin section identical to the 8-bit ISA and one 36-pin section.
Data bus	Total of 16 data lines: first 8 located in 62-pin section, last 8 located in 36-pin section.
Address bus	Total of 24 address lines (16MB of addressable memory): first 20 located in 62-pin section, last 4 located in 36-pin section.

FIGURE 38.6   Typical AT expansion card

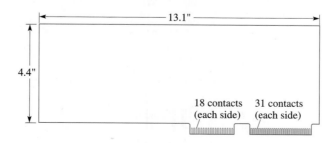

EXPANSION SLOT DESIGNS

Because of the need to meet the requirements of newer microprocessors such as the 80386 and beyond, an expansion slot that could accommodate a 32-bit bus was needed. There were two basic approaches to answer this need. One approach was developed by IBM in its new **micro-channel architecture (MCA)**. Another approach, called the EISA (extended industry standard architecture), was taken by other computer manufacturers. The EISA standard is compatible with the ISA bus. The main difference between the micro-channel and the EISA is that the micro-channel expansion slots are not compatible with the older ISA expansion slots. The EISA expansion slots *are* compatible with the older ISA slots. Figure 38.7 graphically illustrates the differences.

As you can see from Figure 38.7, the micro-channel expansion slots are physically smaller than the older ISA slots; their pin contacts are much closer together. The EISA slots have two rows of connectors. If an older ISA card is used, it will go down only as far as the first row of connectors; the lack of a notch on the card prevents it from going in any farther. On the other hand, if the newer EISA card is used, it will be notched and will go all the way down to the second row of connectors.

There are several different types of micro-channel expansion slots. Some are used to accommodate a 16-bit microprocessor (such as the 80386SX), whereas the others can accommodate the 32-bit microprocessors. Figure 38.8 shows the various kinds of micro-channel expansion slots.

THE MICRO-CHANNEL EXPANSION SLOT

Table 38.4 lists the key features of the 8-bit section of the micro-channel expansion slot.

Table 38.5 lists the key features of the 16-bit extension of the micro-channel expansion slot.

Table 38.6 lists the key features of the 32-bit extension of the micro-channel expansion slot.

Table 38.7 lists the key features of the video extension of the micro-channel expansion slot. This video extension is used for a **video coprocessor**. A video coprocessor is a microprocessor that is dedicated to the display, thus relieving the main system microprocessor of this chore. The result is a more detailed and quicker-responding video display for graphics and animation.

FIGURE 38.7   Three expansion slot designs

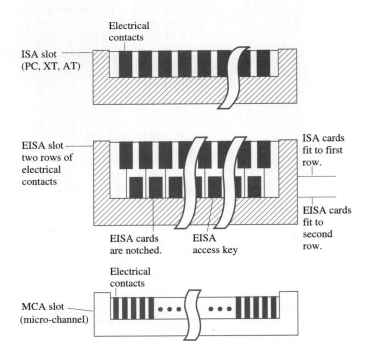

FIGURE 38.8 Various kinds of micro-channel expansion slots

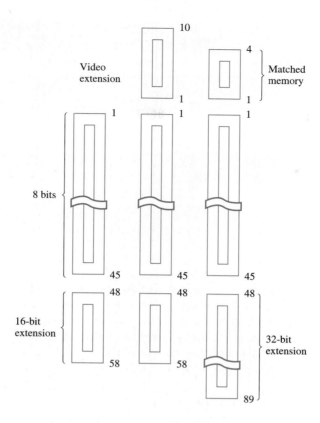

TABLE 38.4 8-bit section of the micro-channel expansion slot

Signals	Comments
Total pins	92 electrical connections
Address lines	24 lines; 16MB of addressable memory
Data lines	8 lines
Audio channel	Single analog audio channel for synthesized voice or music
Power lines	Several +5V DC, +12V DC, and −12V DC lines with ground return (helps reduce noise interference)

TABLE 38.5 16-bit section of the micro-channel expansion slot

Signals	Comments
Total pins	24 electrical connections
Data lines	8 more data lines, increasing the total number of data lines to 16

TABLE 38.6 32-bit section of the micro-channel expansion slot

Signals	Comments
Total pins	62 electrical connections
Data lines	16 more data lines, increasing the total number of data lines to 32
Address lines	16 more address lines, increasing the total number of address lines to 32; 4GB of addressable memory

469

TABLE 38.7 Video extension
section of the micro-channel
expansion slot

Signals	Comments
Total pins	22 electrical connections
Video-control lines	Various video-control lines, such as horizontal and vertical sync, blanking, and video data lines

TABLE 38.8 Matched memory
extension section of the micro-
channel expansion slot

Signals	Comments
Total pins	8 electrical connections
Matched memory lines	Matched memory cycle command, matched memory cycle, matched memory request

Table 38.8 lists the key features of the ***matched memory extension*** section of the micro-channel expansion slot. This extension is used when a higher memory transfer rate can be used by the expansion card. When an internal peripheral is capable of operating at this higher speed, the matched memory provision of the micro-channel can be used. Doing this allows data to be transferred at a 25% increase in speed.

THE LOCAL BUS

As we saw previously, the EISA connector supports 80386, 80486, and Pentium micro-processors by providing a full 32-bit data bus. Three special bus-controlling chips are used to manage data transfers through the EISA connectors. Thus, data that gets transferred between an expansion card and the CPU must go through the bus controller chip set. This effectively reduces the rate at which data can be transferred.

To get around this problem, a new bus architecture was introduced, called the ***local bus***. A local bus connector provides the fastest communication possible between a plug-in card and the machine by bypassing the EISA chip set and connecting directly to the CPU. Local bus video cards and hard drive controllers are popular because of their high-speed data transfer capability.

One initial attempt to define the new local bus was the VESA local bus. VESA stands for Video Electronics Standards Association, an organization dedicated to improving video display and bus technology. VESA local bus cards typically run at 33-MHz speeds, and were originally designed to interface with 80486 signals. VESA connectors are simply add-ons to existing connectors; no special VESA local bus connector exists.

THE PCI BUS

PCI stands for Peripheral Component Interconnect, and it is Intel's offering in the world of standardized buses. The PCI bus uses a *bridge* IC to control data transfers between the processor and the system bus, as indicated in Figure 38.9.

In essence, the PCI bus is not strictly a local bus, since connections to the PCI bus are not connections to the processor, but rather a special PCI-to-host controller chip. Other chips, such as PCI-to-ISA bridges, interface the older ISA bus with the PCI bus, allowing both types of buses on one motherboard, with a single chip controlling them all. The PCI bus is designed to be processor independent, plug-and-play compatible, and capable of 64-bit transfers at 33 MHz and above.

PCI connectors are physically different from all other connectors. Refer to Figure 36.2, which shows four ISA connectors and three PCI connectors. Figure 38.10 shows the pinout for a 32-bit PCI connector.

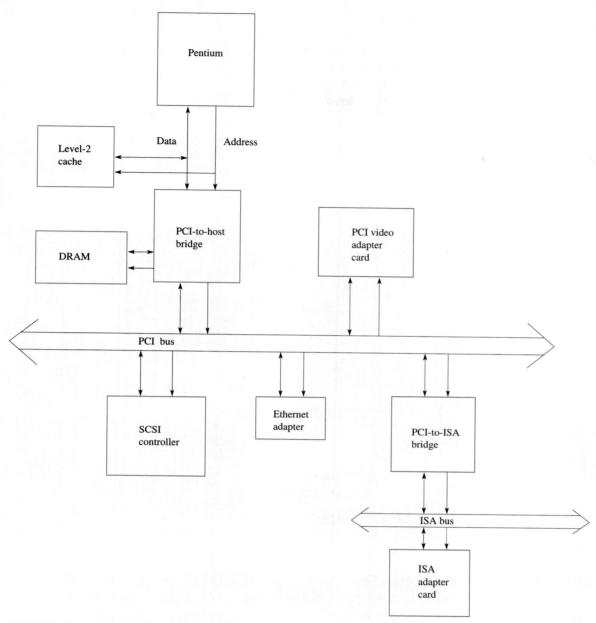

FIGURE 38.9   PCI bridge in a Pentium system

FIGURE 38.10   32-bit PCI connector

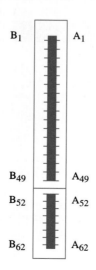

THE PCMCIA BUS

The PCMCIA (Personal Computer Memory Card International Association) bus, now referred to as *PC card bus*, evolved from the need to expand the memory available on early laptop computers. The standard has since expanded to include almost any kind of peripheral you can imagine, from hard drives to LAN adapters and modem/fax cards. Figure 38.11 shows a typical PCMCIA Ethernet card.

The PCMCIA bus supports four styles of cards, as shown in Table 38.9.

All PCMCIA cards allow **hot swapping**, removing and inserting the card with power on.

A type I connector is shown in Figure 38.12. The signal assignments are illustrated in Tables 38.10 and 38.11. The popularity of laptop and notebook computers suggests the continued use of this bus.

FIGURE 38.11   PCMCIA Ethernet card (photograph by John T. Butchko)

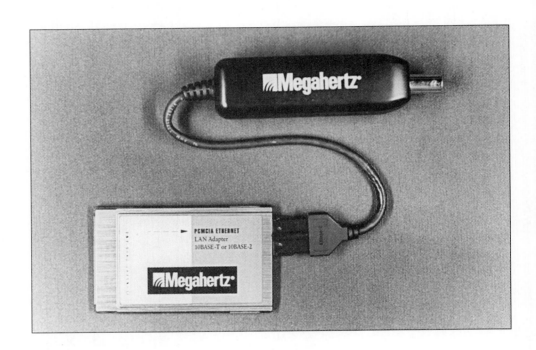

TABLE 38.9   PCMCIA slot styles

Slot Type	Meaning
I	Original standard. Supports 3.3-mm cards. Memory cards only.
II	Supports 3.3-mm and 5-mm cards.
III	Supports 10.5-mm cards, as well as types I and II.
IV	Greater than 10.5 mm supported.

FIGURE 38.12   PCMCIA connector

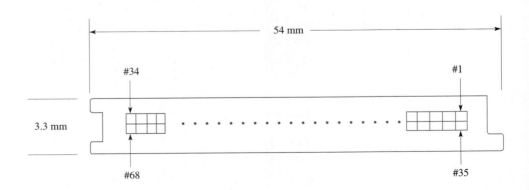

TABLE 38.10   PCMCIA pin
assignments (available at card
insertion)

Pin	Signal	Pin	Signal	Pin	Signal
1	GND	24	A_5	47	A_{18}
2	D_3	25	A_4	48	A_{19}
3	D_4	26	A_3	49	A_{20}
4	D_5	27	A_2	50	A_{21}
5	D_6	28	A_1	51	Vcc
6	D_7	29	A_0	52	Vpp2
7	CE1	30	D_0	53	A_{22}
8	A_{10}	31	D_1	54	A_{23}
9	OE	32	D_2	55	A_{24}
10	A_{11}	33	WP	56	A_{25}
11	A_9	34	GND	57	RFU
12	A_8	35	GND	58	RESET
13	A_{13}	36	CD1	59	WAIT
14	A_{14}	37	D_{11}	60	RFU
15	WE/PGM	38	D_{12}	61	REG
16	RDY/BSY	39	D_{13}	62	BVD2
17	Vcc	40	D_{14}	63	BVD1
18	Vpp1	41	D_{15}	64	D_8
19	A_{16}	42	CE2	65	D_9
20	A_{15}	43	RFSH	66	D_{10}
21	A_{12}	44	RFU	67	CD2
22	A_7	45	RFU	68	GND
23	A_6	46	A_{17}		

TABLE 38.11   PCMCIA signal
differences

Pin	Memory Card	I/O Card
16	RDY/BSY	IREQ
33	WP	IOIS16
44	RFU	IORD
45	RFU	IOWR
60	RFU	INPACK
62	BVD2	SPKR
63	BVD1	STSCHG

AGP

The Accelerated Graphics Port (AGP) is a new technology that improves multimedia performance on Pentium II and III computers. Figure 38.13 shows where the AGP technology fits into the other bus hardware.

The heart of the AGP is the 440LX AGPset hardware, a *quad-ported* data switch that controls transfers between the processor, main memory, graphics memory, and the PCI bus. AGP technology uses a connector similar to a PCI connector.

FIGURE 38.13   AGP interface

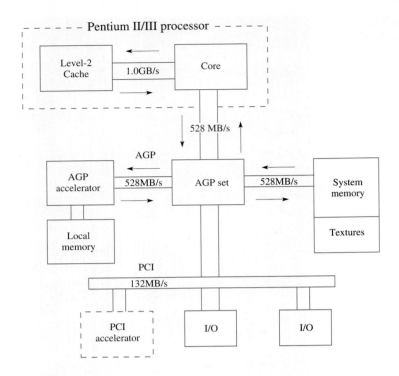

With growing emphasis on multimedia applications, AGP technology sets the stage for improved performance.

USB

The **universal serial bus** is a new peripheral bus designed to make it easier to connect many different types of devices to the PC. These devices include audio players, joysticks, keyboards, scanners, telephones, data gloves, tape and floppy drives, modems, and printers. Motherboards manufactured today have USB support built in (as does Windows 98). USB support can be added to older systems through the use of an adapter card.

Some of the features of the USB are

- Serial bus (simple four-wire cable).
- Up to 127 devices may be connected at the same time.
- 12 Mbps data rate (USB version 2.0 operates at 480 Mbps).
- Devices may be attached with power on.
- Plug-and-play compatibility.
- Data is transmitted in packets.
- USB hubs are used to expand connections to the bus.

Using a USB device eliminates the need to go inside the PC (to install an adapter card) or share COM1 or the printer port with another device.

Figure 38.14 shows three typical USB devices. Pictured are a USB color camera, 4-port USB hub, and USB 10baseT Ethernet adapter.

IEEE 1394 FIREWIRE

The IEEE 1394 FireWire bus is Apple Computer's entry into the world of high-speed bus technology. FireWire is a 400-Mbps plug-and-play serial bus that allows up to 63 devices to be daisy-chained. FireWire is standard on new Apple computers, but it is also available for the PC, with PCI-based FireWire cards providing the necessary hardware interface. Digital video cameras now contain FireWire ports to make it easy to transfer digitized video to your computer.

Table 38.12 provides a comparison of the different expansion slot architectures.

FIGURE 38.14 Three typical USB devices: color camera, 4-port hub, and 10baseT Ethernet adapter

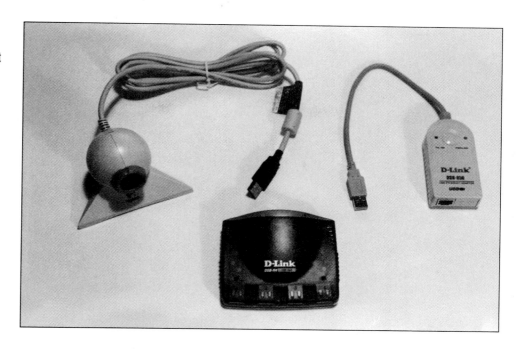

TABLE 38.12 Comparison of expansion slots

Expansion Bus	Data Width	Speed	Addressing	Features
ISA	8	4.77 MHz	1MB	Original PC bus
AT	16	8 MHz	16MB	
MCA	32	10 MHz	16MB	
EISA	32	8 MHz	4GB	
VESA Local	32/64	33/50 MHz	4GB	
PCI	32/64	33	4GB	
PCMCIA	16	8 MHz	64MB	Hot swapping
AGP	32	528MB/s	16–32MB	Video bus only
USB	1	12 Mbps*	127 devices	Serial bus
FireWire	1	400 Mbps	63 devices	Serial bus

*USB version 2.0 operates at 480 Mbps.

TROUBLESHOOTING TECHNIQUES

Good connections between the expansion connector and the adapter card are essential to high-speed digital data communications. Get into the habit of opening your computer chassis from time to time in order to clear the motherboard, fans, and other areas where dust and other contaminants collect. There are many products available for doing cleaning of this kind, from nonconductive spray to mini-vacuum cleaners. Even the individual connector pads on an edge connector can be cleaned by rubbing them gently with a soft eraser.

Performing routine maintenance on your computer will typically save you time, effort, and money in the future.

SUMMARY

In this exercise we discovered that

- There are many types of expansion slots, including ISA, AT, EISA, PCI, AGP, and PCMCIA.
- Expansion slots differ in the size of their data bus, transfer rate, and connector style.

This self-test is designed to help you check your understanding of the background information presented in this exercise.

True/False

Answer *true* or *false*.

1. Expansion slots are generally used to enhance the operation of the computer.
2. The first expansion slot design was the EISA bus.
3. The ISA connector supports 16-bit transfers.
4. All 32-bit system expansion slots are downward compatible with the 16-bit systems.
5. The IBM AT system expansion slots are compatible with the PCI expansion slots.
6. PCMCIA cards are hot swappable.
7. The USB supports 127 devices.

Multiple Choice

Select the best answer.

8. The PCI bus
   a. Uses a PCI-host bridge.
   b. Is not plug-and-play compatible.
   c. Transfers only 32 bits of data.
   d. All of the above.
9. The MCA connector
   a. Has smaller pins than the ISA connector.
   b. Supports 8-, 16-, and 32-bit buses.
   c. Has only one layer of pins.
   d. All of the above.
10. The AT expansion slots
    a. Contain 62 pins that are identical to those of the PC expansion slots.
    b. Have an extra set of 36 pins to accommodate the additional requirements of the 80286.
    c. Contain a total of 98 electrical connections.
    d. All of the above.
11. The reason that the AT expansion slots are divided into two separate sections is
    a. To accommodate PC expansion cards.
    b. To make room for cooling the cards.
    c. Because it allows for better mechanical seating of the cards.
    d. AT expansion slots are not divided into two sections.
12. All computers
    a. Use micro-channel architecture.
    b. Have EISA.
    c. Use ISA.
    d. None of the above.
13. The AGP video connector is controlled by a chip set that is
    a. Single ported.
    b. Dual ported.
    c. Quad ported.
14. The data rate on the USB is
    a. 10 Mbps.
    b. 12 Mbps.
    c. 100 Mbps.

Matching

Match one or more definitions on the right with each term on the left.

15. DMA
16. NMI
17. MCA

   a. IBM's micro-channel.
   b. PC bus system.
   c. Accommodates 32 bits.

18. EISA
19. ISA

    d. Accesses memory directly.
    e. Causes a short interruption.
    f. None of the above.

Completion

Fill in the blank or blanks with the best answers.

20. The AGP chip set requires a(n) _____ microprocessor.
21. The PCMCIA bus is now called the _____ bus.
22. The first connector to support more than 1MB of memory addressing was the _____.
23. PCMCIA cards allow _____ _____.
24. PCI stands for _____ _____ _____.
25. The PCI bus uses a(n) _____ IC to control transfers.
26. There are _____ PCMCIA slot styles.

Open-Ended

27. Explain the purpose of interrupt request lines.
28. Which of the following standards are compatible: ISA, EISA, MCA?
29. State how the EISA expansion slot is made compatible with the ISA expansion card.
30. Explain how micro-channel architecture can accommodate both the 16-bit 80386SX and the 32-bit 80386 microprocessors.
31. What is the distinguishing feature of micro-channel architecture?
32. Why is the local bus faster than all the other bus types?

FAMILIARIZATION ACTIVITY

Your instructor will have several different types of computers available for you to observe. These computers may be accessible to you in the lab or outside the lab. Some of these computers may be located in a retail outlet, where their observation may be part of a homework assignment. In any case, make sure you understand the background information section of this exercise well enough so that you can successfully identify the types of computers presented to you in the performance objective for this exercise.

QUESTIONS/ACTIVITIES

1. What kind of computer systems were available for you to observe as part of this lab exercise?
2. Was an AT-type computer available? If so, how many expansion slots did it have? How many of those slots were made to accommodate a PC expansion card with a "skirt" that came down to the board?
3. Was a computer with an EISA expansion slot available as part of this exercise? If so, what make of computer was it? What kind of microprocessor did it have?
4. As a part of this exercise, was an IBM computer with micro-channel architecture available? If so, what model was it? What kind of microprocessor did it have?
5. In your own words, explain the differences among ISA, EISA, and MCA.
6. If your computer contained a local bus slot, what type of expansion card was plugged into it, if any?

REVIEW QUIZ

Under the supervision of your instructor

1. Identify the type of system, bus structure, and expansion slots for each of at least three computer motherboards and/or expansion cards.
2. Briefly describe the differences among the various bus architectures.

39

Power-On Self-Test (POST)

INTRODUCTION

Joe Tekk pressed the power switch on an old computer for the fifth time. Each time he tried to boot the machine, it beeped a short series of tones through the speaker and halted. Joe thought he had counted correctly but wanted to be sure, so he powered off and on for a sixth time. He heard what he expected: one short beep, then another short beep, then three beeps. He looked up the beep pattern in a diagnostic table. The beeps indicated a CMOS checksum failure.

Joe replaced the CMOS RAM, rebooted, and got the same error message. But then he went into the ROM BIOS setup, adjusted the system settings, saved them, and rebooted again. The checksum error had disappeared.

PERFORMANCE OBJECTIVES

Upon completion of this exercise, you will be able to

Diagnose a computer problem from the results of a POST.

KEY TERMS

Power-on self-tests (POSTs)   Error code

BACKGROUND INFORMATION

DEFINITION OF POST

Every time you turn on a computer, it automatically goes through a series of internal tests. These internal tests are called the *power-on self-tests (POSTs)*. The specific testing done by POST may vary slightly from one computer to the next. However, the essential idea of this kind of automated computer testing during power-up is the same. Understanding what the POST does on any one system will give you enough information to understand other slightly different systems. The POST program is contained in the computer's ROM. Recall that ROM is memory that retains its contents even when the computer is turned off.

What POST does for you, every time you turn on your computer, is to check the major sections of the motherboard, as well as the disk drives, display, and keyboard sections. If no errors are encountered, the POST begins booting up the operating system. The components tested by POST are given in Table 39.1.

TABLE 39.1   POST procedures

Test Name	Action Taken
Basic system	Checks the operation of the microprocessor as well as the area of memory that contains the POST itself and some other areas of memory, including the system buses.
Display	Checks the hardware that operates the video signals. These are the signals that operate the monitor. If there is a second monitor installed in the system, it is not checked.
Memory	Checks all the computer memory. This is accomplished by having data written into the memory and read from it. A number appears in the upper left-hand corner of the screen. This number represents the amount of memory checked at that point in the test. You can watch this number slowly increase to the full amount of memory available in the system.
DMA controller Interrupt logic Programmable timers	Check the operation of all motherboard support circuitry. All components are essential for proper operation.
Keyboard	Checks the circuits that connect the keyboard to the computer. The keyboard is also checked for stuck keys.
Disk drives	Determines which disk drives are installed. If, for any reason, the system does not have any disk drives, this part of the test is omitted.
Adapter cards	Check and configure installed adapter cards. Some of these cards may themselves contain additional ROM chips that have their own POSTs.

POST ERROR MESSAGES

Generally, when the POST encounters an error, it indicates the type of error by a number called a POST *error code*. The POST error codes are numbers whose values indicate the type of problem encountered. Another method used by the POST to indicate errors is an audible signal. Using the system speaker, POST can indicate the type of problem by a series of short beeps.

Table 39.2 gives the general types of indications that occur when a particular type of test encounters an error in the system.

Audible Error Codes

As you saw in Table 39.2, certain errors cause specified types and numbers of beeps when detected by POST. For example, when there is a problem with the display circuitry, one long beep and two short beeps are emitted from the system speaker.

POST also has other beep codes to indicate detected failures in other parts of the system. The audible codes for IBM-BIOS are listed in Table 39.3. Other BIOS manufacturers, such as AMI and Phoenix, use different codes.

Suppose a computer provides no beeps at all when powered-on. What does this mean? Is the power supply or CPU not working? Remember that even getting a series of beeps means some major components are working.

TABLE 39.2 POST error
indications

Type of Failure	Indication
Basic system	A failure in the basic system test is indicated by the system's being halted, with no visible display and no beep. The cursor is left visible on the screen.
Display	A display error causes the system to emit one long beep followed by two short ones. POST continues.
Memory	With this type of error, the screen displays a numeric error code. The value of the error code indicates where the problem is in memory.
Motherboard logic	Most errors cause a beep code to be emitted prior to a system halt.
Keyboard	A keyboard failure causes a number to be displayed on the screen. For a stuck key, the value of that key is displayed.
Disk drive	The error code 601, 1780, or 1781 is displayed on the screen.

TABLE 39.3   IBM-BIOS POST audible error codes

1 short beep	System OK
2 short beeps	POST error displayed on screen
Repeating short beeps	Power supply, system boards
3 long beeps	3270 keyboard card
1 long, 1 short beep	System board
1 long, 2 short beeps	Display adapter (MDA, CGA)
1 long, 3 short beeps	EGA
Continuous beep	Power supply, system board

UNDERSTANDING POST ERROR CODES

In almost all personal computer POSTs, the numeric error codes can be broken into two major parts: a device number and a two-digit error code. The device number followed by two zeros (such as 100 for the system board) indicates that no errors have been detected. Specific error codes should be referenced using the manuals that come with the computer system.

Table 39.4 lists the major error codes and their causes. The table uses numbers such as 4xx. This means that a 4 represents an error in the monochrome display adapter. The xx represents any value from 00 (meaning no problem) to 99, indicating the specific problem in that part of the computer system. Thus, error codes for the monochrome display adapter are between 400 and 499.

TROUBLESHOOTING TECHNIQUES

Windows 95/98 provides a way to examine what kind of hardware problems exist in the current machine configuration. Within the control panel, the Device Manager submenu of the System Properties window is used to examine all the installed hardware components recognized by Windows 95/98. Figure 39.1 shows a typical Device Manager display, indicating a problem with the plug and play BIOS.

Often, the Device Manager provides a clue to determining why the affected hardware is not working. If you do not have a BIOS error code reference, the Device Manager is a good substitute.

TABLE 39.4 Major error codes and their causes

Error Code	Cause
01x	Undetermined problem
02x	Power supply errors
1xx	System board errors
2xx	RAM memory errors
3xx	Keyboard errors
4xx	Monochrome display adapter errors
4xx	On PS/2 systems, parallel port errors
5xx	Color graphics adapter card errors
6xx	Floppy drive or adapter card errors
7xx	Math coprocessor errors
9xx	Parallel printer adapter errors
10xx	Alternate parallel printer adapter errors
11xx	Asynchronous communications adapter errors
12xx	Alternate asynchronous communications adapter errors
13xx	Game adapter control errors
14xx	Printer errors
15xx	Synchronous data link control; communications adapter errors
16xx	Display emulation errors (specifically 327x, 5520, 525x)
17xx	Fixed disk errors
18xx	I/O expansion unit errors
19xx	3270 PC attachment card errors
20xx	Binary synchronous communications adapter errors
21xx	Alternate binary synchronous communications adapter errors
22xx	Cluster adapter errors
24xx	Enhanced graphics adapter (EGA) errors
29xx	Color or graphics printer errors
30xx	Primary PC network adapter errors
31xx	Secondary PC network adapter errors
33xx	Compact printer errors
36xx	General-purpose interface bus (GPIB) adapter errors
38xx	Data acquisition adapter errors
39xx	Professional graphics controller errors
48xx	Internal modem errors
71xx	Voice communications adapter errors
73xx	3.5-inch external disk drive errors
74xx	On PS/2 systems, display adapter errors
85xx	IBM expanded memory adapter (XMA) errors
86xx	PS/2 systems point device errors
89xx	Music card errors
100xx	PS/2 multiprotocol adapter errors
104xx	PS/2 fixed disk errors
112xx	SCSI adapter errors

FIGURE 39.1 Device Manager tab

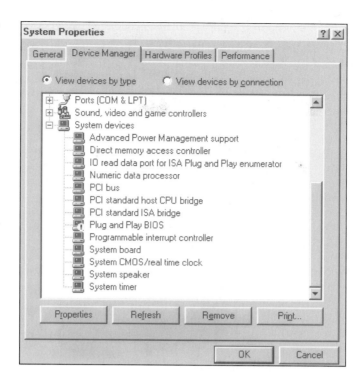

SUMMARY

In this exercise we discovered that

- The personal computer runs power-on self-tests every time it boots.
- POST error messages are numerical and audible (beep codes).
- Windows Device Manager helps identify hardware errors.

SELF-TEST

This self-test is designed to help you check your understanding of the background information presented in this exercise.

True/False

Answer *true* or *false*.

1. Every time you turn on the computer, it automatically goes through a series of internal tests.
2. The test that your computer automatically goes through checks everything except the microprocessor and math coprocessor.
3. All memory except for the memory that contains the test itself is checked by the test through which your computer goes.
4. The POST software is stored in RAM.
5. A series of audible beeps may be used to indicate a POST error.

Multiple Choice

Select the best answer.

6. POST stands for
   a. Power-on self-test.
   b. Passed-on surface-testing.
   c. PreprOgrammed self-test.
   d. None of the above.
7. A keyboard error encountered by POST causes
   a. The POST to stop.
   b. A number to be displayed on the screen and the testing to continue.
   c. The keyboard is not tested by POST.
   d. Only a stuck key is displayed by POST.

483

8. A power supply problem encountered by POST could cause
   a. No beeps at all.
   b. A continuous beep.
   c. A series of short beeps.
   d. Any of the above.
9. Adapter cards
   a. Are not tested during the POST.
   b. May contain their own POST in ROM.
   c. Require POST add-on software from the manufacturer.
10. Windows identifies hardware errors using the
    a. Control Panel.
    b. Device Monitor.
    c. Device Manager.

Matching

Match a problem code on the right with each type of error on the left.

11. RAM memory errors   a. 13xx
12. Keyboard errors   b. 12xx
13. Game adapter control errors   c. 2xx
                                      d. 3xx

Completion

Fill in the blank or blanks with the best answers.

14. The error code 8924 would represent a(n) _____ _____ error.
15. An error code of 728 means a(n) _____ _____ error.
16. Error codes of the form _____ represent problems with the graphics printer.
17. POST may use the system _____ to indicate an audible error code.
18. Numeric error codes contain two parts: _____ number and error code.

FAMILIARIZATION ACTIVITY

Observe and record the following error codes. Make sure you listen for audible error codes as well as look for numerical error codes.

1. Boot a normally working computer. Observe all beeps, the duration (short or long) of the beeps, and how many beeps there are. At the same time, view your screen to observe all numerical values produced by your computer's POST. Record this information here.

2. Turn off the computer and remove the keyboard plug from the computer (so the keyboard is no longer attached to the computer). Now boot the computer. Again, record all of your POST observations.

3. Turn off the computer. Reconnect the keyboard. Now, while booting the computer, hold down a key on the keyboard (this simulates a stuck key). Record your observations here (including the key you held down).

4. Turn off the computer. Remove the monitor cable from the computer case (so the monitor is no longer connected to the computer). Boot the system and record your observations of the POST.

5. If your instructor gives a demonstration of POST messages for specific problems introduced into the computer, record your observations in the following space.

QUESTIONS/ACTIVITIES

1. How long does the POST take for your computer system? (Use a digital watch.)
2. Can you tell how much computer memory your system has from any information given during the POST? Explain.
3. What happens during the POST if you hold down more than one key at the same time (simulating more than one stuck key)?
4. List all the sections that are being tested in your specific computer system during the POST. How do you know this?

REVIEW QUIZ

Under the supervision of your instructor

Diagnose a computer problem from the results of a POST.

40 Motherboard Replacement and Setup

One Saturday afternoon, Joe Tekk was visiting some relatives. Many of the children and young adults were involved in a heated argument about why the figures in the new computer game they were using were not moving around like they were supposed to. Joe knew none of their theories were correct, so he added his two cents to the conversation.

Joe asked, "Does your computer have an MMX processor?"

The group fell silent. Joe continued, "That software uses the new MMX instructions that are part of all of the new microprocessors. The new instructions are designed for multimedia applications." Joe examined the system and showed all who were interested how to identify the processor type. Joe was right, there was no MMX capability.

Before Joe knew what was happening, he was roped into upgrading the PC. Before he left, he determined what he would need to perform the upgrade. The next weekend he returned with all of the necessary parts.

After carefully extracting the old motherboard and inserting the new one, the procedure was a complete success. Soon after that, the screen was filled with moving game characters.

PERFORMANCE OBJECTIVES

Upon completion of this exercise, you will be able to

1. Remove and replace a personal computer motherboard and successfully boot the computer system.
2. Identify the form factor of a motherboard.

KEY TERMS

Form factor   BIOS upgrade

BACKGROUND INFORMATION

In past exercises, you had the opportunity to learn about different aspects of the motherboard. You learned about the various types of microprocessors and coprocessors, types and uses of memory, and motherboard expansion slots.

In this exercise, you will discover how to remove and replace the motherboard of a personal computer properly and safely. The first time you change one should be under supervised laboratory conditions.

OVERALL SAFETY

In every case of motherboard removal and replacement, always observe electrical safety precautions. Make sure the system is turned off and the power cable is *completely* removed.

WHY REPLACE THE MOTHERBOARD?

There are many reasons why a motherboard might be replaced. The first and foremost reason is advances in technology. For someone who purchased one of the early PCs, the fast processing speeds and large hard disk space available on a Pentium would be a dream come true. For someone who already owns a Pentium, a newer Pentium or Pentium II/III/4 with MMX technology might the desired goal. These two situations require different approaches.

Generally, the first thing to do is determine what upgrade paths are available. For someone with an old PC (386 or older), it makes sense to just buy a new system, rather than upgrade. The reason for this is simple—every piece of hardware must be replaced. This includes items like the motherboard, CPU, memory, and so on. That is a complete system. Instead of building a PC from parts (unless you want the experience), simply purchase one of the new under $1000 PCs that contains a Pentium III 733 MMX, 64MB memory, 20GB disk, 48X CD-ROM, 56K modem, and so on, including the monitor.

Another common situation is when a system already contains relatively new memory, a large hard disk, super VGA graphics, a fast CD-ROM drive, and so on, but runs too slowly. These PCs may be candidates for a simple processor and/or memory upgrade. For example, a Pentium 75 can be upgraded to a Pentium 166. The various upgrade options are determined by the capabilities of the motherboard. A simple upgrade might accomplish the desired goal of increased speed at a fraction of the cost of a new PC.

The last common situation involves upgrading the PC to new technology that can still use the existing expansion cards, 72-pin SIMM memory, and disks. For example, let's take the same Pentium 75 as the starting point, but this time add the additional requirement of MMX technology as the desired goal. In this case, the motherboard can be replaced, providing a new Pentium or Pentium II microprocessor with MMX technology while at the same time using the old hard disks, memory, and all other hardware as well. At a cost of about $300 to $400, this upgrade path may provide a cost-effective solution for access to the latest technology.

The rest of this exercise is devoted to the last scenario, in which the motherboard must be replaced. For example, we can assume the upgrade path is from a Pentium without MMX technology to a Pentium, Pentium Pro, or Pentium II/III/4 using MMX technology. Each of these upgrade paths will result in the purchase of different motherboards.

MOTHERBOARD FORM FACTORS

Motherboard *form factors* describe the physical size, layout, and features of a particular motherboard. Currently there are several popular choices available, each of them providing different types of technology. For example, NLX motherboards provide capability for Pentium II/III processors with AGP graphics support. Table 40.1 shows some of the details for the Baby-AT, ATX, LPX, and NLX motherboard form factors.

TABLE 40.1   Common motherboard form factors

Form Factor	Size
Baby-AT	8.5" × 11"
ATX	8.5" × 11"
LPX	9" × 10.6"
NLX	Generic Riser card

TABLE 40.2   Chip set
properties

Property	Types
Memory type	FPM, EDO, SDRAM, ECC, parity
Secondary cache	Burst, pipeine burst, synchronous, asynchronous
CPU type	486, P-24T, P5, P54C/P55C, Pentium Pro, Pentium II/III/4
Maximum memory bus speed	33, 40, 50, 60, 66, 75, 83, 100, 133 MHz
PCI type	32 bit, 64 bit

CHIP SETS

Each motherboard provides a *chip set* designed to control the activity of the system. The chip sets are designed to provide the general features built into the motherboard. This includes items such as the memory controller, EIDE controller, PCI bridge, clock, DMA control, mouse, and keyboard controls. Table 40.2 shows some of the common chip set properties. When shopping for a motherboard, it pays to know as much as possible about your chip set options.

BIOS UPGRADES

BIOS, or the Basic Input Output System, is stored in ROM (read only memory) on many of the older motherboards. In contrast, new motherboards may contain updatable BIOS memory. The ability to update BIOS may be important, as many features of the motherboard are exploited by the BIOS. For example, the new plug-and-play standard Version 1.1A is implemented in BIOS. Without the BIOS plug-and-play feature, the new plug-and-play hardware may not be recognized by the operating system. Using Flash memory, a ***BIOS upgrade*** may be as close as the Internet. The actual procedure may involve the following steps:

1. Write down all current settings.
2. Download the BIOS files.
3. Create a boot floppy containing the BIOS upgrade.
4. Perform the upgrade.
5. Reset system.
6. Verify the new BIOS version.
7. Verify the settings.

Note that if the BIOS upgrade does not execute properly, the system may be left in an unbootable state. Extreme caution must be exercised when upgrading system BIOS. More information on this important activity can be found in Exercise 54.

REMOVAL PROCESS

Often the process of extracting the motherboard can be broken down into several steps. For example, the following steps can be used to remove the motherboard from most PCs:

1. Remove the power cord.
2. Remove all cables.
3. Remove the system cover.
4. Remove the peripheral cards.
5. Label and remove all cables from the motherboard:
   Floppy disk drive
   Hard disk drive (primary and secondary)

CD-ROM drive
Serial and parallel
Power supply
Speaker
Indicator
Reset
Etc.

6. Remove the motherboard.
7. Clean the case, fan, and so on.

If for some reason your PC is a little different, you may need to develop a custom solution.

NEW MOTHERBOARD INSTALLATION

The installation of a new motherboard can also be broken down into several steps, as follows:

1. Insert new motherboard into the case and secure it.
2. Insert microprocessor (and cooling fan).
3. Insert memory modules.
4. Replace all cables:
   Floppy disk drive
   Hard disk drive (primary and secondary)
   CD-ROM drive
   Serial and parallel
   Power supply
   Speaker
   Indicator
   Reset
   Etc.
5. Insert and secure all peripheral cards.
6. Replace the cover.
7. Replace the power cord.
8. Power up the system.

TROUBLESHOOTING TECHNIQUES

This approach to changing a motherboard works very well, but sometimes there is an occasional problem, such as a connector not fitting properly or forgetting to perform one of the steps. Retrace your steps to identify and correct any problems.

It is important to note that changing the motherboard hardware is only half of the installation procedure. The hard drive still contains information about the old motherboard. For example, when booting Windows for the first time after a new motherboard has been installed, Windows will detect the new hardware and begin a process to make the necessary modifications to the Windows operating system to support the new hardware.

During this process, it may be necessary to reboot the computer several times while you view each of the new devices that Windows finds. Note that you may need to supply a Windows installation CD-ROM so that Windows can copy any required files. Follow each of the instructions provided by Windows.

SUMMARY

In this exercise we discovered that

- There are many reasons to replace a motherboard.
- Motherboards come in several form factors.
- It may be possible to upgrade the BIOS on a motherboard.

This self-test is designed to help you check your understanding of the background information presented in this exercise.

True/False

Answer *true* or *false*.

1. Before removing the motherboard, you should make sure all hard disks have been removed from the enclosure.
2. Whatever upgrade path is chosen, a new motherboard must be installed.
3. A Pentium can be upgraded to MMX without purchase of a new motherboard.
4. A chip set can be updated using a software procedure.
5. The form factor describes the physical size, layout, and features of a motherboard.
6. It is not possible to upgrade the BIOS on a motherboard.
7. PXX is a motherboard form factor.

Multiple Choice

Select the best answer.

8. The most common motherboard form factors are
   a. ATX, LPX, and NLX.
   b. A memory controller, EIDE controller, and PCI bridge.
   c. Designed to all provide the same features.
   d. All of the above.
9. Some types of BIOS software
   a. May be updated at a later time.
   b. May be stored in ROM.
   c. Cannot be updated.
   d. All of the above.
10. When updating BIOS software, it is always a good idea to
    a. Reset the computer before making any changes.
    b. Write down all the BIOS settings before making any changes.
    c. Update all files directly.
    d. Get a version from the manufacturer with the best features.
11. Each chip set is designed to
    a. Address data on the hard disks directly.
    b. Coordinate all the activity on the motherboard.
    c. Stop viruses from infecting the system.
    d. Allow for use of the MMX instructions.
12. Immediately after a motherboard has been replaced, the system
    a. Boots normally and runs much faster.
    b. Does not boot and beeps three times.
    c. Detects all new hardware in one easy step.
    d. Detects all new hardware during several reboot operations.
13. A motherboard replacement may be required to
    a. Utilize a newer, faster processor.
    b. Change the operating system of the computer.
    c. Both a and b.
14. The motherboard BIOS may need an upgrade to
    a. Speed up the processor.
    b. Add MMX technology.
    c. Take advantage of plug-and-play features.

Completion

Fill in the blank or blanks with the best answers.

15. Extreme caution must be exercised when performing a BIOS upgrade because the system may be left in a(n) _____ state.

16. The _____ includes support for plug-and-play hardware.
17. AGP graphics are used on the _____ motherboard form factor.
18. Each of the internal _____ cards must be removed prior to motherboard replacement.
19. Both the _____ and _____ disk cables must be reconnected after a motherboard is replaced.
20. Each motherboard uses a particular _____ set.
21. BIOS stored in a(n) _____ memory may be upgraded.

FAMILIARIZATION ACTIVITY

Using the proper procedure, remove the motherboard on your system unit. Once the motherboard has been removed, have your instructor check all the system parts removed, including the motherboard. Then, with your instructor's approval, reinstall the motherboard in the computer system. Have your instructor check out the system before you plug it in and apply power to the system.

QUESTIONS/ACTIVITIES

1. What type of computer system was used by your lab group?
2. Outline the steps required for the removal of the motherboard from your computer system.
3. What difficulties did you encounter, if any, in the removal of the motherboard from your system?
4. Were there other computer systems used in the lab for this exercise that were different from your computer system? Explain.

REVIEW QUIZ

Under the supervision of your instructor

1. Remove and replace a personal computer motherboard and successfully boot the computer system.
2. Identify the form factor of a motherboard.

41

Hard Disk Fundamentals

INTRODUCTION

Joe Tekk drove his car to a local computer store. It was a rainy, windy Saturday. He'd been all over town looking for a specific Maxtor hard drive. Everyone was sold out. By the time Joe walked into the computer store, he was soaking wet and fuming.

He walked to the rear of the store, where the hard drives were kept. Because he was wet, he did not touch anything, but he carefully searched the display cases for the hard drive he was looking for. He found it and his whole mood changed, just as a salesperson came to assist him.

"Can I help you?"

"Sure," Joe replied, his face breaking into a large grin. "I'd like this Maxtor hard drive." Joe pointed to the drive, which had the highest storage capacity of all the drives in the display.

"You are pretty happy for a guy who's soaking wet."

Joe laughed. "I don't care about being wet. I just hate having to delete files when my hard drive is full."

PERFORMANCE OBJECTIVES

Upon completion of this exercise, you will be able to

1. Determine if the hard disk has been set up for the most efficient system operation by looking at the directory structure and the PATH command in the AUTOEXEC.BAT file.
2. Determine the type of hard drive used in your system.

KEY TERMS

Platter	Stripe set
Cylinder	RAID technology
Partitioning	ESDI
Primary partition	IDE
Logical block addressing (LBA)	SCSI
FAT	Caching
FAT32	Sector
Transaction	Cluster
Mirror	

The construction of a hard disk system is much different from that of a floppy disk system. With the floppy disk system, data is stored on each side of the disk, but in a hard disk system, there is usually more than one disk, or ***platter***. Figure 41.1 shows the structure of a two-platter hard disk system, in which there are four sides for storing data.

As you can see from Figure 41.1, the four sides are labeled 0 through 3. Figure 41.2 shows how a floppy disk organizes its data in single concentric tracks, as compared with a hard disk system, which organizes its data in a combination of tracks called ***cylinders***.

PARTITIONS

A hard disk can be formatted so that it acts as two or more independent systems. As an example, it is possible for a hard disk to operate under two entirely different operating systems, such as Windows and UNIX. Doing this is called ***partitioning*** the disk, as shown in Figure 41.3. This is sometimes necessary when a computer is part of a collection of computers connected together over a network.

FIGURE 41.1   Typical hard disk structure

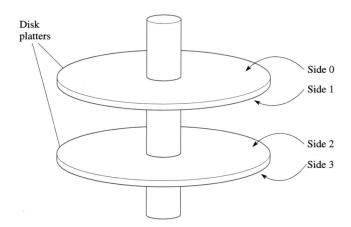

FIGURE 41.2   Floppy and hard disk organization

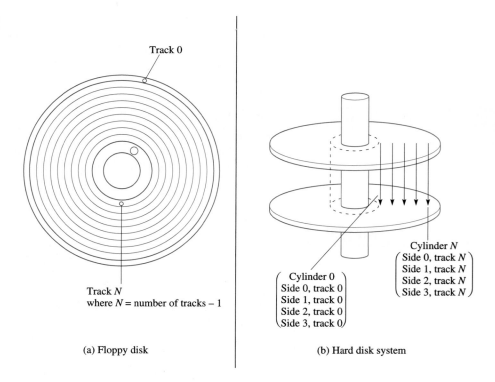

Track 0

Track N
where N = number of tracks − 1

(a) Floppy disk

Cylinder 0
(Side 0, track 0
Side 1, track 0
Side 2, track 0
Side 3, track 0)

Cylinder N
(Side 0, track N
Side 1, track N
Side 2, track N
Side 3, track N)

(b) Hard disk system

494

FIGURE 41.3   Single and multiple partitions of a hard disk

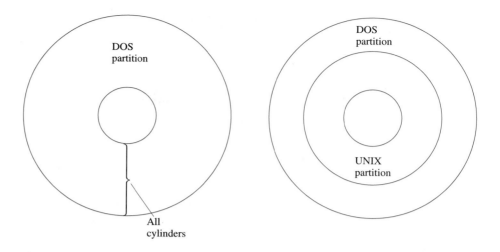

FIGURE 41.4   Windows 98 FDISK utility

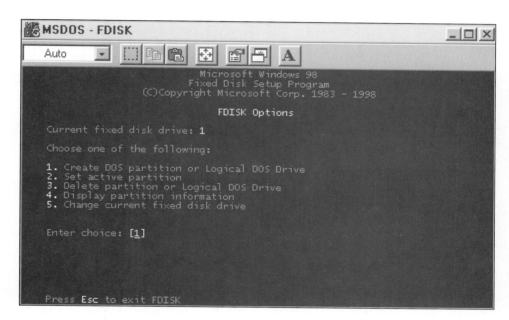

In the early versions of DOS, no disk partition could be larger than 32MB, and no one operating system could access more than one partition. This means that no matter how much data the hard disk could hold, early versions could make use of only 32MB of the disk surface. DOS 4.0 broke the 32MB limitation on hard disks.

When a disk is partitioned, its ***primary partition*** (the one from which it boots) is called the C: drive, and the remaining partitions are referred to as D, E, F, and so on.

Every disk may be partitioned differently, but there are a few rules that must be followed. For example, Windows 95 Version A can recognize disks as large as 2.1GB. Windows 95 Version B and above can address disks as large as 4TB (terabytes). The operating system determines the maximum size that can be handled. Updates to an operating system add capabilities for new technology.

The FDISK tool supplied by DOS and Windows is used to create partitions on the hard disk. Figure 41.4 shows the Windows 98 FDISK menu. It looks very similar to the old DOS FDISK program. FDISK is used to create, modify, or delete partitions on a hard drive. Extreme caution must be observed when working with the FDISK program.

There are also many specialty programs designed to make the disk partitioning process easier and more flexible than the FDISK program. For example, PartitionMagic by Power-Quest allows for a hard drive to be partitioned dynamically, saving time and disk space. Figure 41.5 shows the PartitionMagic main window. Information about the default drive is

FIGURE 41.5   PartitionMagic main window

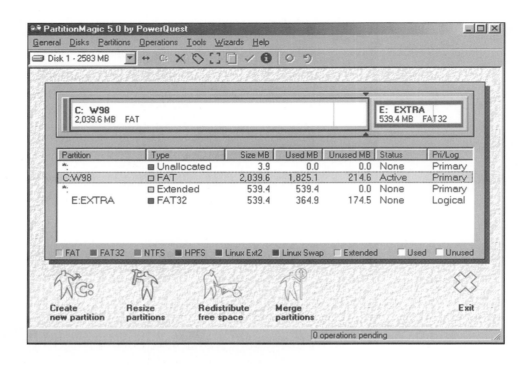

FIGURE 41.6   Disk Usage Partition Information

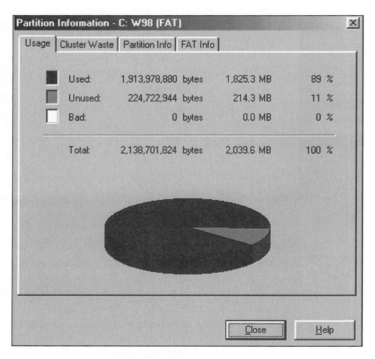

displayed automatically, showing the size of each partition and associated disk format. By selecting Info from the Operations menu, the Partition Information window is displayed showing the default disk usage statistics, as shown in Figure 41.6.

The Cluster Waste tab shows the current amount of disk space that is wasted. This waste is attributed to the smallest amount of disk space that can be allocated by the operating system. For example, if we want to store one character in a file, it will be stored in a 32K chunk of space on a computer with 32K clusters (shown in Figure 41.7). The cluster size is determined by the type of file structure used on the disk, such as FAT16 or FAT32, which will be discussed shortly.

FIGURE 41.7   Cluster Waste Partition Information

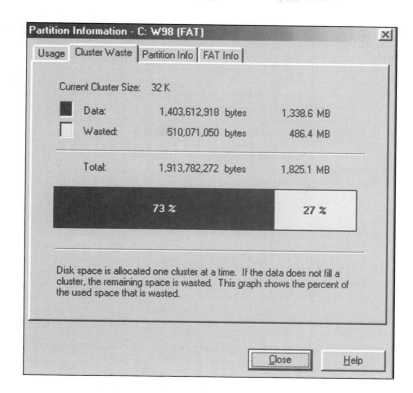

FIGURE 41.8   Physical Partition Information

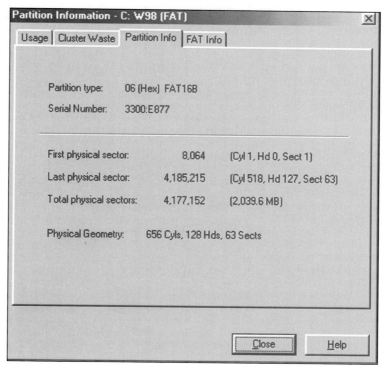

PartitionMagic also can display information about the physical layout of a partition, as shown in Figure 41.8. The first, last, and total physical sectors are displayed, along with the corresponding cylinder and head information. The disk physical geometry is also indicated on the Partition tab.

Details about the FAT are available on the FAT Info tab shown in Figure 41.9. This window contains the details of the FAT structure, such as the number of FATs, root directory capacity, First FAT sector, First Data sector, and other interesting information.

FIGURE 41.9   FAT Info tab

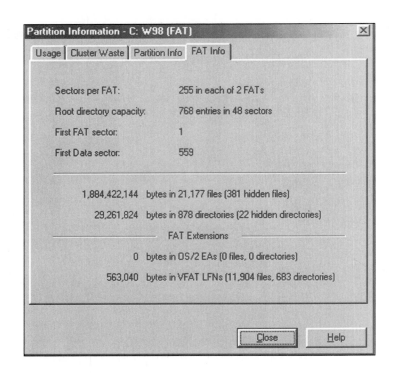

FIGURE 41.10   Cluster analysis window

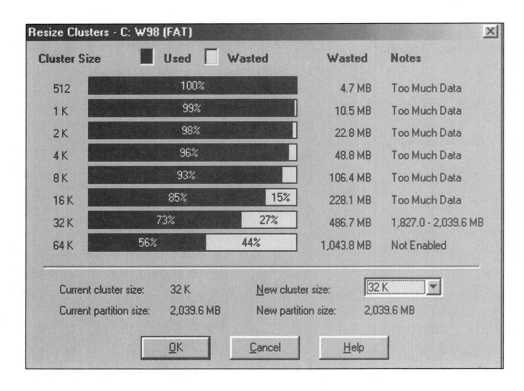

PartitionMagic also can change partition information dynamically, such as changing the cluster size. Figure 41.10 shows the common cluster sizes and associated wasted space. In this case, since the disk is close to capacity, PartitionMagic cannot recommend any type of changes; otherwise, the user may select a new cluster size and change the partition size. If it is necessary to change the disk partition frequently, it may be a good idea to invest in a software package such as PartitionMagic.

FIGURE 41.11   Windows NT Disk Administrator window

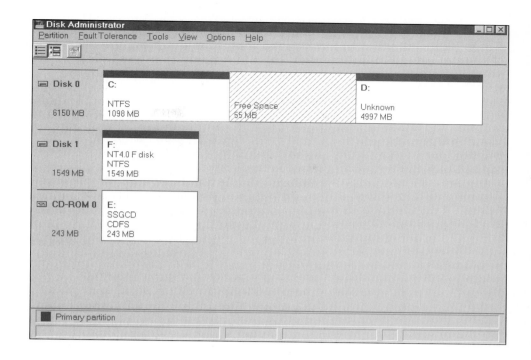

WINDOWS NT DISK ADMINISTRATOR

Windows NT provides a different method to deal with the chore of managing disks and disk partitions. The Disk Administrator utility, located in the Windows NT Administrative Tools (Common) menu, offers many advantages over running the traditional FDISK program in a DOS window. Figure 41.11 shows a graphical display of the physical partitions on a computer system. Notice how Disk 0 contains an NTFS partition at the beginning of the disk labeled "C:" and an unknown partition at the end of Disk 0 labeled "D:" with 55MB of free space remaining. Also note from the figure that Disk 1 contains an NTFS format and a label of "F:" with no free space, and the CD-ROM 0 reports information about the CD currently in the drive. On this computer, the D: disk is actually the Windows 98 operating system, but because the format is FAT32, Windows NT cannot read it. Likewise, Windows 98 cannot see the NTFS partitions.

You are encouraged to explore the capabilities of the Disk Administrator utility.

LBA

Logical block addressing (LBA) is a method to access IDE (Integrated Drive Electronics) hard disk drives. Using LBA, disks larger than 504MB (1024 cylinders) can be partitioned using FDISK or PartitionMagic. Actually, LBA has been around for quite some time now, and has been incorporated into system BIOS on most PCs. Before this, BIOS limitations prevented FDISK from using the entire drive and it was necessary to use custom software, called a Dynamic Drive Overlay. The last option was to simply stay under the limit.

LBA may be implemented in four ways. For example:

1. ROM BIOS support for INT 13h.
2. Hard disk controller support for INT 13h.
3. Use only 1024 cylinders per partition.
4. Real-mode device driver support for geometry translation.

Windows 95 and Windows 98 support the first three methods directly. The last method requires a special version of the Dynamic Drive Overlay software.

IDE disks using the ATA interface also use BIOS INT 13h services. The disk drive identifies itself to the system BIOS specifying the number of cylinders, heads, and sectors per track. The number of bytes in each sector is always 512.

FAT32

As hard drives grew in storage capacity, they quickly reached the maximum size supported by DOS and Windows (initially only 32MB, then 504MB, then 2.1GB). This limitation was based on the number of bits used to store a cluster number. The original FAT12 used a 12-bit *FAT* entry. FAT16 added four more bits, allowing for up to 65,536 clusters. With each cluster representing 16 sectors on the disk, and each sector storing 512 bytes, a cluster would contain 8K. The total disk space available with 65,536 clusters of 8K is 512MB, a small hard drive by today's standards.

One way to support larger partitions is to increase the size of a cluster. Storing 32K in a cluster allows a 2048MB (2.048GB) hard drive, but also increases the amount of wasted space on the hard drive when files smaller than 32K are stored. For example, a file of only 100 bytes is still allocated 32K of disk space when it is created because that is the smallest allocation unit (one cluster). You would agree that most of the cluster is wasted space. Some disk compression utilities reclaim this wasted file space for use by other files. In general, however, large cluster sizes are not the solution to the limitation of the FAT16 file system.

FAT32 uses 32-bit FAT entries, allowing 2200GB hard drives without having to result to using large cluster sizes. In fact, FAT32 typically uses 4K clusters, which helps keep the size of the FAT small and lowers the amount of wasted space. FAT32 is only used by Windows 95 Version B (OEM Service Pack 2) and Windows 98. Several utilities, such as PartitionMagic, are able to convert a FAT16 disk into a FAT32 disk. Windows NT has its own incompatible file system called NTFS.

NTFS

The *NT file system*, or *NTFS*, is used on Windows NT computers. Using NTFS, it is possible to protect individual items on a disk and therefore prevent them from being examined or copied. This is a feature commonly found on multiuser computers such as Windows NT.

NTFS was designed to be more efficient than FAT and to have a higher degree of reliability. Efficiency is improved through the NTFS directory scheme, which uses B-trees instead of linear lists. Searching for a file is much faster with a B-tree than a linear list. The overhead required to support NTFS is larger than FAT, however, preventing the use of NTFS on a floppy.

Reliability is provided through a number of methods. Most significantly, NTFS uses *transactions* to verify that a disk operation was completed successfully. Data protection is provided through fault-tolerant *RAID technology* (see Appendix C for more details) that supports multiple hard drives (up to 32) in arrangements called *mirrors* or *stripe sets*.

Finally, it is possible to convert a FAT file system into an NTFS file system, but not vice versa, and Windows 95 and 98 do not recognize NTFS–formatted partitions.

HARD DRIVE INTERFACES

Many companies manufacture hard drives for personal computers. Even though each company may design and build its hard drives differently, the interface connectors on each drive must conform to one of the accepted standards for hard drive interfaces. These interface standards are illustrated in Figure 41.12.

The first popular hard drive interface scheme was invented by Shugart Technologies. Called ST506, it requires two cables (control and data) between the controller card and the hard drive. This is shown in Figure 41.12(a). Serial data passes back and forth between the controller and hard drive over the data cable. A second hard drive is allowed; the second drive

FIGURE 41.12 Hard drive connections

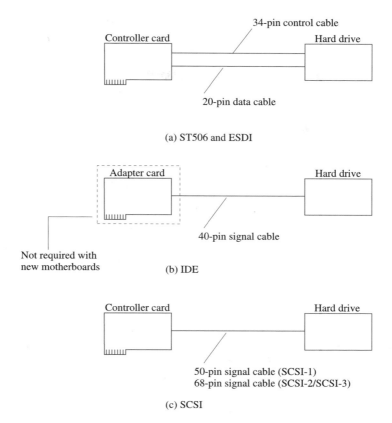

(a) ST506 and ESDI

(b) IDE

(c) SCSI

shares the control cable with the first drive, but has its own data cable. Jumpers must be set on each drive for proper operation, and the last drive needs to contain termination resistors.

An improvement on the ST506 standard was developed by Maxtor, another hard drive manufacturer. Called ESDI (Enhanced Small Device Interface), this interface uses the same two cables as the ST506, but allows data to be exchanged between the controller and hard drive at a faster rate. Although similar in operation to ST506, ESDI is not electrically compatible with it. Thus, ST506 hard drives require ST506 controllers, and ESDI hard drives require ESDI controllers.

The IDE (Integrated Drive Electronics) interface, shown in Figure 41.12(b), has virtually replaced the older ST506 standard. A single cable is used to exchange parallel data between the adapter card and the hard drive. A second hard drive uses the same cable as the first, with a single jumper on each drive indicating if it is the primary drive. A significant difference in the IDE standard is that the controller electronics are located *on the hard drive itself.* In a two-drive system, the primary drive controls itself and the other hard drive as well. The adapter card plugged into the motherboard merely connects the hard drive to the system buses, using parallel data transfers. This allows typical data transfers with an IDE hard drive of up to 8MB per second.

New motherboards have built-in EIDE (enhanced IDE) controllers, which provide signals for four EIDE connectors. This eliminates the need for an adapter card. One pair of connectors are the *primary* connectors, the other pair are the *secondary* connectors. Each pair can support two IDE hard drives (CD-ROM and tape backup drives as well) in a master/slave configuration. Figure 41.13 shows the pin and signal assignments for the EIDE interface, and a typical IDE hard drive.

The earlier IDE interface lacked the upper eight data lines D_8 through D_{15}. The IDE specification also limited hard drive capacity to just 504MB, a small size by today's standards. The EIDE specification increases drive capacity to more than 8GB, expands the maximum number of drives from two to four, and increases the data transfer rate to over 16MB/second.

Right-clicking on My Computer and then choosing Properties allows you to check the system properties. In Device Manager, double-click on the Hard disk controller entry and

FIGURE 41.13 (a) Primary
EIDE connector pin/signal
assignments and (b) typical
IDE hard drive (photograph
by John T. Butchko)

$\overline{\text{RESET}}$	1	2	GND
D_7	3	4	D_8
D_6	5	6	D_9
D_5	7	8	D_{10}
D_4	9	10	D_{11}
D_3	11	12	D_{12}
D_2	13	14	D_{13}
D_1	15	16	D_{14}
D_0	17	18	D_{15}
Ground	19	20	Key (missing pin)
DRQ3	21	22	GND
$\overline{\text{IOW}}$	23	24	GND
$\overline{\text{IOR}}$	25	26	GND
IOCHRDY	27	28	ALE
DACK3	29	30	GND
IRQ14*	31	32	IO16
A_1	33	34	GND
A_0	35	36	A_2
CS0	37	38	CS1
SLV/ACT	39	40	GND

* IRQ15 for the secondary EIDE connector

(a)

(b)

then the primary IDE controller entry. You should get a settings window similar to that shown in Figure 41.14. In Windows NT, this information is available in the Windows NT Diagnostics window.

As indicated in Figure 41.14, the primary IDE controller uses interrupt 14. A handful of I/O ports are required as well, to issue commands and read data from the controller.

FIGURE 41.14   Hardware
settings for primary IDE
controller

Primary IDE controller (single fifo) Properties   ? ✕

General | Resources

Primary IDE controller (single fifo)

Resource settings:

Resource type	Setting
Input/Output Range	01F0 - 01F7
Input/Output Range	03F6 - 03F6
Interrupt Request	14

You cannot modify the resources for this device directly. To
change its resources, change the resources of its parent device.

Parent device: Standard Dual PCI IDE Controller

Conflicting device list:

No conflicts.

OK | Cancel

The SCSI (Small Computer System Interface) standard, shown in Figure 41.12(c),
offers more expansion capability than any of the three previously mentioned standards.
Like the IDE standard, the SCSI standard uses a single cable to control the hard drive. But
whereas all three previous standards allowed at most two hard drives, the SCSI standard allows up to *seven* devices to be daisy-chained on the single 50-pin cable. Each device can be
a hard drive, if necessary. This difference is important to computer users who have large
storage requirements and might require gigabytes (1GB = 1024MB) of hard drive capacity.
SCSI is used to connect a wide variety of devices (usually 8 or 16) together on a shared bus.
For example, several disk drives, a tape drive, and a scanner can be connected to one SCSI
bus. Each device on the bus and the controller card itself requires an address and the end of
the SCSI cable, or the last device on the bus must be terminated. Figure 41.15 shows a
daisy chain of SCSI devices. The last device contains a terminator.

There are several different types of SCSI buses, each allowing for a specific cable type,
transfer rate, bus width, and so on. Table 41.1 shows the different SCSI standards.

FIGURE 41.15   (a) A daisy
chain of SCSI devices

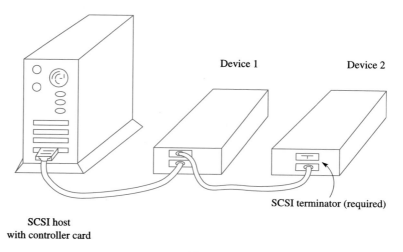

Device 1   Device 2

SCSI terminator (required)

SCSI host
with controller card

(a)

(continued on the next page)

FIGURE 41.15 *(continued)*
(b) SCSI terminators (passive)

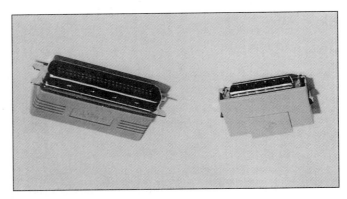

(b)

TABLE 41.1 SCSI bus standards

Standard	Bus Width	Max Transfer Rate (Mbps)	Cable Type
SCSI-1	8	4	Not specified
SCSI-2	8	5	A
	16	10	B
SCSI-3	16	10	P
	32	20	P, Q

TABLE 41.2 SCSI bus lengths

Bus Type	Single-Ended	Differential
SCSI-1	6 meters	25 meters
SCSI-2	6 meters	25 meters
SCSI-3	3 meters	25 meters

SCSI buses also have specific length requirements falling into two categories: single-ended and differential. A single-ended SCSI bus is cheap and fast over short distances. Differential SCSI can be used over longer distances. Table 41.2 shows the SCSI bus length requirements.

There are also different types of connectors used to connect SCSI devices together. Check the individual requirements for each SCSI device to determine the appropriate type. Note that SCSI devices are generally more expensive than non-SCSI devices, but they provide for combinations of devices not possible with standard PC technology.

DATA STORAGE

Although there are differences between IDE hard drives and SCSI hard drives (and all the other types), there is also something in common: each hard drive uses the flux changes of a magnetic field to store information on the hard drive platter surface. A number of different techniques are used to read and write 0s and 1s using flux transitions. Some of the more common techniques are listed in Table 41.3.

DISK CACHING

Because of mechanical limitations (rotational speed of the platters; movement and settling time of the read/write head), the rate at which data can be exchanged with the hard drive is

TABLE 41.3 Data recording techniques

Technique	Meaning/Operation
MFM	Modified Frequency Modulation. Magnetic flux transitions are used to store 0s and 1s.
RLL	Run Length Limited. Special flux patterns are used to store *groups* of 0s and 1s.
Advanced RLL	Advanced Run Length Limited. Permits data to be recorded at higher density than RLL.

limited. It is possible to increase the data transfer rate significantly through a technique called *caching*. A hardware cache is a special high-speed memory whose access time is much shorter than that of ordinary system RAM. A software cache is a program that manages a portion of system RAM, making it operate as a hardware cache. A computer system might use one or both of these types of caches, or none at all.

The main idea behind the use of a cache is to increase the *average* rate at which data is transferred. Let's see how this is done. First, we begin with an empty cache. Now, suppose that a request to the hard drive controller requires 26 sectors to be read. The controller positions the read/write head and waits for the platters to rotate into the correct positions. As the information from each sector is read from the platter surface, a copy is written into the cache. This entire process may take a few *milliseconds* to complete, depending on the drive's mechanical properties. If a future request requires information from the same 26 sectors, the controller reads the copy from the cache instead of waiting for the platter and read/write head to position themselves. This means that data is accessed at the faster rate of the cache (whose access time might be as short as 10 *ns*). This is called a cache *hit*. If the requested data is not in the cache (a *miss*), it is read from the platter surface and copied into the cache as it is outputted, to avoid a miss in the future. The cache uses an algorithm to help maintain a high hit ratio.

The same method is used for writing. Data intended for the hard drive is written into the cache very quickly (8Mbps), and then from the cache to the platter surface at a slower rate (2.5Mbps) under the guidance of the controller.

Many hard drives now come with 256K of onboard hardware cache. In addition, a program called SMARTDRV can be used to manage system RAM as a cache for the hard drive. To use SMARTDRV, a line such as

```
C:\DOS\SMARTDRV.EXE 2048
```

must be added to your AUTOEXEC.BAT file. This command instructs SMARTDRV to use 2MB of expanded or extended memory as a cache. Small programs that are run frequently (DOS utilities stored on the hard drive) load and execute much more quickly with the help of SMARTDRV.

DISK STRUCTURE

The information presented here applies equally well to floppy disks as to hard disks. Figure 41.16 shows how a disk is divided into *sectors*. From the figure, you can see that a sector is a specified pie-slice area on the disk.

Disk sectors and tracks are not physically on the disk, just as data is not physically on the disk. They are simply magnetic patterns placed on the disk by electrical impulses. The number of tracks available on the disk varies. For example, a standard 3.5-inch floppy disk has 80 tracks, whereas a hard disk may contain 650 tracks. The number of sectors a disk has also varies. This is illustrated in Table 41.4.

FIGURE 41.16 Disk sectors

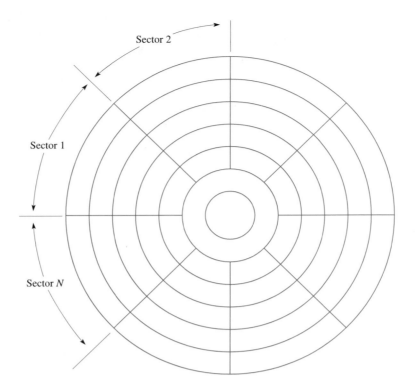

TABLE 41.4 Floppy disk configurations

Disk Size	Disk Type	Tracks/Side	Total Sectors*
5.25	Single-sided—8 sectors per track	40	320
5.25	Single-sided—9 sectors per track	40	360
5.25	Double-sided—8 sectors per track	40	640
3.5	Double-sided—9 sectors per track	40	720
3.5	Quad-density—9 sectors per track	80	1440
5.25	Quad-density—15 sectors per track	80	2400

*Common sector sizes for disks are 128, 256, 512, and 1024 bytes.

DISK STORAGE CAPACITY

You can calculate the storage capacity of a disk as follows:

$$DSC = sides \times tracks \times sectors \times size$$

where

DSC = disk storage capacity
sides = number of disk sides used
tracks = number of disk tracks per side
sectors = number of disk sectors per track
size = size of each sector in bytes (usually 512)

As an example, consider a double-sided, double-density (nine-sector-per-track) disk. From Table 41.4, you can see that such a disk has 40 tracks per side and nine sectors per track, where each sector stores 512 bytes. Thus, for this type of disk, the total disk storage capacity is

$$DSC = 2 \times 40 \times 9 \times 512 = 368,640 \text{ bytes (or 360K)}$$

Recall that each disk contains a boot sector. This boot sector is contained in sector 1, side 0, track 0, as illustrated in Figure 41.17. Table 41.5 lists the information contained in a boot sector.

FIGURE 41.17   Location of the boot sector

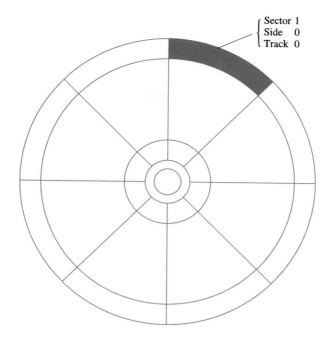

Sector 1
Side   0
Track   0

TABLE 41.5   Information in a boot sector

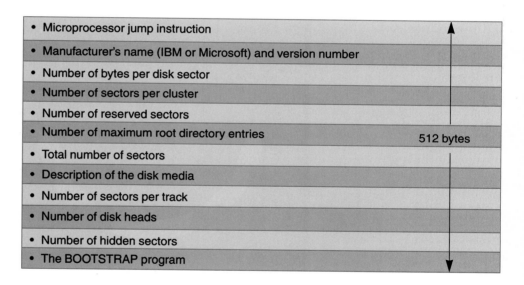

- Microprocessor jump instruction
- Manufacturer's name (IBM or Microsoft) and version number
- Number of bytes per disk sector
- Number of sectors per cluster
- Number of reserved sectors
- Number of maximum root directory entries
- Total number of sectors
- Description of the disk media
- Number of sectors per track
- Number of disk heads
- Number of hidden sectors
- The BOOTSTRAP program

512 bytes

Following the boot sector, there is a file allocation table (FAT). This table is used by DOS to record the number of disk sectors on the disk that can be used for storage as well as bad sectors that must not be used for storage. Several sectors are reserved for the FAT. All hard disks come from the factory with a certain number of bad sectors. During final product testing, these bad sectors are usually found and are usually then labeled on the hard drive unit itself.

In order to ensure reliability, each disk contains a duplicate copy of the FAT. This means that if one copy of the FAT goes bad, the backup copy is available for use.

DISK FRAGMENTATION

Disk fragmentation is the result of one or more disk files being contained in scattered sectors around the disk, as shown in Figure 41.18. Observe that the read/write heads may take more than one revolution of the disk to read all the file information scattered across the various sectors as a result of disk fragmentation.

**FIGURE 41.18   Disk frag-
mentation**

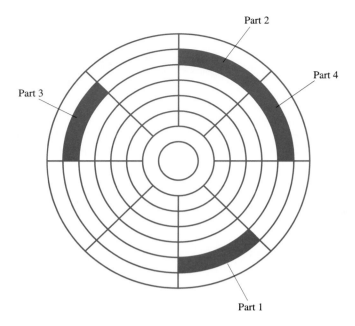

**FIGURE 41.19   Contiguous file
data on the same disk track**

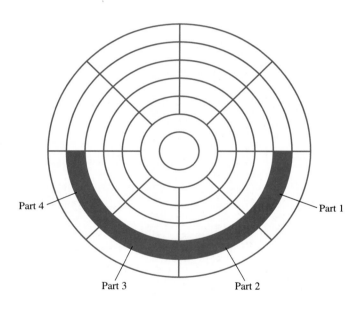

Next, consider the file data distributed in contiguous sectors around the disk, as shown in Figure 41.19. The way the data is distributed here, it is conceivable that it could all be read in one revolution of the disk. The difference between fragmented data and contiguous data is that it takes longer to read fragmented data.

You must keep in mind that disk drives are slow when compared with the rest of the computer system. If information is fragmented over the hard disk, it takes longer to read the disk and significantly slows the entire computer system (because it takes longer to read and write the information). Disk fragmentation occurs when files are repeatedly added and deleted on a disk. Once disk fragmentation occurs (especially on the hard drive), your system will begin to run more slowly when it interacts with the hard drive. To eliminate this, you must use a hard disk utility program to defragment the disk. You will learn how to do this in the next exercise. It is important now to realize that you must do this periodically in order to maximize system performance.

The defragmentation process can be automated in Windows 95/98 using the system agent. During periods of inactivity, the system agent will periodically run the disk utility tools, keeping the disk in reasonably good condition. The defragmentation process can also

FIGURE 41.20   Selecting a disk to defragment

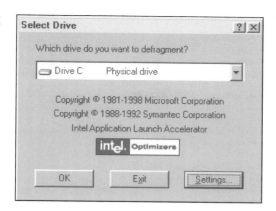

FIGURE 41.21   Disk Defragmenter Settings window

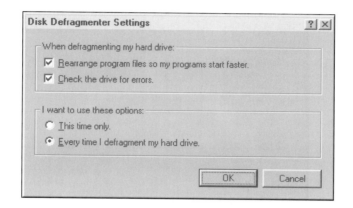

FIGURE 41.22   Defragmentation status window

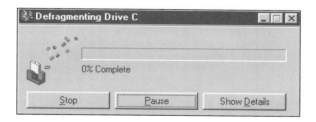

be run on demand by selecting Disk Defragmenter from the System Tools submenu (located under Accessories on the Start menu). Figure 41.20 shows the initial defrag screen presented by Windows.

From this menu, it is very easy to start the defragmentation process, select another drive, check or modify the advanced parameters, or exit. Figure 41.21 shows the Disk Defragmenter Settings window.

Figure 41.22 shows the brief status of the defragmentation process. We can also view the details of the defragmentation process by clicking the Show Details button as shown in Figure 41.23. Notice how each of the disk clusters is presented on the screen. It is interesting to watch the defragmentation process. Depending on the amount of fragmentation, the process may last just a few minutes or as long as a few hours. Select Legend to view the legend shown in Figure 41.24 to help you identify the different types of disk clusters. As you can see, there are many different possible states for a disk cluster.

FILE ALLOCATION

Every time DOS has to get space on the disk for a file, it looks at the FAT for unused disk *clusters*. A cluster is a set of contiguous disk sectors. DOS will always try to minimize

FIGURE 41.23 Details of the defragmentation process

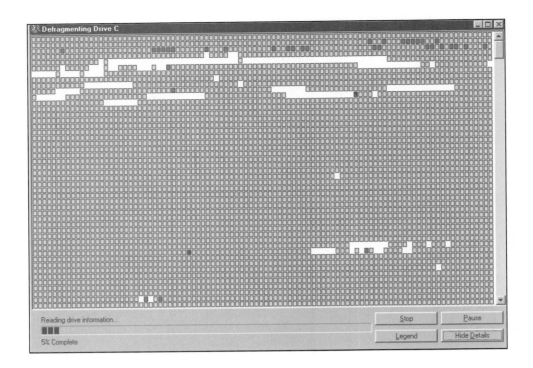

FIGURE 41.24 Defragmentation details legend

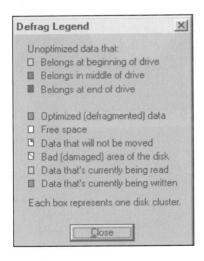

disk fragmentation, if possible. Every file written to the disk has a directory entry that is a 32-byte record, which contains the following information:

- Name of file
- File extension
- Attribute byte
- File time
- File date
- Number of the starting cluster
- Size of the file

DOS allocates disk space for directory entries in a special directory area that follows the FAT. This means that a disk can hold only a certain number of files, depending on the size and density of the disk. To DOS, a subdirectory is treated simply as a standard DOS file. This means that DOS stores subdirectory information in the same manner as it stores file information. Because of this, the number of subdirectories is also limited by the amount of free file space on the disk.

FIGURE 41.25   Windows 98
File System Properties window

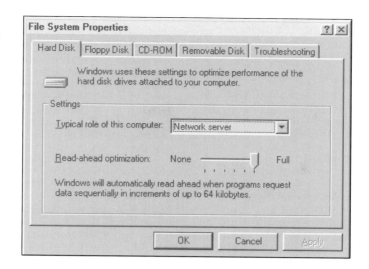

OPTIMIZING DISK PERFORMANCE

When setting up the structure of a hard disk, you should do so in a manner that will optimize the overall performance of the entire system. This means having your files organized to enhance maximum disk performance and to defragment the disk periodically to get rid of disk fragmentation.

For setting up the directory of a hard disk, the root directory should contain only the following:

- Operating system files
- AUTOEXEC.BAT
- CONFIG.SYS

All other files should be placed in subdirectories. For example, all the Windows files should be placed in \WINDOWS. No other files should be added to the root directory. Doing this will improve system performance.

In addition to careful directory planning, you can instruct Windows to use the hard disk in a certain way, depending on the role your machine is playing on the network. Figure 41.25 shows the File System Properties window, accessed through the Performance menu of System Properties. Currently the machine is set up as a network server, enabling a 64K read-ahead on the hard drive to help supply data faster. Other options are desktop computer and mobile docking station.

USING THE PATH COMMAND

To maintain optimum system performance, you should carefully structure the PATH command. The PATH command specifies a series of drives and subdirectory paths that can be searched by the operating system every time a command is not found in the current directory. This command is kept in the AUTOEXEC.BAT file. The form of the PATH command is

$$PATH [[Drive:]path[;[drive:][path] . . .]$$

What happens here is that the operating system first searches in the order given by the PATH command, from left to right, for the path of a particular file that is not in the active directory. Thus, you should organize the PATH command so that the most frequently used paths appear first and the least frequently used paths appear last.

**TROUBLESHOOTING
TECHNIQUES**

For a little historical perspective about the price of 1 bit of hard drive storage, examine the numbers in Table 41.6. Clearly, it is getting cheaper every day to buy a large hard drive, with more than a 600-fold reduction in the cost of a bit since 1995.

TABLE 41.6   Cost of 1 bit of hard drive storage

Hard Drive	Cost	Year Purchased	Cost/Bit
18MB	$2500	1981	4.29¢
30MB	$300	1985	.000953¢
212MB	$375	1989	.000168¢
853MB	$319	1995	.000035¢
1.7GB	$250	1996	.000013¢
5.1GB	$199	1998	.000004¢
8.4GB	$129	1999	.000001¢
40GB	$200	2001	.000000058¢

SUMMARY

In this exercise we discovered that

- Hard disks contain platters, with the tracks on each patter forming a cylinder.
- Hard disks must be partitioned (using FDISK, PartitionMagic, or some other utility).
- File system types include FAT, FAT32, NTFS, and others.
- Hard drive interfaces include ST506, IDE, and SCSI.
- Disk caching improves the read/write performance of a hard drive.
- Disk fragmentation lowers the read/write performance of a hard drive.

SELF-TEST

This self-test is designed to help you check your understanding of the background information presented in this exercise.

True/False

Answer *true* or *false*.

1. The construction of a hard disk system is the same as that of a floppy disk system.
2. A disk in a hard drive system is referred to as a *platter.*
3. A hard disk can be partitioned so that it has more than one operating system.
4. IDE and EIDE hard drives have the same capacity.
5. Each disk sector contains 1024 bytes.
6. Multiple sectors are grouped together as clusters.
7. FAT32 uses 32-bit file entries.

Multiple Choice

Select the best answer.

8. In a hard disk system, a combination of tracks is called
   a. The drive.
   b. A cylinder.
   c. A cluster.
   d. The spindle.
9. In DOS versions 3.3 and earlier, the maximum size of the primary partition was
   a. 640K.
   b. 32MB.
   c. 32K.
   d. Limited by the size of the disk.
10. A disk *sector* is
   a. The part of the disk only written to.
   b. Two or more consecutive tracks.
   c. Used in hard disks, not in floppies.
   d. A pie-slice portion of a disk.

11. SCSI support requires
    a. A special socket.
    b. An SCSI controller card.
    c. The SCSI chip set.
    d. All of the above.
12. The FAT32 file system was introduced with
    a. 32 copies of the FAT for safety.
    b. Windows 95B.
    c. A compressed disk.
    d. All of the above.
13. NTFS stands for
    a. NT File System.
    b. NT Fault Server.
    c. NT File Session.
14. NTFS uses _____ to verify that disk operations are successful.
    a. Heartbeats.
    b. Transactions.
    c. Update profiles.

Matching

Match a statement on the right with each DOS version on the left.

15. 3.2   a. Partition size may be up to 2GB.
16. 4.01   b. Removed the 32MB limit for the primary partition.
17. 5.0   c. Size limit of a single partition is 32MB.
                           d. No size limit on a partition other than the primary one.

Completion

Fill in the blank or blanks with the best answers.

18. The term _____ _____ indicates that files are scattered among many different sectors on the disk.
19. When setting up a hard disk for optimum performance, you should have as few files as possible in the _____ directory.
20. The _____ command determines the order in which directories will be searched for files not in the active directory.
21. SCSI stands for _____ _____ _____ _____.
22. The last device on an SCSI bus must be a(n) _____.
23. NTFS provides RAID support through mirrors and _____ sets.

FAMILIARIZATION ACTIVITY

1. Using the technical manual for the hard disk system in your assigned computer, determine the following:
    a. The number of platters in the hard drive system.
    b. Maximum storage capacity of the hard drive system.
    c. The type of hard disk system used.
2. Turn on your system and determine the following:
    a. Operating system version used.
    b. Structure of the directories.
    c. Structure of the PATH command.
3. Based on the information you obtained in this activity, write a recommendation of how the system performance could be improved; if you don't think it could be improved, state why.

QUESTIONS/ACTIVITIES

1. In your own words, explain the major differences between a hard disk system and a floppy disk system.

2. List and explain the different types of hard disk systems.
3. Sketch what is meant by a cylinder in a hard disk system.
4. Describe what is meant by a disk partition.
5. Determine the disk storage capacity of a hard disk system that has four sides, 160 tracks, nine sectors, and a sector size of 1024 bytes.

Under the supervision of your instructor

1. Determine if the hard disk has been set up for the most efficient system operation by looking at the directory structure and the PATH command in the AUTOEXEC.BAT file.
2. Determine the type of hard drive used in your system.

42

Hard Drive Backup

INTRODUCTION

Joe Tekk had just gotten back from lunch when the phone rang. It was Don, the senior technician. Don described a problem one of RWA's customers was having with their computer.

Don said, "Jim told me that when they turn the computer on, it displays the message 'hard disk 0 failure,' or something like that."

Joe knew this was a bad sign and immediately began wondering if someone was regularly backing up the files on the computer. Joe asked, "Don, did they perform backups?"

Don responded in a very serious tone, "I asked that question, too, and got the response you don't want to hear." He continued, "Joe, please take care of this for me."

Joe decided to make the best out of a bad situation by helping the customer implement a regular backup procedure so this problem will not happen again. Luckily, he was also able to recover many of the files from the defective hard drive. When he returned to the shop, the disk worked fine . . . for about 15 minutes.

PERFORMANCE OBJECTIVES

Upon completion of this exercise, you will be able to

1. Back up and restore a set of files from one disk to another and read the backup log created in the process.
2. Use PKZIP and PKUNZIP (or WINZIP) to back up and restore a compressed directory of files.
3. Back up and restore a set of files from disk to tape and tape to disk.

KEY TERMS

Disk backup
Full backup
Incremental backup
Source disk
Target disk

Log file
Lossless compression
Low-level format
Defect mapping
Disk geometry

BACKGROUND INFORMATION

In this exercise, you have the opportunity to learn how to copy the information from a hard disk onto floppies or tape, partition and format the hard disk, and then restore information back to the hard disk. This process needs to be done whenever a hard disk is replaced.

515

First, you will see how to back up a hard drive. Then you will be presented with the information on how to partition and format the drive using DOS or Windows. Next, you will see how to return the information from the backup disks to the hard drive.

DISK BACKUP

The contents of a hard drive are very important to the overall operation of any computer system. This is where all the frequently used programs, files, and information are kept. A hard drive crash can cause the loss of some or all of this valuable information. The periodic backing up of its files is an important part of using a computer. The hard disk should be backed up periodically, as well as before a computer is shipped.

BACKUP SCHEDULES

Generally, the task of performing disk backups is taken very seriously. A strict schedule of backups should be performed regularly. Since it can take a lot of time to back up a hard disk, usually the backup process is broken down into two types of backups. First, a full backup is used to write all the information from a hard drive to a tape drive, floppy disks, or Zip disks with enough excess capacity to contain the backed-up files. After a full disk backup has been completed, incremental backups can be performed to save any files that have been modified since the last backup date. Table 42.1 shows a typical backup schedule using a combination of both full and incremental saves.

The tapes or disks containing the backed-up data may be kept in a safe place for as long as necessary. For example, a PC containing information collected at a bank, medical, or dental office may need to be saved for some specific period of time as required by law, thus requiring many tapes to be used. In contrast, disk file information maintained by a video store needs to be completely restored from the last full backup and any incremental backups. Each situation may be different, but the same question needs to be answered in each case: If the hard drive crashes, can the data be restored?

WINDOWS 95/98 BACKUP

Windows 95/98 comes with backup software used to write information to a limited number of tape devices. In order to determine whether a compatible tape device is attached, it is necessary to run the backup utility program. Figure 42.1 shows the first screen displayed by the Microsoft Backup program indicating the general steps involved in performing a backup. Generally, performing a backup involves selecting the files to back up, selecting the destination device, and starting the process. It is likely the backup software will confirm any action where data would be overwritten. Answer all questions very carefully.

Next, as shown in Figure 42.2, Backup will automatically create a full backup file set, which can be used to restore a Windows 95/98 boot disk, including all the Registry files. This file set is used only when performing a full backup and restore, and is not used for incremental backups. This backup file set is used to restore files after a catastrophic hard disk failure.

TABLE 42.1   Typical file backup schedule

Day	Backup Type
Monday	Incremental
Tuesday	Incremental
Wednesday	Incremental
Thursday	Incremental
Friday	Full

FIGURE 42.1   Windows Backup system tool

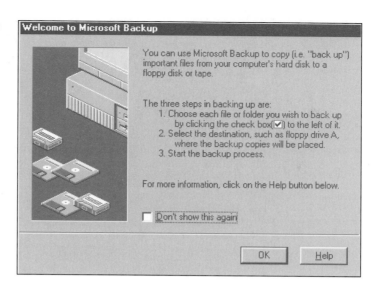

FIGURE 42.2   Backup automatically creates a full backup file set

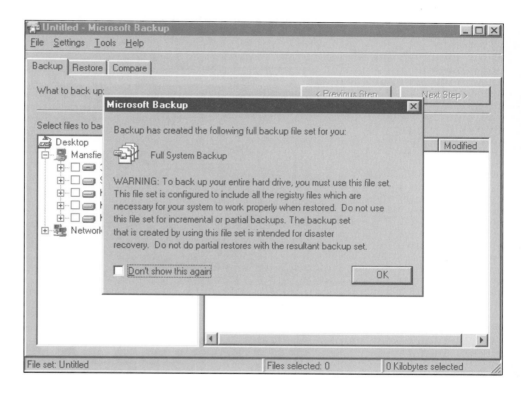

The Options window selected from the Settings pull-down menu allows the individual settings of the Backup program to be modified. The Backup utility contains components used to back up, restore, and compare files. For example, to modify Backup settings, simply select the Backup tab. Figure 42.3 shows the Settings—Options for the Backup operation. Either a full backup or an incremental backup *type* can be selected, and other options such as data compression and tape or floppy erasure settings can be modified as necessary. Care must be exercised when changing any of the Backup settings.

Just as the Backup settings may be changed, the settings of the Restore operation may also be modified as shown in Figure 42.4. Choices can be made as to where files will be restored and what types of action Backup will take when it comes across files with dates from those recorded on the tape. Again, care must be taken when modifying any of these options.

FIGURE 42.3 Backup Settings—Options

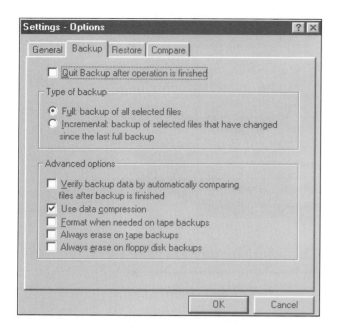

FIGURE 42.4 Restore Settings—Options

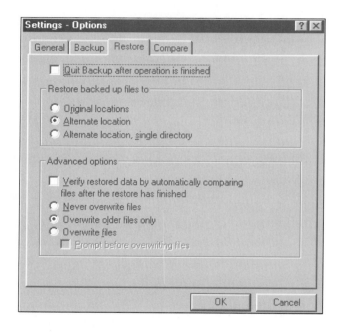

To select files to back up, simply click on the drive or folder and then select the Next Step button as shown in Figure 42.5. It is possible to select any combination of disks or folders.

Unfortunately, as shown in Figure 42.6, Backup did not find a compatible tape drive, and the destination for the backup must be selected from the remaining devices. Sometimes, this is an adequate solution, but usually the cost to store backup information on a hard disk is too expensive when compared with the cost of a tape capable of storing 2, 4, or 8GB of information.

Many times, tape drives come with their own software and do not rely on the built-in Backup utility. The Hewlett-Packard tape drive designed for use under Windows 95/98 comes with its own software that mimics the functions of the built-in Backup utility. Figure 42.7 shows the HP tape drive and several tape cartridges. Figure 42.8 shows a typical backup screen using the Colorado Backup utility. The Settings Options menus are identical to the built-in Backup utility.

FIGURE 42.5   Microsoft Backup file section screen

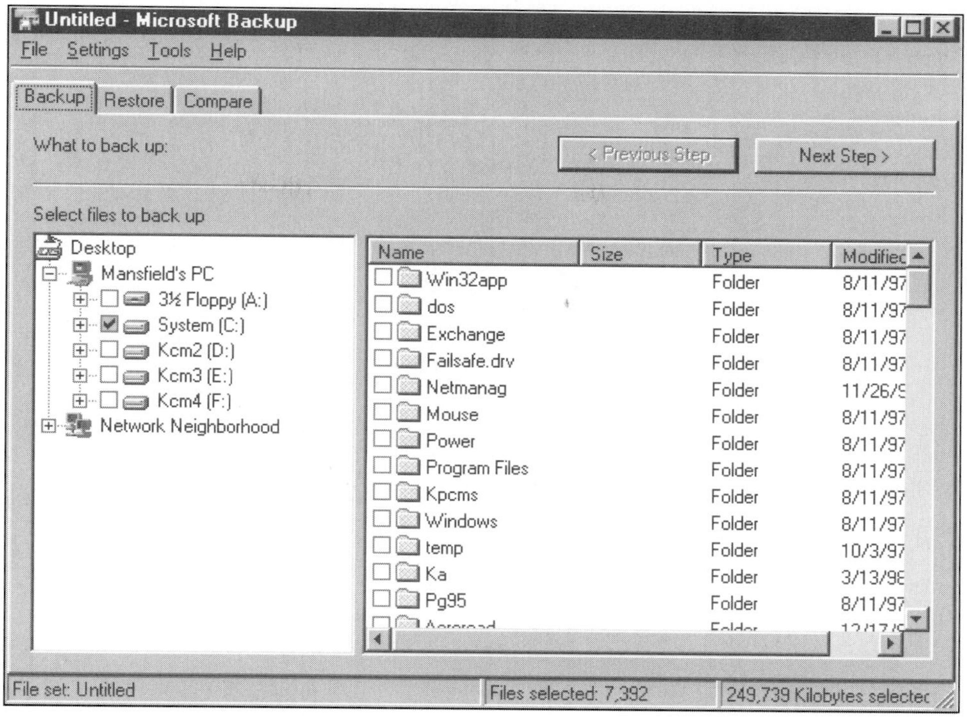

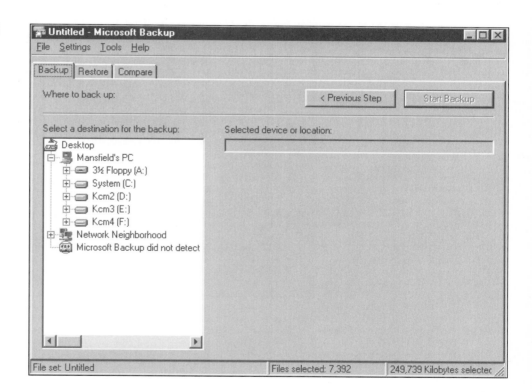

FIGURE 42.6   Backup did not find a tape drive

To begin a backup operation, simply select files to be backed up. This involves left-clicking on the appropriate disks or folders; this causes the backup software to scan for selected files to back up. (The backup software will display a window showing the status of the selection process, as illustrated in Figure 42.9.) Note the file selection operation may take several minutes to complete. Notice how the number of files selected and the size of these files are displayed in the status bar at the bottom of the window. By pressing the Next button, the destination for the backup may be selected. Figure 42.10 shows the list of

FIGURE 42.7 (a) HP tape
backup unit and (b) assorted
tape cartridges

(a)

(b)

devices that can be selected as the destination for the backup. Notice the HP Colorado
T1000e is selected in the list, and the right side of the Backup window shows the current
tape's capacity and its total of free and used space. After selecting the backup device, it is
again necessary to left-click the Next button to finalize any remaining options.

Each backup is appropriately named, usually describing the type and location of the data.
For example, in Figure 42.11, the name "Files from System (C:) and other sources" is automatically displayed. The backup file set can also be password protected to guarantee
privacy. The password, if selected, must be used to access the data on the tape. Without it,
it will be impossible to restore or view the contents of the tape. Obviously, care must be

FIGURE 42.8   Colorado Backup file selection screen

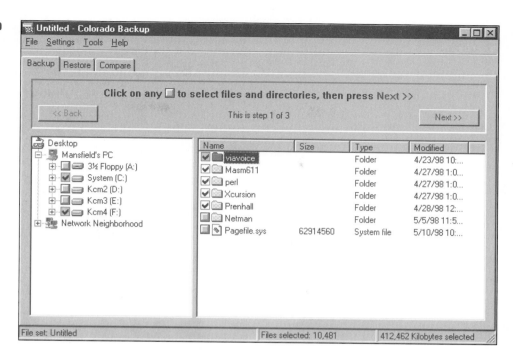

FIGURE 42.9   File Selection status window

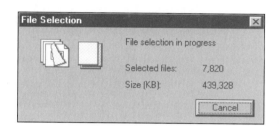

FIGURE 42.10   Selecting the destination device

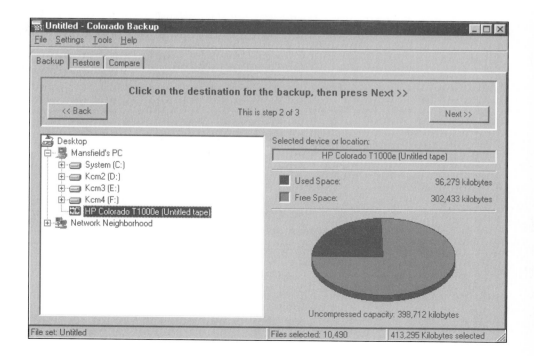

FIGURE 42.11   Backup options window

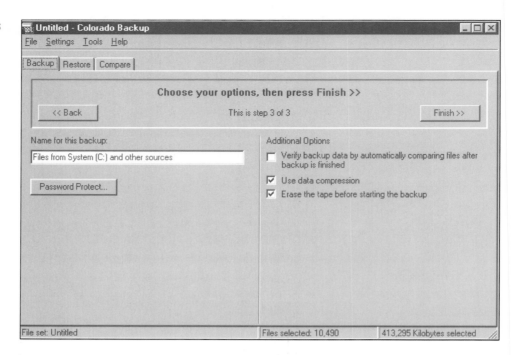

FIGURE 42.12   Remove check mark to erase the tape

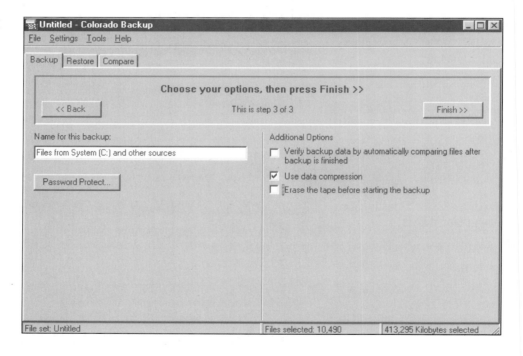

exercised when assigning a password to a file set or changing any of the settings. Notice how "Erase tape before starting the backup" has been deselected in Figure 42.12. This will cause the new backup information to be appended to the end of the tape. When all the settings have been verified, the backup process can be initiated by left-clicking the Finish button.

The Backup program displays the window shown in Figure 42.13 as it prepares to create the new file set. After preparation is complete, Backup will display the current status of the backup as shown in Figure 42.14. When the backup requires a new tape, it will pause until a new tape is inserted in the tape drive and then the process will continue. When the backup is complete, a pop-up window is displayed, as shown in Figure 42.15.

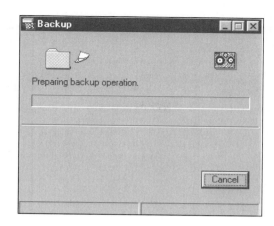

FIGURE 42.13 Backup prepares to write information to tape

FIGURE 42.14 Backup writes information to tape

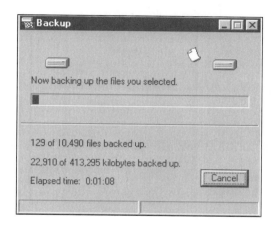

FIGURE 42.15 Confirmation of successful backup operation

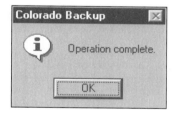

RESTORING FILES

When it becomes necessary to restore files to your system, for whatever reason, it involves selecting the appropriate tape that contains the required information. Then the process involves running the Backup program and selecting the Restore tab from the Backup window. Then it is necessary to select the appropriate device and file set. This is illustrated in Figure 42.16.

The Next button is used to display the file selection screen. Individual files, folders, and disks may be selected for restoration by simply left-clicking on each disk or folder. After the files to restore have been selected and the Next button is pressed, a few additional options shown on Figure 42.17 may be changed if necessary. It is important to use extreme caution when modifying any of these options to prevent possible loss of data.

It is a good idea to spend some time with the Backup program to become familiar and comfortable performing all types of routine backup and restore operations. You are also encouraged to experiment with the compare operation used to perform comparisons between disk and tape files.

FIGURE 42.16 Select the backup source and save set

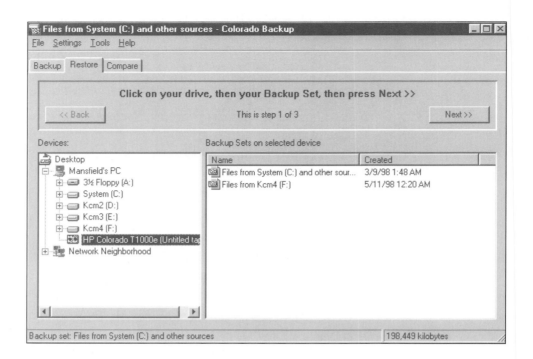

FIGURE 42.17 Selecting the appropriate Restore options

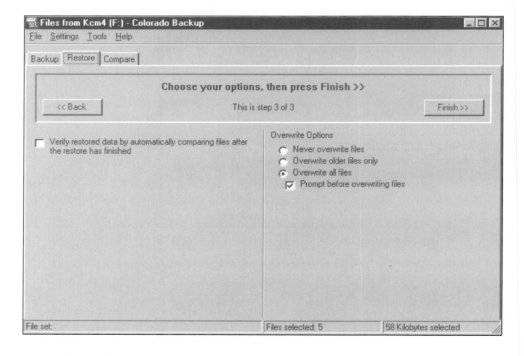

WINDOWS NT BACKUP

In a Windows NT environment, backups are usually considered serious business. A loss of data without replacement can put a company at risk of going out of business. There are two different areas to focus on in a Windows NT domain: the backup requirements of the server itself and the backup requirements of the clients on the network. Basically, all the disks physically connected to the NT server need to be backed up as well as all the disks on the computers in the network. Using the built-in Windows NT backup program or a program from a third party, it is easy to handle this important and otherwise difficult task. You are encouraged to explore the backup program(s) available on your Windows NT computer.

DISK BACKUP USING DOS

DOS contains a command used just for the purpose of backing up the files of a hard disk. The command has the following format:

BACKUP
[drive1:][path][filename][drive2:][/s][/m][/a][/f:size][/d:date][/t:time][/L[[drive:]
[path]filename]]

where   drive1 = the disk drive to be backed up (the *source disk*)
     drive2 = the disk to contain the backed-up files (the *target disk*)
     /s = a command that causes the backing up of subdirectories
     /m = a command that backs up only those files that have been changed since the last backup process
     /a = a command that causes the files that are to be backed up to be appended (as opposed to written over) on the target disk
     /f:size = a command used if the target disk is not formatted before starting the backup process. (It is important to note that the DOS *format* command must be accessible to the backup program for this command to work.)
     /d:date = a command that backs up all files that have been modified on or after *date*
     /t:time = a command that backs up all files that have been modified on or after *time*
     /L:filename = a command that causes a record to be kept in a file of all the files that have been backed up. (This file is called a *log file*. If you do not specify a file name for this log file, DOS will give it the default name of BACKUP.LOG.)

Table 42.2 lists the various values that can be used for target disks during the backup process. These values are used with the /f:size modifier when the target disk has not been formatted.

It is important to note that BACKUP does *not* copy the two hidden DOS files, IO.SYS and MSDOS.SYS. Nor does BACKUP copy the COMMAND.COM file. BACKUP allows you to place a new DOS operating system on the hard disk before restoring the backed-up files to it. Thus, the new IO.SYS and MSDOS.SYS files, as well as the COMMAND.COM file, are not overwritten during the process of transferring the backed-up files to the hard disk.

Be sure not to use an old version of DOS (such as 3.2 or earlier) to restore files backed up by DOS 4.0 or later. If you try to do this, the files will not be properly restored, which will result in the loss of data.

Remember that the backup process will erase any old files already on the target disk unless you use the /a switch with the BACKUP command.

For example, to back up an entire directory to a blank formatted disk (for a directory named \MYSTUFF), you would use

```
C> BACKUP \MYSTUFF A:
```

TABLE 42.2   Values of /f:size for unformatted target disk

Target Disk Size	Allowed Values for /f:size
160K single-sided 5 inch	160, 160K, 160K
180K single-sided 5.25 inch	180, 180K, 180K
320K double-sided 5.25 inch	320, 320K, 320K
360K double-sided 5.25 inch	360, 360K, 360K
720K double-sided 3.5 inch	720, 720K, 720K
1.2M double-sided 5.25 inch	1200, 1200K, 1200K, 1.2, 1.2M, 1.2MB
1.44M double-sided 3.5 inch	1440, 1440K, 1440K, 1.44, 1.44M, 1.44MB
2.88M double-sided 3.5 inch	2880, 2880K, 2880K, 2.88, 2.88M, 2.88MB

RESTORING FILES USING DOS

Once files have been backed up, you can use the DOS RESTORE command to restore the backed-up files from the floppy disk to the hard drive. The RESTORE command has the form

RESTORE drive1:[drive2:][pathname][/s][/p][/b:date][/a:date][/e:time][/m][/n]

where   drive1 = the drive that contains the backup files (source drive)
          drive2 = the drive to which the backed-up files will be restored (target drive)
          /s = a command that lets all subdirectories be restored
          /p = a command that stops the restoring process to let you know that a file to be restored matches an existing file that has been changed since the backup process was done. (It asks if you want to replace that file with the backed-up file.)
      /b:date = a command that causes files to be restored that have been last modified on or before *date*
      /a:date = a command that causes files to be restored that have been modified on or after *date*
      /e:time = a command that causes files to be restored that have been modified at or after *time*
          /m = a command that restores only those files that have been modified since the last backup
          /n = a command that restores only those files that do not exist on the target disk

Remember that the DOS RESTORE command does not restore the system files. It is good practice to check all the restored files with a DIR command to make sure the restoring process has been completed and has accomplished what you expected. To restore the entire directory \MYSTUFF, use the command

```
RESTORE A: \MYSTUFF
```

When the RESTORE process is completed, one of the following exit codes will be displayed:

0 = normal operation
1 = no files found to be restored
2 = restoring process terminated by user
3 = restoring process terminated because of an error

USING PKZIP AND PKUNZIP FOR BACKUPS

The PKZIP and PKUNZIP utilities (available free on the Web) use a digital data compression algorithm to compact the amount of storage space required by a file or group of files. The algorithm is classified as *lossless*, since files must be identical to their original contents when uncompressed.

Figure 42.18 shows the contents of the GRAFX directory. The total space required by the files is 672,033 bytes.

To compress all files in the GRAFX directory, use the DOS commands:

```
C> CD \GRAFX
C> PKZIP GRAFX
```

The PKZIP program compresses each file individually and stores the entire compressed group of files in a new file called GRAFX.ZIP. When this has been done, the display looks like that shown in Figure 42.19. Notice that some files were not compressed. PKZIP decided that they could not be significantly compressed and stored them without change.

Figure 42.20 shows the new GRAFX.ZIP file in the directory listing. The GRAFX.ZIP file takes only 368,971 bytes. Recall that the original directory contained 672,033 bytes. This surely is a significant savings in disk space, and will reduce the required backup time because less information needs to be saved.

FIGURE 42.18 **Contents of the GRAFX directory**

```
Volume in drive D is SUPERSTUFF
Volume Serial Number is 1037-11D2
Directory of D:\GRAFX

.             <DIR>         04-19-94    9:22a
..            <DIR>         04-19-94    9:22a
DA      EXT       1,156 06-22-94    1:11p
GGEN    C        19,093 08-20-94   11:58a
GGEN    EXE      98,402 06-22-94    1:11p
GO      BAT           9 08-20-94    8:13a
OBJS    SFB     150,758 06-22-94    1:11p
PACMAP  DAT     317,049 06-22-94    1:11p
RCT     EXE      85,566 10-16-94    4:00a
        9 file(s)       672,033 bytes
                    160,186,368 bytes free
```

FIGURE 42.19 **Output of the PKZIP program**

```
PKZIP (R)   FAST!   Create/Update Utility   Version 1.1   03-15-90
Copr. 1989-1990 PKWARE Inc.   All Rights Reserved.   PKZIP/h for help
PKZIP Reg. U.S. Pat. and Tm. Off.

Creating ZIP: GRAFX.ZIP
    Adding: DA.EXT       imploding (46%), done.
    Adding: GGEN.C       imploding (77%), done.
    Adding: GGEN.EXE     storing   ( 0%), done.
    Adding: GO.BAT       storing   ( 0%), done.
    Adding: OBJS.SFB     imploding (32%), done.
    Adding: PACMAP.DAT   imploding (66%), done.
    Adding: RCT.EXE      imploding (40%), done.
```

FIGURE 42.20 **GRAFX.ZIP directory entry**

```
Volume in drive D is SUPERSTUFF
Volume Serial Number is 1037-11D2
Directory of D:\GRAFX

.             <DIR>         04-19-94    9:22a
..            <DIR>         04-19-94    9:22a
DA      EXT       1,156 06-22-94    1:11p
GGEN    C        19,093 08-20-94   11:58a
GGEN    EXE      98,402 06-22-94    1:11p
GO      BAT           9 08-20-94    8:13a
OBJS    SFB     150,758 06-22-94    1:11p
PACMAP  DAT     317,049 06-22-94    1:11p
RCT     EXE      85,566 10-16-94    4:00a
GRAFX   ZIP     368,971 11-22-94    9:36a
       10 file(s)     1,041,004 bytes
                    159,793,152 bytes free
```

To get the original files back, use the following commands:

```
C> CD \GRAFX
C> PKUNZIP A:GRAFX
```

(assuming that the GRAFX.ZIP file was saved on drive A:).

WINZIP

The WINZIP utility (shareware) is designed with an easy-to-use graphical interface that zips any file dragged and dropped into its window. You may also navigate through the file system, selecting files to add to the ZIP archive. Figure 42.21 shows the initial WINZIP screen. To create a new ZIP archive, left-click on the New button. You will then have to select a destination folder and drive, and enter a file name for the archive.

After doing this, you are then able to add files to the archive, compressing them as they are added. Files can be removed from the archive without having to start over. Figure 42.22

FIGURE 42.21 Initial WINZIP screen (classical interface)

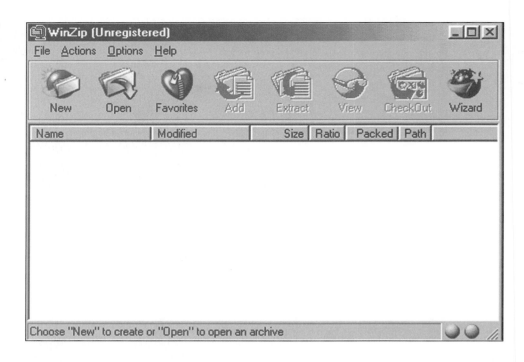

FIGURE 42.22 Adding files to the archive

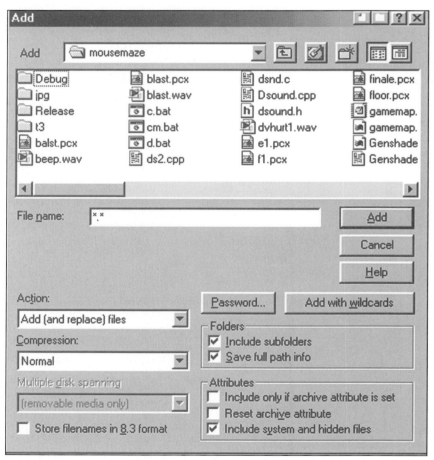

shows WINZIP's Add window, which indicates that files in the MOUSEMAZE folder are being considered for the NETMAZE.ZIP archive. Note the many different options available, such as file name format, what type of files to include, and how subfolders should be handled. With the current options, left-clicking the Add button will cause all files in the MOUSE-MAZE folder *and all of its subfolders* to be archived. Figure 42.23 shows the results.

FIGURE 42.23 Compressed files in the new archive

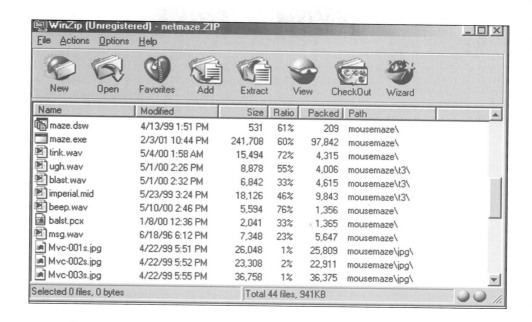

It is interesting to examine the compression results for each file. In addition, it is significant to note that the MOUSEMAZE folder and subfolders contain 941K bytes of information, and the resulting NETMAZE.ZIP archive produced by WINZIP is only 693K.

To extract files from an archive, left-click the Extract button and select the destination folder and files to extract.

WINZIP handles multiple archive types, including Internet file archive types TAR and GZIP. Visit www.winzip.com for more information about WINZIP, or to download a newer version than the one included on the companion CD.

PREPARATION OF A NEW HARD DISK

To prepare a new hard disk for storing programs, you must go through a two-step process:

1. Hard disk partitioning
2. High-level formatting

Once you have performed these two steps, the hard drive is ready to store programs that will be used by the system in which it is installed.

One process has already been performed on the hard drive at the factory. It is called low-level formatting. A *low-level format* causes all the tracks and sectors to be developed on the hard drive. This is the process that causes the outline of these tracks to be written.

Part of the process of low-level formatting is called *defect mapping*. Defect mapping is the process of letting the low-level formatting software know where the defects on the hard drive are located. All new hard drives will have the locations of these defects printed on a label on the drive itself and/or in the accompanying instruction manual. This information is recorded during the low-level formatting process in order to have the formatting program mark these sections so that nothing ever gets written to them.

The first step in preparing a hard disk is to partition the disk. Partitioning is accomplished with the FDISK command. The FDISK command has the following format:

FDISK

The FDISK command configures a hard disk for use with DOS or Windows. Through the use of this command, you can do any of the following:

• Make a primary DOS partition.
• Make an extended DOS partition.

- Change the active partition.
- Show the partition data.
- If the system has more than one hard disk, select the hard disk you want.

If you reconfigure your hard disk using FDISK, you will destroy all existing files. You must make sure that you have backup files of all the hard disk files before you create a partition with this command.

When using FDISK, you must make a primary DOS partition before you can create an extended DOS partition. With DOS 4.0 and later, you will need only one DOS partition, the primary one. For making a primary DOS partition, choose the first selection on the menu; the screen display will then be as shown in Figure 42.24.

If you use the default settings on the display, you will be able to create a primary DOS partition of up to 2GB on the hard drive. After the partitioning is complete, the message shown in Figure 42.25 will be displayed.

After completing the instructions shown on the screen, you must then format the hard drive. Usually you will want to start your system from the hard drive. If so, use the /S option with the FORMAT command. The /S stands for "system" and causes FORMAT to transfer a copy of the boot-up files to the partition after formatting is complete. Therefore, from drive A: (with your DOS system disk in that drive), enter:

```
A> FORMAT C: /S
```

and the high-level formatting process will be executed.

On models 386 and higher, the BIOS setup program must be executed so that new drive parameters can be specified. For example, if an older 406MB hard drive has been replaced with a newer 2.5GB hard drive, the drive parameters specifying the number of heads, cylinders, and sectors per track will surely be different. For example, a new 2.5GB hard drive uses these numbers (which are referred to as the *disk geometry*):

Cylinders	656
Heads	128
Sectors	63

where $656 \times 128 \times 63 \times 512$ bytes per sector equals 2,708,471,808 bytes, or 2.583GB. All three parameters must be saved by BIOS so that the system knows how to access the new hard drive correctly.

After all this, the hard drive is ready to receive programs. Remember that to boot from the hard drive you must make sure that it has the COMMAND.COM file as well as the proper CONFIG.SYS file and, usually, an AUTOEXEC.BAT file.

FIGURE 42.24 Screen display for primary DOS partition

```
Create DOS Partition

Current fixed disk drive: 1

   1. Create Primary DOS Partition
   2. Create Extended DOS Partition
   3. Create Logical DOS Drive(s) in
      the Extended DOS Partition

Enter choice: [1]

Press ESC to return to Fdisk Options
```

FIGURE 42.25 Screen display after partitioning

```
Create DOS Partition

Current fixed disk drive: 1

Do you wish to use the maximum size
for a DOS partition and make the DOS
partition active (Y/N).........?  [Y]

Press ESC to return to Fdisk Options
```

Many problems associated with hard drive backups are usually a result of the connection between the computer and the tape drive. The parallel port is used to communicate with tape devices (because the data can be transferred faster using parallel data lines). The parallel port in newer computers can be configured in BIOS in different ways, such as standard or EPP. If the setting is incorrect, the tape device will not work properly. It may be necessary to modify the BIOS setting to change the parallel port from EPP to standard, or vice versa. Many times, the documentation provided by the manufacturer can provide the answer to this problem as well as many other types of common problems.

Other problems and questions arise when the C: drive has been replaced. In this case, assuming we have a *good* backup, it is necessary to reinstall the operating system on the new hard disk before any tape backup files can be restored. After all the files have been restored, a reboot is necessary to complete the process.

SUMMARY

In this exercise we discovered that

- Disk backups are important and should be performed using regular schedules (incremental and full).
- New hard drives must be partitioned and formatted before they can be used.
- PKZIP, PKUNZIP, and WINZIP are utilities designed for compressing and uncompressing files.

SELF-TEST

This self-test is designed to help you check your understanding of the background information presented in this exercise.

True/False

Answer *true* or *false*.

1. The Windows built-in backup program works with all types of tape devices.
2. A full disk backup using Windows will automatically include all of the Registry files.
3. You need to use only the FORMAT command on a hard disk before placing programs on it.
4. The standard method used to back up a hard drive is with the COPY or XCOPY command.
5. In the standard course of normal computer operations, you should periodically back up the programs on a hard disk.
6. Defect mapping locates problems with a disk at the factory.
7. The compression algorithms used by PKZIP and WINZIP are lossless.

Multiple Choice

Select the best answer.

8. You should back up a hard drive
   a. Just before shipping the computer unit.
   b. Before reformatting the hard drive.
   c. As a periodic routine of normal computer use.
   d. All of the above.
9. Before you can start the backup process with a hard drive, you must
   a. Have a set of formatted disks on which to store the information.
   b. Make sure there are no subdirectories, because these will not be backed up.
   c. Create a record of those files to be backed up, so you will be able to restore them later.
   d. None of the above.
10. In DOS 4.0 the BACKUP command will
    a. Copy the two hidden DOS files IO.SYS and MSDOS.SYS.
    b. Not copy the two hidden files IO.SYS and MSDOS.SYS.
    c. Work with the RESTORE command for DOS 3.2 and lower.
    d. None of the above.

11. A recommended backup schedule includes
    a. A complete backup of the entire system each day.
    b. An initial complete backup of the entire system, followed by daily incremental backups.
    c. Incremental backups of the hard disk followed by a complete backup.
    d. Requirements that all files be backed up monthly.
12. When modifying the Windows Backup settings
    a. There is absolutely no risk of losing information.
    b. Extreme caution must be exercised to prevent accidental loss of data.
    c. Windows will automatically prevent accidental loss of data.
    d. A special password must be used to prevent accidental loss of data.
13. The /S in FORMAT /S stands for
    a. Shutdown.
    b. System.
    c. Supervisor.
14. Low-level formatting is performed
    a. At the factory.
    b. Every time the drive is partitioned.
    c. At boot time.

Matching

Match an item on the right with each phrase on the left.

15. Low-level format
16. Partition a disk
17. High-level format

a. Performed at the factory
b. FORMAT /S
c. RESTORE
d. FDISK

Completion

Fill in the blanks with the best answers.

18. It is necessary to create a(n) _____ on each hard disk before the drive can be formatted.
19. The disk geometry settings for each hard drive are stored in the system _____ settings.
20. The DOS RESTORE command does not restore the _____ files.
21. When preparing a newly installed hard drive, the first software preparation is creating a _____.
22. All data on a hard disk is _____ when the disk is formatted.
23. The parameters of a hard drive (cylinders, heads, etc.) are referred to as its _____.
24. FDISK is used to create a _____ on a hard disk.

Your system may have an A: and a C: drive or an A: and a B: drive. To do this activity, you will use the A: drive as the target drive (the drive that will contain the backed-up files) and the other drive as the source drive (the drive that contains the files to be backed up).

1. Make sure both disks (source and target) are formatted.
2. Change the volume name of the source disk to SOURCE and the volume name of the target disk to TARGET.
3. Clearly label both disks: the source disk as SOURCE and the target disk as TARGET.
4. On the SOURCE disk, create the directory and file structure shown in Figure 42.26.
5. Place the target disk in drive A:. Place the source disk in drive B: or make drive C: the active drive.
6. Make the MYSTUFF directory the active directory on the source disk.

FIGURE 42.26 Directory and
file structure for SOURCE disk

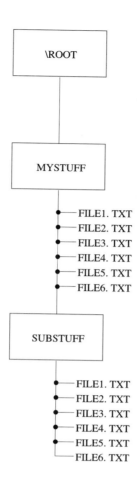

7. You will now back up all the files and subdirectories on the source disk to the target
   disk. From the source drive (with **MYSTUFF** as the active directory), enter

   ```
   B:> BACKUP B: A: /L/S
   ```

8. After the backup process is completed, look at the directory of the target disk and
   record what you see (include the volume name as well as the names of the files).
9. Try to erase the files on the target disk by using the DOS ERASE command. Record the
   message you get on the screen when you try this.
10. Use the DOS ATTRIB command to check the attributes of the files on the target disk.
    Record what you observe.
11. Go to the root directory of the source disk. Observe the new file there called
    BACKUP.LOG. Using the DOS TYPE command, observe and record the contents of
    this file.
12. Repeat this process using PKZIP and PKUNZIP, and/or WINZIP.

QUESTIONS/ACTIVITIES

1. What message appeared on the screen just before the backup process started when you
   first backed up the source disk with the BACKUP B: A: /L/S? Explain why this message was there.
2. Did the volume name of the target disk change after the backup process? Why do you
   think this happened?
3. If the backup process requires more than one target disk (because of the number and
   size of the files to be backed up), what do you think will be the volume name of the
   second target disk? What do you think will be the file names on the second target disk?
4. What are the attributes of a backed-up file created by the DOS BACKUP process?

5. In what directory is the BACKUP.LOG file placed? What are the contents of this file?
6. What happens if you try to PKZIP a ZIP file? Does it compress even more?

Under the supervision of your instructor

1. Back up and restore a set of files from one disk to another and read the backup log created in the process.
2. Use PKZIP and PKUNZIP (or WINZIP) to back up and restore a compressed directory of files.
3. Back up and restore a set of files from disk to tape and tape to disk.

43

Hard Disk Replacement and File Recovery

INTRODUCTION

Joe Tekk examined the old computer system sitting on his desk. A client had brought it in, requesting more hard drive space. The system contained an 850MB hard drive, which was full and already compressed. Joe grabbed a spare 2GB hard drive and connected it as a slave. He powered up the system and ran the BIOS setup program to update the drive information. After the system booted up, the new 2GB drive was available as drive D:. Joe called the customer to say the computer was ready.

Later, after thinking about the installation, Joe recalled the first time he replaced a hard drive. Everything went wrong. The system would not boot with the new drive installed, and eventually a number of files were corrupted on the original drive. After much trial and error, Joe was able to get both drives operating, having learned a great deal about hard drives in the process. Each new installation teaches me something else, *Joe thought to himself.*

PERFORMANCE OBJECTIVES

Upon completion of this exercise, you will be able to do one or all of the following, depending on your lab situation:

1. Remove and replace a hard drive, making sure the files are backed up.
2. Install a new hard drive, partitioned and formatted.
3. Use the DOS RECOVER command on a single file and on a directory of files. Show the contents of one of the recovered directory files and rename it with a descriptive name.
4. Show how to recover a file from the Recycle Bin.

KEY TERMS

IDE hard drive
SCSI hard drive

RECOVER

BACKGROUND INFORMATION

In Exercises 41 and 42, you had the opportunity to learn about hard disk fundamentals and backup. In this exercise, you are shown how to replace a hard disk. You also have the opportunity to learn how to recover lost or damaged files using the DOS RECOVER command

and Recycle Bin. Both of these skills are important for the repair and maintenance of micro-computer systems.

In all the following examples of a hard drive replacement, it is assumed that you have already removed the system cover. In all cases, observe electrical safety precautions, which include having the power cord disconnected from the system as well as from any electrical outlet. It is also assumed that you have completely backed up the hard drive before practicing any removal and replacement procedures. See Exercise 42 for hard drive backup procedures if you need a review.

Most hard drives are mounted in their metal cage by two or more screws on each side of the drive. It is not difficult to physically remove or install the hard drive, but a great deal of preparation is required to ensure the drive is replaced/installed properly.

PHYSICAL CONSIDERATIONS

Several factors must be considered when replacing a hard drive (or adding a new one). Is the drive an old 5.25-inch unit that we want to replace with a new 3.5-inch drive? Is a spare power connector available, along with room in the hard drive bay? Is the power supply powerful enough to support another piece of hardware? Does the motherboard have inte-grated-drive electronics built-in? How much money do you want to spend? Are you willing to mix one manufacturer's hard drive with another's, such as a Maxtor drive with a Connor, Seagate, or Western Digital? Does your BIOS support large hard drives (LBA performed by BIOS)? If not, do you have memory to spare for the dynamic drive overlay software that loads at boot time to control the drive?

Every user will have his or her own reasons for replacing or adding a hard drive. Let's examine some common scenarios that are typical of how systems are upgraded. Keep in mind that hard drive installations generally encounter some kind of glitch that adds more time to the job than originally planned. A wise person will set aside additional time up front, just in case.

Scenario #1: Adding a Second IDE Hard Drive (as a Slave)

Perhaps the easiest thing to do when you've run out of disk space is to buy a new drive to use as a second hard drive, a slave to the original hard drive, which is the master. An IDE hard drive contains its own onboard controller, which is disabled for slave operation. The IDE controller on the master drive controls both drives. This is illustrated in Figure 43.1.

Adding a second drive as a slave involves changing the jumpers on the hard drive to in-dicate slave operation. Figure 43.2 shows the master/slave jumper settings for a Maxtor IDE hard drive.

The manufacturer provides technical details on the hard drive, its jumper settings, log-ical parameters (such as number of cylinders and sectors), and operational characteristics. Looking through the technical details is time well spent. You can learn useful information about BIOS, DOS, Windows, and hard drive technology from the installation directions that come with the hard drive.

To add a new, second hard drive on a machine with a BIOS that supports LBA, do the following:

1. Set the jumpers on the new drive for slave operation.
2. Connect the IDE ribbon cable to the new drive and attach power. The colored stripe on the ribbon cable indicates which side of the IDE connector contains pin 1.
3. Turn on power and enter the system BIOS setup program.
4. Use the hard drive autodetect feature to find the new drive and load its parameters. Typically the BIOS will recommend the best mode of operation (such as LBA and Normal) for the drive.
5. Save the new settings and boot the machine.
6. Use FDISK to create a partition on the new drive.
7. Reboot and format the new partition.

Note that the drive parameters can also be entered by hand, skipping autodetect.

Be aware that adding any type of new drive may cause drive letters to change.

FIGURE 43.1 Adding a second IDE disk drive

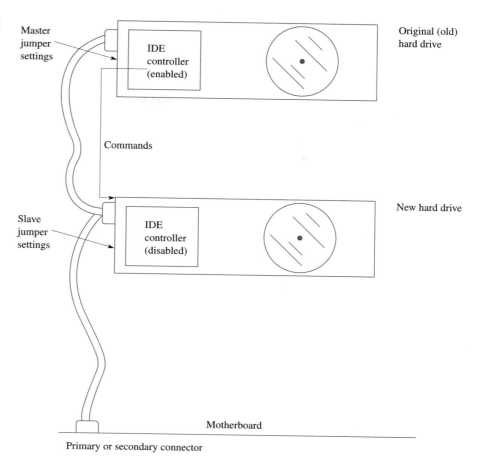

Master jumper settings

IDE controller (enabled)

Original (old) hard drive

Commands

Slave jumper settings

IDE controller (disabled)

New hard drive

Motherboard

Primary or secondary connector

FIGURE 43.2 Disk drive jumper settings

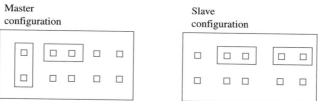

Master configuration

Slave configuration

Scenario #2: Adding a Second IDE Hard Drive (as a Master)

In this scenario, the new drive becomes the master, with the old drive changing from master to slave. This will require the new drive to be formatted as a system disk, so that the operating system can boot from it at power-up.

If the old master drive contains Windows, you will have to transfer all associated files to the new drive (using XCOPY or PartitionMagic's copying feature). Alternately, you could install Windows on the new drive and leave everything else on the old drive. This option may require you to edit various setup parameters, since the old drive will be assigned a new drive letter when it is changed to slave operation. At the worst, you may need to reinstall other applications to repair the drive letter change.

Assuming that we will reinstall the operating system, here is what we need to do:

1. Set the jumpers on the new drive for master operation.
2. Connect the IDE ribbon cable to the new drive and attach power.
3. Change the old hard drive to slave mode.
4. Turn on power and enter the system BIOS setup program.
5. Use the hard drive autodetect feature to determine the parameters for each drive.
6. Save the new settings and boot the machine.
7. Use FDISK to create a partition on the new drive.

8. Insert the Windows CD-ROM and follow the installation procedure.
9. When the installation is complete, check all applications on the old drive (which should now be drive D:) to see which ones require path and/or setting changes.
10. Delete the old operating system from the old drive.

Windows NT is able to boot from drives other than C: (such as the E partition on a multiple-partition hard drive), which adds a small amount of flexibility to the installation.

Scenario #3: Replacing an Old IDE Drive

It is often the case with old hard drives that their capacity is so small compared with the new drive (200MB versus 5GB) that we do not bother to keep the old drive. This brings up an interesting problem: How can we transfer an exact copy of the old drive's data to the new drive? One way to do this would involve the following:

1. Add the second drive as a slave (as in Scenario #1).
2. Format the new drive with the /S option to transfer the operating system to it. The operating system on the new drive will match the operating system on the old drive.
3. Use the XCOPY command to copy all files from the old drive (the master) to the new drive. Try XCOPY *.* /S D:.
4. Shut the system down and remove the old drive.
5. Change the jumpers on the new drive to make it a master.
6. Power-up and set the drive parameters in the BIOS setup program.
7. Save the BIOS parameters and boot the system.

One problem with the XCOPY command is that it does not copy hidden files, so this method is not preferred. Instead, you could simply load the drive with a copy of a recent backup made of the old drive. Or, better yet, use PartitionMagic to make a *clone*, an exact copy of the old drive's partition, and store it on the new drive.

Scenario #4: Adding an SCSI Drive

When adding an SCSI drive to a PC, you also add an SCSI controller card. The SCSI controller card can be plugged into a PCI slot. The SCSI controller card can then be connected to the SCSI disk, using either the internal or external connectors available on the controller card. Plug-and-play SCSI adapters are supported by Windows.

SCSI buses can support multiple devices and each device must be assigned a unique address on the bus (called a *device ID*). This is done automatically during the installation process provided by the manufacturer.

Internal drives can be mounted into any suitable location.

THE DOS RECOVER COMMAND

There are times when your disk may develop a bad sector or two. If these bad sectors are part of the contents of a file, you will lose the parts of the file that were contained on those now-bad sectors.

DOS provides a command to help restore the file. The DOS RECOVER command works by reading the file to another file and omitting the part of the file that contains the bad sector. This means that the part of the file that had the bad sector is still lost, but, if this is a text file, you can use a text editor to reinsert the lost text. If, on the other hand, this is an .EXE or .COM file, it probably will not be usable because of the data lost from the bad sector.

It should be noted that this DOS command does not recover erased files. In MS-DOS 5.0 and later, a new command called UNDELETE can be used to attempt to restore a deleted program from a disk (provided the sectors that stored this program have not been written over by another program).

The DOS command for attempting to recover a damaged file is

RECOVER[drive:][path]filename

where   drive = the drive that contains the file(s) to be recovered
        path = the path that contains the file(s) to be recovered
   filename = the name of the file (wild cards may be used) to be recovered. (Using wild cards will recover only the first matching file, not the rest.)

To use the RECOVER command to recover a file called STORY.TXT on drive A: in the root directory, you would enter

```
A> RECOVER STORY.TXT
```

The recovered file now appears in the root directory. You will find that all recovered files are *always placed in the root directory.*

If you need to recover a group of files in a directory, such as a directory called MYSTUFF, all the recovered files are placed in the root directory and named as follows:

FILE000N.REC

For example, if the files to be recovered in the MYSTUFF directory are

```
TEXT1.TXT
TEXT2.TXT
TEXT3.TXT
TEXT4.TXT
```

and you enter the DOS RECOVER command

```
A> RECOVER \MYSTUFF
```

after the recovery process, you will find the following files in your root directory:

```
FILE0001.REC
FILE0002.REC
FILE0003.REC
FILE0004.REC
```

You can also attempt to recover a whole disk of files on the A: drive, for example, by entering

```
A> RECOVER A:
```

This will place *all* the recovered files in all the directories on the disk into the root directory. Recall that the root directory can hold only a certain number of files. Because of this, you should attempt to recover only one file at a time, use a text editor to replace lost data, rename the file by a descriptive name, and place the file in a subdirectory (say, a directory called \RECOVER).

You can use the DOS CHKDSK command to check a file for bad sectors. If any are encountered, use the DOS RECOVER command.

The UNDELETE command provides a list of files it will attempt to undelete. You must supply the first letter of the undeleted file.

THE RECYCLE BIN

Recall from Exercise 12 that files deleted while inside Windows are sent to the Recycle Bin. They are not deleted from the hard drive until the Recycle Bin is emptied. Figure 43.3 shows the contents of the Recycle Bin after several files were deleted. Any or all of the files may be undeleted (recovered) using the Restore option from the File menu. You must select the files to recover.

Note that files deleted inside DOS are not sent to the Recycle Bin.

FIGURE 43.3   Contents of the Recycle Bin

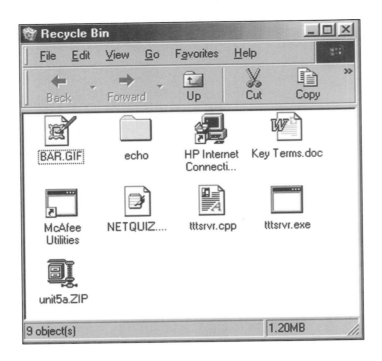

One of the most frightening error messages you may encounter reads like this:

Missing operating system

That simple message says it all. This error is an indication that the system cannot find a bootable partition when powered on.

If you have a start-up disk, boot the system with it and use FDISK to examine the partition information for the hard drive. It is possible that the partition exists and contains valid information (including the missing operating system), but has not been made *active*. FDISK can be used to set the partition active, as can any other disk utility, such as PartitionMagic. As shown in Figure 43.4, PartitionMagic clearly indicates the type and size of each partition and the active partition. Always use FDISK or an equivalent utility to check for the existence of a partition before taking drastic measures, such as starting over from scratch.

FIGURE 43.4   PartitionMagic drive information

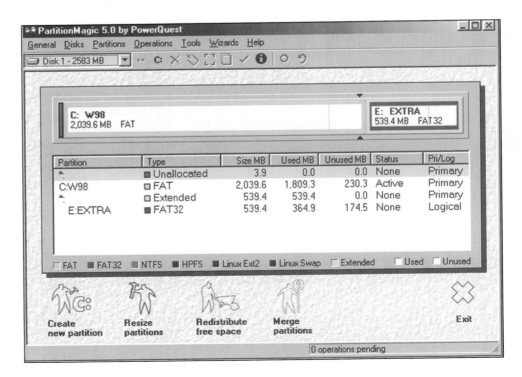

SUMMARY

In this exercise we discovered that

- There are many scenarios when replacing a hard drive.
- It is important to have backup copies of your files.
- DOS files may possibly be recovered using the RECOVER command.
- Deleted Windows files may be restored from the Recycle Bin.

SELF-TEST

This self-test is designed to help you check your understanding of the background information presented in this exercise.

True/False

Answer the following questions *true* or *false*.

1. If you lose any files on a hard drive as a result of its removal and replacement, you need only use the DOS RECOVER command to get them back.
2. Because of the DOS RECOVER command, it is no longer necessary to waste time backing up files on a hard drive.
3. A hard drive may have multiple partitions.
4. When a new hard drive is added to a system, it must be the master.
5. A maximum of two IDE hard drives may be installed on a system.
6. Files may be recovered from the Recycle Bin.
7. Hard drive space is recovered as soon as a file is sent to the Recycle Bin.

Multiple Choice

Select the best answer.

8. When replacing the hard drive, you must make sure that the
   a. Drive is never dropped or mishandled.
   b. Mounting screws of the correct size, thread, and number are used.
   c. System power cable is unplugged.
   d. All of the above.
9. The odd-colored stripe on the hard drive ribbon cable
   a. Represents the ground connection.
   b. Must go to the last pin of the connector.
   c. Means "do not remove."
   d. Represents pin 1 of the connector.
10. SCSI hard drives can be added to
    a. Any system containing an SCSI controller.
    b. Only systems that have no IDE hard drives.
    c. Only systems that do have IDE hard drives.
11. The onboard IDE controller is _____ during slave operation.
    a. Disabled.
    b. Enabled.
    c. There is no onboard controller.
12. The BIOS _____ feature may be used to identify hard drives.
    a. Autodetect.
    b. Drive-scan.
    c. Drive-detect.

Matching

Match a statement on the right with each DOS command on the left.

13. RECOVER MYFILE
14. RECOVER A:
15. RECOVER \MYSTUFF
16. UNDELETE

a. Recovers only one file.
b. Recovers a directory of files.
c. Recovers a whole disk of files.
d. Recovers every file ever deleted.
e. May recover recently deleted files.

Completion

Fill in the blanks with the best answers.

17. The DOS RECOVER command rewrites all the files except for the part contained in any bad _____.
18. Using the DOS RECOVER command for an entire disk may overload the _____ directory.
19. Recovered files from a directory are rewritten in the form of _____ , where N is the file number.
20. Device IDs are used with _____ hard drives, printers, scanners, and other devices.
21. IDE hard drives are jumpered as master or _____.
22. One _____ on a hard drive must be active.

FAMILIARIZATION ACTIVITY

The familiarization activity for this exercise is divided into three parts. The first part deals with the removal and replacement of a hard disk drive. The second part deals with the DOS RECOVER command. Depending on your lab situation, you may not have access to a computer that can have its hard drive removed. Instead, the hard drive removal and replacement activity may be given as a classroom demonstration. The third part of the activity covers the use of the Windows Recycle Bin feature.

Hard Drive Removal and Replacement

Using the procedure appropriate to your computer system, remove and replace its hard drive system. Make sure you use all the proper safety procedures during this process. This procedure may involve partitioning, formatting, and reinstallation of applications.

The DOS RECOVER Command

Using your floppy disk, create a subdirectory called MYSTUFF and place five text files inside it. Make one text file in the root directory.

1. Use the RECOVER command to recover the file in the root directory. Explain how you did this and the results you got.
2. Next, use the RECOVER command to recover all the text files in the MYSTUFF directory. Again, explain how you did this.
3. What new files do you now have in the root directory? List them below.

4. Are the contents of these files the same as those of the original files? How did you determine this?

The Recycle Bin

Create a file using a word processor (EDIT or Microsoft Word). Save it somewhere on drive C:. Use Windows Explorer to locate the file and delete it. Then open the Recycle Bin.

1. Is your file in the Recycle Bin?
2. Are other files in the Recycle Bin?
3. Does the hard drive free space change when you empty the Recycle Bin? To determine this, use the DIR command from a DOS shell on the C: drive both before and after the file is emptied from the Recycle Bin.

1. Who was the manufacturer of the hard drive that you removed and replaced in the lab?
2. Was it necessary to back up the files on the hard drive before removing them? Explain.
3. Which scenario covered in this exercise comes closest to your lab situation?
4. Where are the files recovered using the DOS RECOVER command stored?
5. Why should you be cautious about using the DOS RECOVER command to recover a whole disk of files at the same time?

Do one or all of the following, depending on your lab situation:

1. Remove and replace a hard drive system, making sure the files are backed up.
2. Install a new hard drive, partitioned and formatted.
3. Use the DOS RECOVER command on a single file and on a directory of files. Show the contents of one of the recovered directory files and rename it with a descriptive name.
4. Show how to recover a file from the Recycle Bin.

44

Video Monitors and Video Adapters

Joe Tekk had been using his new office PC for some time now. He was happy with it, but he was very curious about the new technology being marketed touting the Pentium 4 and AGP graphics. He decided to buy one of the new PCs for his home. He thought it might be delivered today since he chose the rush delivery option. All he could think about was the new high-speed Pentium 4 processor, huge hard drive, AGP video card with 8MB of video memory, and a 17-inch monitor.

It was just after noon when he got a phone call from Donna, the secretary. "Two big boxes were just delivered with your name on them, Joe. Can you please pick them up?" she asked. Joe quickly got the equipment cart and rushed to the office.

Within a few minutes of getting home, Joe had all of the boxes unpacked. First, he removed the system cover to examine the hardware components. Both the processor and the video card were of great interest to Joe since he had spent a lot of time researching exactly what to buy. Joe satisfied his curiosity and replaced the system cover. He was eager to see the computer in action.

Joe booted the system and began to run some of the demonstration programs that came preinstalled. While he viewed a video, he noticed immediately the high-quality output of the display. The processor and AGP graphics card displayed the full motion video file without any delay or jitter as he had seen on many other computers. For a moment, Joe forgot he was looking at a PC display.

Upon completion of this exercise, you will be able to

1. Identify the type of monitor and display adapter being used by a given computer system and ensure that they have been properly installed and adjusted.
2. Discuss the differences between CRT monitors and flat panel displays.

Video monitor
Video adapter card
Additive color mixing
Pixel
Aspect ratio

Dot pitch
Text mode
Graphics mode
Extended character set
Alphanumeric mode

Video graphics array   Thin-film transistor (TFT)
Multisync monitor   VGA adapter
Flat panel display   SVGA adapter
Liquid crystal display (LCD) technology   Graphics accelerator
Dual-scan twisted nematic ((DSTN)   AGP adapter

OVERVIEW

The computer display system used by your computer consists of two separate but essential parts: the *video monitor* and the *video adapter card* as shown in Figure 44.1. Note from the figure that the monitor does not get its power from the computer; it has a separate power cord and its own internal power supply.

The video adapter card [Figure 44.1(b)] interfaces between the motherboard and the monitor. This card processes and converts data from the computer and allows you to see all the things you are used to seeing displayed on the screen.

FIGURE 44.1 The two essential parts of a computer display system   (a) Video adapter card with companion monitor and (b) SVGA graphics accelerator card *(photograph by John T. Butchko)*

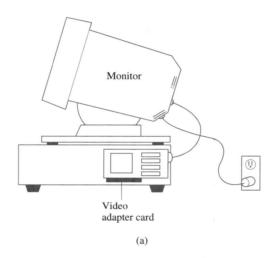

(a)

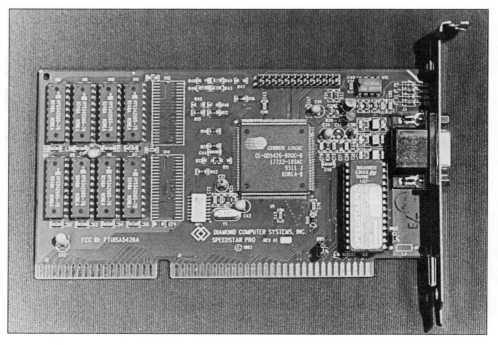

(b)

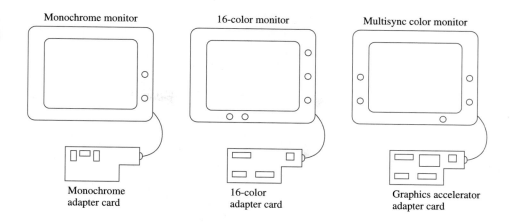

FIGURE 44.2   Necessity of each computer monitor having its own matching adapter

Monochrome monitor

16-color monitor

Multisync color monitor

Monochrome adapter card

16-color adapter card

Graphics accelerator adapter card

It is very important to realize that there are many different types of monitors and that each type of monitor essentially requires its own special video adapter card, as shown in Figure 44.2. Connecting a monitor to an adapter card not made for it can severely damage the monitor or adapter card, or both.

MONITOR SERVICING

Very seldom is the computer user expected to repair a computer monitor. Computer monitors are very complex devices that require specialized training to repair. These instruments contain very high and dangerous voltages that are present even when no power is being applied. The servicing of the monitor itself is, therefore, better left to those who are trained in this specialty.

What you need to know is what kinds of monitors are available, their differences, and how they interface with the computer. Then you need to know enough about hardware and software in order to tell if a problem that appears on the monitor is in the monitor itself, its adapter card, the computer, or the monitor cabling—or is simply a result of the customer's lack of understanding about how to operate the computer.

MONITOR FUNDAMENTALS

All monitors have the basic sections shown in Figure 44.3. Table 44.1 lists the purpose of each of the major sections of a computer monitor.

FIGURE 44.3   Major sections of a computer monitor

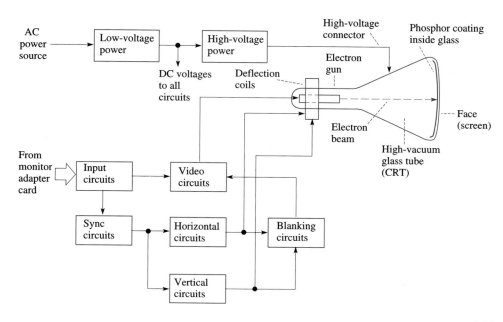

TABLE 44.1 Major sections of a computer monitor

Section	Purpose
Glass CRT	The cathode-ray tube (CRT) creates the image on the screen. It is so named because the source of electrons is called the cathode, and the resulting stream of electrons is called its rays (cathode rays).
Electron gun	Generates a fine stream of electrons that are attracted toward the glass face of the CRT (the screen) by the large positive voltage applied there.
Phosphor coating	A special kind of material that emits light when struck by an electron beam.
High-voltage power source	Supplies the large positive voltage required by the CRT to attract the electrons from the electron gun.
Deflection coils	Generate strong magnetic fields that move the electron beam across the face of the CRT.
Horizontal circuits	Generate waveforms applied to the deflection coils, causing the electron beam to sweep horizontally across the face of the CRT from left to right.
Vertical circuits	Generate waveforms applied to the deflection coils, causing the horizontal sweep of the electron beam to move vertically across the face of the CRT from top to bottom and creating a series of horizontal lines.
Blanking circuits	Cause the electron beam to be cut off from going to the face of the CRT so that it isn't seen when the electron beam is retracing from right to left or from bottom to top. (This is similar to what you do when writing. You lift your pen from the surface of the paper after you finish a line and return to the left side of the paper to begin a new line just below it.)
Video circuits	Control the intensity of the electron beam that results in the development of images on the screen. The intensity of the beam is varied as the beam is swept from left to right. An entire screen of lines forms the image.
Sync circuits	Electrical circuits that help synchronize the movement of the electron beam across the screen.
Low-voltage power supply	Supplies the operating voltages required by the various circuits inside the monitor.

MONOCHROME AND COLOR MONITORS

One of the differences between a monochrome (single-color) monitor and a color monitor is in the construction of the CRT. The differences are illustrated in Figure 44.4.

As shown in the figure, the color CRT contains a triad of color phosphor dots. Even though this consists of only three color phosphors, all the colors you see on a color monitor are produced by means of these three colors (including white, which is produced by controlling the intensity of the three colors: red is 30%, green is 59%, and blue is 11%). This process, called *additive color mixing*, is illustrated in Figure 44.5.

The other differences between monochrome and color monitors are the circuits inside these systems as well as their adapter cards. Some of these differences are the high voltages in a color monitor that are several times higher than those found in a monochrome monitor.

FIGURE 44.4 Monochrome and color CRTs

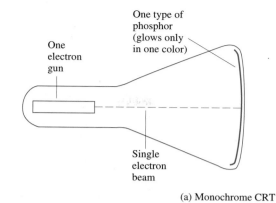

(a) Monochrome CRT

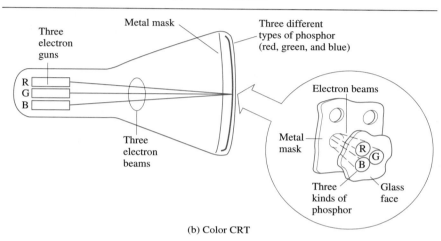

(b) Color CRT

FIGURE 44.5 Additive color mixing

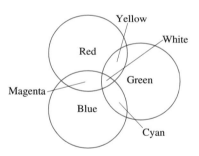

Usually, these voltages are on the order of 30,000V or more. You should note that this high voltage can be stored by the color CRT and still be present even when the set is unplugged from the AC outlet. A special probe is used to discharge the CRT.

ENERGY EFFICIENCY

Energy-efficient PCs are designed with energy efficiency in mind. The system BIOS, monitor video card, and other hardware must support either the Advanced Power Management (APM) or VESA BIOS extensions for power management (VBE/PM) standards. Some computers may support limited power management or energy saving features.

It is estimated by the U.S. Environmental Protection Agency (EPA) that the average office desktop computer or workstation uses around $105 of electrical power annually. When all desktops are considered, the total consumption adds up to around 5% of all electrical energy consumed in the United States. The EPA estimates that by using energy-efficient equipment, as much as $90 a year per computer can be saved.

TABLE 44.2   Major video controls

Control	Purpose
Contrast	A gain control for the circuits that determine the strength of the signal used to place images on the screen. It affects the amount of difference between light and dark.
Brightness	Controls the amount of high voltage applied to the CRT, which controls the strength of the beam. The higher the voltage, the stronger the beam and the brighter the picture.
Vertical size	Controls the output of the vertical circuit, changing the amount of the vertical sweep of the CRT and thus changing the vertical size of the displayed image.
Horizontal size	Controls the output of the horizontal circuit, changing the amount of horizontal sweep of the CRT and thus changing the horizontal size of the displayed image.
Vertical hold	Helps adjust the synchronous circuits so the image is stable in the vertical direction.
Horizontal hold	Helps adjust the synchronous circuits so the image is stable in the horizontal direction.

The EPA has proposed a set of guidelines for energy-efficient use of computers, work stations, monitors, and printers. The EPA *Energy Star* program requires the computer and monitor to use less than 30 watts each when they are not being used (for a total of 60 watts including both the system unit and the monitor). Personal computers adhering to the Energy Star recommendations are also called *green* PCs.

Each computer can be set up to automatically reduce energy usage using the standby and sleep modes. The standby mode is activated after a user-specified period of inactivity. The sleep mode is automatically activated after the standby time has expired. If the computer is used during the standby energy-saving mode, it takes just a short period of time before the monitor is usable. Sleep mode is similar to a power-down of the monitor and requires some additional time before the monitor is usable.

The EPA Web site located at http://www.epa.gov/energy_star maintains a list of all energy-efficient computer products. Look for the Energy Star trademark on product packaging and the marketing materials supplied by most manufacturers.

VIDEO CONTROLS

Table 44.2 lists some of the major video controls and their purposes.

PIXELS AND ASPECT RATIO

Figure 44.6 illustrates two important characteristics of computer monitors. As shown in the figure, a *pixel* (or pel) is the smallest area on the screen whose intensity can be controlled. The more pixels available on the screen, the greater the detail that can be displayed. The number of pixels varies among different types of monitors; the more pixels, the more expensive the monitor. The *aspect ratio* indicates that the face of the CRT is not a perfect square. It is, instead, a rectangle. This is important to remember, especially if you are developing software for drawing squares and circles; you may wind up with rectangles and ellipses. The size of a pixel is referred to as its *dot pitch* and is a function of the number of pixels on a scan line and the distance across the display screen.

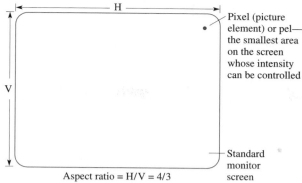

FIGURE 44.6 Pixels and aspect ratio

Pixel (picture element) or pel—the smallest area on the screen whose intensity can be controlled

Standard monitor screen

Aspect ratio = H/V = 4/3

FIGURE 44.7 Text and graphics modes

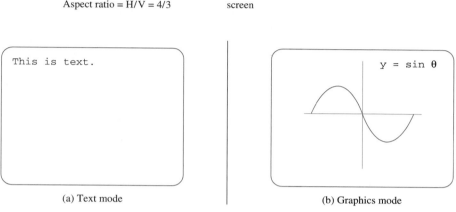

This is text.

$y = \sin \theta$

(a) Text mode

(b) Graphics mode

MONITOR MODES

There are two fundamental modes in which the monitor communicates with the computer: the **text mode** and the **graphics mode**. Figure 44.7 illustrates the difference.

In the text mode, the CRT display gets its information from a built-in ROM chip referred to as the *character ROM*. This may not be a separate ROM chip but part of another, larger one. This ROM contains all the characters on your keyboard, plus many more. This group of characters is known as the **extended character set** and may, among other things, be used in combination to form squares, boxes, and other shapes while your computer is still in the text mode. To get any of these extended characters on the screen (or to get *any* character on the screen), simply hold down the Shift and Alt keys at the same time and then type in the character number. For example, to get the character ö, hold down the Shift and Alt keys and type in 148 on the keypad; when you lift up on the Shift and Alt keys, the character appears (and can also appear on output from the printer, depending on the type of printer).

The advantage of the text mode is that it doesn't take much memory and the visual results are predictable and easy to achieve (you need only to press a key on the keyboard). The size of the text screen is 80×25 or 40×25. The text screen is sometimes referred to as the **alphanumeric mode**.

When the monitor and its circuits are in the graphics mode, an entirely different use of memory is required. RAM is used because a program has complete control over the intensity and (in the case of color) the color of each pixel. The more pixels available on the monitor, the more memory required. In order to display detailed graphics, you must have a big and powerful machine, which means a more expensive system as well as a more expensive monitor.

Just to give you an idea of the memory requirements for graphics, if your monitor has 640 horizontal pixels and 480 vertical pixels, the total number of pixels that must be addressed by RAM is $640 \times 480 = 307{,}200$, which is more than a third of a megabyte for just one screen. If color is not used in the graphics mode, less memory is required (because the computer needs to store less information about each pixel).

TYPES OF MONITORS

In order to understand the differences among the most common types of computer monitors, you must first understand the definitions of the terms used to describe them. Table 44.3 lists the major terms used to distinguish one monitor from another.

Now that you know the definitions of some of the major terms used to distinguish one monitor from another, you can be introduced to the most common types of monitors in use today. Table 44.4 lists the various types of monitors and their distinguishing characteristics.

TABLE 44.3 Computer monitor terminology

Term	Definition
Resolution	The number of pixels available on the monitor. A resolution of 640 × 480 means that there are 640 pixels horizontally and 480 pixels vertically.
Colors	The number of different colors that may be displayed at one time in the graphics mode. For some color monitors, more colors can be displayed in the text mode than in the graphics mode. This is possible because of the reduced memory requirements of the text mode.
Palette	A measure of the full number of colors available on the monitor. However, not all the available palette colors can be displayed at the same time (again, because of memory requirements). You can usually get a large number of colors with low resolution (fewer pixels) or a smaller number of colors (sometimes only one) with much higher resolution—again, because of memory limitations.
Display (digital or analog)	There are basically two different types of monitor displays, *digital* and *analog*. Some of the first computer monitors used poor-quality analog monitors. Then digital monitors, with their better overall display quality, became more popular. Now, however, the trend is back to analog monitors because of the increasing demand for high-quality graphics, in which colors and shades can be varied continuously to give a more realistic appearance.

TABLE 44.4 Common types of computer monitors

Type	Resolution*	Colors	Palette	Display
Monochrome composite	640 × 200	1	1	Analog
Color composite	640 × 200	4	4	Analog
Monochrome display	720 × 350	1	1	Digital
RGB (CGA)	640 × 200	4	16	Digital
EGA	640 × 350	16	64	Digital
PGA	640 × 480	Unlimited	Unlimited	Analog
VGA	640 × 480	256	262,144	Analog
SVGA	1280 × 1024	Varies	Varies	Digital/ analog
Multiscan	Varies	Unlimited	Unlimited	Digital/ analog

*In general, the higher the resolution, the higher the scan frequency. For example, the typical scan frequencies of EGA and VGA monitors are 21.5 KHz and 31.5 KHz, respectively.

Monochrome Composite Monitor

The first computers used *monochrome composite monitors*, which actually were television sets. Because of the way a standard television set works, only 40 characters could be displayed across the face of the CRT. Because of the poor resolution of this type of monitor, it is almost never used with computers today.

Color Composite Monitor

The *color composite monitor* was similar to the monochrome composite monitor except that it could show color. It had even poorer resolution than the monochrome composite monitor and, as a result, is almost never used with computers today.

Monochrome Display Monitor

The *monochrome display monitor* was the standard monitor used by IBM when the PC was first introduced. This monitor, with very good resolution and a green screen, is the single-color (monochrome) monitor you usually see with many older office PC and XT systems.

RGB Monitor

The *red-green-blue (RGB) monitor* is a technologically out-of-date monitor that was one of the first popular color monitors to be used with the IBM PC. On this monitor, which had poor resolution, it was extremely difficult to read either text or any kind of graphics.

EGA Monitor

The *enhanced graphics array (EGA) monitor* is the answer to the RGB monitor. It is a very popular medium-priced color monitor. Its resolution is not quite as good as that of the monochrome display monitor.

PGA Monitor

The *professional graphics array (PGA) monitor*, which has a better resolution than that of the EGA, has not been very popular because of its high price. It is not as popular as the next color monitor, the VGA.

VGA Monitor

The *video graphics array (VGA) monitor* is one of the most popular color monitors; it provides high color resolution at a reasonable price. More and more software with graphics is making use of this type of monitor. The associated cards have a high scanning rate, resulting in less eye fatigue both in text and in graphics modes.

SVGA Monitor

Higher screen resolution and new graphics modes make the *Super VGA (SVGA) monitor* even more popular than the VGA monitor.

Multiscan Monitor

The *multiscan monitor* was one of the first monitors that could be used with a wide variety of monitor adapter cards. Since this type of monitor can accommodate a variety of adapter cards, it is sometimes referred to as the *multidisplay* or *multisync monitor*.

Flat Panel Displays

An alternative to the CRT monitor is the *flat panel display*. This device does not use a CRT to display pixels, but instead controls all the pixels by individually turning them on or off. Instead of an electron beam hitting a phosphor-coated display, the flat panel display uses liquid crystal technology to develop an image. The *liquid crystal display (LCD)* screen is

FIGURE 44.8   CRT and flat
panel displays

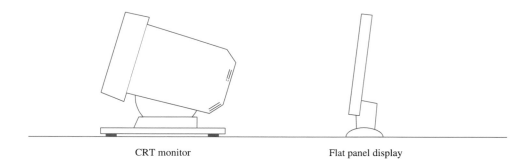

CRT monitor   Flat panel display

composed of pixel cells containing liquid crystals that change orientation when a small voltage is applied, affecting the way light passes through the crystal. Flat panel displays do not require deflection coils, as the pixel cells are controlled electronically. Thus, flat panel displays use less power than CRT monitors, and are physically much smaller and lighter, as indicated in Figure 44.8.

The two major types of LCD technology are ***dual-scan twisted nematic (DSTN)*** and ***thin-film transistor (TFT)***. DSTN displays are also called *passive matrix* displays, and are not as bright as TFT displays and have a smaller viewing angle. TFT displays use transistors to control each pixel cell, and are called *active matrix* displays. TFT displays are thus brighter than DSTN displays, have a larger viewing angle and resolution, and create more accurate colors.

Even though the flat panel display has several advantages over the CRT monitor, there are also some disadvantages. First, the maximum resolution of a flat panel display is fixed. If, for example, the maximum resolution is 800×600, smaller resolutions, such as 640×480, are displayed by using a smaller area of the screen. The size of the pixel cells cannot be changed (as they can in a CRT monitor by adjusting the scanning frequency).

Second, since the pixel cells are controlled individually, a malfunction in a cell causes it to be permanently on or off, slightly reducing the quality of the image.

Third, flat panel displays are much more expensive than CRT monitors. So, you must balance the desire to have a small, lightweight, low-power display with some imperfections against a large, heavy, power-hungry CRT monitor with a different set of imperfections (flicker, focusing).

DISPLAY ADAPTERS

As previously stated, a computer monitor must be compatible with its adapter card. If it is not, damage to the monitor or adapter card, or both, could result.

MDA Adapter

The *MDA* or *monochrome display adapter* is an adapter card that contains both a 9-pin D-shell connector and a parallel printer connector. Figure 44.9 shows the connections for this type of adapter. This is purely a text mode adapter, offering no graphics capabilities. This adapter was used by the most popular green-screen monochrome monitors.

CGA Adapter

The *CGA* or *color graphics array adapter* was one of the first color adapters. It can generate 16 different colors (including black, dark gray, light gray, and white). It has several different graphic modes, with different levels of resolution. This adapter will also operate monochrome displays, but not the IBM monochrome designed for the MDA adapter. This adapter allows for three levels of resolution, the *low-*, *medium-*, and *high-resolution modes*. The low-resolution mode allows all 16 colors to be displayed at the same time, but the resolution is only 200×160 pixels. This results in such a poor image that this mode is not supported by any IBM system. In the medium-resolution mode, there are 200×320 pixels, which allow for only four colors to be displayed at the same time from four possible palettes. The first two palettes are red, green,

FIGURE 44.9   Pin diagram for an MDA adapter

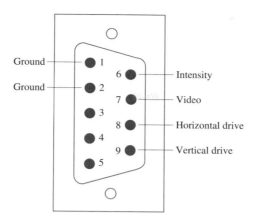

FIGURE 44.10   Pin diagram for CGA adapter

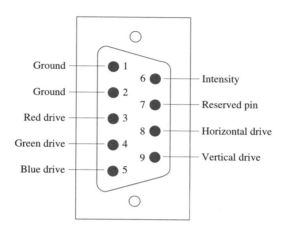

brown, and black; and white, magenta, cyan, and black (or a software-selected background color). The last two palettes are identical to the first two, but with high-intensity colors.

In the high-resolution mode, you can get only one color at a time. However, the trade-off is a much higher resolution: 640×200 pixels. This adapter card comes with its own extra 16K of memory for the displays. The pin diagram for this adapter is shown in Figure 44.10.

MGA Adapter

A small company in Berkeley, California, called Hercules Technology created an adapter card that worked with the IBM monochrome system. This card, called the "Hercules Graphics Card," was sometimes called the *MGA* or *monochrome graphics adapter*. For the first time there was a low-cost, reliable adapter card that could provide graphics capabilities for the IBM monochrome screen. True, there was only one color, but you could now have graphics as well as text. Many other companies emulated the design of this card and the associated software that went with it. It has a resolution of 720×348 pixels, which, with its 64K of onboard memory, allows each pixel to be either ON or OFF.

EGA Adapter

The *EGA* or *enhanced graphics adapter* can operate an RGB, EGA, or multiscan monitor. It was the first color graphics adapter put out by IBM. The pin diagram for this adapter is shown in Figure 44.11.

VGA Adapter

The *VGA (video graphics array card)* was the fastest-growing graphics card in terms of popularity until the SVGA card became available. The VGA adapter card uses a 15-pin high-density pin-out, as shown in Figure 44.12. The VGA 15-pin adapter can be wired to fit the standard 9-pin graphics adapter, as shown in Figure 44.13.

FIGURE 44.11 Pin diagram for EGA adapter

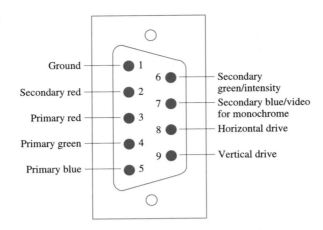

FIGURE 44.12 Pin diagram for VGA adapter

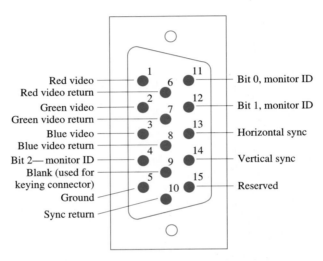

FIGURE 44.13 Nine-pin adapter cable for VGA

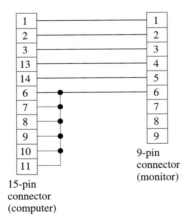

SVGA Adapter

The Super VGA (SVGA) graphics interface uses the same connector that VGA monitors use. However, more display modes are possible with SVGA than with VGA.

8514 Display Adapter

The *8514 display adapter* card provides a resolution of 1024 × 768 pixels and enough memory (about 0.5MB) to allow 16 colors to be used at the same time. This adapter is made for the IBM PS/2 line of computers. In order to take full advantage of this adapter, the IBM 8514 display should also be used.

MCGA Adapter

The *MCGA (multicolor graphics array) adapter* is integrated into the motherboard of some IBM PS/2 model computers. This adapter integration supports the CGA mode if a suitable analog display card is added. It has an available palette of 262,144 colors, with 64 shades of gray available at the same time. It has two graphics modes, 640 × 480 and 320 × 200 pixels.

VESA

The *VESA* (Video Electronics Standards Association) specification has been developed to guide the operation of new video cards and displays beyond VGA. New BIOS software that supports the VESA conventions is contained in an EPROM mounted on the display card. The software also supports the defined VESA video modes. Some of these new modes are 1024 × 768, 1280 × 1024, and 1600 × 1200, with up to 16 million possible colors.

GRAPHICS ACCELERATOR ADAPTERS

A ***graphics accelerator*** is a video adapter containing a microprocessor designed specifically to handle the graphics processing workload. This eliminates the need for the system processor to handle the graphics information, allowing it to process other instructions (nongraphics related) instead.

Aside from the graphics processor, there are other features offered by graphics accelerators. These features include additional video memory, which is reserved for storing graphical representations, and a wide bus capable of moving 64 or 128 bits of data at a time. Video memory is also called VRAM and can be accessed much faster than conventional memory.

Many new multimedia applications require a graphics accelerator to provide the necessary graphics throughput in order to gain realism in multimedia applications. Table 44.5 illustrates the settings available for supporting many different monitor types and refresh rates.

Most graphics accelerators are compatible with the new standards such as Microsoft DirectX, which provides an application programming interface, or API, to the graphics subsystem. Usually, the graphics accelerators are also compatible with OpenGL for the Windows NT environment. Figure 44.14 shows two graphics accelerator display adapters.

AGP ADAPTER

The *Accelerated Graphics Port (AGP)* is a new interface specification developed by Intel. The ***AGP adapter*** is based on the PCI design but uses a special point-to-point channel so that the graphics controller can directly access the system main memory. The AGP channel is 32 bits wide and runs at 66 MHz. This provides a bandwidth of 266 Mbps as opposed to the PCI bandwidth of 133 Mbps.

TABLE 44.5   Common monitor support

Resolution	Colors	Memory	Refresh Rates
640 × 480	256	2MB	60, 72, 75, 85
	65K	2MB	
	16M	2MB	
800 × 600	256	2MB	56, 60, 72, 75, 85
	65K	2MB	
	16M	2MB	
1024 × 768	256	2MB	43 (interlaced), 60, 72, 75, 85
	65K	2MB	
	16M	4MB	
1280 × 1024	256	2MB	43 (interlaced), 60, 75, 85
	65K	4MB	
	16M	4MB	

FIGURE 44.14 (a) PCI graphics accelerator and (b) AGP graphics accelerator

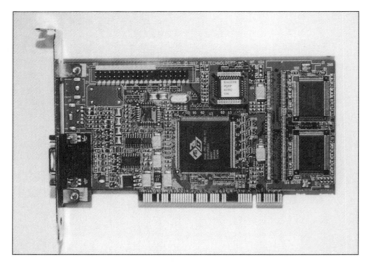

(a)

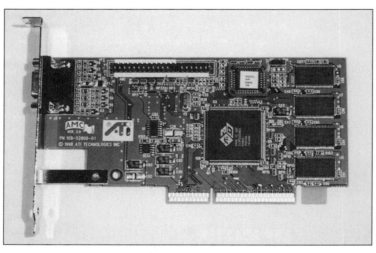

(b)

AGP optionally supports two faster modes, with throughput of 533Mbps and 1.07Gbps. Sending either one (AGP 1X), two (AGP 2X), or four (AGP 4X) data transfers per clock cycle accomplishes these data rates. Table 44.6 shows the different AGP modes. Other optional features include AGP texturing, sideband addressing, and pipelining. Each of these options provides additional performance enhancements.

AGP graphics support is provided by the new NLX motherboards, which also support the Pentium II microprocessor (and above). It allows for the graphic subsystem to work much closer with the processor than previously available by providing new paths for data to flow between the processor, memory, and video memory. Figure 44.15 shows this relationship.

TABLE 44.6 AGP graphics modes

Mode	Throughput (Mbps)	Data Transfers per Cycle
1x	266	1
2x	533	2
4x	1066	4

FIGURE 44.15   AGP
configuration

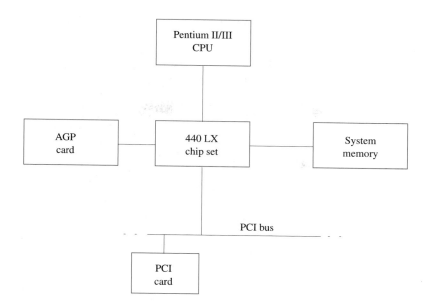

AGP offers many advantages over traditional video adapters. You are encouraged to become familiar with the details of the AGP adapter.

**TROUBLESHOOTING
TECHNIQUES**

Most monitors produced today can be controlled by software. In order to take advantage of this feature, the monitor must be recognized by the operating system. Windows 95/98 will display the monitor's specific information when the Change Display Type button is selected from the Display Properties screen as shown in Figure 44.16.

The Change Display Type window shows the current settings of the display adapter and the monitor type, as illustrated in Figure 44.17. Note the additional check box setting used to inform Windows the monitor is Energy Star compliant. If enabled, the monitor may be

**FIGURE 44.16   Access to
change the display type**

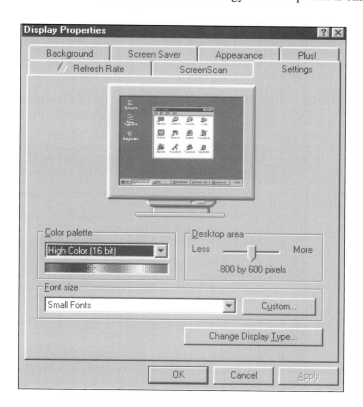

FIGURE 44.17 Setting the
monitor type

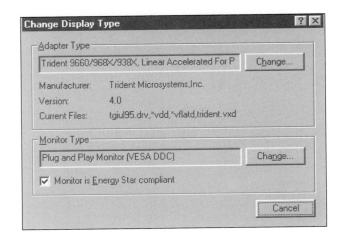

FIGURE 44.17 Setting the
monitor type

shut down during a period of inactivity. Be sure to verify the system, video card, and monitor can support Energy Star features before enabling them. Windows NT provides similar screens to accomplish the same tasks.

SUMMARY

In this exercise we discovered that

- CRT monitors form an image by sweeping an electron beam horizontally and vertically across a phosphor-coated surface.
- There are many types of CRT monitors and their associated display adapters, such as EGA, VGA, and SVGA.
- Graphics accelerators and the AGP adapter are designed to increase graphics performance.
- Flat panel displays use liquid crystals to illuminate pixels.

SELF-TEST

This self-test is designed to help you check your understanding of the background information presented in this exercise.

True/False

Answer *true* or *false*.

1. The standard computer monitor gets its electrical power directly from the computer.
2. A computer monitor requires special circuits such as those found on a video adapter card to process the information from the computer.
3. There are only two types of monitors used with computers: the color monitor and the monochrome monitor.
4. Any color monitor will work with any adapter card.
5. Connecting a monitor to an adapter card not made for it can severely damage the monitor.
6. TFT is a type of CRT monitor.
7. Flat panel displays use less power than CRT monitors.

Multiple Choice

Select the best answer.

8. Computer monitors are
   a. Easy to service.
   b. Normally serviced by computer technicians.
   c. Never in need of servicing.
   d. None of the above.

9. A computer monitor contains
   a. Very high and dangerous voltages.
   b. The same low voltages as does the computer.
   c. Easy-to-service circuits.
   d. Very few circuits, because most of the circuits are contained in the computer itself.
10. The high voltage in a computer monitor
    a. Is not dangerous because it lasts for only a short time.
    b. May still be present even when its power is disconnected.
    c. Is automatically discharged through the computer when the power is turned off.
    d. Exists only in the TV-type monitor, not the computer-type monitor.
11. The actual glass tube used in the computer monitor is referred to as the
    a. ABC.
    b. Glass tube (GT).
    c. Computer monitor (CM).
    d. Cathode-ray tube (CRT).
12. The image on the monitor is formed by
    a. Spraying the screen with electrons all at the same time.
    b. Developing a series of horizontal lines that are made across the screen from top to bottom.
    c. Developing a series of vertical lines that are made across the screen from left to right.
    d. Having the electron rays strike pixels at the proper point in time.
13. Flat panel displays control each pixel cell
    a. Electronically.
    b. With an electron beam.
    c. Magnetically.
14. The AGP bus design is similar to
    a. ISA.
    b. MCA.
    c. PCI.

Matching

Match a definition on the right with each term on the left.

15. Contrast	a.	Smallest element on the screen that can be controlled.
16. Pixel	b.	Uses a character set contained in ROM.
17. Aspect ratio	c.	Affects the difference between light and dark on the screen.
18. Pel	d.	The width of the monitor in comparison with the height.
19. Text mode	e.	Controls how bright the image is on the screen.
	f.	None of the above.

Completion

Fill in the blanks with the best answers.

20. There are basically two monitor modes, the _____ mode and the graphics mode.
21. The number of pixels available on the monitor is called the _____.
22. The _____ is the measure of the full number of colors available on the monitor.
23. The VGA display has a maximum resolution of _____.
24. There are basically two different types of monitor displays: one is _____ and the other is analog.
25. DSTN stands for Dual-Scan _____ Nematic.
26. The TFT display has a(n) _____ viewing angle than a DSTN display.

Open-Ended

27. Explain what the term *multisync* means.

28. What is the fastest-growing and most popular graphics adapter card?
29. What is the maximum resolution of the SVGA adapter card?
30. What is a green PC?
31. Why change the monitor type?

FAMILIARIZATION
ACTIVITY

1. Determine the type of monitor used by your computer.
2. Determine the type of display adapter card used by your computer.
3. Is your monitor energy efficient?
4. Does your monitor support a graphics mode?
5. What is the resolution of your monitor? Are there different modes of operation of your monitor that will allow for different amounts of resolution?
6. Sketch the connector used to interface your monitor and the computer. List the purpose of each connection. How did you determine this information?
7. What are the controls used on your monitor? State what each of the controls does.
8. How many different color palettes are available with your monitor?
9. How many different colors can be displayed at the same time by your monitor?
10. How much memory is installed on your display adapter?

QUESTIONS/ACTIVITIES

1. List the most popular types of monitors used by computers.
2. Explain the operation of a CRT used in a color monitor.
3. Show, by a diagram, the major sections of a monitor.
4. Explain the major differences between a monochrome CRT and a color CRT.
5. State the difference between a color monitor's palette and its ability to display several colors at the same time.

REVIEW QUIZ

Under the supervision of your instructor

1. Identify the type of monitor and display adapter being used by a given computer system and ensure that they have been properly installed and adjusted.
2. Discuss the differences between CRT monitors and flat panel displays.

45

The Computer Printer

On Friday afternoon, Joe Tekk was planning to leave work a little bit early. He was just finishing up a report that he had been working on all week. Just a few more minutes, *he thought to himself.*

He sent the final copy of the document to the laser printer and waited a few minutes at his desk before he went to the shared network printer to retrieve the output. As he picked up the stack of papers, he noticed a large black streak running down the middle of the first page. He leafed through the papers—every single page had the same dark streak. Not today, *he thought to himself.*

Joe checked the printer's status display and then proceeded to open the cover to examine the components inside. He removed the toner cartridge and performed a careful visual inspection. He cleaned up a small pile of toner close to the fuser unit. He then gently shook the toner cartridge to evenly distribute the toner inside and replaced it.

As Joe put all of the pieces back together, he thought to himself, That didn't take too much time. I still can get out early. *He reset the printer and waited for a test sheet to print to see if the problem was fixed. Everything looked fine. He went back to his desk to print another copy of his document. Again, he waited at his desk while the printer completed the print job.*

When Joe picked up the papers, he studied the first sheet very carefully. It looked great. Then he leafed through the rest of the pages. A few of them were faded, and a few didn't print at all. The status indicator on the printer read "Add toner." Joe looked for a new cartridge but could not find one anywhere. When he asked Donna, the secretary, where the toner cartridges were stored, she smiled and said, "Oh, they are on backorder. Sorry!"

Joe reluctantly decided to select a different printer. When he looked at the output from the new printer, he noticed the formatting was completely different and he would have to edit the document again. He spent the rest of the afternoon making changes.

Upon completion of this exercise, you will be able to

1. Perform a printer self-test.
2. Produce a printer test page.
3. From the user's manual and an inspection of the printer, determine whether the printer is in need of any periodic maintenance.

Impact printer
Nonimpact printer
Dot-matrix printer
Overstrike
Ink-jet printer
Bubble-jet printer

Laser printer
Electrophotographic process
ASCII code
Printer-control code
Escape code

There are two fundamental types of printers used with personal computers: the *impact printer* and the *nonimpact printer*. The impact printer uses some kind of mechanical device to impart an impression to the paper through an inked ribbon. The nonimpact printer uses heat, a jet of ink, electrostatic discharge, or laser light. Nonimpact printers form printed images without making physical contact with the paper. These two types of printers are illustrated in Figure 45.1.

IMPACT PRINTERS

The most common type of impact printers is the *dot-matrix printer*. The dot-matrix printer makes up its characters by means of a series of tiny mechanical pins that move in and out to form the various characters printed on the paper.

FIGURE 45.1 Two fundamental types of printers (a) Impact (dot-matrix illustration) and (b) Nonimpact (laser illustration)

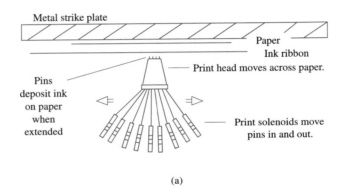

Metal strike plate

Paper
Ink ribbon

Print head moves across paper.

Pins deposit ink on paper when extended

Print solenoids move pins in and out.

(a)

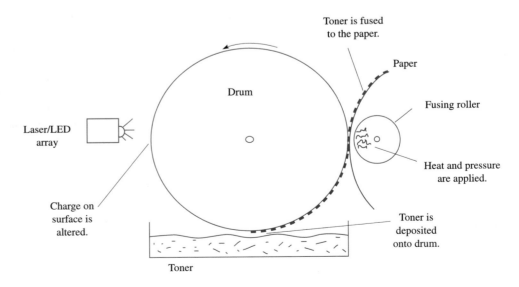

Toner is fused to the paper.

Paper

Drum

Fusing roller

Laser/LED array

Heat and pressure are applied.

Charge on surface is altered.

Toner is deposited onto drum.

Toner

(b)

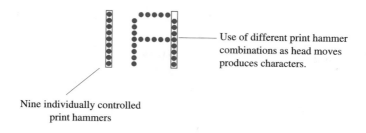

FIGURE 45.2 9-pin dot-matrix print head

Use of different print hammer combinations as head moves produces characters.

Nine individually controlled print hammers

The Dot-Matrix Printer

The dot-matrix printer, one of the most popular types of printers, uses a mechanical printing head that physically moves across the paper to be printed. This mechanical head consists of tiny movable wires that strike an inked ribbon to form characters on the paper. There are two popular kinds of dot-matrix print heads: one consists of 9 pins (the movable wires) and the other consists of 24 pins. A 9-pin print head is shown in Figure 45.2.

The 24-pin dot-matrix printer is more expensive than the 9-pin model. However, because both types have modes of operation that allow for an *overstrike* of the image (with the head moving slightly and the image being struck again), the 9-pin model can produce close to what is known as letter-quality printing. The 24-pin model can produce an even sharper character when operated in the same overstrike mode. Because of the manner in which characters are formed in this type of printer, the printing of graphic images is possible.

NONIMPACT PRINTERS

The most popular nonimpact printers are the *ink-jet printer*, *bubble-jet printer*, and the *laser printer*. The ink-jet printer uses tiny jets of ink that are electrically controlled. The laser printer uses a laser to form characters. The laser printer resembles an office photo-copying machine.

Ink-Jet and Bubble-Jet Printers

An ***ink-jet printer*** uses electrostatically charged plates to direct jets of ink onto paper. The ink is under pressure and is formed by a mechanical nozzle into tiny droplets that can be deflected to make up the required images on the paper. A ***bubble-jet printer*** uses heat to form bubbles of ink. As the bubbles cool, they form the droplets applied to the paper. Ink-jet and bubble-jet printers cost more than impact printers but are quieter and can produce high-quality graphic images.

Laser Printers

Laser printers use the six-step ***electrophotographic process*** to produce high-quality text and graphics, including images. The six steps are as follows:

1. Charging, in which the photoreceptor drum is charged with static electricity.
2. Exposure, in which the print image is exposed onto the photoreceptor drum (by a laser or an LED array).
3. Developing, in which the toner is applied to the exposed portions of the photoreceptor drum.
4. Transfer, in which the toner is transferred from the photoreceptor drum to the paper.
5. Fusing, in which the toner is heated and bonded to the paper.
6. Cleaning, in which the photoreceptor drum is cleaned to prepare it for the next exposure.

This process is illustrated in Figure 45.1(b).

Because of their high-quality output, laser printers find wide application in desktop publishing, computer-aided design, and other image-intensive computer applications. Laser printers are usually at the high end of the price range for computer printers.

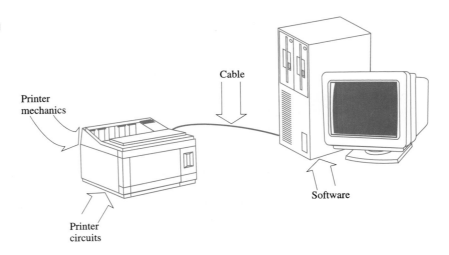

FIGURE 45.3 Problem areas in computer printers

Cable

Printer mechanics

Software

Printer circuits

TECHNICAL CONSIDERATIONS

Most printer problems are caused by software. What this means is that the software does not match the hardware of the printer. This is especially true when the printing of graphics is involved. Software troubleshooting for printers is presented later in this section. When there is a hardware problem with printers, it is usually the interface cable that goes from the printer adapter card to the printer itself. This is illustrated in Figure 45.3.

Printer Cables

The interface cable is used to connect the printer to the computer. Previously, there was limited communication between the printer and computer. The computer received a few signals from the printer such as the online or offline indicator, the out-of-paper sensor, and the print buffer status. As long as the printer was sending the correct signals to the computer, the computer would continue to send data.

Advances in printer technology now require a two-way communication between the computer and printer. As a result of these changes, a new bidirectional printer cable is required to connect most new printers to the computer. The bidirectional cables may or may not adhere to the new IEEE standard for Bidirectional Parallel Peripheral Interface. The IEEE 1284 Bitronic printer cable standard requires 28 AWG construction, a Hi-flex jacket, and dual shields for low EMI emissions. The conductors are twisted into pairs to reduce possible cross talk. Some new printers also come equipped with a USB connector, requiring a USB cable.

Check the requirements for each printer to determine the proper cable type. Figure 45.4 shows two popular parallel printer cable styles. Many different lengths of printer cables are available. It is usually best to use the shortest cable possible in order to reduce the possibility of communication errors.

Printer Hardware

A printer requires periodic maintenance. This includes vacuuming out the paper chaff left inside the printer. A soft dry cloth should be used to keep the paper and ribbon paths clean. It is a good idea to use plastic gloves when cleaning a printer, because the ink or toner is usually difficult to remove from the skin. With dot-matrix printers, be careful of the print heads. These heads can get quite hot after extended use. Make sure you do not turn the print platen rollers when the power is on because a stepper motor is engaged when power is applied. This little motor is trying to hold the print platen roller in place. If you force it to move, you could damage the stepper motor. When replacing the ink cartridge on a color printer, be sure to run the manufacturer's printhead alignment program.

Laser Hardware

Essentially, laser printers require very little maintenance. If you follow the instructions that come with the printer, the process of changing the toner cartridge (after about 3500 copies) also performs the required periodic maintenance on the printer.

FIGURE 45.4 Typical printer cables

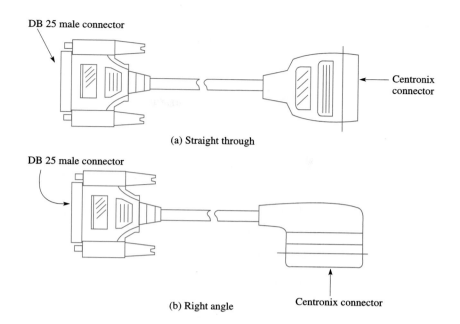

DB 25 male connector

Centronix connector

(a) Straight through

DB 25 male connector

Centronix connector

(b) Right angle

When using a laser printer, remember that such a machine uses a large amount of electrical energy and thus produces heat. So make sure that the printer has adequate ventilation and a good source of reliable electrical power. This means that you should not use an electrical expansion plug from your wall outlet to run your computer, monitor, and laser printer. Doing so may overload your system.

When shipping a laser printer, be sure to remove the toner cartridge. If you don't remove it, it could open up and spill toner (a black powder) over the inside of the printer, causing a mess that is difficult to clean up.

TESTING PRINTERS

When faced with a printer problem, first determine if the printer ever worked at all or if this is a new installation that never worked. If it is a new installation and has never worked, a careful reading of the manual that comes with the printer is usually required to make sure that the device is compatible with the printer port. Table 45.1 lists some of the most direct methods for troubleshooting a computer printer.

SYSTEM SOFTWARE

Many types of commands are sent to the printer while it is printing. Some of these commands tell the printer what character to print; others tell the printer what to do, such as performing a carriage return, making a new line, or doing a form feed. This is all accomplished by groups of 1s and 0s formed into a standard code that represents all the printable characters and the other commands that tell the printer what to print and how to print it.

The ASCII Code

Table 45.2 lists all the printable characters for a standard printer. The code used to transmit this information is called the ***ASCII code*** (see Appendix A). ASCII stands for American Standard Code for Information Interchange.

As you can see in Table 45.2, each keyboard character is given a unique number value. For example, a space is number 32 (which is actually represented by the binary value 0010 0000 when transmitted from the computer to the printer). The number values that are less than 32 are used for controlling the operations of the printer. These are called ***printer-control codes***, or simply ***control codes***. These codes are shown in Table 45.3.

567

TABLE 45.1 Printer troubleshooting methods

Checks	Comments
Check if printer is plugged in and turned on.	The printer must have external AC power to operate.
Check if printer is online and has paper.	Printers must be *online*, meaning that their control switches have been set so that they will print (check the instruction manual). Some printers will not operate if they do not have paper inserted.
Print a test page.	Select the Print Test Page option as shown in Figure 45.5. Confirm that the page printed properly (see Figure 45.6). Figure 45.7 shows the printer test page output.
Do a printer self-test.	Most printers have a self-test mode. In this mode, the printer will repeat its character set over and over again. You must refer to the documentation that comes with the printer to see how this is done.
Do a PrintScreen.	If the printer self-test works, then with some characters on the computer monitor, hold down the Shift key and press the PrintScrn key at the same time. What is on the monitor should now be printed. Do not use a program (such as a word-processing program) because the software in the program may not be compatible with the printer.
Exchange printer cable.	If none of the above tests work, the problem may be in the printer cable. At this point, you should swap the cable with a known good one.
Replace the printer adapter card.	Try replacing the printer adapter card with a known good one. Be sure to refer to the printer manual to make sure you are using the correct adapter card.
Check parameters for a serial interface.	If you are using a serial interface printer from a serial port, make sure you have the transmission rate set correctly, along with the parity, number of data bits, and number of stop bits. Refer to the instruction manual that comes with the printer and use the correct form of the DOS MODE command.
Check the configuration settings.	Check all the configuration settings available on the printer.
Check the software installation.	When software is installed (such as word-processing and spreadsheet programs), the user may have had the wrong printer driver installed (the software that actually operates the printer from the program).

The definitions of the control code abbreviations are as follows:

ACK	Acknowledge	GS	Group separator
BEL	Bell	HT	Horizontal tab
BS	Backspace	LF	Line feed
CAN	Cancel	NAK	Negative acknowledge
CR	Carriage return	NUL	Null
DC_1–DC_4	Device control	RS	Record separator
DEL	Delete idle	SI	Shift in
DLE	Data link escape	SO	Shift out
EM	End of medium	SOH	Start of heading
ENQ	Enquiry	SP	Space
EOT	End of transmission	STX	Start text
ESC	Escape	SUB	Substitute
ETB	End of transmission block	SYN	Synchronous idle

FIGURE 45.5 Print Test Page option

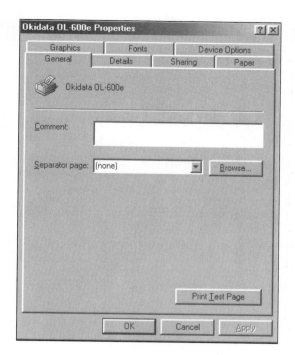

FIGURE 45.6 Printer test page confirmation

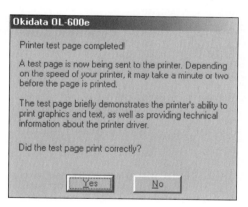

FIGURE 45.7 Printer test page output

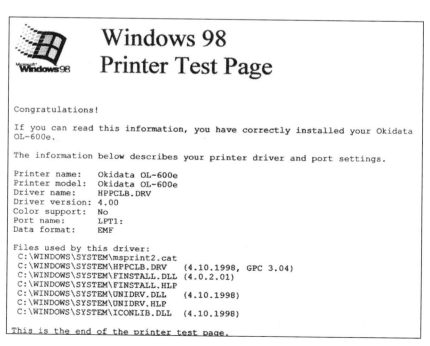

TABLE 45.2 Standard printable ASCII code

Dec Hex Char	Dec Hex Char	Dec Hex Char
32 20 SP	64 40 @	96 60 '
33 21 !	65 41 A	97 61 a
34 22 "	66 42 B	98 62 b
35 23 #	67 43 C	99 63 c
36 24 $	68 44 D	100 64 d
37 25 %	69 45 E	101 65 e
38 26 &	70 46 F	102 66 f
39 27 '	71 47 G	103 67 g
40 28 (	72 48 H	104 68 h
41 29)	73 49 I	105 69 i
42 2A *	74 4A J	106 6A j
43 2B +	75 4B K	107 6B k
44 2C ,	76 4C L	108 6C l
45 2D –	77 4D M	109 6D m
46 2E .	78 4E N	110 6E n
47 2F /	79 4F O	111 6F o
48 30 0	80 50 P	112 70 p
49 31 1	81 51 Q	113 71 q
50 32 2	82 52 R	114 72 r
51 33 3	83 53 S	115 73 s
52 34 4	84 54 T	116 74 t
53 35 5	85 55 U	117 75 u
54 36 6	86 56 V	118 76 v
55 37 7	87 57 W	119 77 w
56 38 8	88 58 X	120 78 x
57 39 9	89 59 Y	121 79 y
58 3A :	90 5A Z	122 7A z
59 3B ;	91 5B [	123 7B {
60 3C <	92 5C \	124 7C	
61 3D =	93 5D]	125 7D }
62 3E >	94 5E ^	126 7E ~
63 3F ?	95 5F _	

ETX	End text	US	Unit separator
FF	Form feed	VT	Vertical tab
FS	Form separator		

The way you can enter an ASCII control code is by holding down the Alt key and the Shift key at the same time and using the numeric keypad to enter the ASCII control code. You will see the corresponding character on the screen when you release the Alt and Shift keys. (The top row of numbers on your keyboard will not work—use only the ones on the numeric keypad portion of the keyboard for this.)

TABLE 45.3 ASCII control codes

Dec	Hex	Char	Code
0	0	^@	NUL
1	1	©	SOH
2	2	●	STX
3	3	♥	ETX
4	4	♦	EOT
5	5	♣	ENQ
6	6	♠	ACK
7	7	•	BEL
8	8	◙	BS
9	9	○	HT
10	A	■	LF
11	B	♂	VT
12	C	♀	FF
13	D	♪	CR
14	E	♫	SO
15	F	○	SI
16	10	►	DLE
17	11	◄	DC$_1$
18	12	↕	DC$_2$
19	13	‼	DC$_3$
20	14	¶	DC$_4$
21	15	§	NAK
22	16	▬	SYN
23	17	↨	ETB
24	18	↑	CAN
25	19	↓	EM
26	1A	→	SUB
27	1B	←	ESC
28	1C	∟	FS
29	1D	↔	GS
30	1E	▲	RS
31	1F	▼	US

For example, to create a text file that will cause the printer to eject a sheet of paper (a form feed), using an editor, you would press

Alt+Shift+12

What will appear on the screen when you release the Alt+Shift keys is

^L

This may not be exactly what you expected, but it is your monitor's way of interpreting what you have just entered. If you put this into a text file (say it is called FORMFED.TXT), then when it is sent to the printer, it will be interpreted as a form feed, and the printer will feed a new sheet of paper. You can then make a batch file that uses this form-feed text file (called FORMFED.BAT) that contains this command:

COPY FORMFED.TXT PRN

To test the printer's form feed, you would simply enter

A> FORMFED

and the printer (if on and ready) should then feed a sheet of paper through. You could also make up your own custom batch files to perform other types of printer tests or put your name in them (so the name appears printed on the sheet).

EXTENDED ASCII CODES

If you set up your printer to act as a graphics printer (by setting the appropriate configuration; refer to the user manual that comes with the printer), you can extend the character set to include many other forms of printable characters. These characters are shown in Table 45.4.

You can write these extended character codes to your printer by creating text files. To do this, again hold down the Alt and Shift keys and type the number code into the numeric keypad on your keyboard. For example, to get the Greek letter Σ, simply press Alt+Shift+228; when you lift up on the Alt+Shift keys, a Σ will appear on the monitor. If you include this in a text file (or do a PrintScreen), you can transfer it to the printer. The extended characters 176 through 223 are used for creating boxes, rectangles, and other shapes on the monitor or printer while it is still in the text mode. If you can't get these extended characters on the printer, it is either because you have not set the printer to the graphics mode or your printer simply can't perform the functions required by this mode.

OTHER PRINTER FEATURES

Recall that most printers allow you to get different kinds of text (such as 80 or 132 characters of text across the page) or change the page orientation from portrait to landscape. You can also create batch files to test the capabilities of the printer to print the following:

- Bold text
- Underscores
- Overscores
- Superscripts
- Subscripts
- Compressed or expanded text
- Italics

For example, a new printer does not do bold text. You could move the printer to a different computer to see if the problem is in the printer, or you could have a batch file you created that will quickly test if the printer really is capable of producing bold text. If the printer can do it, the problem is probably in the software, because a new printer driver needs to be installed (which comes from the software vendor). To do this, you need to understand what printer manufacturers do in order to get their printers to create features such as bold text, subscripts, and superscripts.

PRINTER ESCAPE CODES

The ESC (escape) character is used by printer manufacturers as a preface. It is an easy way for them to get a whole new set of printer commands. The character ESC generally doesn't do anything by itself; what it does is to tell the printer that the character or set of

TABLE 45.4 Extended ASCII character set

Dec	Hex	Char	Dec	Hex	Char	Dec	Hex	Char	Dec	Hex	Char
128	80	Ç	160	A0	á	192	C0	└	224	E0	α
129	81	ü	161	A1	í	193	C1	┴	225	E1	β
130	82	é	162	A2	ó	194	C2	┬	226	E2	Γ
131	83	â	163	A3	ú	195	C3	├	227	E3	π
132	84	ä	164	A4	ń	196	C4	─	228	E4	Σ
133	85	à	165	A5	Ń	197	C5	┼	229	E5	σ
134	86	å	166	A6	ª	198	C6	╞	230	E6	μ
135	87	ç	167	A7	º	199	C7	╟	231	E7	τ
136	88	ê	168	A8	¿	200	C8	╚	232	E8	φ
137	89	ë	169	A9	⌐	201	C9	╔	233	E9	θ
138	8A	è	170	AA	¬	202	CA	╩	234	EA	Ω
139	8B	ï	171	AB	½	203	CB	╦	235	EB	δ
140	8C	î	172	AC	¼	204	CC	╠	236	EC	∞
141	8D	ì	173	AD	¡	205	CD	═	237	ED	Ø
142	8E	Ä	174	AE	"	206	CE	╬	238	EE	∈
143	8F	Å	175	AF	"	207	CF	╧	239	EF	∩
144	90	É	176	B0	░	208	D0	╨	240	F0	≡
145	91	æ	177	B1	▒	209	D1	╤	241	F1	±
146	92	Æ	178	B2	▓	210	D2	╥	242	F2	≥
147	93	ô	179	B3	│	211	D3	╙	243	F3	≤
148	94	ö	180	B4	┤	212	D4	╘	244	F4	⌠
149	95	ò	181	B5	╡	213	D5	╒	245	F5	⌡
150	96	û	182	B6	╢	214	D6	╓	246	F6	÷
151	97	ù	183	B7	╖	215	D7	╫	247	F7	≈
152	98	ÿ	184	B8	╕	216	D8	╪	248	F8	°
153	99	Ö	185	B9	╣	217	D9	┘	249	F9	∙
154	9A	Ü	186	BA	║	218	DA	┌	250	FA	·
155	9B	¢	187	BB	╗	219	DB	█	251	FB	√
156	9C	£	188	BC	╝	220	DC	▄	252	FC	η
157	9D	¥	189	BD	╜	221	DD	▌	253	FD	²
158	9E	₧	190	BE	╛	222	DE	▐	254	FE	■
159	9F	ƒ	191	BF	┐	223	DF	▀	255	FF	

characters that follows is to be treated in a special way. As an example, an <ESC> E means to begin bold text and <ESC>F means to end the bold text. The exact escape sequence is different for different printer manufacturers, and you need to find the sequence for your printer in the user's manual.

You could have a batch file calling a text file that tests for bold printing, such as

Mickey Brown's Printer test:
This is normal text.
<ESC>E
This is now bold text.
<ESC>F
This is back to normal text.

The problem in creating this kind of text file is to actually enter the Esc key into it (just pressing the Esc key doesn't do it). The secret to doing this is to enter a Ctrl+V (hold down the Ctrl key while pressing the V key) and then follow it with the [(left bracket). So when you see the text

<ESC>E

it really means Ctrl+V[E, which will start boldface printing. Remember, for the printer you are using, the *escape code* may be different. All you need to do is to use the operator's manual that comes with the printer to determine the proper escape code for each printer's unique features.

MULTIFUNCTION PRINT DEVICES

It is becoming more and more common to see printers bundled with other common products, like a fax machine. For example, a fax machine usually prints any faxes received. With some modifications, it can print data received from a computer. These types of printers generally use either ink-jet or bubble-jet printer technology.

Similarly, when sending a fax, the image or text that is sent must be scanned. Again, by making some additional modifications, the scanner can provide the scanned data to a computer instead of a fax. These three features—printing, faxing, and scanning—are available on most multifunction printers. Other features such as an answering machine may also be included. Multifunction devices can save a lot of money while offering the convenience of many products in one package.

ENERGY EFFICIENCY

Like computers and monitors, printers can waste a tremendous amount of energy. This is because printers are usually left on 24 hours a day but are active only a small portion of the time. The EPA Energy Star program recommends that a printer automatically enter a sleep mode when not in use. In sleep mode, a printer may consume between 15 and 45 watts of power. This feature may cut a printer's electricity use by more than 65%.

Other efficiency options recommended by the EPA include printer sharing, duplex printing, and advanced power management features. Printer sharing reduces the need for an additional printer. Power management features can reduce the amount of heat produced by a printer, contributing to a more comfortable workspace and reduced air-conditioning costs. Consider turning off a printer at night, on weekends, or during extended periods of inactivity.

TROUBLESHOOTING TECHNIQUES

The most common problems with printers usually involve the quality of the output. Many problems are associated with the supply of ink or toner. Printers also contain many mechanical components that are a common point of failure. In the case of a dot-matrix printer, the ribbon may need to be replaced, the printhead may need to be replaced, or the pin feeds may occasionally require some adjustment. For an ink-jet printer, the ink cartridge may become clogged with dried ink and may need to be cleaned to restore the print quality. There is no set schedule for these events to occur.

The best course of action is to be prepared for common problems that can be encountered. For example, it is a good idea to keep printer supplies on hand, so when a problem occurs, it can be remedied quickly. Table 45.5 contains a list of items that should be kept on hand. Remember, many of these items have a certain shelf life. Rotate the stock regularly.

TABLE 45.5 Common printer
types and supplies

Printer Type	Supplies
Dot matrix	Ribbons, pin-feed paper
Ink-jet and bubble-jet	Black ink cartridges, color ink cartridges, single-sheet ink-jet paper
Laser	Toner cartridges, single-sheet laser-quality paper
Color laser	Cyan, yellow, magenta, and black toner cartridges, single-sheet color laser-quality paper

SUMMARY

In this exercise we discovered that

- Printers come in two types: impact and nonimpact.
- Impact printers use a mechanical printhead to strike dots onto the paper to form characters and graphics.
- Nonimpact printers use static electricity to deposit ink or toner onto the paper.
- The electrophotographic process contains six steps: charging, exposure, developing, transfer, fusing, and cleaning.
- ASCII control codes can be used to control a printer.

SELF-TEST

This self-test is designed to help you check your understanding of the background information presented in this exercise.

True/False

Answer *true* or *false*.

1. There are basically two types of printers: impact printers and nonimpact printers.
2. A bubble-jet printer is capable of producing graphics.
3. A dot-matrix printer cannot do text, but it can do graphics.
4. An IEEE 1284 cable is used to support serial printer communications.
5. Most printer problems may be corrected by simply turning the printer off and on.
6. The electrophotographic process contains six steps.
7. The parallel cable connector is called the Centronix connector.

Multiple Choice

Select the best answer.

8. A nonimpact printer is the
   a. Dot-matrix printer.
   b. Ink-jet printer.
   c. Laser printer.
   d. Both b and c.
9. Most printer problems are caused by
   a. Hardware in the printer.
   b. The printer interface card.
   c. Software.
   d. The printer cable.
10. A quick way of checking a printer/computer interface is to
    a. Do a PrintScreen with some text on the monitor.
    b. Run it from the user's word-processing program.
    c. Swap the printer with a known good one.
    d. None of the above.

11. A printer test page
    a. Is performed automatically.
    b. Tests the connection to the printer.
    c. Tests the printer driver.
    d. Both b and c.
12. Printer features include
    a. The ASCII character set.
    b. Extended ASCII characters.
    c. Various output choices.
    d. None of the above.
13. Which is not an electrophotographic step?
    a. Charging.
    b. Transducing.
    c. Transfer.
14. Printer escape codes begin with which ASCII character?
    a. CR.
    b. ESC.
    c. LF.

Matching

Match a definition on the right with each term on the left.

15. ASCII code
16. ESC
17. Extended character set

   a. A printer code prefix.
   b. Standard code used for communicating between the computer and printer.
   c. Letters of the Greek alphabet.
   d. None of the above.

Completion

Fill in the blank or blanks with the best answers.

18. For a printer to print special graphics-type symbols (in text mode), it must be set to the _____ _____ mode.
19. The printer code for beginning bold text is called a(n) _____ sequence.
20. To enter an <ESC> in a text file, you must press Ctrl _____ followed by the _____ symbol.
21. Page orientation can be either _____ or _____.
22. ASCII characters less than 32 are called _____ _____.
23. During the _____ step, the print-image is applied to the photoreceptor drum.
24. The electrophotographic process uses _____ electricity.

FAMILIARIZATION ACTIVITY

Using the printer available with your system

1. Determine whether the printer has a graphics or similar setting. Refer to the owner's manual.
2. If the printer does have such a setting, describe what is needed to put the printer into this setting (the switches and positions of these switches).
3. Using the owner's manual, determine what you must do in order to perform a printer self-test. Then actually run a printer self-test, and attach one sheet of the results to this exercise.
4. From the owner's manual, determine and list the special printing characteristics available (such as boldface), along with their <ESC> sequences.
5. Develop a batch file or files using the <ESC> sequences for your printer that will test all the printing characteristics of your particular printer.

1. What kind of printer did you use for this exercise?
2. If you need to get a new ribbon or ink cartridge for this printer, where would you get it and for what would you ask?
3. Does the printer you used in this exercise have the ability to print in color? Explain.
4. Does your printer use a serial or a parallel interface? How did you determine this?

REVIEW QUIZ

Under the supervision of your instructor

1. Perform a printer self-test.
2. Produce a printer test page.
3. From the user's manual and an inspection of the printer, determine whether the printer is in need of any periodic maintenance.

46 Keyboards and Mice

Joe Tekk's phone rang just as Joe walked into his office. It was his friend Ken Koder.

"Hey, Joe, you remember that problem I had with my keyboard when we were trying to use pcANYWHERE? Well, it showed up in other places. I would press a key and a different character would appear on the screen."

Joe recalled the problem. "Sure, I remember. You tried to enter '1' and kept getting an exclamation point. What did you find out?"

Ken explained, "Out of desperation I deleted my keyboard using Device Manager. When I rebooted, Windows found my keyboard and reinstalled drivers for it. Everything is fine now."

Joe thought about that approach. "I never would have tried that, Ken."

Ken laughed. "I don't know why I did it; it was just something else to try."

PERFORMANCE OBJECTIVES

Upon completion of this exercise, you will be able to

1. Ensure that the computer keyboard has been properly connected to the system.
2. Explain how the keyboard should be maintained.
3. Ensure that the computer mouse has been properly connected and adjusted.
4. Check keyboard/mouse properties under Windows.

KEY TERMS

Keyboard
Character
Character code

Keyboard interface connector
Trackball

BACKGROUND INFORMATION

One of the most common ways of getting information into a computer is through a *keyboard.* A computer keyboard consists of separate keys that, when tapped, send specific codes to the computer. Essentially, such a code tells the computer that a key is depressed, what key it is that is being depressed, and when the key is no longer depressed.

Another device used for getting information into a computer is a computer mouse. A mouse is simply a device that moves a cursor to any desired area of the screen. The computer always knows at what position on the screen the cursor is located. On the mouse

FIGURE 46.1 Actions of a keyboard and a mouse

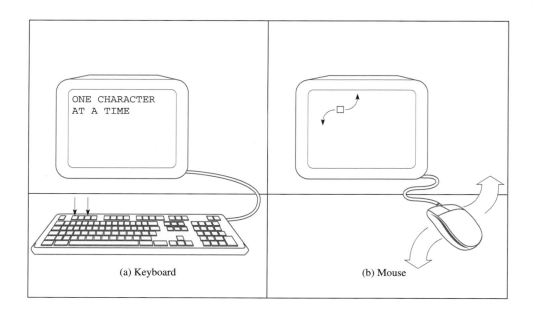

(a) Keyboard

(b) Mouse

itself, there are buttons (usually two or three). When a button is depressed, this—along with the position of the cursor on the screen—gives the computer specific information. Usually the screen contains information as to what that particular area of the screen means to the mouse user. For instance, it could mean to begin or terminate a process. Figure 46.1 shows actions of a keyboard and a mouse.

As you can see from Figure 46.1, both of these devices are input devices. The major disadvantages of the keyboard are the typing skill required to use it and the need to know specific key sequences to initiate computer actions (such as DIR in order to get the directory listing of a disk).

The disadvantages of the keyboard are overcome through the use of a mouse. Using the mouse does not require any typing skills or knowledge of special key sequences (such as the DOS commands). Windows essentially requires only the use of the mouse for system interaction. Windows allows you to execute multiple programs simultaneously, and quickly change from one application to another, with a single mouse click.

THE KEYBOARD

IBM and compatible computers have gone through an evolution of keyboards. The first keyboard was the 83-key PC keyboard that was also used with some of the first XT systems. This keyboard had 10 function keys along the left side, with an Esc key at the upper left and a combination numeric keypad–cursor control section on the right (see Figure 46.2).

The 84-key AT keyboard became the next in the line of keyboards and the standard for all AT-model computers. It was similar in layout to the 83-key PC keyboard. One of the

FIGURE 46.2 The original 83-key PC keyboard

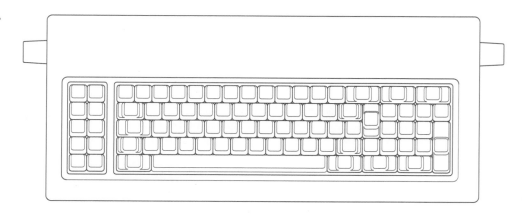

FIGURE 46.3   Enhanced
keyboard

changes was moving the Esc key to the top of the numeric keypad section. Three LED status lights were added to indicate when the Caps Lock, Num Lock, and Scroll Lock functions of the keyboard were enabled.

IBM later came out with an 84-key space-saving keyboard. This keyboard has its function keys along the top with no numeric keypad–cursor control section on the right.

A new keyboard, used with most PCs, contains 101 keys. The function keys are located horizontally along the top of the keyboard, where there are now 12 of them. The Esc key is at the upper left. The keyboard contains duplicate cursor-movement and other similar keys. This is sometimes referred to as the enhanced keyboard (Figure 46.3).

Identifying Keys

There are four ways of identifying a key on a keyboard: by the *character* on the cap of the key, by the *character code* associated with each key-cap character, by the scan code of the key, and by the decimal key-location number. These are illustrated in Figure 46.4.

During the POST, the first part of the keyboard scan code is displayed if there is a problem with that particular key. Figure 46.5 shows the scan codes for a 101-key keyboard, and Figure 46.6 shows the key-location numbers for the same keyboard.

As shown in Figure 46.6, each key is assigned a decimal number that is used as a key-location reference on most drawings. These numbers are used only as convenient guides for the physical location of the various keys and bear no relationship to the actual characters generated by the corresponding keys.

Keyboard Servicing

Outside of routine cleaning of the keyboard, there is little you can do to service it. In many cases, the keyboard assembly is a sealed unit. The major hazards to a keyboard are spilled liquids. Periodically you can use a chip puller to pull the keytops off the keyboard. (Be sure to have a similar keyboard to use as a reference when replacing these key caps.) Then hold the keyboard upside down and blow it out with compressed air.

The keyboard is connected to the computer through a cable to the *keyboard interface connector*. This connector is shown in Figure 46.7.

FIGURE 46.4   Four ways of identifying a key on an IBM keyboard

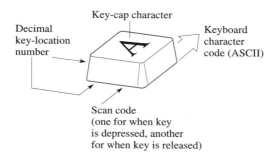

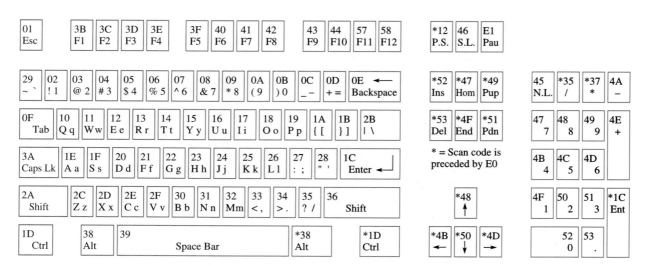

FIGURE 46.5   Scan codes for 101-key keyboard

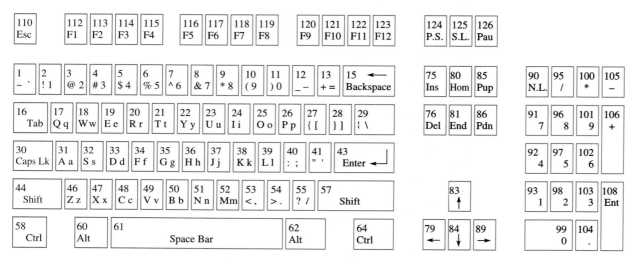

FIGURE 46.6   Key-location numbers for a 101-key keyboard

FIGURE 46.7   Keyboard interface connector (socket)

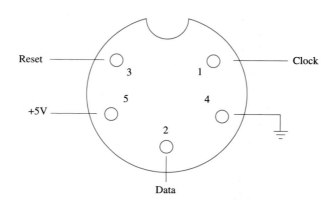

FIGURE 46.8 Microsoft
Natural keyboard

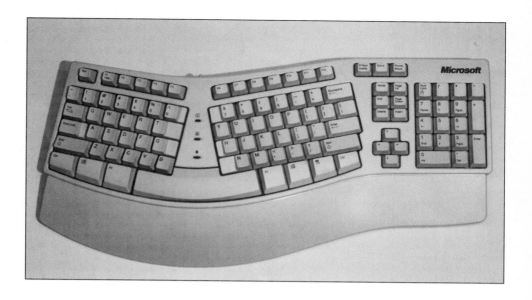

You can use a voltmeter to test the operation of the keyboard interface connector. Voltages between all pins and pin 4 of the connector should be in the range of 2V to 5.5V DC. If any of these voltages are wrong, the problem is usually in the computer's system board. If these voltages are correct, the problem may be in the keyboard or its connector cable. Some keyboards have one or more switches (on the bottom side) to make them compatible with the computer to which they are connected. Check these switch settings as well as the documentation that comes with the keyboard.

If you find that only one key is malfunctioning, you can check the small spring on the key. Simply remove the key cap, under which you will see a small spring. Try pulling the spring slightly and then replace the key cap. You can check the cable continuity by carefully removing the bottom plate of the keyboard and observing how the cable interfaces with the computer.

Because new keyboards are so inexpensive, it is usually cheaper to replace a bad keyboard.

Today, you can purchase custom keyboards that have themes (a *Star Trek* keyboard), keyboards with infrared transmitters and no cables, or a keyboard with the keys arranged into two groups (as shown in Figure 46.8), one for each hand. You can even buy a keyboard with a scanner built in.

THE MOUSE

The mouse interfaces with the computer through a serial port. In the PC, the mouse is either connected to the 9-pin male plug on the COM1 serial port, a USB port, or to its own dedicated port. A 9- to 25-pin adapter is available if the serial port has a 25-pin connector. Figure 46.9 shows a typical computer mouse.

Figure 46.10 illustrates the internal operation of a mouse. Two optical encoding disks (one horizontal, one vertical) spin as the mouse is moved. Optical sensors mounted on the disks detect the direction and speed of the movement and relay the information to a mouse controller IC. The controller generates a serial signal to the computer that represents the encoded mouse movement.

It is a good idea to periodically clean the mouse by removing dust and small fibers that accumulate around the shaft of the rollers. Otherwise, mouse movement may become erratic in one or both directions.

TRACKBALLS

A *trackball* is similar to a mouse except the device does not move. Instead, the user pushes a round trackball around inside its case, allowing the same movement as a mouse but not

FIGURE 46.9   Typical computer mouse

FIGURE 46.10   Internal operation of a mouse

Vertical
motion
detectors

Roller

Light source

Encoding
disk:

Horizontal motion
detectors

Rubber-coated
metal ball

Optical
encoding disk

Slots in
disk allow
light to pass
through

(a) Horizontal and vertical encoders

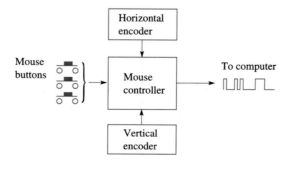

Horizontal
encoder

Mouse
buttons

Mouse
controller

To computer

Vertical
encoder

(b) Block diagram

requiring a mousepad or large surface for movement. Many laptop and notebook computers have trackball mouse devices built in.

TROUBLESHOOTING TECHNIQUES

Usually the biggest problems with a mouse-to-computer connection is improper installation of the mouse software. To correct this problem, read the literature that comes with the mouse. In order for the mouse to interface effectively with the computer, the software that comes with the mouse must be installed in the system as directed by the manufacturer. If you are running DOS, you also want to make sure that the system's CONFIG.SYS and AUTOEXEC.BAT files have been properly set up so that the mouse driver is automatically installed each time the system is booted up. Again, this information is included in the literature that comes with the mouse. The important point here is that you read the literature and follow the directions.

Usually, there is an optional utility program with the software that comes with the mouse that helps you test and adjust the mouse interface (by means of software). These utilities are perhaps one of the best tests of mouse performance.

Windows supplies its own mouse drivers, eliminating the need for any setup in CONFIG.SYS and AUTOEXEC.BAT, and provides a great amount of control over how the mouse appears and operates.

Clicking on the Mouse icon in Control Panel brings up the Mouse Properties window shown in Figure 46.11. Here you can adjust such important parameters as the double-click speed and left/right-handed operation. Figures 46.12 and 46.13 show two additional control windows dealing with the appearance of the mouse pointer. If the mouse is not responding correctly, it may be necessary to change its driver or hardware properties. In Control Panel, double-clicking the System icon and then selecting the Device Manager tab will allow you to double-click Mouse and check the driver information. This information is illustrated in Figure 46.14. Mouse information is available under the Devices icon in the Windows NT Control Panel. Figure 46.15 shows the associated Windows NT mouse information.

SUMMARY

In this exercise we discovered that

- Each key on a keyboard has a character, character code, decimal key-location number, and scan code.
- The mouse uses dual optical encoders to sense movement in the horizontal and vertical directions.

FIGURE 46.11 Initial Mouse Properties window

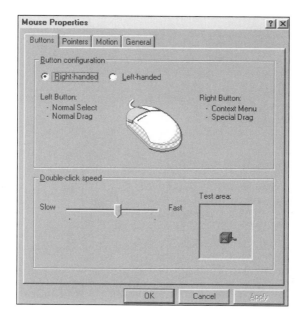

FIGURE 46.12   Mouse pointer types

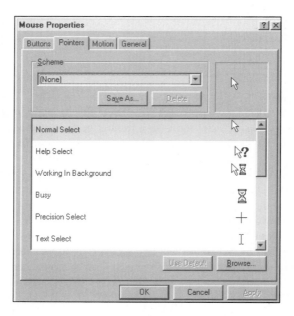

FIGURE 46.13   Additional pointer controls

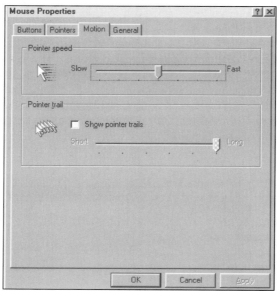

FIGURE 46.14   Examining the Mouse Driver information

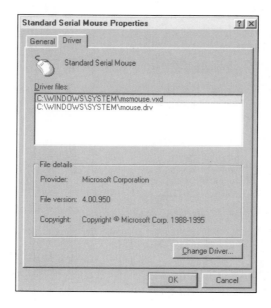

FIGURE 46.15 Windows NT Mouse Driver status

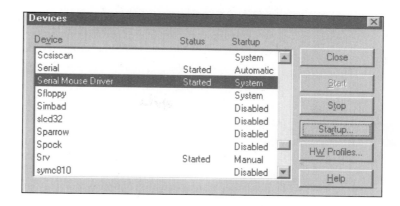

This self-test is designed to help you check your understanding of the background information presented in this exercise.

True/False

Answer *true* or *false*.

1. One of the most common ways of getting information into the computer is through the keyboard.
2. Generally speaking, a mouse is easier to use than a keyboard.
3. Scan codes are the same as ASCII codes.
4. The mouse is a serial device.
5. The mouse uses three optical encoding disks to sense movement.

Matching

Match one or more descriptions on the right with each term on the left.

6. 101-key keyboard
7. 83-key keyboard
8. 84-key keyboard

a. The enhanced keyboard.
b. Used in the AT line.
c. First IBM keyboard.
d. None of the above.

Completion

Fill in the blanks with the best answers.

9. The _____ code is associated with each key-cap character on the keyboard.
10. The key-location number is used as a convenient guide to the _____ location of the actual key on the keyboard.
11. The mouse controller IC generates a _____ waveform.
12. Windows provides its own mouse _____.
13. A(n) _____ is similar to a mouse, but does not move when used.
14. The _____ icon in Control Panel provides access to the mouse properties.
15. A software _____ is required to use a mouse in Windows.

1. Using the information provided in the background information of this exercise, determine the type of keyboard used by your computer.
2. Sketch the connector that the keyboard uses on the rear of the computer.
3. Where are the function keys located on your keyboard? What purposes do the keys serve?
4. Does your keyboard have any switches (usually located on the bottom of the keyboard)? If so, determine and record the purposes of these switches.
5. If your system has a mouse, what software has been added to the system so that the mouse can function?

6. If there is a utility program for checking the operation of the mouse with your system, practice using it. This is one of the requirements of the performance objectives. Otherwise, check and experiment with the Mouse Properties under Windows.
7. Sketch the connector used by your system mouse at the rear of the computer.

1. Explain the major differences between a keyboard and a mouse.
2. Which do you think is easier to use, a keyboard or a mouse? Why?
3. If your keyboard has a Num Lock key, what purpose does it serve? How did you determine this?
4. Is there a separate utility that comes with the mouse for testing its operation? If so, what exactly does the utility do?

Under the supervision of your instructor

1. Ensure that the computer keyboard has been properly connected to the system.
2. Explain how the keyboard should be maintained.
3. Ensure that the computer mouse has been properly connected and adjusted.
4. Check keyboard/mouse properties under Windows.

47 Telephone Modems

INTRODUCTION

Joe Tekk was looking through his storage boxes in the attic of his parents' home. In one box he found his old 300-baud acoustic modem. He pulled it out and looked at it: shiny white plastic with black rubber couplers for the telephone handset. Toggle switches on the control panel selected different modes of operation. Joe recalled connecting to a college mainframe for the first time and endlessly dialing his old rotary telephone until the single line to the college was available.

Joe thought it was funny that back then 300 baud seemed really fast. Today, modems as small as credit cards operate almost 200 times faster and still do not seem fast enough to satisfy the growing needs of personal and business computing. Joe wondered exactly when the change had occurred.

PERFORMANCE OBJECTIVES

Upon completion of this exercise, you will be able to

1. Connect to the Internet using a modem.
2. Successfully use a computer modem to transmit and receive a text file.
3. Explain the various MNP and CCITT standards.
4. Describe several encoding methods.

KEY TERMS

Modulate
Demodulate
Data terminal equipment
Data communications equipment
Baud
AT command set
Hayes compatible
Simplex
Duplex

Full duplex
Half duplex
Echo
Local echo
Remote echo
Quadrature modulation
Trellis modulation
MNP standards
CCITT standards

This exercise has to do with communications between computers. The user of one computer may interact with another computer—which may be located thousands of miles away—as if it were sitting right in the same room.

In order for computers to communicate in this manner, four items must be available, as shown in Figure 47.1.

As shown in Figure 47.1, there must be some kind of link between the computers. The most convenient link to use is the already-established telephone system lines. Using these lines and a properly equipped computer allows communications between any two computers that have access to a telephone. This becomes a very convenient and inexpensive method of communicating between computers.

There is, however, one problem. Telephone lines were designed for the transmission of the human voice, not for the transmission of digital data. Therefore, in order to make use of these telephone lines for transmitting computer data, the ONs and OFFs of the computer must first be converted to an analog signal, sent over the telephone line, and then reconverted from analog back to the ONs and OFFs that the computer understands. This concept is shown in Figure 47.2.

THE MODEM

The word *modulate* means to change. Thus an electronic circuit that changes digital data into analog data can be called a *modulator.* The word *demodulate* can be thought of as

FIGURE 47.1 Four items necessary for computer communications

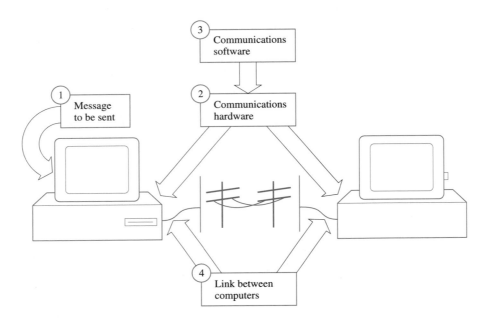

FIGURE 47.2 Basic needs for the use of telephone lines in computer communications

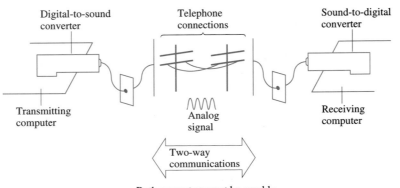

meaning "unchange," or restore to an original condition. Any electronic circuit that converts the analog signal used to represent the digital signals back to the ONs and OFFs understood by a computer can, therefore, be called a demodulator. Since each computer must be capable of both transmission and reception, each computer must contain an electrical circuit that can modulate as well as demodulate. Such a circuit is commonly called a <u>modu</u>lator/ <u>dem</u>odulator, or *modem.*

For personal computers, a modem may be an internal or an external circuit—both perform identical functions.

THE RS-232 STANDARD

The EIA (Electronics Industries Association) has published the EIA *Standard Interface Between Data Terminal Equipment Employing Serial Binary Data Interchange*—specifically, EIA-232-C. This is a standard defining 25 conductors that may be used in interfacing **data terminal equipment** (DTE, such as your computer) and **data communications equipment** (DCE, such as a modem) hardware. The standard specifies the function of each conductor, but it does not state the physical connector that is to be used. This standard exists so that different manufacturers of communications equipment can manufacture devices that can communicate with each other. In other words, the RS-232 standard is an example of an interface, essentially an agreement among equipment manufacturers on how to allow their equipment to communicate.

The RS-232 standard is designed to allow DTEs to communicate with DCEs. The RS-232 uses a DB-25 connector; the male DB-25 goes on the DTEs and the female goes on the DCEs. The RS-232 standard signals are shown in Figure 47.3.

The RS-232 is a digital interface designed to operate at no more than 50 feet with a 20,000-bps bit rate. The **baud**, named after J. M. E. Baudot, actually indicates the number of *discrete* signal changes per second. In the transmission of binary values, one such change represents a single bit. What this means is that the popular usage of the term *baud* has become the same as bits per second (bps). Table 47.1 shows the standard set of baud rates.

TELEPHONE MODEM SETUP

The most common problem with telephone modems is the correct setting of the software. There are essentially six distinct areas to which you must pay attention when using a telephone modem:

1. Port to be used
2. Baud rate
3. Parity
4. Number of data bits
5. Number of stop bits
6. Local echo ON or OFF

FIGURE 47.3 The RS-232 standard signals

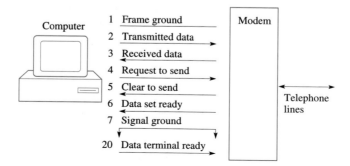

TABLE 47.1   Standard baud rates

Low Speed	High Speed
300	
600	
1200	14,400
2400	28,800
4800	33,600
9600	56K

Most telephone modems have a default setting for each of these areas. However, as a user you should understand what each of these areas means. You will have to consult the specific documentation that comes with the modem in order to see how to change any of the settings in the list. For now it is important that you understand the idea behind each of these areas.

Port to Be Used

The most common ports to be used are COM1 and COM2. Other possible ports are COM3 and COM4. The port you select from the communications software depends on the port to which you have the modem selected. On most communications software, once you set the correct port number, you do not need to set it again.

Baud Rate

Typical values for the baud rate are between 9600 and 56K. Again, these values can be selected from the communications software menu. It is important that both computers be set at the same baud rate.

Parity

Parity is a way of having the data checked. Normally, parity is not used. Depending on your software, there can be up to five options for the parity bit, as follows:

Space: Parity bit is always a 0.
  Odd: Parity bit is 0 if there is an odd number of 1s in the transmission and is a 1 if there is an even number of 1s in the transmission.
 Even: Parity bit is a 1 if there is an odd number of 1s in the transmission and is a 0 if there is an even number of 1s in the transmission.
 Mark: Parity bit is always a 1.
 None: No parity bit is transmitted.

Again, what is important is that both the sending and receiving units are set up to agree on the status of the parity bit.

Number of Data Bits

The number of data bits to be used is usually set at 8. There are options that allow the number of data bits to be set at 7. It is important that both computers expect the same number of data bits.

Number of Stop Bits

The number of stop bits used is normally 1. However, depending on the system, the number of stop bits may be 2. Stop bits are used to mark the end of each character transmitted. Both computers must have their communications software set to agree on the number of stop bits used.

WINDOWS MODEM SOFTWARE

Windows has built-in modem software, accessed through the Control Panel. Clicking on the Modem icon displays the window shown in Figure 47.4. Notice that Windows indicates the presence of an external Sportster modem. To test the modem, click the Diagnostics tab. Figure 47.5 shows the Diagnostics window.

Selecting COM2 (the Sportster modem) and then clicking More Info will cause Windows to interrogate the modem for a few moments, and then display the results in a new window, shown in Figure 47.6.

Specific information about the modem port is displayed, along with the responses to several AT commands. The *AT command set* is a standard set of commands that can be sent to the modem to configure, test, and control it. Table 47.2 lists the typical *Hayes*

FIGURE 47.4 Modems Properties window

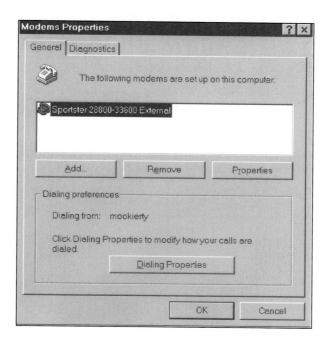

FIGURE 47.5 Modems Diagnostics window

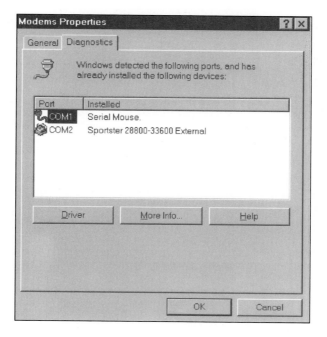

FIGURE 47.6 Modem diagnostic information

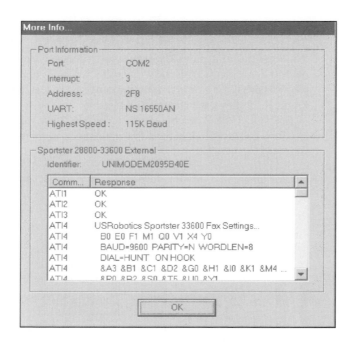

TABLE 47.2 Selected AT commands

Command	Function	Command	Function
A/	Repeat last command	Xn	Result code type
A	Answer	Yn	Long space disconnect
Bn	Select CCITT or Bell	Zn	Recall stored profile
Cn	Carrier control option	&Cn	DCD option
D	Dial command	&Dn	DTR option
En	Command echo	&F	Load factory defaults
Fn	Online echo	&Gn	Guard tone option
Hn	Switch hook control	&Jn	Auxiliary relay control
In	Identification/checksum	&M0	Communication mode option
Kn	SRAM buffer control	&Pn	Dial pulse ratio
Ln	Speaker volume control	&Q0	Communication mode option
Mn	Speaker control	&Sn	DSR option
Nn	Connection data rate control	&Tn	Self-test commands
On	Go online	&Vn	View active and stored configuration
P	Select pulse dialing	&Un	Disable Trellis coding
Qn	Result code display control	&Wn	Stored active profile
Sn	Select an S-register	&Yn	Select stored profile on power-on
Sn=x	Write to an S-register	&Zn=x	Store telephone number
Sn?	Read from an S-register	%En	Auto-retrain control
?	Read last accessed S-register	%G0	Rate renegotiation
T	Select DTMF dialing	%Q	Line signal quality
Vn	Result code form	-Cn	Generate data modem calling tone

compatible commands (first used by Hayes in its modem products). An example of an AT command is

$$ATDT\ 778\ 8108$$

which stands for AT (attention) DT (dial using tones). This AT command causes the modem to touch-tone dial the indicated phone number. Many modems require an initial AT command string to be properly initialized. This string is automatically output to the modem when a modem application is executed.

TELEPHONE MODEM TERMINOLOGY

In using technical documentation concerning a telephone modem, you will encounter some specialized terminology. Figure 47.7 illustrates some of the ideas behind some basic communication methods. As you can see from the figure, *simplex* is a term that refers to a communications channel in which information flows in one direction only. An example of this is a radio or a television station.

Duplex

The *duplex* mode refers to two-way communication between two systems. This term is further refined as follows. *Full duplex* describes a communication link that can pass data in two directions at the same time. This mode is analogous to an everyday conversation between two people either face-to-face or over the telephone. *Half duplex* refers to a communication link that allows data to pass in both directions, one at a time. An example is two people talking over CB radios.

Echo

Terminology used here has to do with how the characters you send to the other terminal are displayed on your monitor screen. The term *echo* refers to the method used to display characters on the monitor screen. First, there is a *local echo*. A local echo means that the sending modem immediately returns or echoes each character back to the screen as it is

FIGURE 47.7   Some basic communication methods

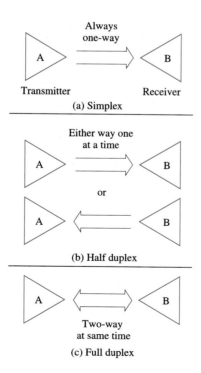

(a) Simplex

(b) Half duplex

(c) Full duplex

FIGURE 47.8   Echo modes

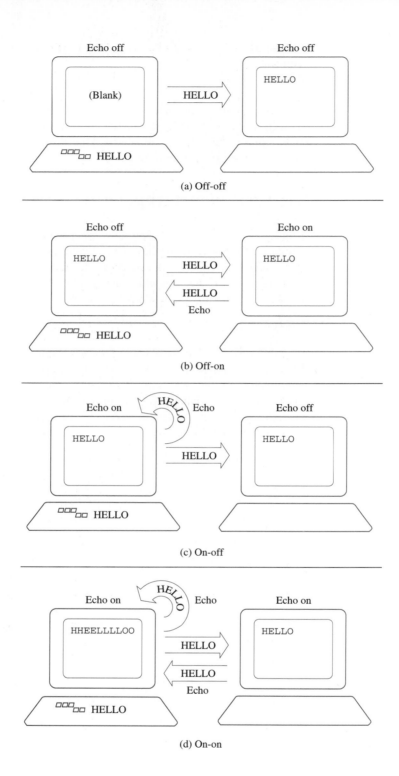

(a) Off-off

(b) Off-on

(c) On-off

(d) On-on

entered into the keyboard. This mode is required before transmission, so that you can see what instructions you are giving the communications software. Next there is *remote echo*. Remote echo means that during the communications between two computers, the remote computer (the one being transmitted to) sends back the character it is receiving. The character that then appears on your screen is the result of a transmission from the remote unit. This is a method of verifying what you are sending. To use the remote-echo mode, you must be in the full duplex mode. This idea is illustrated in Figure 47.8.

MODULATION METHODS

Many different techniques are used to encode digital data into analog form (for use by the modem). Several of these techniques are

- AM (amplitude modulation)
- FSK (frequency shift keying)
- Phase modulation
- Group coding

Figure 47.9 shows how the first three of these techniques encode their digital data.

To get a high data rate (in bits per second) over ordinary telephone lines, group coding techniques are used. In this method, one cycle of the transmitted signal encodes two or more bits of data. For example, using *quadrature modulation*, the binary patterns 00, 01, 10, and 11 encode one of four different phase shifts for the current output signal. Thus, a signal that changes at a rate of 2400 baud actually represents 9600 bps!

Another technique, called *Trellis modulation*, combines two or more other techniques, such as AM and quadrature modulation, to increase the data rate.

MNP STANDARDS

MNP (Microcom Networking Protocol) is a set of protocols used to provide error detection and correction, as well as compression, to the modem data stream. Table 47.3 lists the MNP classes and their characteristics.

FIGURE 47.9   Modulation techniques

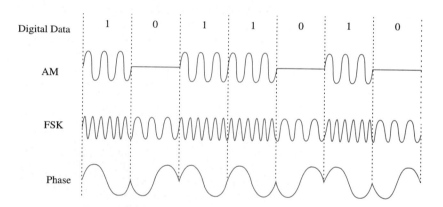

TABLE 47.3   MNP standards

Class	Feature
1	Asynchronous, half duplex, byte oriented
2	Asynchronous, full duplex, byte oriented
3	Synchronous, full duplex, byte oriented
4	Error correction, packet oriented
5	Data compression
6	Negotiation
7	Huffman data compression
9	Improved error correction
10	Line monitoring

Note: There is no MNP-8 standard.

TABLE 47.4   CCITT standards

Standard	Data Rate (bps)
V.22	1200
V.22 bis	2400
V.32	9600
V.32 bis	14,400
V.32 terbo	19,200
V.34	28,800/33,600
V.90	56K

MNP classes 4 and above are used with newer, high-speed modems. When two modems initially connect, they will negotiate the best type of connection possible, based on line properties and the features and capabilities of each modem. The CCITT standards supported by the modem are also part of the negotiation. Let's look at these standards as well.

CCITT STANDARDS

CCITT (French abbreviation for International Telegraph and Telephone Consultive Committee) standards define the maximum operating speed (as well as other features) available in a modem (which is a function of the modulation techniques used). Table 47.4 lists the CCITT standards.

Earlier, low-speed standards not shown are the Bell 103 (300 bps using FSK) and Bell 212A (1200 bps using quadrature modulation). V.22 is similar in operation to Bell 212A and is more widely accepted outside the United States.

The V.90 standard, finalized in early 1998, outlines the details of modem communication at 56Kbps, currently the fastest speed available for regular modems. Fax modems have their own set of standards.

ISDN MODEMS

ISDN (Integrated Services Digital Network) is a special connection available from the telephone company that provides 64Kbps digital service. An ISDN modem will typically connect to a *basic rate ISDN* (BRI) line, which contains two full duplex 64Kbps B channels (for voice/data) and a 16Kbps D channel (for control). This allows up to 128Kbps communication. ISDN modems are more expensive than ordinary modems and require you to have an ISDN line installed before you can use it.

CABLE MODEMS

One of the most inexpensive, high-speed connections available today is the *cable modem*. A cable modem connects between the television cable supplying your home and a network interface card in your computer. Two unused cable channels are used to provide data rates in the hundreds of thousands of bits per second. For example, downloading a 6MB file over a cable modem takes less than 20 seconds (during several tests of a new cable modem installation). That corresponds to 2,400,000 bps! Of course, the actual data rate available depends on many factors, such as the speed the data is transmitted from the other end and any communication delays. But unlike all other modems, the cable modem has the capability to be staggeringly fast, due to the high bandwidth available on the cable. In addition, a cable modem is typically part of the entire package from the cable company, and is returned when you terminate service. The cost is roughly the same as the cost of basic cable service.

FAX/DATA MODEMS

It is difficult to find a modem manufactured today that does not have fax capabilities built into it. Since fax/data modems are relatively inexpensive, it does not make sense to purchase a separate fax machine (unless it is imperative that you be able to scan a document before transmission). Word-processing programs (such as WordPerfect) now support the use of a fax/data modem, helping to make the personal computer almost an entire office by itself.

PROTOCOLS

A *protocol* is a prearranged communication procedure that two or more parties agree on. When two modems are communicating over telephone lines (during a file transfer from a computer bulletin board or an America Online session), each modem has to agree on the technique used for transmission and reception of data. Table 47.5 shows some of the more common protocols. The modem software that is supplied with a new modem usually allows the user to specify a particular protocol.

Figure 47.10 shows the results of running the WINIPCFG utility to get information about a modem connection. Under Dial-Up Networking, Windows will establish a PPP (Point-to-Point Protocol) connection using the modem (called the PPP Adapter). PPP allows the use of TCP/IP over a serial link (provided by the modem). Note the Adapter Address 44-45-53-54-00-00. The hexadecimal values may look familiar to someone who knows the ASCII character set. The values 44-45-53-54 represent the letters DEST, as in destination. It is interesting to discover hidden things like this.

TABLE 47.5 Modem communication protocols

Protocol	Operation
Xmodem	Blocks of 128 bytes are transmitted. A checksum byte is used to validate received data. Bad data is retransmitted.
Xmodem CRC	Xmodem using Cyclic Redundancy Check to detect errors.
Xmodem-1K	Essentially Xmodem CRC with 1024-byte blocks.
Ymodem	Similar to Xmodem-1K. Multiple files may be transferred with one command.
Zmodem	Uses 512-byte blocks and CRC for error detection. Can resume an interrupted transmission from where it left off.
Kermit	Transmits data in packets whose sizes are adjusted to fit the needs of the other machine's protocol.

FIGURE 47.10 WINIPCFG information for a modem connection

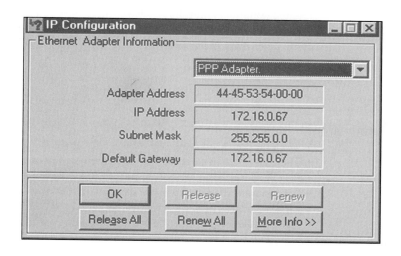

TABLE 47.6   Common telephone modem problems

Symptom	Possible Cause(s)
Can't connect	Usually this means that your baud rates or numbers of data bits are not matched. This is especially true if you see garbage on the screen, especially the { character.
Can't see input	You are typing in information but it doesn't appear on the screen. However, if the person on the other side can see what you are typing, it means that you need to turn your local echo on. In this way, what you type will be echoed back to you, and you will see it on your screen.
Get double characters	Here you are typing information and getting double characters. This means that if you type HELLO, you get HHEELLLLOO; at the same time, what the other computer is getting appears normal. This means that you need to turn your local echo off. In this way, you will not be echoing back the extra character. With some systems *half duplex* refers to local echo on, whereas *full duplex* refers to local echo off.

TROUBLESHOOTING TECHNIQUES

Table 47.6 lists some of the most common problems encountered in telephone modems. As you will see, most of the problems are software related.

Other common problems encountered involve very simple hardware considerations. For example, telephone modems usually come with two separate telephone line connectors.

The purpose of the phone input is to connect a telephone, not the output line from the modem, to the telephone wall jack. The phone input is simply a convenience. It allows the telephone to be used without having to disconnect a telephone line from the computer to the wall telephone jack. If you mistakenly connect the line from the wall telephone jack to the phone input, you will be able to dial out from your communications software, but your system will hang up on you. Make sure that the telephone line that goes to the telephone wall jack comes from the *line output* and not the *phone output* jack of your modem.

Another common hardware problem is a problem in your telephone line. This can be quickly checked by simply using your phone to get through to the other party. If you cannot do this, then neither can your computer.

A problem that is frequently encountered in an office or school building involves the phone system used within the building. You may have to issue extra commands on your software in order to get your call out of the building. In this case you need to check with your telecommunications manager or the local phone company.

Sometimes your problem is simply a noisy line. This may have to do with your communications provider or it may have to do with how your telephone line is installed. You may have to switch to a long-distance telephone company that can provide service over more reliable communication lines. Or you may have to physically trace where your telephone line goes from the wall telephone jack. If this is an old installation, your telephone line could be running in the wall right next to the AC power lines. If this is the case, you need to reroute the phone line.

SUMMARY

In this exercise we discovered that

- Modems convert digital data into an analog signal for use over the telephone.
- The data rate of a serial connection is referred to as the baud rate.
- RS-232 describes a serial communication standard.
- A communication link may be simplex, half duplex, or full duplex.
- MNP standards provide error detection and correction, compression, and negotiation.
- CCITT standards define the various operating speeds (baud rates) for modems.

This self-test is designed to help you check your understanding of the background information presented in this exercise.

True/False

Answer *true* or *false*.

1. The most convenient link to use in order to communicate between computers at long distances is the two-way radio.
2. Telephone lines easily allow for the transmission of the computer's ON and OFF signals.
3. The word *modulate* means "to change."
4. 9600 bps is the same as 9600 baud.
5. MNP standards define the maximum baud rates available.

Multiple Choice

Select the best answer.

6. The word *modem* comes from
   a. The first manufacturer of the device.
   b. An FCC regulation.
   c. A combination of the words "modulate" and "demodulate."
   d. None of the above.
7. The RS-232 standard
   a. Sets the transmission rate.
   b. Specifies the function of each conductor used in data communications.
   c. Was an old standard set by the FCC for communications between computers.
   d. Both b and c are correct.
8. Full duplex mode means
   a. Simultaneous communications between two systems.
   b. One-way communication from one system to another.
   c. Communication to more than one system.
   d. None of the above.
9. Which is not a modem property?
   a. Efficiency.
   b. Baud rate.
   c. Parity.
10. In the AT command AT DT, the DT means
    a. Device timeout.
    b. Dial using DTMF tones.
    c. Demodulate.

Matching

Match a corrective action on the right with each symptom on the left.

11. Nothing appears on the screen when you type.
12. You get double characters on the screen when you type.
13. You see a bunch of incorrect characters on the screen.

a. Check baud rate and number of data bits.
b. Turn local echo on.
c. Turn local echo off.
d. None of the above.

Completion

Fill in the blanks with the best answers.

14. The purpose of the _____ connection on the modem is as a convenience for connecting a telephone.
15. The purpose of the _____ connection on the modem is for the purpose of connecting the modem to the telephone lines.

16. The _____ rate is a common unit of measurement of the number of discrete changes per second.
17. A prearranged communication method is called a(n) _____.
18. In quadrature modulation, one cycle encodes _____ bits of data.
19. ISDN stands for Integrated Services _____ Network.

FAMILIARIZATION ACTIVITY

1. Read the documentation that accompanies the modem used by your system.
2. If one is not already present, create a text file on your disk that can be sent from one computer to another over the telephone modem.
3. If it has not already been done, properly connect your computer to a designated telephone outlet.
4. Your instructor will assign a telephone number for you to dial. Using your modem software, dial the assigned number and set parameters to an agreed-to setting.
5. Transmit the text file on your disk to the remote station.
6. Hang up and then verify (by telephone) that your text file has been received.
7. Now set your modem software so that your computer is in the receive mode. An assigned remote terminal will now contact you and send you a text file. Make sure that the received text file is stored on your floppy disk.
8. By telephone, verify to the sender that the text file has been properly received.
9. Experiment with the AT commands. Use the reference provided with your modem or find additional AT command information on the Web.

QUESTIONS/ACTIVITIES

1. Explain, through the use of diagrams, what is needed in order to have two computers communicate with each other over standard voice telephone lines.
2. What is a modem? What does it do? What is the origin of the term "modem"?
3. State what is meant by the RS-232 standard.
4. Define baud rate. How is this term used?
5. Explain how the echo mode is properly used between computers.
6. What is the purpose of a modem protocol?
7. Check your modem properties under Windows. Also, run WINIPCFG during a Dial-Up Networking session and examine the Adapter Address. Is it 44-45-53-54-00-00?
8. Attempt to measure the connection speed of your modem using a line-speed testing Web site or a program such as AnalogX's Netstat. Search the Web for "line speed test" and try several different sites. Netstat can be downloaded from www.analogx.com/contents/download/network/nsL.htm.

REVIEW QUIZ

Under the supervision of your instructor

1. Connect to the Internet using a modem.
2. Successfully use a computer modem to transmit and receive a text file.
3. Explain the various MNP and CCITT standards.
4. Describe several encoding methods.

48

CD-ROM and Sound Card Operation

INTRODUCTION

Joe Tekk was in his office at RWA Software, speaking slowly and clearly into a microphone. "Notepad . . . notepad . . . notepad," he was saying as Don, the senior technician, walked in. Don waited until Joe was finished speaking and asked, "What are you doing?"

"I'm training a speech recognizer to understand a bunch of different words. I want to use it to let me do simple things in Windows, such as open Notepad or Calculator, or shut down automatically whenever I say, 'Goodnight.'"

"Why did you say each word three times?"

Joe turned and picked up the speech board manual lying open on his desk. "It says you have to say each word more than once to build an accurate recognition envelope." Joe handed the manual to Don. "Pretty cool, huh?"

"Just beautiful, Joe," Don replied. "I think I'd like to teach it a few words of my own."

PERFORMANCE OBJECTIVES

Upon completion of this exercise, you will be able to

1. Discuss what is required to operate a CD-ROM drive and a sound card.
2. Explain what it means to be MPC compliant.
3. Discuss the various CD-ROM formats.

KEY TERMS

MPC compliant
Pits
High Sierra format
ISO-9660
Multisession CD-ROM
CLV
CAV
ATAPI

Digital-to-analog converter
Low-pass filter
Digital signal processor
MIDI
Controller
Sequencer
Wave table synthesis

BACKGROUND INFORMATION

The term *multimedia* is now generally applied to personal computers equipped with CD-ROM drives and sound cards. Entire encyclopedias are now available on CD-ROM. Access the subject of spacecraft, and you get live-action video of an *Apollo* moon shot, complete with the accompanying audio. Hundreds of software packages are available on CD-ROM, with more appearing every day.

A computer is said to be **MPC (Multimedia PC) compliant** if it contains the following hardware:

- 386SX-16 processor (or better)
- 4MB of RAM (or more)
- 40MB hard drive (or more)
- A color VGA display (or better)
- A mouse
- A single-spin CD-ROM (or faster)

Single-spin (1x) CD-ROM drives transfer data at a maximum rate of 150KB/second. A double-spin (2x) CD-ROM drive transfers data at 300KB/second, and so on.

A computer is MPC-2 compliant if it contains the following updated hardware:

- 486SX-25 processor (or better)
- 8MB of RAM (or more)
- 160MB hard drive (or more)
- A color VGA display (or better)
- A mouse
- A double-spin (2x) CD-ROM (or faster)

MPC-3 compliant hardware requires a 75-MHz Pentium and a 540-MB hard disk. The CD-ROM must be at least a 4x.

Note that current technology far exceeds the minimum requirements established by MPC-3.

In this exercise, we will see what is necessary to install a CD-ROM drive and sound card in a machine.

CD-ROM OPERATION

A CD-ROM stores binary information in the form of microscopic *pits* on the disk surface. The pits are so small that a CD-ROM typically stores more than 650MB of data. This is equivalent to more than 430 1.44MB floppies. A laser beam is shone on the disk surface and either reflects (no pit) or does not reflect (pit), as you can see in Figure 48.1.

These two light states (reflection and no reflection) are easily translated into a binary 0 and a binary 1. Since the pits are mechanically pressed into a hard surface and only touched by light, they do not wear out or change as a result of being accessed.

FIGURE 48.1   Reading data from a CD

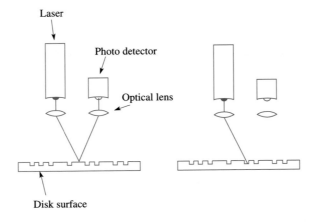

(a) Laser light reflects.   (b) Laser light gets trapped in a pit.

**FIGURE 48.2   Compact disk
(a) Dimensions and
(b) structure**

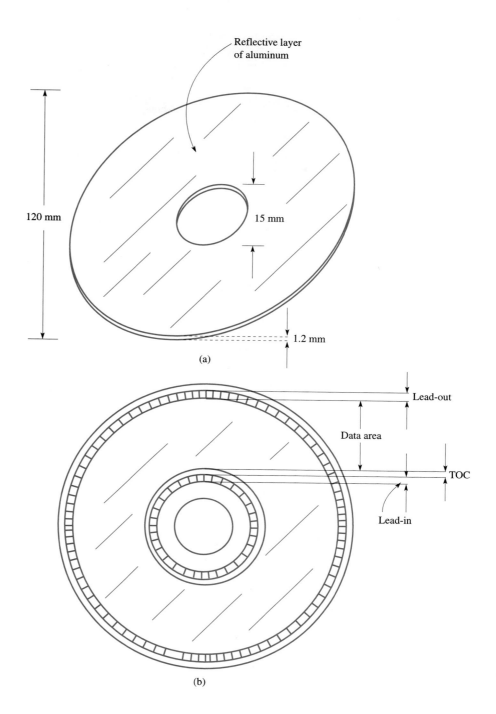

(a)

(b)

PHYSICAL LAYOUT OF A COMPACT DISK

Figure 48.2 shows the dimensions and structure of a compact disk. The pits previously described are put into the reflective aluminum layer when the disk is manufactured. Newer recordable CDs use a layer of gold instead of aluminum so that they can be written to using a low-power laser diode.

THE HIGH SIERRA FORMAT

The High Sierra format specifies the way the CD is logically formatted (tracks, sectors, directory structure, file name conventions). This specification is officially called ISO-9660 (International Standards Organization).

TABLE 48.1 CD-ROM standards

Book	Feature
Red	CD-DA. Digital audio. PCM encoding.
Green	CD-I. Interactive (text, sound, and video). ADPCM and MPEG-1 encoding.
Orange	CD-R. Recordable.
Yellow	CD-ROM. Original PC CD-ROM format. 150KB/s transfer rate.

CD-ROM STANDARDS

The evolution of the CD-ROM is documented in several **books**. These are described in Table 48.1. The red book describes the method used to store digital audio on the CD-ROM. Pulse code modulation (PCM) is used to sample the audio 44,100 times/second with 16-bit sampling.

The green book specifies the CD interactive format typically used by home video games. Text, audio, and video are interleaved on the CD-ROM, and the MPEG-1 (Motion Picture Entertainment Group) video encoding method requires special hardware inside the player to support real-time video. The CD-ROM-XA (extended architecture) format is similar to CD-I.

The orange book provides the details on recordable CD-ROM drives. Gold-based disks are used to enable data to be written to the CD-ROM, up to 650MB. A **multisession CD-ROM** allows you to write to the CD more than once, and requires a multisession CD-ROM drive. Compact disks that only allow one recording session are known as WORM drives, for write once, read mostly.

The yellow book describes the original PC-based CD-ROM format (single spin), which specifies a data transfer rate of 150KB/s. 2x CD-ROMs transfer at 300KB/s. 4x CD-ROMs transfer data at 600KB/s. Currently there are 52x CD-ROM drives on the market, with faster ones coming.

CLV Versus CAV

There are two methods for spinning the CD in a CD-ROM drive:

- **CLV** (constant linear velocity)
- **CAV** (constant angular velocity). Newer than CLV.

In CLV, the CD is rotated at different speeds so that the data transfer rate remains constant. For example, the CD rotates slower when the read head is over the outer tracks and faster when the head is over the inner tracks.

In CAV, the CD spins at a constant rate. Thus, data is transferred faster from the outer tracks than it is from the inner tracks.

Some drives use a combination of CLV and CAV technology.

PHOTO CD

Developed by Kodak, the photo CD provides a way to store high-quality photographic images on a CD (using recordable technology) in the CD-I format. Each image is stored in several different resolutions, from 192×128 to 3072×2048, using 24-bit color. This allows for around 100 images on one photo CD.

ATAPI

ATAPI stands for AT Attachment Packet Interface. ATAPI is an improved version of the IDE hard drive interface and uses *packets* of data during transfers. The ATAPI specification supports CD-ROM drives, hard drives, tape backup units, and plug-and-play adapters.

CD-ROM INSTALLATION

The CD-ROM drive, like the floppy drive, mounts with a few screws in an empty drive bay and is connected to the computer by a ribbon cable plugged into an adapter card or an IDE connector on the motherboard. A second cable reserved for left and right audio signals is connected between the CD-ROM drive and the sound card. This allows music CDs to be played with the CD-ROM drive. As a matter of fact, most CD-ROM drives come with application software that turns the color display into a large CD control panel for a stereo.

The CD-ROM drive is equipped with the same 4-pin power connector found on hard and floppy drives.

There are two styles of CD-ROM drives: those that have manual cartridge loading and those that have automatic cartridge loading. As shown in Figure 48.3, both drives contain an activity indicator, an audio jack, and a volume control on the front panel. In the manual drive, the user must pull the cartridge out and push it in. In the automatic drive, the user presses a button to load or eject a small platter that holds the CD.

Once the CD-ROM drive has been installed, two pieces of software must be loaded to make the drive operational. The first is a manufacturer-specific driver that manages the hardware interface between the CD-ROM drive and the computer. The second required program, which is the same for all CD-ROM drives, is the MSCDEX.EXE (Microsoft CD-ROM Extensions) file that comes with DOS and Windows. This program is used to access the CD-ROM drive as if it were an ordinary floppy drive, with sectors, tracks, a file allocation table, and so on. The first program needs to be loaded from the CONFIG.SYS file. MSCDEX.EXE gets loaded during processing of the AUTOEXEC.BAT file. Most CD-ROM drive manufacturers provide an automated software installation program with their drives that makes the necessary changes in both CONFIG.SYS and AUTOEXEC.BAT, and copies the necessary files to the hard disk.

FIGURE 48.3   Two styles of CD-ROM drives   (a) Manual cartridge opening/closing performed by user and (b) automatic cartridge opening/closing when user presses Load/Eject button

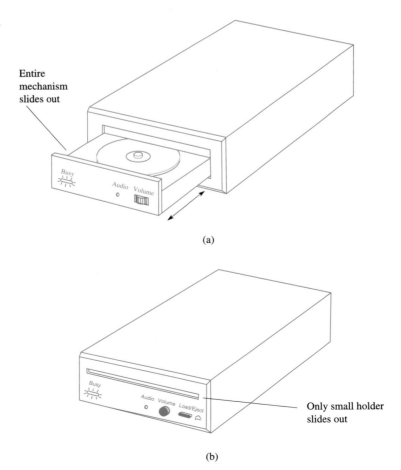

Entire mechanism slides out

(a)

Only small holder slides out

(b)

During the software installation, you are usually offered a choice between software transfer and DMA transfer of CD-ROM data. The DMA transfer option is faster than software transfer, but requires a spare DMA channel and an unused IRQ line. Generally, the setup parameters of the CD-ROM drive must be tweaked to get the optimal performance out of the drive.

A sample line from CONFIG.SYS for a Mitsumi CD-ROM drive is as follows:

DEVICE=C:\MTMCDS.SYS /D:CDROM-1 /L:R /P:300 /A:1 /M:1 /I:3

where   MTMCDS.SYS = the name of the software transfer driver
/D:CDROM-1 = the device name of the CD-ROM drive
/P:300 = the base address of the adapter card
/A:1 = the audio play mode
/M:1 = the number of memory buffers to use
/I:3 = the IRQ signal used by the CD-ROM drive

The corresponding line from AUTOEXEC.BAT is as follows:

C:\DOS\MSCDEX.EXE /D:CDROM-1 /L:R

where   MSCDEX.EXE = the required DOS interface program
/D:CDROM-1 = the same device name used in the CONFIG.SYS file
/L:R = the drive letter (R) of the CD-ROM

CD-ROM PROPERTIES IN WINDOWS

Right-clicking on the CD-ROM icon in the My Computer window displays the Properties window shown in Figure 48.4. The label box indicates that there is a disk in the CD-ROM drive (the Windows CD-ROM). Note that there is no free space indicated in the pie chart.

SOUND CARD OPERATION

Along with CD-ROM drives, sound cards for PCs have also increased in popularity. Currently, 128-bit sound cards are available that provide multiple audio channels and FM-quality sound, and are compatible with the MIDI (Musical Instrument Data Interface) specification. Figure 48.5 shows a typical PCI sound card.

FIGURE 48.4   CD-ROM Properties window

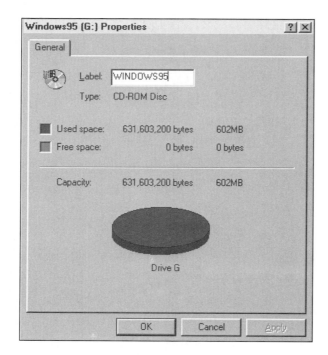

FIGURE 48.5   PCI sound card

FIGURE 48.6   How binary data is converted into an analog waveform

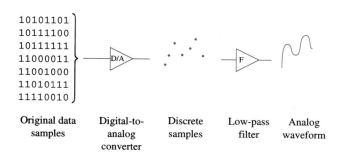

```
10101101 ⎫
10111100 ⎪
10111111 ⎪
11000011 ⎬
11001000 ⎪
11010111 ⎪
11110010 ⎭
```

Original data   Digital-to-   Discrete   Low-pass   Analog
samples   analog   samples   filter   waveform
                     converter

The basic operation of the sound card is shown in Figure 48.6. Digital information representing samples of an analog waveform are inputted to a ***digital-to-analog converter***, which translates the binary patterns into corresponding analog voltages. These analog voltages are then passed to a ***low-pass filter*** to smooth out the differences between the individual voltage samples, resulting in a continuous analog waveform. All of the digital/analog signal processing is done in a custom ***digital signal processor*** chip included on the sound card.

Sound cards also come with a microphone input that allows the user to record any desired audio signal.

MIDI

MIDI stands for Musical Instrument Digital Interface. A MIDI-capable device (electronic keyboard, synthesizer) will use a MIDI-in and MIDI-out serial connection to send messages between a ***controller*** and a ***sequencer***. The PC operates as the sequencer when connected to a MIDI device. MIDI messages specify the type of note to play and how to play it, among other things. Using MIDI, a total of 128 pitched instruments can generate 24 notes in 16 channels. This can be accomplished in a PC sound card by using frequency modulation or ***wave table synthesis***, the latter method utilizing prerecorded samples of notes stored in a data table. The output of a note is controlled by several parameters. Figure 48.7 illustrates the use of attack, delay, and release times to shape the output waveform envelope. Each of the four parameters can be set to a value from 0 to 15.

FIGURE 48.7   Note envelope

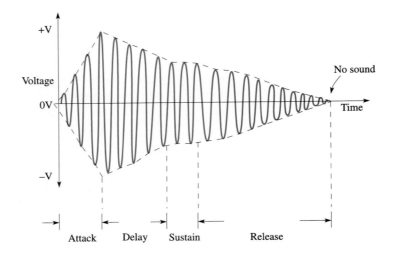

SOUND CARD INSTALLATION

Once the sound card has been plugged into the motherboard, it is necessary to connect a set of external speakers to it. Sound cards do not use the internal speaker of the PC.

Like the CD-ROM drive, sound cards require a software driver program in order to work properly. Once again, the manufacturer of the sound card usually provides an automated setup program that will make the necessary changes in the file. Even so, it is necessary to pick the correct combination of I/O ports and IRQ lines so that both the CD-ROM drive and the sound card work together. A connector for a CD-ROM is provided on the sound card, which helps simplify installation and setup.

The CONFIG.SYS lines for a Creative Labs 16 MultiCD Sound Blaster are as follows:

```
DEVICE=C:\SB16\DRV\CTSB16.SYS /UNIT=0 /BLASTER=A:220 /I:7 /D:1 /H:5
DEVICE=C:\SB16\DRV\CTMMSYS.SYS
```

where   CTSB16.SYS and CTMMSYS.SYS = the device drivers
/BLASTER=A:220 = the base port address
/I:7 = IRQ 7
/D:1 = the low-DMA channel 1
/H:5 = the high-DMA channel 5

System DMA channels are used to transfer data to and from the sound card and must be properly picked so that they do not conflict with DMA channels being used by other devices (such as the CD-ROM drive).

No software is loaded from the AUTOEXEC.BAT file, but a few environment variables are defined as follows:

```
SET SOUND=C:\SB16
SET BLASTER=A220 I7 D1 H5 P330 T6
SET MIDI=SYNTH:1 MAP:E
```

These environment variables are examined by the Sound Blaster software when it needs to know vital settings after the machine has been booted.

The software supplied with sound cards is capable of using many different types of audio file formats. These include the .WAV files used by Windows, as well as .VOC (voice), .CMF (Creative Music File), and MIDI file formats.

When combined with a CD-ROM drive, the sound card provides a complete multimedia environment on the personal computer.

One of the most common reasons a new CD-ROM drive or sound card does not work has to do with the way its interrupts and/or DMA channels are assigned.

Figure 48.8 shows the location of the sound card in the hardware list provided by Device Manager. The AWE-32 indicates that the sound card is capable of advanced wave effects using 32 voices.

Figure 48.9 shows the interrupt and DMA assignments for the sound card. Typically, interrupt 5 is used (some network interface cards also use interrupt 5), as well as DMA

FIGURE 48.8   Selecting the sound card

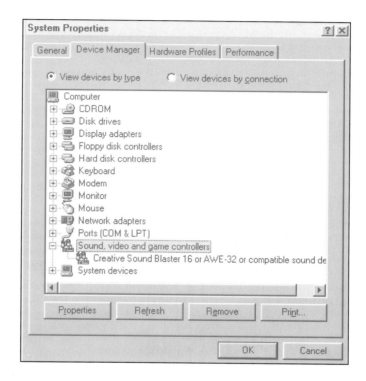

FIGURE 48.9   Sound card settings

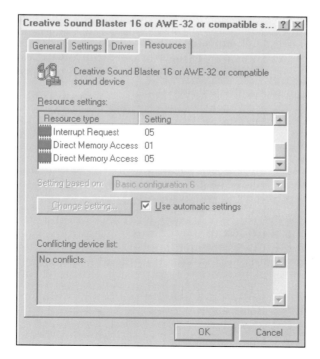

channels 1 and 5. If the standard settings do not work, you need to experiment until you find the right combination.

SUMMARY

In this exercise we discovered that

- There are levels of MPC compliancy.
- CDs stored more than 650MB of data using microscopic pits on the CD surface.
- CD-ROM standards are documented in different *books*.
- CD-ROM drives use CLV and CAV technology to spin the CD.
- Sound cards use wave table synthesis and digital-to-analog conversion to create audio sounds.

SELF-TEST

This self-test is designed to help you check your understanding of the background information presented in this exercise.

True/False

Answer *true* or *false*.

1. Music CDs can be played on a CD-ROM drive.
2. Only MSCDEX.EXE is needed to operate a CD-ROM drive.
3. The CONFIG.SYS file loads the CD-ROM device driver.
4. The CONFIG.SYS file loads the sound card device driver.
5. The CD-ROM drive and sound card work with any IRQ line.
6. A 32x CD-ROM drive transfers data at a rate of 32 times 150 KB/s.
7. MIDI is used for synthesized speech.

Multiple Choice

Select the best answer.

8. In the term *MPC compliant*, MPC stands for
   a. Master Peripheral Controller.
   b. Multi-Purpose CD-ROM.
   c. Multimedia Personal Computer.
9. Digital information is stored on the surface of a CD as
   a. Microscopic bar codes.
   b. Microscopic pits.
   c. Microscopic 1s and 0s.
10. Sound cards
    a. Require their own device driver.
    b. Share a device driver with the CD-ROM drive.
    c. Do not require a device driver.
11. Which format supports recordable CDs?
    a. CD-I.
    b. CD-XA.
    c. CD-R.
12. This technology spins the CD at a constant rate:
    a. CLV.
    b. CAV.
    c. Both a and b.
13. CAV stands for
    a. Constant amplitude voltage.
    b. Constant angular velocity.
    c. Constant audio velocity.

Completion

Fill in the blank or blanks with the best answers.

14. Digital data is stored through the use of _____ on the CD surface.
15. A double-spin CD-ROM drive transfers data at the maximum rate of _____.
16. Sound cards use a _____ chip to generate audio signals.
17. AWE stands for _____ _____ _____.
18. A standard musical serial interface is called _____.
19. A MIDI controller sends data to a MIDI _____.
20. ATAPI stands for AT attachment _____ interface.

FAMILIARIZATION ACTIVITY

1. Put a software CD into a CD-ROM drive and do a complete directory listing of its contents with the DOS command

```
DIR /S
```

How many files are stored on the CD? How many bytes are stored? Does the directory listing indicate the free space on the CD? Should there be any free space?
2. Put an audio CD into a CD-ROM drive. Try to do another directory listing. What type of error do you get?
3. Use whatever sound card software you have to record three .WAV files: one that is one second long, one that is five seconds long, and one that is 10 seconds long. Compare the sizes of these files.

QUESTIONS/ACTIVITIES

If a sound card is capable of recording speech from a microphone, then it should be possible to issue spoken commands to your computer. How do you think that a computer program might be used to recognize speech?

REVIEW QUIZ

Under the supervision of your instructor

1. Discuss what is required to operate a CD-ROM drive and a sound card.
2. Explain what it means to be MPC compliant.
3. Discuss the various CD-ROM formats.

49

Multimedia Devices

INTRODUCTION

Joe Tekk went with a friend—Marilyn Jayne, a photographer—to the camera store to look for a new digital camera. She invited Joe because she knew he was familiar with the lingo associated with digital cameras and with computers in general. She was interested in professional-style equipment that would allow her to perform high-quality work on her PC.

She decided to make the switch to digital because many of her competitors were already using the new technology. She was impressed with the quality of their work.

The salesperson showed them every camera in the store, commenting on each special feature. Some of them use DOS-formatted disks and store the images as .JPEG files.

She settled on a reasonably priced model that was easy to use and had enough features to offer tremendous flexibility.

As they left the store, Joe replied, "I love to spend other people's money!"

PERFORMANCE OBJECTIVES

Upon completion of this exercise, you will be able to

1. Identify different types of multimedia devices.
2. Recognize how each device is interfaced to the personal computer.
3. Discuss a method of troubleshooting multimedia devices.

KEY TERMS

Digital camera
Charge-coupled device (CCD)
Scanner

Optical character recognition (OCR)
Digital versatile disk (DVD)
Bar code reader

BACKGROUND INFORMATION

Multimedia devices are part of a growing portion of the computer industry. New technology is making it possible to interface more and more electronic devices to the personal computer. Recall that the new MMX and 3DNow! instructions are specially designed to speed up multimedia applications. Multimedia devices include scanners, bar code readers, cameras, and television and video add-on cards, just to name a few. Other multimedia applications build on the multimedia devices already installed in the system, such as voice recognition software, which uses the existing sound card and microphone as input to the voice recognition application software.

Multimedia devices involve three major areas of product development:

- Hardware
- Interface
- Software

Generally, advances in multimedia devices come about from advances in related technological areas, such as changes in data compression algorithms, data storage devices, data transports, and new interfaces, such as the universal serial bus (USB). Figure 49.1 shows several different types of multimedia devices. Let's examine a few of them.

CAMERAS

The *digital camera* stores images digitally (in the camera memory) as opposed to recording them on film. After a picture has been taken, it can be downloaded to a computer and manipulated with a graphics program. Unlike film, which has high resolution, digital cameras are limited by the amount of available memory and the resolution of the digitizing mechanism.

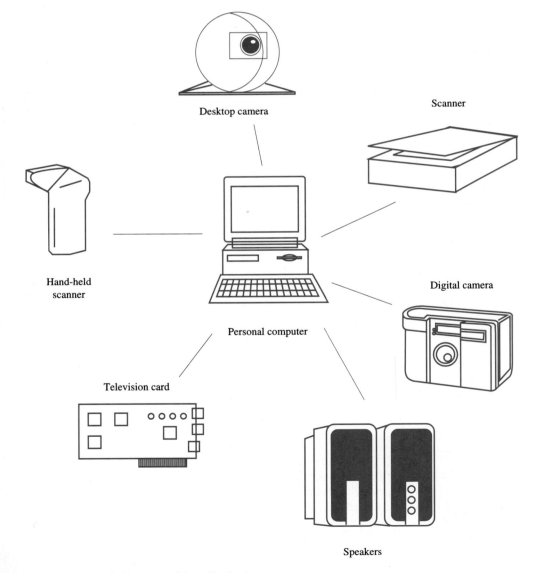

FIGURE 49.1   Different multimedia devices

FIGURE 49.2 USB color camera

Ultimately, the resolution is determined by the output device on which it is displayed, for example, a graphics display at 800×600, or a laser printer with a resolution of 600×600.

A digital camera uses either **charged-coupled devices (CCDs)** or CMOS chips. More expensive cameras use the CCD method, whereas the CMOS chips are found in cheaper cameras. In either case, the big advantage of digital cameras is the reduced cost of obtaining the images, and the speed, because there is no traditional film processing.

There are many other types and styles of digital cameras. Some of them are designed to sit on top of a monitor, so a picture of the computer user can be stored, displayed on the screen, or sent out on the Internet. Figure 49.2 shows a color camera that connects to the universal serial bus.

SCANNERS

Scanners are a common multimedia device used to work with both text and graphical images. There are many different types of scanners available on the market, such as flatbed, image, and the single-sheet feed scanners commonly found on multipurpose machines. Most devices are *Twain compliant* (the de facto interface standard for scanners), meaning that they are compatible with standard Twain device drivers and are supported by all Twain compatible software packages.

OCR, or *optical character recognition*, is performed by examining a scanned image that contains text. The software intensively examines the scanned image to match characters with the graphical representation. There is always an element of error introduced, depending on the quality of the input source, orientation, and quality of the hardware. Most software packages perform a spell check of any information that has been read to identify where the scanner had difficulty identifying the characters. Figure 49.3 shows OmniPage Pro recognizing some scanned text.

TELEVISION CARDS

One of the newest multimedia add-on cards is the television adapter. With the addition of cable television wire, television signals can be displayed directly on the desktop. An alternate source of programming is offered through the Web.

FIGURE 49.3 Using OCR to recognize text (OmniPage and OmniPage Pro are registered trademarks of ScanSoft, Inc., www.scansoft.com)

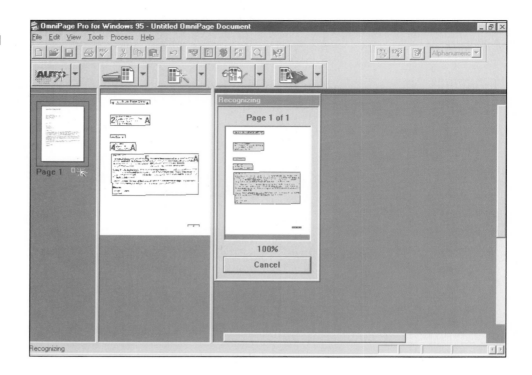

FIGURE 49.3 Using OCR to recognize text (OmniPage and OmniPage Pro are registered trademarks of ScanSoft, Inc., www.scansoft.com)

Television cards support the various recognized standards, such as NTSC, PAL, and SECAM, and many of them offer full-motion MPEG capture capability. Optional features include MTS stereo, remote control, and radio features. The tuners can usually display many channels on the screen simultaneously and switch between channels very easily.

DVD

Digital Versatile Disk (DVD) CDs and CD-ROM drives make use of advancements in laser technology to significantly increase data storage capacity. Compare an ordinary CD's 650MB with a DVD CD's 4.7GB to 17GB of storage. Typically, two disks are bonded back-to-back, and may contain two data layers on each side.

DVD CDs support digital audio (surround-sound format), up to four hours of video (MPEG-2 video compression) per side, and the DVD-ROM file system. Table 49.1 shows the various DVD prerecorded and recordable formats.

Do not confuse DVD with DIVX (Digital Video Express), a different video technology used in pay-per-view systems. DIVX disks are not DVD disks, but a DIVX player will play a video DVD disk.

TABLE 49.1 DVD storage capacity

Prerecorded DVD		
Format	Capacity	Sides/Layers
DVD-5	4.7GB	1/1
DVD-9	8.5GB	1/2
DVD-10	9.4 GB	2/1
DVD-18	17.0GB	2/2
Recordable DVD		
Format	Capacity	
DVD-R	3.8 GB/side	
DVD-RAM	2.6 GB/side	

BAR CODE READERS

A **bar code reader** is one of the most common multimedia devices. Bar codes are used to easily identify many different objects, such as books in a library, pieces of equipment that must be inventoried, letters being mailed through the postal service, or a host of other applications.

There are many different bar codes, each with a different purpose or capability. Some of the most common bar codes are

- Code 39

  Code 39 is variable-length and can encode the following characters:

  0123456789ABCDEFGHIJKLMNOPQRSTUVWXYZ-.*$+%

- Code 93

  Code 93 is variable length and can encode the complete 128-character ASCII code. It also provides a higher character density than code 39.

- Code 128

  Code 128 is a variable-length, high-density code capable of 106 different bar and space patterns, each of which can have one of three different meanings.

- POSTNET

  POSTNET (POSTal Numeric Encoding Technique) is a five-, nine-, or 11-digit numeric code used by the U.S. Postal Service to encode ZIP code information. POSTNET is unlike other bar codes because the data is encoded in bar height instead of bar width and spaces. Most standard bar code readers cannot decode the POSTNET code.

- UPC codes

  UPC codes fall into three categories: UPC-A, UPC-E, and UPC supplemental. UPC-A codes consist of 11 data digits, whereas the UPC-E can encode only six digits. A supplemental two- or five-digit number can be appended to the main code.

- EAN (European Article Numbering, also called JAN in Japan)

  EAN coding uses the same size requirement and coding mechanism as the UPC codes, changing only the meaning of the digital data. For example, EAN-13 is similar to UPC-A, except for the addition of a country code, which is encoded into positions 12 and 13.

- Codabar

  The Codabar code is variable length, encoding the following characters:

  0123456789-$:/.+ABCD

- PDF417

  PDF417 is a high-density, two-dimensional code capable of encoding the entire ASCII character set. The PDF417 code can encode as many as 2725 characters in a single bar code.

INTERFACING MULTIMEDIA DEVICES

Every type of multimedia device must be interfaced to the PC. Usually, the interface is in the form of a cable connection to a serial, parallel, or USB port, but connections can also be of the wireless nature, using infrared or radio signals.

For communication to occur between the PC and the multimedia device, it is necessary for the interfacing to be working properly. If problems occur, all three items are suspect: the multimedia device, the cable, and the PC interface. Check the cabling and the status indicators on all equipment to help isolate any problems or errors.

TABLE 49.2   Common Internet multimedia file types

File Extension	File Type	Features
.GIF	Image	256 colors, animation, lossless compression
.JPG	Image	16M colors, lossy compression
.WAV	Audio	Mono/Stereo, no compression
.MP3	Audio	Good compression, high-quality audio
.MID	Audio	High-quality instrumental music, low data rate
.MPG	Video+Audio	Lossy video compression, 16M colors, high-resolution, CD-quality audio
.AVI	Video+Audio	256-color, low-resolution, low-quality audio
.PDF	WYSIWYG document	Printable (from PDF viewer) document containing text and images

DEVICE DRIVERS

Each multimedia device will probably contain its own device driver to control each device. There are many times when a device driver is out of date or contains an error that causes a problem. In such cases, it is best to make sure the most recent version of the device driver is on hand. All manufacturers provide access to the latest set of device drivers for their products on the World Wide Web. Search their Web site or check the documentation provided with each device for a specific way to obtain the latest drivers.

Note that multimedia vendors constantly fix minor problems and enhance their products. You may want to visit the manufacturer's device drivers Web page just to see if there are any reported problems. If there are, they can be fixed by downloading the latest driver. Be sure to follow all directions provided with the software.

INTERNET MULTIMEDIA FILES

There are many different file types in use on the Internet, supporting images, audio and video, and documents. Table 49.2 summarizes several of these file types.

The Internet contains many free applications capable of creating, editing, and viewing or playing each type of file. Simply search for the desired file type and spend some time looking through the search results. You may find such applications as

- Paint Shop Pro, for images
- NETTOOB, for playing/editing audio and video
- Adobe Acrobat, for viewing PDFs

OTHER MULTIMEDIA APPLICATIONS

Other multimedia applications, such as voice recognition software, use the existing hardware already installed on the personal computer, such as a sound card, speakers, and a microphone. If a computer does not have a sound card, for example, it must be purchased separately before the voice recognition application will run. Some of the voice recognition products include a headset for use with the software as shown in Figure 49.4. The voice recognition software must be *trained* to recognize the speech patterns of each user. Commonly a set of words or sentences is read aloud and the computer stores the characteristic information.

Video editing is another multimedia application commonly being performed on a PC. Most products require special hardware to save and compress the video in real time. During playback, no special hardware is required to display the output.

FIGURE 49.4 Headset with earphone and microphone

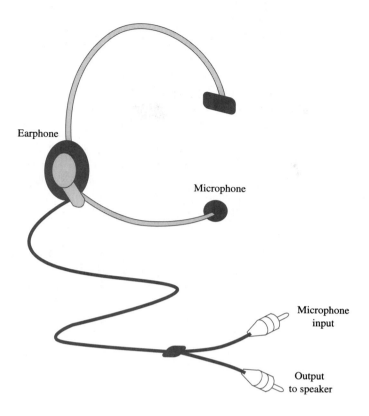

Earphone

Microphone

Microphone input

Output to speaker

Many different problems can be encountered with every multimedia device. Since there are three potential sources of any problem, it may be difficult to determine the cause. Use the following list of techniques when trying to resolve specific problems:

- Make sure your system meets all the requirements described by the product documentation.
- Make sure all cable connections are secure.
- Restart your computer to see if the problem is still present.
- Make sure you have the latest device drivers.
- Uninstall and reinstall the software.
- Refer to the customer support offered by the product manufacturer.

SUMMARY

In this exercise we discovered that

- There are many multimedia devices, including cameras, scanners, DVD CDs, and CD-ROM drives.
- Each multimedia device requires a device driver.
- There are several Internet multimedia file types, including .GIF, .JPG, .MPG, and .PDF.

SELF-TEST

This self-test is designed to help you check your understanding of the background information presented in this exercise.

True/False

Answer *true* or *false*.

1. Digital cameras require traditional photo processing techniques.
2. OCR is used to store the image in a digital camera.
3. Every multimedia device may be interfaced to the PC.
4. Multimedia applications take advantage of MMX and 3DNow! instructions.

5. Every bar code reader can read every type of bar code.
6. DVDs store more data than ordinary CD-ROMs.
7. A multimedia device may require a device driver.

Multiple Choice

Select the best answer.

8. Multimedia devices commonly involve three areas of product development:
    a. Hardware, software, and the interface.
    b. Input, output, and indicator lights.
    c. Microphones, speakers, and pointing devices.
9. Bar code readers can be used to
    a. Differentiate between many different items.
    b. Identify multiple objects with the same code.
    c. Both a and b.
10. A device driver is used to
    a. Specify a multimedia device operation.
    b. Make sure all cable connections are secure.
    c. Both a and b.
11. A headset contains
    a. A speaker and a microphone.
    b. A speaker only.
    c. A microphone only.
12. CCD cameras are
    a. More expensive than CMOS cameras.
    b. Cheaper than CMOS cameras.
    c. Approximately the same price.
13. CCD stands for
    a. Charge-coupled device.
    b. Capacitive-coupled device.
    c. Charge-controlled device.
14. Which is not a multimedia file type?
    a. MPEG.
    b. MUL.
    c. MP3.

Matching

Match a description of bar code on the right with each item on the left.

15. Code 128
16. Code 93
17. EAN
18. POSTNET
19. PDF417

    a. High density, two-dimensional code encoding 2725 characters.
    b. Encodes the complete 128-character ASCII code.
    c. High density, using 106 different bar and space patterns.
    d. Encodes five, nine, or 11 digit numeric codes.
    e. Encodes a country code.

Completion

Fill in the blank or blanks with the best answers.

20. Each multimedia device requires a(n) _____ _____ to control the multimedia device.
21. _____ communication uses infrared light and radio signals.
22. Before installing a multimedia device, make sure your computer meets the _____ requirements as specified by the manufacturer.
23. The latest version of the _____ _____ can usually be found on the manufacturer's Web site.
24. Voice recognition software requires _____ in order to increase the recognition accuracy.
25. The _____ image file type only allows 256 colors.
26. An MPG file provides 16M _____.

1. Take a trip to a local office supply store (such as Staples or OfficeMax) and examine the multimedia equipment available. Check the prices on the following:
   - Color scanner
   - Color camera (with internal floppy)
   - Bar code reader/printer
   - Polaroid picture scanner
   - Business card scanner
   - Video telephone
2. Visit a local desktop publishing house and ask for a demonstration of its multimedia software.
3. Look up the definition of each multimedia file type shown in Table 49.2.

1. What multimedia devices are missing from this exercise?
2. What multimedia devices would you most like to have and use? Why?
3. Why put a TV card in your computer when you can just watch a real television?

Under the supervision of your instructor

1. Identify different types of multimedia devices.
2. Recognize how each device is interfaced to the personal computer.
3. Discuss a method of troubleshooting multimedia devices.

UNIT VI Selected Topics

50

An Introduction to Intel Microprocessor Architecture

INTRODUCTION

Joe Tekk was sitting at his desk, drawing on a sheet of paper. Don, the senior technician, asked him what he was drawing.

Joe replied, "I'm making a diagram of the internal registers in the Pentium processor. I have to meet with a group of Boy Scouts in the networking seminar room in 15 minutes. They are are visiting local high-technology companies to learn about careers."

Don had some free time to spare. "Do you need any help with them?"

"Sure, Don. That would be great."

Don turned to leave. "I'll go get a motherboard to show them." Joe finished sketching the register model. Then he scanned it and saved the scanned image of his register model as a .JPG image file. He copied the image file to his Web directory, and created a Web page for it that displayed the file and a short description of the register model. Don arrived with a motherboard just as Joe was finishing. He looked at what Joe had done in the six minutes he was gone.

"Nice work, Joe. What took you so long?"

PERFORMANCE OBJECTIVES

Upon completion of this exercise, you will be able to

1. Outline the software model of the 80x86 microprocessor family.
2. Discuss the various instruction groups and addressing modes that are available.
3. Explain how a 20-bit address is formed from two 16-bit numbers.

KEY TERMS

Binary number
Bit
Hexadecimal number
Real mode
Protected mode
General-purpose register
Segment register
Flag register
Byte

Word
Double word
Byte-swapping
Little endian format
Instruction
Addressing mode
Physical address
Segment
Megabyte

Beginning with the 8088 and 8086 microprocessors, Intel started a microprocessor family that grew to include the 80186, 80286, 80386, 80486, and finally the new Pentium series. Since all processors in the series run the same basic instructions, they are collectively referred to as the 80x86 family. In this exercise, we examine the real-mode architecture of the 80x86. In Exercise 53 we will look at another mode of operation, called *protected mode*, in which the full 32-bit power of the processor is available.

The introduction of the 8088 and 8086 into the arena of microprocessors came at a time when we were reaching the limits of what an 8-bit machine could do. With their restricted instruction sets and addressing capabilities, it was obvious that something more powerful was needed. The 80x86 contains instructions previously unheard of in 8-bit machines, a very large address space, many different addressing modes, and an architecture that easily lends itself to multiprocessing or multitasking (running many programs simultaneously).

First, let's look at the types of numbers we must work with when we deal with the microprocessor at its own level.

BINARY NUMBERS

Working with microprocessors requires a good grasp of the binary and hexadecimal number systems. Both are related to the number system we use every day, the decimal number system.

A decimal number is composed of one or more digits chosen from a set of 10 digits (0, 1, 2, 3, 4, 5, 6, 7, 8, 9), as shown in Figure 50.1. Each digit in a decimal number has an associated *weight* that is used to give the digit meaning. For example, the decimal number 357 contains three 100s, five 10s, and seven 1s. The weight of the digit 3 is 100, the weight of the digit 5 is 10, and the weight of the digit 7 is 1. The weight of each digit in a decimal number is related to the *base* of the number. Decimal numbers are base-10 numbers. Thus, the weights are all multiples of 10. Look at our example decimal number again:

Digits:	3	5	7
Weights as powers of 10:	10^2	10^1	10^0
Actual weight values:	100	10	1
Components of number:	300	50	7

Notice that the weights are all powers of 10, beginning with 0. The components of the number are found by multiplying each digit value by its respective weight. The number itself is found by adding the individual components. This technique applies to numbers in *any* base.

A *binary number* is a number composed of digits (called *bits*) chosen from a set of only two digits (0, 1), as shown in Figure 50.2. Base 2 is used for binary numbers because there

FIGURE 50.1   Three-digit decimal number

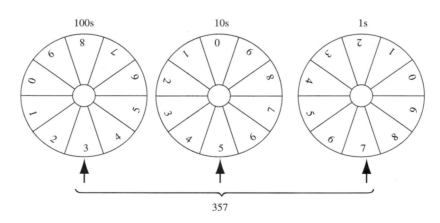

FIGURE 50.2   Five-bit binary number

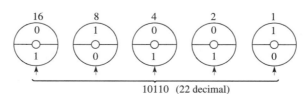

are only two legal digits in a binary number. This means that the weights of the bits in a binary number are all multiples of 2. Consider the binary number 10110. The associated weights are as follows:

Bits:	1	0	1	1	0
Weights as powers of 2:	2^4	2^3	2^2	2^1	2^0
Actual weight values:	16	8	4	2	1
Components of number:	16		4	2	

The components are again found by multiplying each bit in the number by its respective power of 2. The individual components add up to 22. Thus, 10110 binary equals 22 decimal. We now have a technique for determining the decimal value of any binary number.

Going from one base to another requires a *conversion*. As we just saw, going from 10110 to 22 required us to perform a *binary-to-decimal* conversion. How do we go the other way? For example, what binary number represents the decimal number 37? This requires a *decimal-to-binary* conversion. One way to do this conversion is as follows:

$$37/2 = 18 \text{ with 1 left over}$$
$$18/2 = 9 \text{ with 0 left over}$$
$$9/2 = 4 \text{ with 1 left over}$$
$$4/2 = 2 \text{ with 0 left over}$$
$$2/2 = 1 \text{ with 0 left over}$$
$$1/2 = 0 \text{ with 1 left over}$$

The number is repeatedly divided by 2 and the remainder recorded. When we get to "1/2 = 0 with 1 left over," we are done dividing. The binary result is found by reading the remainder bits from the bottom up. So, 37 decimal equals 100101 binary. To check this result, we use the binary-to-decimal conversion technique:

1	0	0	1	0	1
32	16	8	4	2	1
32			4		1

The sum of 32, 4, and 1 is 37—the original number.

Here are some common binary and decimal numbers:

Binary	Decimal
1010	10
1111	15
1100100	100
10000000	128
11111111	255
1111101000	1000

Clearly, a good understanding of the various powers of 2 is a valuable aid in performing conversions. The first 20 powers of 2 are shown in Table 50.1.

TABLE 50.1   The first 20 powers of 2

Power		Value	Power		Value
2^0	=	1	2^{10}	=	1024
2^1	=	2	2^{11}	=	2048
2^2	=	4	2^{12}	=	4096
2^3	=	8	2^{13}	=	8192
2^4	=	16	2^{14}	=	16,384
2^5	=	32	2^{15}	=	32,768
2^6	=	64	2^{16}	=	65,536
2^7	=	128	2^{17}	=	131,072
2^8	=	256	2^{18}	=	262,144
2^9	=	512	2^{19}	=	524,288

HEXADECIMAL NUMBERS

It is not easy to remember large binary numbers. For instance, examine the 20-bit binary number shown below for five seconds. Then close your eyes and try to repeat it.

$$10101111011010110011$$

Were you able to do it? Most people cannot, because their short-term memory is not capable of storing so many bits of information. It is possible, however, to remember shorter sequences of characters. Try this character sequence:

AF6B3

Were you able to remember it? Those who have difficulty with the 20-bit example can usually remember this five-character example easily. Here is where the trick comes in: the 20-bit binary number *has the same value* as the five-character sequence. The sequence AF6B3 is a ***hexadecimal number*** (base 16). This number is derived from the 20-bit binary number as follows:

1. Separate the number into groups of 4 bits each:

   1010 1111 0110 1011 0011

2. Find the individual decimal equivalent of each group:

   1010 1111 0110 1011 0011

    10   15   6   11   3

3. Replace each decimal value from 10 to 15 with the corresponding letter from A to F. Thus, the 10 becomes an A, the 11 becomes a B, and the 15 becomes an F.

   1010 1111 0110 1011 0011

    10   15   6   11   3

    A   F   6   B   3

This technique makes it possible to work with large binary numbers by using their hexadecimal equivalents.

The word *hexadecimal* refers to "6" and "10," or "16." The hexadecimal system contains numbers composed of digits and letters chosen from the set (0, 1, 2, 3, 4, 5, 6, 7, 8, 9, A, B, C, D, E, F). The decimal numbers 10, 11, 12, 13, 14, and 15 are represented in hexadecimal as A, B, C, D, E, and F, respectively. Each digit or letter in a hexadecimal number represents 4 binary bits (as we have seen). The binary patterns associated with the 16 hexadecimal symbols are shown in Table 50.2.

It is much more convenient (and easier on the memory) to use two hexadecimal symbols than to use 8 binary bits. For instance, 3EH (the H stands for hex) means 00111110B (B for binary). Larger binary numbers prove this point even better (as our AF6B3H example showed). Since microprocessor address and data lines commonly use 8, 16, or even 32 bits of data, the two-, four-, or eight-symbol hexadecimal equivalents are easier to deal with.

TABLE 50.2   Binary equivalents of hexadecimal symbols

Hex	Binary	Hex	Binary
0	0 0 0 0	8	1 0 0 0
1	0 0 0 1	9	1 0 0 1
2	0 0 1 0	A	1 0 1 0
3	0 0 1 1	B	1 0 1 1
4	0 1 0 0	C	1 1 0 0
5	0 1 0 1	D	1 1 0 1
6	0 1 1 0	E	1 1 1 0
7	0 1 1 1	F	1 1 1 1

This brief discussion should have familiarized you with the types of numbers that we encounter when dealing with microprocessors. Now let's see what microprocessors do with them.

THE REAL-MODE SOFTWARE MODEL OF THE 80X86

The real-mode software model describes the internal organization of the 80x86 from the perspective of a programmer, who needs to know such things as how many internal registers there are, how many bits are stored in each register, and the names of the registers. The term *real mode* refers to the original 8086 (and 8088) internal organization. With the release of the 80286 microprocessor, another mode of operation called *protected mode* was introduced, which allows virtual addressing, multitasking, and protected memory accesses. All advanced 80x86 processors initially operate in real mode and may be switched into protected mode by special software. Information on the advanced Intel microprocessors and protected mode is provided in Exercise 53.

The 80x86 microprocessor (operating in real mode) contains four data registers, referred to as AX, BX, CX, and DX. All are 16 bits wide, and may be split up into two halves of 8 bits each. Figure 50.3 shows how each half-register is referred to by the programmer. Five other 16-bit registers are available for use as pointer or index registers. These registers are the *stack pointer* (SP), *base pointer* (BP), *source index* (SI), *destination index* (DI), and *instruction pointer* (IP). None of the five may be divided up in a manner similar to the data registers. AX, BX, CX, DX, BP, SI, and DI are referred to as *general-purpose registers*.

A major difference between the 80x86 and many other CPUs on the market has to do with the next group of registers, the *segment registers*. Four segment registers are used by the processor to control all accesses to memory and I/O, and must be maintained by the programmer. The *code segment* (CS) is used during instruction fetches, the *data segment* (DS) is most often used by default when reading or writing data, the *stack segment* (SS) is used during stack operations such as subroutine calls and returns, and the *extra segment* (ES) is used for anything the programmer wishes. All segment registers are 16 bits long.

Finally, a 16-bit *flag register* is used to indicate the results of arithmetic and logical instructions. Included are zero, parity, sign, and carry flags, plus a few others. Together, these 14 registers make an impressive set.

FIGURE 50.3 Real-mode software model of the 80x86

	15	8 7	0	
AX	AH	AL		Accumulator
BX	BH	BL		Base
CX	CH	CL		Count
DX	DH	DL		Data

	15	0	
SP	Stack pointer		
BP	Base pointer		
SI	Source index		
DI	Destination index		
IP	Instruction pointer		

	15	0
CS	Code segment	
DS	Data segment	
SS	Stack segment	
ES	Extra segment	

	15	0
	Flags	

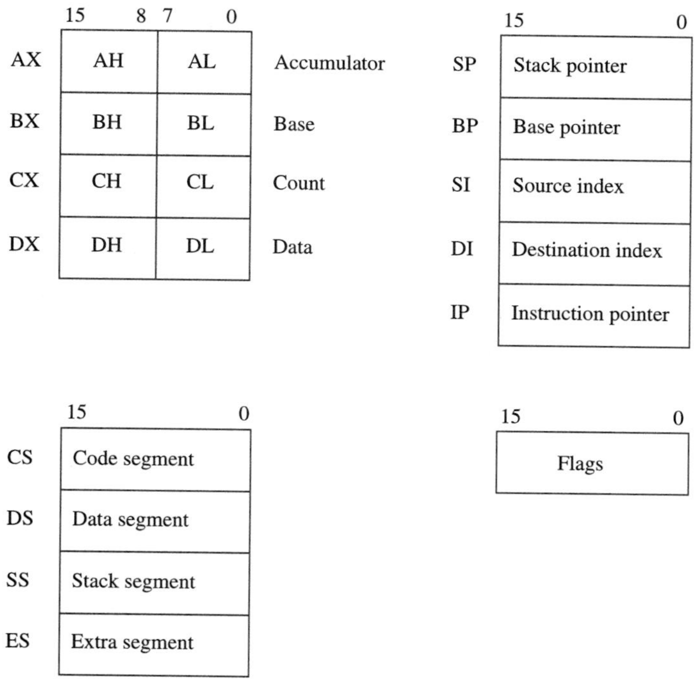

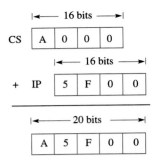

FIGURE 50.4 Generating a 20-bit address in the 80x86

80X86 PROCESSOR REGISTERS

The 16-bit registers just introduced in Figure 50.3 are combined in an interesting way to form the necessary 20-bit address required to access memory. If you recall, there were no 20-bit registers shown in the software model. How then does the 80x86 generate 20-bit real-mode addresses?

Segment Registers

The four segment registers, CS, DS, SS, and ES, are all 16-bit registers that are controlled by the programmer. A segment, as defined by Intel, is a 64K block of memory, starting on any 16-byte boundary. Thus, 00000, 00010, 00020, 20000, 8CE90, and E0840 are examples of valid segment addresses. The information contained in a segment register is combined with the address contained in another 16-bit register to form the required 20-bit address. Figure 50.4 shows how this is accomplished. In this example, the code segment register contains A000 and the instruction pointer contains 5F00. The 80x86 forms the 20-bit address A5F00 in the following way: First, the data in the code segment register is shifted 4 bits to the left. This has the effect of turning A000 into A0000. Then the contents of the instruction pointer are added, yielding A5F00. All external addresses are formed in a similar manner, with one of the four segments used in each case. Each segment register has a default usage. The 80x86 knows which segment register to use to form an address for a particular application (instruction fetch, stack operation, and so on). The 80x86 also allows the programmer to specify a different segment register when generating some addresses.

General-Purpose Registers

The seven general-purpose registers available to the programmer (AX, BX, CX, DX, BP, SI, and DI) can be used in many different ways; they also have some specific roles assigned to them. For instance, the accumulator (AX) is used in multiply and divide operations and in instructions that access I/O ports. The count register (CX) is used as a counter in loop operations, providing up to 65,536 passes through a loop before termination. The lower half of CX, the 8-bit CL register, is also used as a counter in shift/rotate operations. Data register DX is used in multiply and divide operations and as a pointer when accessing I/O ports. The last two registers are the source index and destination index (referred to as SI and DI, respectively). These registers are used as pointers in string operations.

Even though these registers have specific uses, they may be used in many other ways simply as general-purpose registers, allowing for many different 16-bit operations.

Flag Register

Figure 50.5 shows the nine flag assignments within the 16-bit flag register. The flags are divided into two groups: *control flags* and *status flags*. The control flags are IF (*interrupt enable flag*), DF (*direction flag*), and TF (*trap flag*). The status flags are CF (*carry flag*), PF (*parity flag*), AF (*auxiliary carry flag*), ZF (*zero flag*), SF (*sign flag*), and OF (*overflow flag*). Most of the instructions that require the use of the ALU affect these flags.

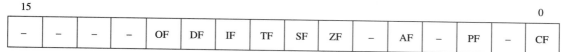

FIGURE 50.5   Real-mode 80x86 flag register

80X86 DATA ORGANIZATION

The 80x86 microprocessor has the capability of performing operations on many different types of data. In this section, we will examine what some of the more common data types are and how they are represented and used by the processor. The 80x86 contains instructions that directly manipulate single bits and other instructions that use 8-, 16-, and even 32-bit numbers. By common practice, 8-bit binary numbers are referred to as *bytes*. Processor register halves AL, BH, and CL are examples of where bytes might be stored and utilized.

Sixteen-bit numbers are known as *words* and may require an entire processor register for storage. Registers DX, BP, and SP are used to hold word data types. In register DX, DH contains the upper 8 bits of the number and DL holds the lower 8 bits.

Some instructions (particularly multiply and divide) allow the use of 32-bit numbers. These data types are called *double words* (or long words). In this case, the 32-bit number is stored in registers DX and AX, with DX holding the upper 16 bits of the number.

It is important to keep track of the data type being used in an instruction, because incorrect or undefined types may lead to incorrect program assembly or execution.

One of the differences between the Intel line of microprocessors and those made by other manufacturers is Intel's way of storing 16-bit numbers in memory. A method that began with the 8080 and has been used on all upgrades, from the 8085 to the Pentium, is a technique called *byte-swapping*. This technique is sometimes confusing for those unfamiliar with it, but it becomes clear after a little exposure. When a 16-bit number must be written into the system's byte-wide memory, the low-order 8 bits are written into the first memory location and the high-order 8 bits are written into the second location. Figure 50.6 shows how the 2 bytes that make up the 16-bit hexadecimal number 2055 are written into locations 18000 and 18001, with the low-order 8 bits (55) going into the first location (18000). This is what is known as byte-swapping. The lower byte is always written first, followed by the high byte. This method of storing numbers is also called *little-endian format*.

Reading the 16-bit number out of memory is performed automatically by the processor with the aid of certain instructions. The 80x86 knows that it is reading the lower byte first and puts it in the correct place. Programmers who manipulate data in memory must remember to use the proper practice of byte-swapping or discover that their programs do not give the correct, or expected, results.

80X86 INSTRUCTION TYPES

The 80x86 instruction set is composed of six main groups of instructions. Examining the instructions briefly here will provide a good overall picture of the capabilities of this processor.

FIGURE 50.6   Word storage using Intel byte-swapping

Address	Data	
18000	55	Low byte of word
18001	20	High byte of word

Data Transfer Instructions

Fourteen data transfer instructions are used to move data among registers, memory, and the outside world. Also, some instructions directly manipulate the stack, whereas others may be used to alter the flags.

The data transfer instructions are

MOV	Move byte or word (2 bytes) to register/memory
PUSH	Push word onto stack
POP	Pop word off stack
XCHG	Exchange byte or word
XLAT	Translate byte
IN	Input byte or word from port
OUT	Output byte or word to port
LEA	Load effective address
LDS	Load pointer using data segment
LES	Load pointer using extra segment
LAHF	Load AH from flags
SAHF	Store AH in flags
PUSHF	Push flags onto stack
POPF	Pop flags off stack

Arithmetic Instructions

Twenty instructions make up the arithmetic group. Byte and word operations are available on almost all instructions. A nice addition are the instructions that multiply and divide. Previous microprocessors did not include these instructions, forcing the programmer to write subroutines to perform multiplication and division when needed. Addition and subtraction of both binary and binary-coded decimal (BCD) operands are also allowed.

The arithmetic instructions are

ADD	Add byte or word
ADC	Add byte or word plus carry
INC	Increment byte or word by 1
AAA	ASCII adjust for addition
DAA	Decimal adjust for addition
SUB	Subtract byte or word
SBB	Subtract byte or word and carry
DEC	Decrement byte or word by 1
NEG	Negate byte or word
CMP	Compare byte or word
AAS	ASCII adjust for subtraction
DAS	Decimal adjust for subtraction
MUL	Multiply byte or word (unsigned)
IMUL	Integer multiply byte or word
AAM	ASCII adjust for multiply
DIV	Divide byte or word (unsigned)
IDIV	Integer divide byte or word
AAD	ASCII adjust for division
CBW	Convert byte to word
CWD	Convert word to double word

Bit Manipulation Instructions

Thirteen instructions capable of performing logical, shift, and rotate operations are contained in this group. Many common Boolean operations (AND, OR, NOT) are available in the logical instructions. These, as well as the shift and rotate instructions, operate on bytes or words. No single-bit operations are available.

The bit manipulation instructions are

NOT	Logical NOT of byte or word
AND	Logical AND of byte or word
OR	Logical OR of byte or word
XOR	Logical Exclusive-OR of byte or word
TEST	Test byte or word
SHL	Logical shift left byte or word
SAL	Arithmetic shift left byte or word
SHR	Logical shift right byte or word
SAR	Arithmetic shift right byte or word
ROL	Rotate left byte or word
ROR	Rotate right byte or word
RCL	Rotate left through carry byte or word
RCR	Rotate right through carry byte or word

String Instructions

Nine instructions are included to specifically deal with string operations. String operations simplify programming whenever a program must interact with a user. User commands and responses are usually saved as ASCII strings of characters, which may be processed by the proper choice of string instruction.

The string instructions are

REP	Repeat
REPE (REPZ)	Repeat while equal (zero)
REPNE (REPNZ)	Repeat while not equal (not zero)
MOVS	Move byte or word string
MOVSB (MOVSW)	Move byte string (word string)
CMPS	Compare byte or word string
SCAS	Scan byte or word string
LODS	Load byte or word string
STOS	Store byte or word string

Program Transfer Instructions

This group of instructions contains all jumps, loops, and subroutine (called *procedure*) and interrupt operations. The great majority of jumps are *conditional*, testing the processor flags before execution.

The program transfer instructions are

CALL	Call procedure (subroutine)
RET	Return from procedure (subroutine)
JMP	Unconditional jump
JA (JNBE)	Jump if above (not below or equal)
JAE (JNB)	Jump if above or equal (not below)
JB (JNAE)	Jump if below (not above or equal)
JBE (JNA)	Jump if below or equal (not above)
JC	Jump if carry set
JE (JZ)	Jump if equal (zero)
JG (JNLE)	Jump if greater (not less or equal)
JGE (JNL)	Jump if greater or equal (not less)
JL (JNGE)	Jump if less (not greater or equal)
JLE (JNG)	Jump if less or equal (not greater)
JNC	Jump if no carry
JNE (JNZ)	Jump if not equal (not zero)
JNO	Jump if no overflow
JNP (JPO)	Jump if no parity (parity odd)

JNS	Jump if no sign
JO	Jump if overflow
JP (JPE)	Jump if parity (parity even)
JS	Jump if sign
LOOP	Loop unconditional
LOOPE (LOOPZ)	Loop if equal (zero)
LOOPNE (LOOPNZ)	Loop if not equal (not zero)
JCXZ	Jump if CX equals zero
INT	Interrupt
INTO	Interrupt if overflow
IRET	Return from interrupt

Processor Control Instructions

This last group of instructions performs small tasks that sometimes have profound effects on the operation of the processor. Many of these instructions manipulate the flags.

The processor control instructions are

STC	Set carry flag
CLC	Clear carry flag
CMC	Complement carry flag
STD	Set direction flag
CLD	Clear direction flag
STI	Set interrupt enable flag
CLI	Clear interrupt enable flag
HLT	Halt processor
WAIT	Wait for TEST in activity
ESC	Escape to external processor
LOCK	Lock bus during next instruction
NOP	No operation

80X86 ADDRESSING MODES

The 80x86 offers the programmer a wide number of choices when referring to a memory location. Many people believe that the number of *addressing modes* contained in a microprocessor is a measure of power. If that is so, the 80x86 should be counted among the most powerful processors. Many of the addressing modes are used to generate a *physical address* in memory. Recall from Figure 50.4 that a 20-bit real-mode address is formed by the sum of two 16-bit address registers. One of the four segment registers will always supply the first 16-bit address. The second 16-bit address is formed by a specific addressing mode operation. We will see that there are several different ways in which the second part of the address may be generated. An acceptable notation that represents both 16-bit halves of the address in Figure 50.4 is A000:5F00. We will make use of this addressing format in a subsequent exercise.

Real-Mode Addressing Space

All addressing modes eventually create a physical address that resides somewhere in the 00000 to FFFFF real-mode addressing space of the processor. Figure 50.7 shows a brief memory map of the 80x86 addressing space, which is broken up into 16 blocks of 64K each. Each 64K block is called a *segment*. A segment contains all the memory locations that can be reached when a particular segment register is used. For example, if the data segment contains 0000, then addresses 00000 through 0FFFF can be generated using the data segment. If, instead, register DS contains 1800, then the range of addresses becomes 18000 through 27FFF. It is important to see that a segment can begin on *any* 16-byte boundary. So 00000, 00010, 00020, 035A0, 10800, and CCE90 are all acceptable starting addresses for a segment.

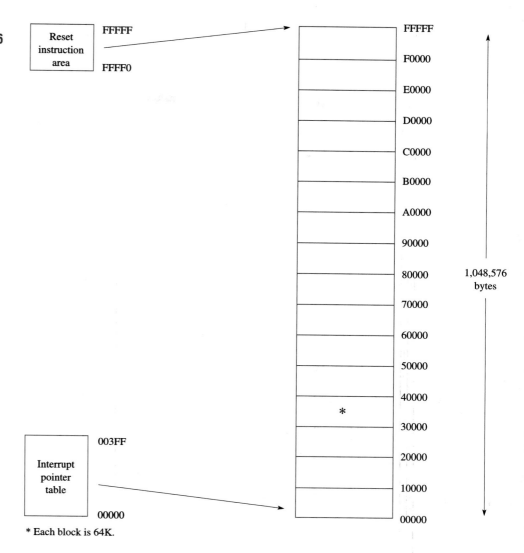

FIGURE 50.7 Real-mode addressing space of the 80x86

Reset instruction area — FFFFF, FFFF0

Interrupt pointer table — 003FF, 00000

FFFFF
F0000
E0000
D0000
C0000
B0000
A0000
90000
80000
70000
60000
50000
40000
*
30000
20000
10000
00000

1,048,576 bytes

* Each block is 64K.

Altogether, 1,048,576 bytes can be accessed by the processor. This is commonly referred to as 1 ***megabyte***. Small areas of the addressing space are reserved for special operations. At the very high end of memory, locations FFFF0 through FFFFF are assigned the role of storing the initial instruction used after a RESET operation. At the low end of memory, locations 00000 through 003FF are used to store the addresses for 256 interrupt vectors (although not all are commonly used in actual practice). This dedication of addressing space is common among processor manufacturers, and may force designers to conform to specific methods or techniques when building systems around the 80x86. For instance, EPROM is usually mapped into high memory, so that the starting execution instructions will always be there at power-on.

Addressing Modes

The simplest addressing mode is known as *immediate addressing*. Data needed by the processor is actually included in the instruction. For example

```
MOV CX,1024
```

contains the immediate data value 1024. This value is converted into binary and included in the code of the instruction.

637

When data must be moved between registers, *register addressing* is used. This form of addressing is very fast, because the processor does not have to access external memory (except for the instruction fetch). An example of register addressing is

```
ADD AL,BL
```

where the contents of registers AL and BL are added together, with the result stored in register AL. Notice that both operands are names of internal 80x86 registers.

The programmer may refer to a memory location by its specific address by using *direct addressing*. Two examples of direct addressing are

```
MOV AX,[3000]
```

and

```
MOV BL,COUNTER
```

In each case, the contents of memory are loaded into the specified registers. The first instruction uses square brackets to indicate that a memory address is being supplied. Thus, 3000 and [3000] are allowed to have two different meanings. The second instruction uses the symbol name COUNTER to refer to memory. COUNTER must be defined somewhere else in the program in order to be used in this way.

When a register is used within the square brackets, the processor uses *register indirect addressing*. For example

```
MOV BX,[SI]
```

instructs the processor to use the 16-bit quantity stored in the SI (source index) register as a memory address. A slight variation produces *indexed addressing*, which allows a small offset value to be included in the memory operand. Consider the instruction

```
MOV BX,[SI + 10]
```

The location accessed by this instruction is the sum of the SI register and the offset value 10.

When the register used is the base pointer (BP), *based addressing* is employed. This addressing mode is especially useful when manipulating data in large tables or arrays. An example of based addressing is

```
MOV CL,[BP + 4]
```

Including an index register (SI or DI) in the operand produces *based-indexed addressing*. The address is now the sum of the base pointer and the index register. An example might be

```
MOV [BP + DI],AX
```

When an offset value is also included in the operand, the processor uses *based-indexed with displacement addressing*. An example is

```
MOV [BP + SI + 2]
```

It is easily seen that the 80x86 intends the base pointer to be used in many different ways. Other addressing modes are used when string operations must be performed.

The 80x86 processor is designed to access I/O ports, as well as memory locations. When *port addressing* is used, the address bus contains the address of an I/O port instead of a memory location. I/O ports may be accessed in two different ways. The port may be specified in the operand field, as in

```
IN AL,80H
```

or indirectly, by means of the address contained in register DX, as in

```
OUT DX,AL
```

Using DX allows a port range from 0000 to FFFF, or 65,536 individual I/O port locations. Only 256 locations (00 to FF) are allowed when the port address is included as an immediate operand.

A great amount of material was presented in this exercise regarding the Intel 80x86 architecture. It would be helpful to really commit some of the most basic material about the 80x86 to memory. As a minimum, you should be able to do the following without much thought:

- Name all the processor registers, their bit size, and whether they can be split into 8-bit halves.
- Be familiar with several architectural features, such as the processor's addressing space (1MB in real mode), interrupt mechanism, and I/O mechanism.
- Show what is meant by Intel byte-swapping.
- List the names and meanings of the most common flags, such as zero, carry, and sign.
- Describe the method used to form a 20-bit address in real mode (combining a segment register with an offset).
- Show how an instruction is composed of a operation, a set of operands, and a particular addressing mode.

Knowing these basics thoroughly will assist you in further exploration of the Intel 80x86 architecture.

SUMMARY

In this exercise we discovered that

- Two common ways to represent numbers are binary and hexadecimal.
- The real mode of the 80x86 microprocessor contains seven general-purpose registers and four segment registers, as well as a flag register.
- The gerneral-purpose registers are AX, BX, CX, DX, BP, SI, and DI.
- The segment registers are CS, DS, ES, and SS.
- All registers are 16 bits wide.
- Intel byte-swapping stores the low byte of a 2-byte number in the first memory location (little endian format).
- There are many different instruction types and addressing modes.
- The real mode addressing space is 1MB.

SELF-TEST

This self-test is designed to help you check your understanding of the background information presented in this exercise.

True/False

Answer *true* or *false*.

1. Binary numbers must always contain at least one zero.
2. Only binary numbers can be represented with hexadecimal notation.
3. A four-symbol hexadecimal number has an equivalent 16-bit binary number.
4. Only segment registers are used to make an address.
5. The 80486 runs the same basic instructions as the Pentium.
6. The two halves of register AX are AL and AH.
7. The two halves of register SI are SL and SH.

Multiple Choice

Select the best answer.

8. The general-purpose registers are
   a. AX, BX, CX, and DX.
   b. CS, DS, ES, and SS.
   c. AX, BX, CX, DX, BP, SI, and DI.

9. The processor registers are able to hold
   a. 8-bit numbers.
   b. 16-bit numbers.
   c. Both 8- and 16-bit numbers.
   d. 20-bit numbers.
10. An example of immediate addressing is
    a. MOV AL,BL.
    b. AND AL,2FH.
    c. NOP.
11. The real mode addressing space is
    a. 640K.
    b. 1MB.
    c. 16MB.
12. The number 1234H is stored in memory as
    a. 12, 34.
    b. 34, 12.
    c. Both a and b.

Completion

Fill in the blanks with the best answers.

13. CS, DS, SS, and ES are called _____ registers.
14. A(n) _____ contains 2 bytes of information.
15. Reversing the order of the upper and lower bytes of a 16-bit number during memory access is called _____.
16. 80x86 processors can operate in either _____ mode or _____ mode.
17. The binary number 10001 equals _____ decimal.
18. The hexadecimal number 35 equals _____ decimal.

FAMILIARIZATION ACTIVITY

Examine the following short list of instructions:

```
AND  AL,7
ADD  CX,5000
STC
INT  20H
RET
MOV  BL,[SI]
MOV  [DI+4],DX
```

What addressing modes are used for both source and destination operands (if any) in each instruction? What type of immediate data is used (if any), and are 8 or 16 bits needed for the data?

QUESTIONS/ACTIVITIES

1. How many different instructions are there in the 80x86 instruction set?
2. If an instruction allows four addressing modes for its destination operand and six addressing modes for its source operand, how many different instructions are possible?
3. What kinds of instructions might you like to see in an upgrade of the 80x86? Can you describe them?

REVIEW QUIZ

Under the supervision of your instructor

1. Outline the software model of the 80x86 microprocessor family.
2. Discuss the various instruction groups and addressing modes that are available.
3. Explain how a 20-bit address is formed from two 16-bit numbers.

51

Computer Languages

INTRODUCTION

Don, the senior technician, handed Joe Tekk his next assignment. "I'm sorry to do this to you, Joe, but the customer wants his custom digital camera interfaced to a parallel input/ output port. You'll have to write assembly language code to control the data transfer between the camera and the port." Don paused, and then continued in a nervous voice, "He needs it in one week."

Joe took the customer's specification and examined it for a few moments. "One week?" he asked. Then he smiled at Don and said, "Thanks, Don! This is so cool!" He left the room in a hurry.

Don sat in his chair, wondering what Joe was so happy about. Thirty minutes later Joe called Don and said, "Come to my office. I got the camera working."

Don recalled a moment from Joe's job interview when Joe mentioned that he knew "a little assembly language."

That boy does not like to brag, he thought to himself.

PERFORMANCE OBJECTIVES

Upon completion of this exercise, you will be able to

1. Enter an 80x86 assembly language source file.
2. Assemble the source file using MASM.
3. Link the object file using LINK.
4. Run the resulting 80x86 program.
5. Discuss the various statements in a C++ source file.
6. Describe the difference between an assembler and a compiler.
7. Explain the basic difference between an executable program and an interpreted BASIC program.

KEY TERMS

Assembly language
Machine language
Source file
Macro assembler
List file
Object file

High-level language
BASIC interpreter
C++ compiler
Object-oriented programming
Run-time error

In order to use the maximum power of the personal computer, it is necessary to understand how to control and use the hardware on the motherboard and the software capabilities of the processor. In this exercise, we will examine how the 80x86 microprocessor is programmed in its own unique language, which is called *assembly language*. This exercise will expose you to assembly language and give you an appreciation for what happens inside the PC. To learn more about the details of assembly language programming, search the Web or read any of the large number of books on the subject. In addition, we will look briefly at two popular high-level languages, BASIC and C++, and see how they differ from each other and from assembly language.

MACHINE LANGUAGE VERSUS ASSEMBLY LANGUAGE

Our spoken language is one of words and phrases. The processor's language is a string of 1s and 0s. For example, the instruction

```
ADD   AX,BX
```

contains the word ADD, which means something to us. Apparently, we are adding AX and BX, two of the general-purpose registers. So, even though we might be unfamiliar with the details of the 80x86 instruction set, ADD AX,BX means something to us.

If instead we were given the binary string

```
0000 0001 1101 1000
```

or the hexadecimal equivalent

```
01 D8
```

and asked its meaning, we might be hard-pressed to come up with anything. We associate more meaning with ADD AX,BX than we do with 01 D8, which is the way the instruction is actually represented. All programs for the 80x86 are simply long strings of binary numbers.

Because of the processor's internal logic, different binary patterns represent different instructions. Here are a few examples to illustrate this point:

```
01 D8      ADD   AX,BX      ;add BX to AX, result in AX
29 D8      SUB   AX,BX      ;subtract BX from AX, result in AX
21 D8      AND   AX,BX      ;AX equals AX and BX
40         INC   AX         ;add 1 to AX
4B         DEC   BX         ;subtract 1 from BX
8B C3      MOV   AX,BX      ;copy BX into AX
```

Can you guess the meaning of each instruction just by reading it? Do the hexadecimal codes for these instructions mean anything to you? What we see here is the difference between *machine language* and assembly language. The machine language for each instruction is represented by the hexadecimal code. This is the binary language of the machine. The assembly language is represented by the wordlike terms that mean something to us. Putting groups of these wordlike instructions together is how a program is constructed. The format of an assembly language instruction is basically as follows:

```
<opcode>   <destination-operand>,<source-operand>
```

where <opcode> is an instruction from the 80x86 instruction set, and both <destination-operand> and <source-operand> are register names or numbers representing data or memory addresses. For example, in

```
ADD   AX,BX
```

the <opcode> is ADD, the <destination-operand> is register AX, and the <source-operand> is register BX. An instruction may have zero, one, or two operands.

Let's see how an assembly language program is written, converted into machine language, and executed.

THE NUMOFF PROGRAM

When the PC is first turned on, instructions in the start-up software turn the NUM-LOCK indicator on. This indicator is located near the NUM-LOCK button on the keyboard. It is annoying to have to push NUM-LOCK manually every time the PC is turned on (or even rebooted). Luckily, there is a single bit stored in a specific memory location used by DOS that controls the state of the NUM-LOCK indicator. We are about to see that it is possible to write a program called NUMOFF to manipulate the NUM-LOCK status bit.

Using a word processor or text editor, enter the following text file exactly as you see it. Save the ASCII text file under the name NUMOFF.ASM.

```
NUMOFF  SEGMENT  PARA 'CODE'
        ASSUME   CS:NUMOFF,DS:NOTHING
START:  MOV      AX,40H                ;set AX to 0040H
        MOV      DS,AX                 ;load data segment with 0040H
        MOV      SI,17H                ;load SI with 17H
        AND      BYTE PTR [SI],0DFH    ;clear NUM-LOCK bit
        MOV      AH,4CH                ;terminate program function
        INT      21H                   ;exit to DOS
NUMOFF  ENDS
        END      START
```

These 10 lines of code constitute a *source file*, the starting point of any 80x86-based program. Thus, NUMOFF.ASM is a source file.

To convert NUMOFF.ASM into a group of hexadecimal bytes that represent the corresponding machine language, we make use of two additional programs: MASM and LINK. MASM is a *macro assembler*, a program that takes a source file as input and determines the machine language for each source statement. As illustrated in Figure 51.1, MASM creates two new files—the *list file* and the *object file*. The list file contains all the text from the source file, plus additional information, as we will soon see. The object file contains only the machine language.

To assemble NUMOFF.ASM, enter the following command at the DOS prompt:

```
MASM NUMOFF,,;
```

FIGURE 51.1 Operation of the MASM program

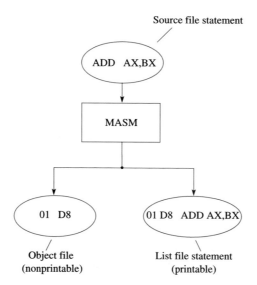

Source file statement

ADD AX,BX

MASM

01 D8

Object file (nonprintable)

01 D8 ADD AX,BX

List file statement (printable)

643

This instructs MASM to assemble NUMOFF.ASM and create NUMOFF.LST (the list file) and NUMOFF.OBJ (the object file). The list file created by MASM looks like this:

```
0000                 NUMOFF SEGMENT PARA 'CODE'
                            ASSUME  CS:NUMOFF,DS:NOTHING
0000 B8 0040   START:  MOV    AX,40H            ;set AX to 0040H
0003 8E D8             MOV    DS,AX             ;load data segment with 0040H
0005 BE 0017           MOV    SI,17H            ;load SI with 17H
0008 80 24 DF          AND    BYTE PTR [SI],0DFH ;clear NUM-LOCK bit
000B B4 4C             MOV    AH,4CH            ;terminate program function
000D CD 21             INT    21H               ;exit to DOS
000F           NUMOFF ENDS
                       END    START
```

In this file it is obvious that MASM has determined the machine language for each source statement. The first column is the set of memory locations in which the instructions are stored. The second column is the group of machine language bytes that represent the actual 80x86 instructions.

The object file created by MASM is *not* executable. Its internal structure is used by another program, called LINK, to create an executable file. LINK uses NUMOFF.OBJ to create NUMOFF.EXE. The DOS command to do this is

LINK NUMOFF,,;

The LINK program creates the executable NUMOFF.EXE file so that it conforms to a structure that DOS knows how to work with and that can be run from the DOS prompt. Recall that any .BAT, .EXE, or .COM file can be run from the DOS prompt.

Right now we have our first working 80x86 program, NUMOFF.EXE. To test it, press the NUM-LOCK button on the PC's keyboard until the NUM-LOCK light goes on. Then execute NUMOFF.EXE by entering:

NUMOFF

at the DOS prompt. The NUM-LOCK light should go off. This is what NUMOFF.EXE does. If you want the NUM-LOCK indicator to be turned off automatically each time your PC is powered up (or rebooted), add the statement

NUMOFF.EXE

to your AUTOEXEC.BAT file.

MASM and LINK have kept up with Intel's advancements in microprocessor architecture and now handle instructions for all machines up through the Pentium. However, for large programming applications, it is usually better to use a *high-level language*, such as C/C++, to write application programs. When the program is compiled, each statement is converted into one or more machine language instructions, and an executable file is generated that contains all the instructions. Thus, everything depends on the machine language instructions fed to the processor. Let us now examine two high-level languages, BASIC and C++.

HIGH-LEVEL LANGUAGES

Rather than using assembly language, it would be easier to give instructions and get answers back using symbols that we already understand. The easiest set of characters that we can use is the English alphabet. It would be very easy to program if you could give an instruction to the computer such as

Figure out my income tax for me and let me know when you're finished so I can tell you where to mail it.

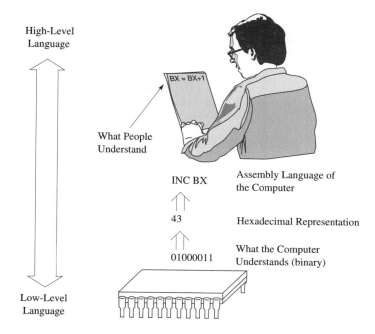

FIGURE 51.2 Different levels of computer languages

High-Level Language

BX = BX+1

What People Understand

INC BX   Assembly Language of the Computer

43   Hexadecimal Representation

01000011   What the Computer Understands (binary)

Low-Level Language

Although computers can't quite yet follow those instructions, a program can be written in words that people easily understand. When the computer is programmed using symbols and letters that the operator can read, this computer language is called a ***high-level language***. The language presented here, BASIC, is a high-level language.

Figure 51.2 illustrates the different levels of computer languages. Any language that is higher in level than machine language requires a program to convert that language into machine language. For example, the next level above the 1s and 0s understood by the microprocessor is the *hexadecimal number system*, a number system containing 16 symbols: the digits 0–9 and the letters A–F. Using the hexadecimal number system, the instruction

```
01000011
```

could be entered into the computer as

```
43
```

—two keystrokes instead of eight!

Programming a computer this way is still called machine-language programming by many programmers. However, it is very difficult to write complete programs in machine language. It is easier to enter a BASIC statement, such as

```
BX = BX + 1
```

than to try and remember the correct machine language to enter. For this reason, a special program called a ***BASIC interpreter*** is used to automatically convert the high-level BASIC language into the low-level machine language understood by the microprocessor. The BASIC interpreter looks at each BASIC statement and converts it into the machine language required by the microprocessor. Luckily, most desktop computers come with a BASIC interpreter. If you want to program the computer in another language, such as C++, you'll need to purchase a special program called a ***compiler***, which will convert the C++ statements into the 1s and 0s that the microprocessor understands. The main difference between an interpreter and a compiler is that the interpreter is used every time the program is executed, whereas the compiler is used only once, to create an *executable* machine-language program. The compiler creates an entirely new program containing only machine language. The BASIC interpreter must determine the correct machine language for every BASIC statement

FIGURE 51.3   Action of an interpreter

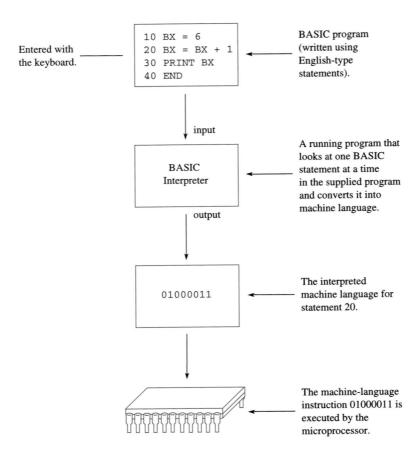

Entered with _____ the keyboard.

```
10 BX = 6
20 BX = BX + 1
30 PRINT BX
40 END
```

BASIC program (written using English-type statements).

↓ input

BASIC Interpreter

A running program that looks at one BASIC statement at a time in the supplied program and converts it into machine language.

↓ output

01000011

The interpreted machine language for statement 20.

The machine-language instruction 01000011 is executed by the microprocessor.

TABLE 51.1   Some programming languages and their advantages

Programming Language	Reason for Use
APL	(A Programming Language) A very powerful language for scientific work.
Assembly language	A low-level programming language that allows direct control of the microprocessor.
BASIC	Easy to learn.
COBOL	(Common Business-Oriented Language) For business-oriented programming.
LOGO	A simple programming language originally designed to teach children how to use computers.
Pascal	Teaches good programming habits.
FORTRAN	(Formula Translation) Used to solve mathematical formulas.
C/C++	More control over computer than other high-level languages, and built-in machine-language interface.

each time the program is executed. It is for this reason that compiled programs run much faster than interpreted programs.

The action of an interpreter is illustrated in Figure 51.3.

BASICA, GWBASIC, and QBASIC are three examples of interpreted BASIC. A package called Visual BASIC is capable of creating executable machine-language programs and thus performs the actions of a compiler.

There are many different programming languages available. Table 51.1 lists some of the more common languages and their particular strengths.

FIGURE 51.4   Computing
circle area

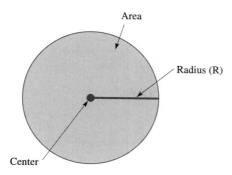

Area

Radius (R)

Center

FIGURE 51.5   Output of circle
area program

```
This program finds the area of a circle.
Enter the value of the radius and
the program will do the rest.
? 6
The area in square units is 113.0973
```

SAMPLE BASIC PROGRAM

The program presented here will compute the area of a circle for a given radius. You will recall that the formula for the area of a circle is

$$Area = \pi R^2$$

where   Area = The area of the circle in square units
π = The constant pi, which is approximately equal to 3.14159
R = The radius of the circle

This is illustrated in Figure 51.4. The BASIC program that solves for the circle area is as follows:

```
10 REM Area of a Circle Program
20 LET PI = 3.14159
30 CLS
40 PRINT "This program finds the area of a circle."
50 PRINT "Enter the value of the radius and"
60 PRINT "the program will do the rest."
70 INPUT R
80 LET Area = PI * R * R
90 PRINT "The area in square units is "; Area
100 END
```

This program will compute the area of a circle for a given radius. When the program is executed, the output on the screen will appear as shown in Figure 51.5.

What the Program Does

As you can see from Figure 51.5, when the program is executed it clears the screen and tells the program user what the program will do and how it is to be used. Then the program waits for the user to enter a number that represents the radius of a circle for which the user wants to find the area. When the user enters the value of the radius and presses the Enter key, the program computes the area of the circle and displays the result.

Look again at the program. You see that it consists of 10 lines, each line beginning with a number. Each line contains a BASIC statement. This idea is illustrated in Figure 51.6.

FIGURE 51.6 Typical BASIC statement structure

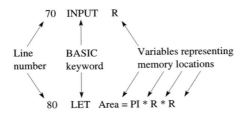

In BASICA and GWBASIC, line numbers are required for each BASIC statement in the program. This helps the editor keep the program statements in order. The line number specified by the programmer tells the editor where the statement belongs in the program.

In QBASIC, line numbers are not required, but may be used if desired. The built-in editor available with QBASIC is called a *screen editor*, which does not require line numbers to keep statements in order. Editing is performed by using the arrow keys (or mouse) to position the cursor at the location where a new statement will be inserted, or a current statement edited or deleted. In general, statements in QBASIC execute in the order that they appear on the screen. Table 51.2 lists the definitions of each of the BASIC statements used in the Area program.

TABLE 51.2 BASIC statements in the Area program

BASIC Statement	What It Does	Use in the Program
REM	This is a remark. It does not cause the BASIC program to do anything.	The REM statement is used by the programmer for comments and notes within the program.
LET	This is the LET statement. It signifies the beginning of a computation or logic statement.	The LET statement sets the value of the variable PI to the value of 3.14159. Now, whenever PI is used in the program, it will have the value of 3.14159.
CLS	This statement clears the monitor screen and starts the cursor at the upper-left corner of the monitor.	The CLS clears the BASIC program itself from the screen. This is done to allow only the program output to appear.
PRINT	This statement causes the characters enclosed inside double quotation marks to be displayed (printed) on the screen when the program is executed.	The PRINT statements cause the instructions to be displayed on the monitor as well as the results of the calculation.
INPUT	This statement causes the program to wait for the user to input a value and press the Enter key.	The INPUT statement causes the program to display a question mark and wait for the user to input the value of the radius of the circle.
*	The asterisk (*) indicates multiplication.	The * is used to indicate multiplication. In this case, it is used to calculate the area of a circle: $$Area = \pi R^2$$ $$Area = PI * R * R$$
LINE NUMBERS	Indicates the order in which each BASIC statement will be executed. Program execution goes from the lowest line number to the largest. Line numbers may be in the range from 0 to 65529.	The program starts with line 10 and continues sequentially to line 100. Usually BASIC line numbers are numbered in units of 10. This is done in case the program needs modification and a new statement is required between two existing ones.

The C++ Environment

The C++ environment contains an *editor, compiler, include files, library files, linker,* and much more. The functions of these components are as follows:

- **Editor**   Allows you to enter and modify your C++ source code.
- **Compiler**   A program that converts the C++ program you have developed into a code understood by the computer.
- **Include Files**   Files that consist of many separate definitions and instructions that may be useful to a programmer in certain instances.
- **Library Files**   Previously compiled programs that perform specific functions. These programs can be used by you to help you develop your C++ programs. For example, the C++ function that allows you to display text on your screen (the `cout` function) is not in the C++ language. Instead, its code is in a library file. The same is true of many other functions, such as graphics, sound, and working with the disk and printer. You can also create your own library files of routines that you use over and over again in different C++ programs. By doing this, you can save hours of programming time and prevent programming errors.
- **Linker**   Essentially, the linker combines all of the necessary parts (such as library files) of your C++ program to produce the final executable code. Linkers play an important and necessary role in all of your C++ programs. In larger C++ programs, it is general practice to break the program down into smaller parts, each of which is developed and tested separately. The linker will then combine all of these parts to form your final executable program code.

You will also need some kind of a disk operating system to assist you in saving your programs. The main parts of the C++ environment are shown in Figure 51.7.

Elements of C++

This section lays the ground rules for the fundamental elements of all C++ programs. In this section you will see many new definitions. These definitions will set the stage for the example program that follows.

- **Characters**   To write a program in C++, you use a set of characters. This set includes the uppercase and lowercase letters of the English alphabet, the 10 decimal digits of the Arabic number system, and the underscore (_) character. Whitespace characters (such as the spaces between words) are used to separate the items in a C++ program, much the same as they are used to separate words in this book. These whitespace characters also include the tab and carriage return, as well as other control characters that produce white spaces.
- **Tokens**   In every C++ source program, the most basic element recognized by the compiler is a single character or group of characters known as a token. Essentially, a token is source program text that the compiler will not break down any further—it is treated as a fundamental unit. As an example, in C++ `main` is a token; so is the required opening brace ({) as well as the plus sign (+).
- **C++ Keywords**   Keywords are predefined tokens that have special meanings to the C++ compiler. Their definitions cannot be changed; thus, they cannot be used for anything else except the intended action they have on the program in which they are used. The most common keywords are as follows:

```
auto       double     int        struct     break      else       public
long       switch     case       enum       register   typedef    private
char       extern     return     union      const      float      protected
short      unsigned   continue   for        signed     void
default    goto       sizeof     volatile   do         if
static     while      cin        cout       class      virtual
```

- **Types of Data**   The C++ language allows three major types of data: numbers, characters, and strings. A *character* is any item from the set of characters used by C++. A *string* is a combination of these characters.
- **Numbers Used by C++**   C++ uses a wide range of numbers. Numbers used by C++ fall into two general categories: *integer* (whole numbers) and *float* (numbers with decimal points). These two main categories can be further divided as shown in Table 51.3.

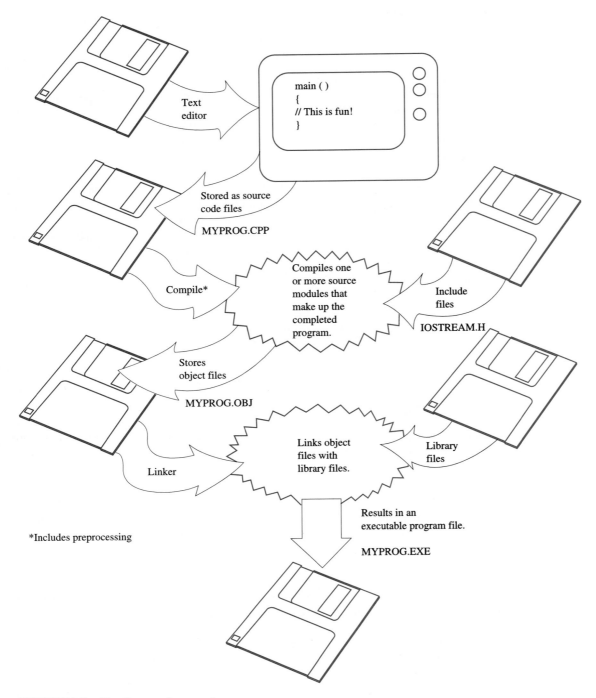

FIGURE 51.7   The C++ environment

As you can see in Table 51.3, C++ offers a rich variety of data types. Generally speaking, the larger the value range of the data type, the more computer memory it takes to store it. As a general rule, you want to use the data type that conserves memory and still accomplishes the desired purpose. As an example, if you needed a data type for counting objects—such as the number of resistors for a parts order—the type `int` would probably do the job. However, if you were writing a program for as much precision as possible, you might consider the type `double`, which can give you 15-digit accuracy.

Examine the sample C++ program shown in Figure 51.8. Can you spot the different tokens and keywords used by the programmer?

TABLE 51.3 Subdivisions of C++ data types

Type Identifier	Meaning	Range of Values
char	character	−128 to 127
int	integer	−32,768 to 32,767
short	short integer	−32,768 to 32,767
long	long integer	−2,147,483,648 to 2,147,483,647
unsigned char	unsigned character	0 to 255
unsigned	unsigned integer	0 to 65,535
unsigned short	unsigned short int	0 to 65,535
unsigned long	unsigned long int	0 to 4,294,967,295
enum	enumerated	0 to 65,535
float	floating point	3.4E +/− 38 (7 digits)
double	double floating point	1.7E +/− 308 (15 digits)
long double	long double floating point	1.7E +/− 4932 (40 digits)

OBJECT-ORIENTED PROGRAMMING

One of the main differences between C and C++ is the support of *objects* in the C++ environment. An object contains both code and data and interacts with the C++ environment in certain, predefined ways. Objects may be duplicated, hidden from each other, and protected from access by other objects. Objects, unlike ordinary functions and variables, have *lifetimes*, or periods of time during execution where they are created (instantiated), used, and then destroyed (in the interest of memory management). AREA is the name of the class in the sample program in Figure 51.8 that is used to instantiate the mycircle object.

In general, working with objects in a C++ program allows the programmer to skip many of the cumbersome chores of data management that would be required in an ordinary C program. The same steps are now performed automatically during an object's lifetime.

TROUBLESHOOTING TECHNIQUES

You may think it premature to discuss troubleshooting techniques when we have been exposed to so little of the assembly language and high-level programming. Even so, we have already seen a number of places where errors can occur, and it would be worthwhile to discuss them. For example, the NUMOFF.ASM source file could have contained one or more *typographical* errors, such as a misspelled instruction (MVO for MOV), or a missing comma, or a comma where a semicolon was expected. Generally, when errors such as these are present in a source file, the assembler will report them with a brief error message.

Even if the source file does not have any typographical errors, we could still run into trouble. We could enter the command to invoke MASM or LINK incorrectly, or not use the correct options.

When the source file correctly assembles and links, and an executable program has been created, there is still the possibility of a ***run-time error*** in the program. Run-time errors are typically caused by incorrect sequences of instructions and incomplete or faulty logicial thinking. The same problems may be encountered when working with BASIC and C++ progams.

To avoid a loss of time and effort, it is good to keep these common stumbling blocks in mind. Paying attention to the details will really pay off as you learn to create a working program with a minimum of time and effort.

```
Preprocessor          #include <iostream.h>
Directives    ─────▶  #include <math.h>
                      #define PI 3.14159
                      #define SQUARE(x) ((x) * (x))

                      /*
                      Program:  Circle Area
                      Developed by: A. G. Programmer

                      Description:  This program will solve for the area of a circle.  The
                                    programmer need only enter the radius.  Value returned
                                    is in square units.

Programmer's          Variables:
Block
                          radius      = Radius of the circle.
                          area        = Area of the circle.

                      Constants:

                          PI = 3.14159
                      */

                      // Class Definition
                                                        Explanation of member
                      class AREA                        variables and functions
                      {
                      public:
                          float radius;                 // Radius of the circle.
Class                 float area;                       // Area of the circle.
Definition
                          AREA(void);                   // Explain program to user.
                          void get_value(void);         // Get radius from user.
                          void circle_area(void);       // Compute area of circle.
                          ~AREA();                      // Display results.
                      };

                      void main()
                      {
Main                  AREA mycircle;                    // Instantiate object.
Function   ─────▶
                          mycircle.get_value();         // Get radius from user.
                          mycircle.circle_area();       // Compute the circle area.
                      }

                      void AREA::AREA()                 // Explains the program.
                      {
Constructor  ─────▶       cout << "This program calculates the area of a circle.\n";
                          cout << "Just enter the value of the radius and press ENTER\n";
                          cout << "\n";                 // Put in a blank line.
                      }

                      void AREA::get_value()            // Gets radius from user.
                      {
                          cout << "Value of the radius ==> ";
                          cin >> radius;
Member                }
Functions  ─────▶
                      void AREA::circle_area()          // Compute the circle area.
                      {
                          area = PI * SQUARE(radius);
                      }

                      void AREA::~AREA()                // Display the answer.
Destructor            {
                          cout << "\n\n";               // Print two blank lines.
                          cout << "The area of the circle is " << area << " square units.\n";
                      }
```

FIGURE 51.8   Structure of a C++ program

In this exercise we discovered that

- Machine language is not the same as assembly language.
- Machine language consists of binary patterns that represent instructions and data.
- Assembly language consists of wordlike terms that represent instructions and other items.
- An assembler converts assembly language (from a source file) into machine language (saved in an object file).
- BASIC and C++ are high-level languages.
- BASIC is an interpreted language.
- C++ is a compiled language that supports objects.

SELF-TEST

This self-test is designed to help you check your understanding of the background information presented in this exercise.

True/False

Answer *true* or *false*.

1. Machine language and assembly language are the same thing.
2. Only MASM is needed to make an executable file.
3. Both the list and object files may be printed.
4. Executable programs can be written only in assembly language.
5. BASIC is a compiled language.
6. C++ supports the use of objects.
7. It is good to have typographical errors in a source file.

Multiple Choice

Select the best answer.

8. The extension reserved for assembly language files is
   a. MLF (machine language file).
   b. SRC (source file).
   c. ASM (assembly).
9. The two files created by the assembler are
   a. MASM and LINK.
   b. Source and target.
   c. Source and object.
   d. List and object.
10. The instruction in the NUMOFF program that actually clears the NUM-LOCK bit is the
    a. MOV instruction.
    b. AND instruction.
    c. INT instruction.
11. PRINT (in BASIC) and cout (in C++) are both used to
    a. Input data from the user.
    b. Output data to the user.
    c. Perform input and output operations.
12. A program that converts assembly language into machine language is an
    a. Assembler.
    b. Compiler.
    c. Interpreter.
13. In the instruction ADD AX,BX, registers as AX and BX are called
    a. Opcodes.
    b. Operands.
    c. Subinstructions.

Completion

Fill in the blanks with the best answers.

14. A text file written in assembly language is called a(n) _____ file.
15. The linker creates an executable file using information from the _____ file.
16. Executable programs may also be written in a(n) _____ language such as C/C++.
17. A(n) _____ converts C++ statements into executable code.
18. Binary patterns are used to represent _____ language.
19. A program is either assembled, compiled, or _____.

FAMILIARIZATION ACTIVITY

1. Use EDIT to enter the NUMOFF.ASM source file. Deliberately misspell one of the MOV instructions (use MVO instead).
2. Assemble NUMOFF using MASM, by means of

   ```
   MASM  NUMOFF,,;
   ```

3. Did MASM discover the error? How can you tell? To see the details of the error, type out the contents of NUMOFF.LST.
4. Edit NUMOFF.ASM so that the instructions are all correct, and reassemble the program.
5. Use the DIR command to view all of the NUMOFF files.
6. Create the NUMOFF executable with the linker using this command:

   ```
   LINK NUMOFF,,;
   ```

7. Use the DIR command to verify that NUMOFF.EXE has been created.
8. Run NUMOFF a few times to verify that it actually turns the NUM-LOCK indicator off.
9. Verify the operation of the BASIC sample program using your own BASIC system.
10. Search the Web for free C++ compilers. Are any available?

QUESTIONS/ACTIVITIES

1. Examine the contents of the NUMOFF.LST file. Do the machine codes for all the MOV instructions have anything in common?
2. Can you think of a way to do the opposite of NUMOFF? That is, can you change NUMOFF so that it turns NUM-LOCK on?
3. Can you think of a way to toggle the NUM-LOCK indicator? (This would cause NUM-LOCK to alternate between ON and OFF.)
4. Explain why an assembler can be thought of as a language translator.
5. Explain why interpreted programs run more slowly than compiled programs.

REVIEW QUIZ

Under the supervision of your instructor

1. Enter the following assembly language source file:

   ```
   MTXT     SEGMENT PARA 'DATA'
   MSG      DB   '80x86 assembly language!$'
   MTXT     ENDS
   PGM      SEGMENT PARA 'CODE'
            ASSUME CS:PGM,DS:MTXT
   START:   MOV  AX,MTXT      ;set up message segment
            MOV  DS,AX
            LEA  DX,MSG        ;set up pointer to message
            MOV  AH,9          ;display string function
   ```

```
        INT   21H                ;DOS call
        MOV   AH,4CH             ;terminate program function
        INT   21H                ;exit to DOS
PGM     ENDS
        END   START
```

2. Assemble the source file using MASM.
3. Link the object file using LINK.
4. Run the resulting 80x86 program.
5. Discuss the various statements in a C++ source file.
6. Describe the difference between an assembler and a compiler.
7. Explain the basic difference between an executable program and an interpreted BASIC program.

52

Hardware and Software Interrupts

INTRODUCTION

Joe Tekk was busy with the installation of a network interface card. However, every time he booted the machine, the network software failed to initialize. All the other hardware in the machine, including the sound card, worked fine.

Eventually, after trial and error, Joe discovered that removing the sound card and commenting out the sound drivers in the CONFIG.SYS and AUTOEXEC.BAT files allowed the network software to load. Joe read through the manual for the sound card and found that its preassigned interrupt was the same as the interrupt used by the network interface card.

The sound card manual explained how to change the interrupt number. Joe made the necessary changes and rebooted the machine. The conflict had disappeared.

PERFORMANCE OBJECTIVES

Upon completion of this exercise, you will be able to

1. Explain how hardware and software interact through interrupts.
2. Discuss the operation of the interrupt vector table.
3. Use DEBUG to view the interrupt vector table.
4. Find the address associated with DOS INT 21H.

KEY TERMS

Software interrupt
Hardware interrupt
Interrupt handler

Interrupt service routine
Interrupt vector table

BACKGROUND INFORMATION

The operation of the personal computer involves cooperation between the hardware of the system and the software running it. Essentially, there are software events that cause the hardware to respond, and hardware events that trigger a response from the software. For example, Figure 52.1 shows how the execution of the DIR command causes the disk drive to turn on so that directory information can be read.

The DIR command routine in this example uses a *software interrupt* to activate the file I/O routines. There are many of these software interrupts reserved for use by BIOS and DOS. All of them operate through the processor's INT instruction. For example, a very

FIGURE 52.1   How software
affects hardware

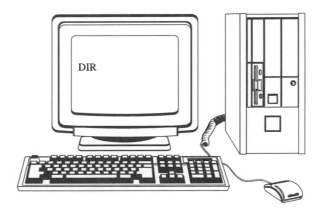

(a)   User enters DIR command.

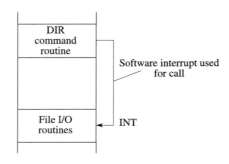

(b)   DIR subroutine calls for file input.

(c)   File routines activate hardware on the disk drive.

useful DOS interrupt is INT 21H, which is capable of performing disk, keyboard, and display I/O; memory management; time/date, printer, and directory functions; and more. We will examine the INT instruction shortly.

Figure 52.2 shows how a hardware event causes a response from the software. Every time a key is pressed on the keyboard, its scan code is transmitted to keyboard logic on the motherboard. The keyboard logic generates a ***hardware interrupt*** to signal the processor that the key code needs to be read. The hardware interrupt causes the processor to stop whatever it is doing and run the keyboard input routine (called an ***interrupt handler*** or ***interrupt service routine***). The processor takes this action because the hardware interrupt is translated into the software interrupt INT instruction. BIOS INT 9 is used to handle the keyboard on the PC. Note that similar events occur in the Windows environment, where interrupts are used and managed much like their DOS counterparts.

THE INTERRUPT VECTOR TABLE

All types of interrupts, whether hardware or software generated, point to a single entry in the processor's ***interrupt vector table***. This table is a collection of 4-byte addresses (2 for CS and 2 for IP) that indicate where the 80x86 should jump to execute the associated interrupt service

FIGURE 52.2 How hardware triggers a software response

(a) User presses a key on the keyboard.

(b) Keyboard transmits key code to keyboard logic on motherboard.

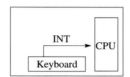

(c) Keyboard logic issues a hardware interrupt.

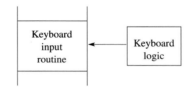

(d) Keyboard input subroutine reads key code.

routine. Since 256 interrupt types are available, the interrupt vector table is 1024 bytes long. The 1K block of memory reserved for the table is located in the address range 00000 to 003FFH. BIOS and Windows (or DOS) automatically initialize the vector table at boot time.

Figure 52.3 shows the organization of the interrupt vector table. Each 4-byte entry consists of a 2-byte IP register value followed by a 2-byte CS register value. Notice that some of the vectors are predefined. Vector 0 has been chosen to handle division-by-zero errors. Vector 1 helps to implement single-step operations. Vector 2 is used when NMI is activated. Vector 3 (breakpoint) is normally used when troubleshooting a new program. Vector 4 is associated with the overflow interrupt. Vectors 5 through 31 are reserved by Intel for use in their products. This does not mean that these interrupt vectors are unavailable to us, but we should refrain from using them in an Intel machine unless we know how they have been assigned.

BIOS and DOS, as well as Windows, use specific interrupt numbers when performing their respective operations. Some of the more common BIOS interrupts are listed in Table 52.1. Table 52.2 lists the more common DOS INT 21H interrupt functions. A programmer using assembly language has a great deal of power available through the use of these interrupts.

VIEWING THE INTERRUPT VECTOR TABLE WITH DEBUG

The contents of the interrupt vector table can be displayed in hexadecimal format through the use of a DOS utility program called *DEBUG*. DEBUG contains a command called

659

FIGURE 52.3 Interrupt vector table

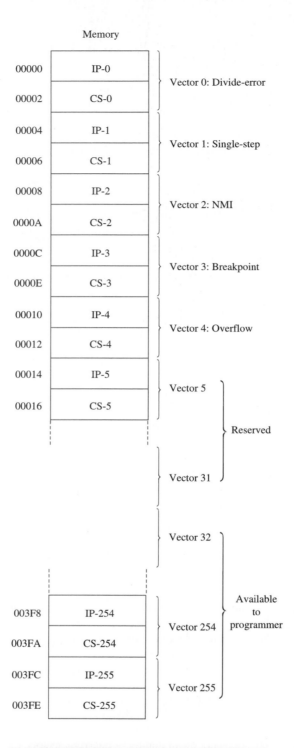

TABLE 52.1 Common BIOS interrupts

Interrupt	Function
INT 09H	Read keyboard
INT 10H	Video services
INT 13H	Disk services
INT 14H	Serial communication services
INT 16H	Keyboard services
INT 17H	Parallel printer services
INT 1AH	Time-of-day services

660

TABLE 52.2 Some of the DOS INT 21H interrupt functions

Function	Operation
01H	Console input with echo
02H	Display output
05H	Printer output
08H	Console input without echo
09H	Display string
1CH	Get drive information
25H	Set interrupt vector
2AH	Get date
2CH	Get system time
30H	Get DOS version number
31H	Terminate and stay resident
35H	Get interrupt vector
39H	Create subdirectory
3BH	Set current directory
3CH	Create file
3DH	Open file with handle
3EH	Close file with handle
3FH	Read from file
40H	Write to file
41H	Delete file
48H	Allocate memory
49H	Free allocated memory
4AH	Modify allocated memory blocks
4CH	Terminate program

dump that displays a range of memory locations on the screen. The user specifies the address of memory where the display will begin. To start DEBUG, enter

```
C> DEBUG
```

DEBUG responds with its own prompt, a dash (–). Enter the DEBUG command

```
-D 0:0
```

which instructs DEBUG to display the first 128 locations of memory, beginning at segment 0, address 0. The result will look similar to this:

```
0000:0000  FE 1A 4B 1E F4 06 70 00-16 00 C9 06 F4 06 70 00   ..K...p.......p.
0000:0010  F4 06 70 00 15 00 A4 15-4C E1 00 F0 6F EF 00 F0   ..p.....L...o...
0000:0020  46 0C 41 0B 23 19 B9 DF-6F EF 00 F0 6F 00 C9 06   F.A.#...o...o...
0000:0030  4D 11 62 E7 82 0E 41 0B-B7 00 C9 06 96 08 91 D6   M.b...A.........
0000:0040  FA 05 62 E7 4D F8 00 F0-41 F8 00 F0 3C 0F 17 09   ..b.M...A...<...
0000:0050  C6 0F 17 09 66 0F 17 09-2E E8 00 F0 01 00 A4 15   ....f...........
0000:0060  D0 E3 00 F0 19 10 17 09-6E FE 00 F0 EE 06 70 00   ........n.....p.
0000:0070  7E 13 85 C9 A4 F0 00 F0-22 05 00 00 5F 5B 00 C0   ~......."..._[..
```

The address for the BIOS keyboard INT 09H is highlighted in boldface. The four hexadecimal values 23 19 B9 DF correspond to a memory address of DFB9:1923. This is the address of the routine that reads the keyboard. The address on your machine will probably be different, since it was not configured at boot time the same as the one used for this example. Recall from Figure 52.1 that the DIR command caused hardware in the disk drive to become activated through the use of BIOS and DOS interrupts. When the INT instruction issued by DOS is processed, the address of the interrupt service routine is loaded from the interrupt vector table. Thus, DOS does not have to know the exact address of the disk routine, just the number of the software interrupt that uses it. These DOS routines are simulated in the Windows environment, which provides support for DOS applications.

The entire interrupt vector table occupies memory from address 0:0 to address 0:3FF. Recall from Figure 52.3 that each vector requires four memory locations of storage. Thus, the address for a specific INT instruction equals four times the value of the interrupt number used in the instruction. For instance, the vector for INT 21H is located at address 0:84, since 21H (33 decimal) times 4 is 84H (132 decimal). To see only the INT 21H vector, use the following DEBUG command:

```
-D 0:84 L 4
```

HARDWARE INTERRUPT ASSIGNMENTS

The hardware interrupt circuitry on the motherboard is responsible for managing the interrupts generated by the various hardware devices used in the computer. These devices include the real-time clock, disk controllers, keyboard logic, and serial and parallel ports. A special chip called a *programmable interrupt controller* is used to do all the work. This chip is programmed during boot time so that it translates one of 16 interrupt requests (IRQ0 through IRQ15) into specific INT instructions. Table 52.3 shows the hardware interrupt request assignments for a typical PC.

Refer to Figure 52.2 once again. The interrupt signal generated by the keyboard logic when it has received a key code is IRQ1. The programmable interrupt controller translates this request into an INT 09H instruction, which forces the processor to suspend what it is doing and process the keyboard interrupt. The suspended main program will resume after the interrupt service routine does its job.

The hardware interrupt logic was extended to 16 request lines in the AT computer. A sample assignment of these interrupt request lines is shown in Figure 52.4. This information is the

TABLE 52.3   Interrupt assignments

IRQ Number	Purpose
0	System timer
1	Keyboard
2	Cascade from IRQ9
3	COM2, COM4
4	COM1, COM3
5	Parallel port 2
6	Floppy disk controller
7	Parallel port 1
8	Real-time clock
9	Sound card
10	Network card
11	Video card
12	Free
13	Coprocessor
14	Hard disk controller (primary)
15	Hard disk controller (secondary)

FIGURE 52.4   Sample output from MSD utility

```
IRQ   Address     Description        Detected             Handled by
---   ---------   ----------------   ------------------   ----------------
  0   0B41:0C46   Timer Click        Yes                  SCH
  1   DFB9:1923   Keyboard           Yes                  Block Device
  2   F000:EF6F   Second 8259A       Yes                  BIOS
  3   06C9:006F   COM2: COM4:        COM2:                Default Handlers
  4   E762:114D   COM1: COM3:        COM1: Serial Mouse   MOUSE.COM
  5   0B41:0E82   LPT2:              No                   SCH
  6   06C9:00B7   Floppy Disk        Yes                  Default Handlers
  7   D691:0896   LPT1:              Yes                  CTSOUND0
  8   06C9:0052   Real-Time Clock    Yes                  Default Handlers
  9   F000:ECF5   Redirected IRQ2    Yes                  BIOS
 10   F000:EF6F   (Reserved)                              BIOS
 11   15A4:0015   (Reserved)                              REDIR5
 12   F000:EF6F   (Reserved)                              BIOS
 13   F000:F0FC   Math Coprocessor   Yes                  BIOS
 14   06C9:0117   Fixed Disk         Yes                  Default Handlers
 15   F000:EF6F   (Reserved)                              BIOS
```

output of the MSD.EXE (Microsoft Diagnostics) program that comes with DOS. Other, similar utilities allow system information to be examined. Even the system BIOS on new machines provides this type of information. It is useful to examine hardware settings to get an overall feel for the system hardware.

Often, when a new piece of hardware has just been installed, the setup software will examine the current IRQ assignments of the system and assign an IRQ that does not conflict with the others. This process is not foolproof, and sometimes the reason behind a mysterious hardware problem is the result of conflicting hardware interrupts.

TROUBLESHOOTING TECHNIQUES

It pays to know the details of interrupt processing when troubleshooting a program. Many times, the fault of erratic execution in an 80x86-based system is a poorly written or incomplete interrupt handler. Understanding the basic principles can help eliminate some of the more obvious problems.

- A valid stack must exist to save all the information required to support the interrupt handler.
- A typical interrupt pushes the current flags and return address (CS:IP).
- Vector addresses are equal to four times the vector number.
- In the interrupt vector table, the handler address is stored in byte-swapped form, with IP as the first word and CS as the second.
- It may be necessary to save and restore registers (via PUSH/POP) in the interrupt handler.
- Use IRET (return from interrupt) to return from an interrupt handler. RET does not work properly with interrupts.
- For an interrupt to work, its vector must be loaded with the starting address of the handler, and the handler code must be in place as well.

These tips should come in handy if you try to write an interrupt handler of your own, or if you are looking at the code for someone else's handler, to determine how it works.

Windows 95/98 reports the current interrupt assignments when you right-click on the Computer icon in the Device Manager window. Figure 52.5(a) shows the initial set of interrupts and their assignments.

Figure 52.5(b) shows the Windows NT Diagnostics Resources display of interrupt resources used by the system. Note that I/O ports, DMA, and Memory are also considered resources used by the system. Their assignments are displayed in a fashion similar to the IRQ display.

FIGURE 52.5 (a) Interrupt
assignments in Windows 95/98
and (b) Windows NT IRQ
Resources

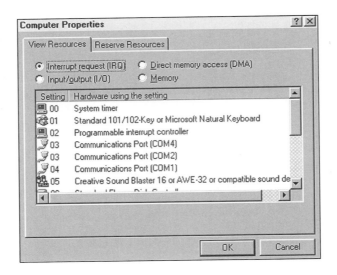

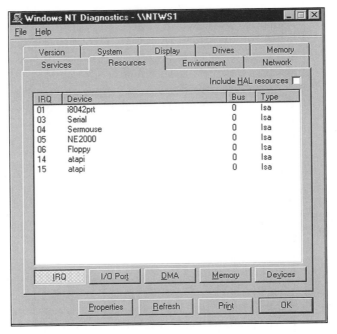

SUMMARY

In this exercise we discovered that

- The operation of the personal computer involves the use of hardware and software interrupts.
- Software interrupts are managed through the interrupt vector table.
- Hardware interrupts are managed by a programmable interrupt controller.
- INT 21H is a common DOS-based software interrupt.

SELF-TEST

This self-test is designed to help you check your understanding of the background information presented in this exercise.

True/False

Answer *true* or *false*.

1. Hardware and software operate independently in the PC.
2. The interrupt vector table is initialized during boot time.

3. The keyboard generates an interrupt once every second.
4. All devices use the same IRQ signal when they need service.
5. The interrupt table contains room for 256 vectors.
6. An interrupt vector is represented by a 16-bit number.

Multiple Choice

Select the best answer.

7. Software interrupts are used by
   a. BIOS only.
   b. Operating system only.
   c. Both BIOS and operating system.
8. An interrupt service routine is located at address 2000:1E00. The segment address of the routine is
   a. 2000.
   b. 1E00.
   c. 3E00.
   d. 20001E00.
9. Interrupt vector 0 is predefined by Intel for
   a. Out-of-memory errors.
   b. Divide-by-0 errors.
   c. Breakpoints.
   d. DOS and BIOS.
10. A useful DOS-based software interrupt is
    a. INT 21.
    b. INT 21H.
    c. INT 000E:0021.
11. A DOS utility capable of examining the interrupt vector table is
    a. IVTVIEW.
    b. VECTORS.
    c. DEBUG.

Completion

Fill in the blank or blanks with the best answers.

12. The keyboard logic generates a(n) _____ interrupt when a key is received.
13. _____ and _____ both use interrupts to control the personal computer hardware.
14. _____ is a utility that allows the user to display the contents of memory.
15. The device responsible for handling hardware interrupts is called a(n) _____ _____ _____.
16. All interrupt vectors require _____ bytes of storage.
17. Interrupt vectors consist of an IP address and a(n) _____ address.

FAMILIARIZATION ACTIVITY

1. Start the DEBUG program by entering

   ```
   C> DEBUG
   ```

2. Display the contents of the first portion of the interrupt vector table with

   ```
   -D 0:0
   ```

3. Are there any interrupt vectors that equal 0000:0000?
4. Find the interrupt vector location for INT 09H and read the associated service routine address.
5. Find the service routine address for the BIOS interrupts listed in Table 52.1.

6. Examine the rest of the interrupt vector table by using the D command three more times, as in

```
-D
```

You will get a new block of 128 bytes each time. Are there any vectors equal to 0000:0000? Are any of the vectors more popular than the others; that is, does one address appear more frequently than the others?

7. If you have MSD.EXE (or some other system utility program), display the configuration of the hardware interrupts for your system.

QUESTIONS/ACTIVITIES

1. DEBUG can be used to view the assembly language associated with an interrupt with its built-in U (unassemble) command. For example, if the address of the keyboard interrupt service routine is DFB9:1923, use the U command as follows:

```
-U DFB9:1923
```

You will get a display similar to this:

```
DFB9:1923 9C              PUSHF
DFB9:1924 FA              CLI
DFB9:1925 2E              CS:
DFB9:1926 833EB02300      CMP     WORD PTR [23B0],+00
DFB9:192B 7510            JNZ     193D
DFB9:192D 2E              CS:
DFB9:192E FE06BA23        INC     BYTE PTR [23BA]
DFB9:1932 2E              CS:
DFB9:1933 FF1EC823        CALL    FAR [23C8]
DFB9:1937 2E              CS:
DFB9:1938 FE0EBA23        DEC     BYTE PTR [23BA]
DFB9:193C CF              IRET
DFB9:193D 50              PUSH    AX
DFB9:193E 06              PUSH    ES
DFB9:193F 2E              CS:
DFB9:1940 8E062119        MOV     ES,[1921]
```

Try this on your own computer.

2. Examine the hardware manual for any card plugged into the motherboard of your computer. What interrupt request line (or lines) does the card use, if any? What is it (are they) for?

REVIEW QUIZ

Under the supervision of your instructor

1. Explain how hardware and software interact through interrupts.
2. Discuss the operation of the interrupt vector table.
3. Use DEBUG to view the interrupt vector table.
4. Find the address associated with DOS INT 21H.

53

The Advanced Intel Microprocessors

INTRODUCTION

Joe Tekk had an old case of floppies that contained programs written in an assembly language class many years ago. He picked one of the floppies at random and put it into the disk drive. A directory showed several files, one of which was an executable for a blackjack game. Joe typed in the name of the program and ran it. It worked just as it was designed to work, even though it was written for an original 8088 PC, and Joe was now running it on his Pentium-based office machine.

Don, the senior technician, walked up and asked Joe what he was doing.

"I'm running a program I wrote years ago, and it still works. Even though Intel has changed and improved the internal hardware architecture of the 80x86 family, all the processors still run code from the original 8086. Even the Pentium III still comes up in real mode at power-on."

"So all the advanced Intel processors initially act like superfast 8086 machines?"

"That's right," Joe replied. "Really fast 8086s. But the way the hardware does it has changed."

PERFORMANCE OBJECTIVES

Upon completion of this exercise, you will be able to

1. Discuss the improvements offered by each new 80x86 processor.
2. Explain the basic operation of an instruction pipeline.

KEY TERMS

Multitasking
Gigabyte
Virtual addressing
Segment descriptor
Page fault
Cache

Cache hit/miss
Pipelining
CISC
RISC
Branch prediction
MMX technology

BACKGROUND INFORMATION

In this exercise we will survey the advanced Intel microprocessors. Most importantly, we will see that every processor in the 80x86 family runs programs written for the first machines in the series, the 8088 and 8086. Thus, even though new personal computers use more advanced Intel microprocessors than the original PC did, they are fully software-compatible with the older machines.

A SUMMARY OF THE 80186

Intel made great efforts to make future processors compatible with the 8086, while at the same time offering additional enhancements and features. The 80186 High-Integration 16-bit Microprocessor can be thought of as a super 8086. Its instruction set is compatible with the 8086, which allows programs written for the 8086 to run on the 80186. Ten additional instructions are included, some of which control the additional 80186 hardware features.

Unlike the 8086, which comes in a 40-pin dual inline package (DIP), the 80186 contains 68 pins and comes housed in a variety of different packages. Plastic leaded chip carrier (PLCC), ceramic pin grid array (PGA), and ceramic leadless chip carrier (LCC) are the three types of packages that the 80186 may be found in. Both the PLCC and LCC packages have 17 pins on each of their four sides. The pins on the PGA come out of the bottom of the package. These new types of packages occupy less space on printed circuit boards and are thus very popular with designers.

Many of the additional pins are needed for signals used and generated by the new hardware features of the 80186. A *programmable interrupt controller* has been added, which supervises the operation of five hardware interrupt lines, including a nonmaskable interrupt signal. These interrupt lines can be programmed for different modes of operation. In *fully nested* mode, the four maskable interrupts are used to generate internal interrupt vectors on a prioritized basis. In *cascade* mode, the four interrupt lines become handshaking signals for an external interrupt controller, greatly increasing the interrupt capability of the processor.

Three 16-bit *programmable timers* have also been added. Two of the timers interface with the outside world and can be programmed for many different operations, such as counting and timing external events and generating waveforms (e.g., square waves and pulses). The third timer is used for internal operations. The timers can be programmed to cause an interrupt when a certain count (called a *terminal count*) is reached. All timers are clocked at a frequency equal to one-quarter of the CPU's internal clock.

A *programmable DMA unit* connected to the internal processor bus allows two independent DMA channels to operate under processor control. The DMA channels are especially useful when transferring large blocks of data between the processor, memory, and I/O devices. Both 8-bit and 16-bit transfers are allowed, and may be transferred as fast as 2.5MB per second. Priorities may be assigned to the two DMA channels in two ways: with one channel having a higher priority than the other, or with both channels having the same priority. Each channel can be programmed to interrupt the processor when it has completed its data transfer.

The *bus interface unit* is similar to the one found in the 8086, with additional signals providing the use of synchronous and asynchronous bus transfers. Finally, a *chip-select unit* has been added, which contains programmable outputs that can be used to select banks of memory or I/O devices. This task was previously accomplished by additional chip-select logic outside the processor. Putting this logic inside, and making it programmable, further reduces the amount of hardware needed to operate the microcomputer system. Wait states can also be programmed to be automatically inserted into bus cycles to allow for slow memory or I/O devices.

From a software standpoint, the 80186 is compatible with the 8086, using the same register set. Many additional hardware registers have been added for the purpose of controlling the new services provided by the 80186. These registers are contained in a dedicated area called the *peripheral control block* (PCB). The PCB is automatically placed at the top of the processor's I/O space (port addresses FF00H to FFFFH) whenever the CPU is reset, but it can be moved to a different location by changing the contents of the processor's *relocation register*. The PCB contains interrupt control registers, timer control registers for all three timers, chip-select control registers, and DMA descriptors for both channels. Programming all the internal hardware enhancements is done through these registers.

Several new instructions have also been added. PUSHA and POPA deal with the stack and are used to push or pop *all* the 80186's registers (AX, BX, CX, etc.). The integer multiply instruction IMUL has been enhanced to allow immediate data as an operand. Also enhanced are the shift and rotate instructions, which now allow a count value to be included in the instruction. For I/O operations, two new instructions allow 8-bit and 16-bit data to be inputted or outputted between an I/O device and a memory location (instead of the usual use of the accumulator). These instructions are INS and OUTS. For byte operations, INSB and OUTSB are used, and INSW and OUTSW are used for word operations. A special form of the instruction allows a *string* of bytes or words to be transferred. Two other instructions are used to manipulate the stack area (with the help of the BP register). These instructions are ENTER and LEAVE, and have been included to assist in the implementation of procedure calls in high-level languages such as C. Finally, the BOUND instruction is included to help in the partitioning of memory for multiuser environments. BOUND checks the contents of a specified register against an allowable range and generates an interrupt if the register is "out of bounds."

All in all, the 80186 offers many significant improvements over the 8086, while still being compatible. For those designers still determined to use an 8-bit data bus, Intel offers the 80188. This processor is an exact internal copy of the 80186, but differs in its use of an external 8-bit data bus.

A SUMMARY OF THE 80286

The next improvement in Intel's line of microprocessors was the 80286 High-Performance Microprocessor with Memory Management and Protection. Unlike the 80186, the 80286 does not contain the internal DMA controllers, timers, and other enhancements. Instead, the 80286 concentrates on the features needed to implement *multitasking*, an operating system environment that allows many programs or tasks to run seemingly simultaneously. In fact, the 80286 was designed with this goal in mind. A 24-bit address bus gives the processor the capability of accessing 16MB of storage. The internal memory management feature increases the storage space to 1 *gigabyte* of virtual address space—more than *one billion* locations of virtual memory. *Virtual addressing* is a concept that has gained much popularity in the computing industry. Virtual memory allows a large program to execute in a smaller physical memory. For example, if a system using the 80286 contained 8MB of RAM, memory management and virtual addressing would permit the system to run a program containing 12MB of code and data, or even multiple programs in a multitasking environment, *all of which* could be larger than 8MB.

To implement the complicated addressing functions required by virtual addressing, the 80286 has an entire functional unit dedicated to address generation. This unit is called the *address unit*. It provides two modes of addressing: 8086 real address mode and protected virtual address mode. The 8086 real address mode is used whenever an 8086 program executes on the 80286. The 1MB addressing space of the 8086 is simulated in the 80286 by the use of the lower 20 address lines. Processor registers and instructions are totally compatible with the 8086.

Protected virtual address mode uses the full power of the 80286, providing memory management, additional instructions, and protection features, while at the same time retaining the ability to execute 8086 code. The processor switches from 8086 real address mode to protected mode when a special instruction sets the protection enable bit in the machine's status word. Addressing is more complicated in protected mode, and is accomplished through the use of *segment descriptors* stored in memory. The segment descriptor is the device that really makes it possible for an operating system to control and protect memory. Certain bits within the segment descriptor are used to grant or deny access to memory in certain ways. A section of memory may be write-protected, or made execute-only, by the setting of proper bits in the access rights byte of the descriptor. Other bits are used to control how the segment is mapped into virtual memory space and whether the descriptor is for a code segment or a data segment. Special descriptors, called *gate descriptors*, are used for other functions. Four types of gate descriptors are call gates, task gates, interrupt gates, and trap gates. They are used to change privilege levels (there are four), switch tasks, and specify interrupt service routines.

The instruction set of the 80286 is identical to that of the 80186, with an additional 16 instructions thrown in to handle the new features. Many of the instructions are used to load and store the different types of descriptors found in the 80286. Other instructions are used to manipulate task registers, change privilege levels, adjust the machine status word, and verify read/write accesses. Clearly, the 80286 differs greatly from the 80186 in the services it offers, while at the same time filling a great need for designers of operating systems.

A SUMMARY OF THE 80386

Intel continued its 8086-compatible trend with the introduction of the 80386 High Performance 32-bit CHMOS Microprocessor with Integrated Memory Management. Software written for the 8088, 8086, 80186, 80188, and 80286 will also run on the 386. A 132-pin pin grid array (PGA) package houses the 80386, which offers a full 32-bit data bus and 32-bit address bus. The address bus is capable of accessing more than 4GB of physical memory. Virtual addressing pushes this to more than 64 *trillion* bytes of storage.

The register set of the 80386 is compatible with earlier models, including all general-purpose registers, plus the four segment registers. Although the general-purpose registers are 16 bits wide on all earlier machines, they can be extended to 32 bits on the 80386. Their new names are EAX, EBX, ECX, and so on. Two additional 16-bit segment registers, FS and GS, are included. These registers are illustrated in Figure 53.1. Note that the original 8086 registers may still be used as 8- or 16-bit registers (AH, AL, AX, for example).

Like the 80286, the 80386 has two modes of operation: real mode and protected mode. In real mode, segments have a maximum size of 64K. In protected mode, a segment can be as large as the entire physical addressing space of 4GB. The new extended flags register contains status information concerning privilege levels, virtual mode operation, and other flags concerned with protected mode. The 80386 also contains three 32-bit control registers. The first, called the *machine control register*, contains the machine status word and additional bits dealing with the coprocessor, paging, and protected mode. The second, *page fault linear address*, is used to store the 32-bit address that caused the last page fault. In a virtual memory environment, physical memory is divided into several fixed-size *pages*. Each page will at some time be loaded with a portion of an executing program or other type of data. When the processor determines that a page it needs to use has not been loaded into memory, a ***page fault*** is generated. The page fault instructs the processor to load the missing page into memory. Ideally, a low page-fault rate is desired.

The third control register, page directory base address, stores the physical memory address of the beginning of the page directory table. This table is up to 4K in length and

FIGURE 53.1   Software model of the 80386

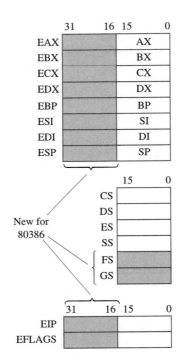

may contain up to 1024 page directory entries, each of which points to another page table area, whose information is used to generate a physical address.

The segment descriptors used in the 80286 are also used in the 80386, as are the gate descriptors and the four levels of privilege. Thus, the 80386 functions much the same as the 80286, except for the increase in physical memory space and the enhancements involving page handling in the virtual environment.

The computing power of each of the processors that have been presented can be augmented with the addition of a floating-point coprocessor. All sorts of mathematical operations can be performed with the coprocessors with 80-bit binary precision. The 8087 coprocessor is designed for use with the 8088 and 8086, the 80287 with the 80286, and the 80387 with the 80386.

A SUMMARY OF THE 80486

This processor is the next in Intel's line of compatible 80x86 architectures. Surprisingly, there are only a few differences between the 80486 and the 80386, but these differences create a significant improvement in performance.

Like the 80386, the 80486 is a 32-bit machine containing the same register set as the 80386 and all of the 80386's instruction set, with a few additional instructions. The 80486 has a similar 4GB addressing space using the same addressing features.

The first improvement over the 80386 is the addition of an 8K **cache** memory. A cache is a very high speed memory, with an access time usually 10 times faster than that of conventional RAM used for external processor memory. The 80486's internal cache is used to store both instructions and data. Whenever the processor needs to access memory, it first looks for it in the cache. If the data is found in the cache, it is read out much faster than if it had to come from external RAM or EPROM. This is known as a *cache hit*. If the data is not found in the cache, the processor must then access the slower external memory. This is called a *cache miss*. The processor tries to keep the cache's hit ratio as high as possible. Consider the following example:

$$\text{RAM access time} = 70 \text{ ns}$$
$$\text{Cache access time} = 10 \text{ ns}$$
$$\text{Hit ratio} = 0.85$$
$$\text{Average memory access time} = 0.85 \times (10 \text{ ns}) \text{ Hit}$$
$$+ (1 - 0.85) \times (10 \text{ ns} + 70 \text{ ns}) \text{ Miss}$$
$$= 20.5 \text{ ns}$$

The average memory access time for a hit ratio of 0.85 is less than 21 ns. The reason for this is as follows: If data is found in the cache (85% of the time), the access time is only 10 ns. If data is not found (15% of the time), the access time is 80 ns (the cache access time plus the RAM access time), because the processor had to read the cache to find out that the data was *not* there.

If you consider that a large portion of a program (or even an *entire* program) might fit within the 8K cache, you will agree that the program will execute very quickly, because most instruction fetches will be for code already in the cache. This architectural improvement significantly increases the processing speed of the 80486. Some of the new 80486 instructions are included to help maintain the cache.

The 80486 has two other improvements. Although it executes the same instruction set as the 80386, the 80486 does so with a redesigned internal architecture. This new design allows many 80486 instructions to execute with fewer clock cycles than required by the 80386. This reduction in clock cycles adds additional speed to the 80486's execution. Also, the 80486 comes with an on-chip coprocessor. You might recall that the 80386 can be connected to an external 80387 coprocessor to enhance performance. The 80486 has the equivalent of an 80387 built right into it. Moreover, since the coprocessor is closer to the CPU, data is transferred more quickly, which adds another performance boost.

Thus, although the 80386 and 80486 share many similarities, the 80486's differences create a much more powerful processor.

PIPELINING

Before we cover the Pentium processor, we will take a short look at the technique of *pipelining*. A pipelined processor executes instructions faster than a nonpipelined processor, as you can see in Figure 53.2. The nonpipelined processor, illustrated in Figure 53.2(a), is designed to execute instructions in three clock cycles. The first clock cycle is the Fetch cycle (F), the second clock cycle is the Decode cycle (D), and the third clock cycle is the Execute cycle (E). A sequence of four instructions executed on this machine requires 12 clock cycles to execute. When the first instruction has been fetched, decoded, and executed, the second instruction begins, and so on. This is an inefficient way of executing instructions, since each stage of execution is idle for two clock cycles.

Figure 53.2(b) shows how pipelining reduces the number of clock cycles it takes to execute the same four instructions. The main difference is that in the pipelined processor, the Fetch, Decode, and Execute operations *overlap*. For example, during the third clock cycle, I_3 is being fetched, I_2 is being decoded, and I_1 is executing. By the end of the sixth clock cycle, four instructions have made it through the Execute unit, and two more instructions have been partially completed. On a much larger scale, suppose that we execute 1000 instructions on the nonpipelined computer. This will require 3000 clock cycles of execution time. In contrast, the pipelined processor will need three clock cycles for the first instruction and *one* clock cycle for each of the remaining 999, for a total of 1002 clock cycles. This is a significant improvement over the nonpipelined result. In effect, an instruction pipeline is capable of executing one instruction every clock cycle.

Pipelining is made possible by the use of latches between consecutive pipeline stages, as shown in Figure 53.3. The latches make it possible for the three pipeline stages to operate in parallel. The result of one stage is latched for use by the next stage. This allows each stage to work on something different each clock cycle.

Intel uses pipelined computer architecture in its 80x86 family of microprocessors. With the design of the Pentium processor, the pipeline was given special attention to boost performance. In the next section you will see what Intel did with the Pentium's pipeline.

FIGURE 53.2 Instruction execution in nonpipelined and pipelined processors

F	D	E	F	D	E	F	D	E	F	D	E	
I_1	I_1	I_1	I_2	I_2	I_2	I_3	I_3	I_3	I_4	I_4	I_4	•••
1	2	3	4	5	6	7	8	9	10	11	12	

(a) Nonpipelined: 12 clock cycles required

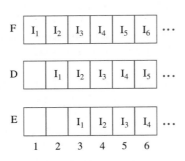

(b) Pipelined: six clock cycles required

FIGURE 53.3 A three-stage pipeline

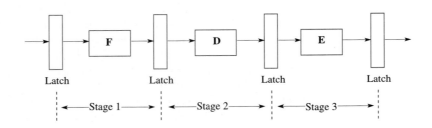

FIGURE 53.4 Architectural layout of the Pentium microprocessor

A SUMMARY OF THE PENTIUM

The Pentium, Pentium II, and Pentium III microprocessors all followed the 80486, paving the way to faster computing with new architectures. In this section we will look at the first processor in the Pentium series and how its main features compare with earlier 80x86 machines. The Pentium will run all programs written for any machine in the 80x86 line, although it does so at speeds many times that of the fastest 80486. Figure 53.4 shows a die shot of the Pentium.

There are two major computer architectures in use: *CISC* and *RISC*. CISC stands for Complex Instruction Set Computer. RISC stands for Reduced Instruction Set Computer. All the 80x86 machines prior to the Pentium can be considered CISC machines. The Pentium itself is a mixture of both CISC and RISC technologies. The CISC aspect of the Pentium provides for upward compatibility with the other 80x86 architectures. The RISC aspects lead to additional performance improvements. Some of these improvements are separate 8K data and instruction caches, dual integer pipelines, and branch prediction.

As Figure 53.5 shows, the Pentium processor is a complex machine with many inter-locking parts. At the heart of the processor are the two integer pipelines, the U pipeline and the V pipeline. These pipelines are responsible for executing 80x86 instructions. A floating-point unit is included on the chip to execute instructions previously handled by the external 80x87 math coprocessors. During execution, the U and V pipelines are capable of executing two integer instructions at the same time, under special conditions, or one floating-point instruction.

The Pentium communicates with the outside world via a 32-bit address bus and a 64-bit data bus. The bus unit is capable of performing *burst* reads and writes of 32 bytes to memory and, through bus-cycle pipelining, allows two bus cycles to be in progress simultaneously.

An 8K instruction cache is used to provide quick access to frequently used instructions. When an instruction is not found in the instruction cache, it is read from the external data

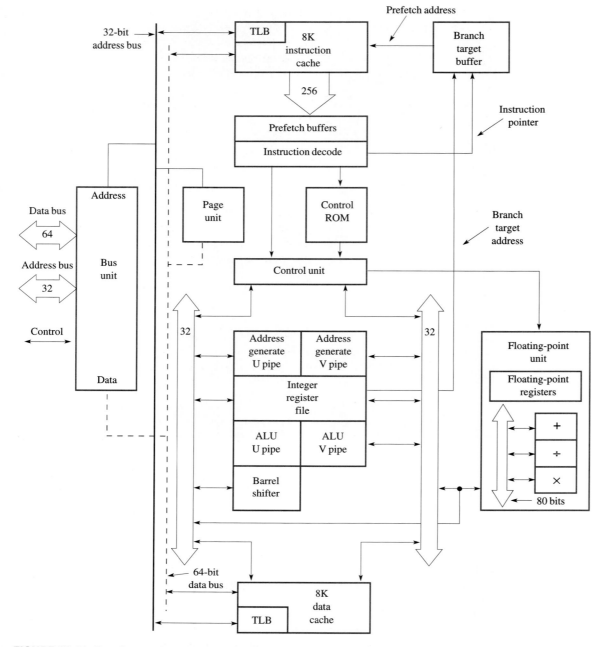

FIGURE 53.5 **Pentium architecture block diagram**

bus and a copy is placed into the instruction cache for future reference. The branch target buffer and prefetch buffers work together with the instruction cache to fetch instructions as fast as possible. The prefetch buffers maintain a copy of the next 32 bytes of prefetched instruction code, and can be loaded from the cache in a single clock cycle, due to the 256 bit-wide data output of the instruction cache.

The Pentium uses a technique called ***branch prediction*** to maintain a steady flow of instructions into the pipelines. To support branch prediction, the branch target buffer maintains a copy of instructions in a different part of the program located at an address called the *branch target*. For example, the branch target of a CALL XYZ instruction is the address of the subroutine XYZ. Certain instructions, such as CALL, may cause the processor to jump to an entirely different program location, instead of simply proceeding to the instruction in the next location. So, just in case the code from the target address is needed, the branch target buffer maintains a copy of it and feeds it to the instruction cache.

A separate 8K data cache stores a copy of the most frequently accessed memory data. Because memory accesses are significantly longer than processor clock cycles, it pays to keep a copy of memory data in a fast-reading cache. The data and instruction caches may both be enabled/disabled with hardware or software. Both also use a translation looka-side buffer, which converts logical addresses into physical addresses when virtual memory is used.

The floating-point unit of the Pentium maintains a set of floating-point registers and provides 80-bit precision when performing high-speed math operations. This unit has been completely redesigned from the one used inside the 80486 and is also pipelined. The floating-point unit uses hardware in the U and V pipelines to perform the initial work during a floating-point instruction (e.g., fetching a 64-bit operand), and then uses its own pipeline to complete the operation. Since both integer pipelines are used, only one floating-point instruction may be executed at a time.

Altogether, the Pentium processor includes many features designed to increase performance over earlier 80x86 machines. This was possible by blending CISC and RISC technologies together. The benefit to us as programmers lies in the fact that all Intel processors from the 8086 up, including the Pentium, run the same basic instruction set, but they do it faster and faster.

TROUBLESHOOTING TECHNIQUES

The advanced nature of the Pentium microprocessor requires us to think differently about the nature of computing. As we have already seen, exotic techniques such as branch prediction, pipelining, and superscalar processing have paved the way for improved performance. Let us take a quick look at some other improvements from Intel:

- Intel has added **MMX technology** to its line of Pentium processors (Pentium, Pentium Pro, Pentium II, Pentium III, and Pentium 4). A total of 57 new instructions enhance the processor's ability to manipulate audio, graphic, and video data. Intel accomplished this major architectural addition by *reusing* the 80-bit floating-point registers in the FPU. Using a method called SIMD (single instruction multiple data), one MMX instruction is capable of operating on 64 bits of data stored in an FPU register.
- The Pentium Pro processor (and the Pentium II) use a technique called *speculative execution*. In this technique, multiple instructions are fetched and executed, possibly out of order, to keep the pipeline busy. The results of each instruction are speculative until the processor determines that they are needed (based on the result of branch instructions and other program variables). Overall, a high level of parallelism is maintained.
- First used in the Pentium Pro, a new bus technology called Dual Independent Bus architecture uses two data buses to transfer data between the processor and main memory (including the level-2 cache). One bus is for main memory, the second for the level-2 cache. The buses may be used independently or in parallel, significantly improving the bus performance over that of a single-bus machine.
- The five-stage Pentium pipeline was redesigned for the Pentium Pro into a *superpipelined* 14-stage pipeline. By adding more stages, less logic can be used in each stage, which allows the pipeline to be clocked at a higher speed. In addition, improvements in microchip manufacturing allow increases in operating frequency. This is easily verified by the more than 1000-MHz processors currently available. Although there are drawbacks to super-pipelining, such as bigger branch penalties during an incorrect prediction, its benefits are well worth the price.
- The Pentium III has several new features that make it a worthy successor to the Pentium II. Streaming SIMD extensions for multimedia applications, several low-power modes, and an Intel processor serial number are three new additions to the powerful Intel Pentium architecture. Examine the literature available on Intel's Web site for additional details.

Every aspect of computing must be studied in order to fully understand how to develop improved methods, software, and hardware for increased performance. Invest some time trying to think of an improvement of your own, as if you were designing a new microprocessor. Look through recent issues of computer architecture journals, or search the Web for information. You will find that a lot of other people are thinking about improvements, too.

In this exercise we discovered that

- All 80x86 microprocessors support the same basic instruction set, using different internal architectures.
- Beginning with the 80286, protected mode operation allows the use of virtual memory and multitasking.
- Beginning with the 80486, internal cache is used to store frequently used instructions and data.
- Pipelining is a technique used to execute instructions in parallel.
- There are two major types of computer architectures: CISC and RISC.

This self-test is designed to help you check your understanding of the background information presented in this exercise.

True/False

Answer *true* or *false*.

1. A program written for the 8088 will run on the 80386.
2. A program written for the Pentium will run on the 80386.
3. All the 80x86 microprocessors support multitasking.
4. In real mode, an 80486 operates like a very fast 8086.
5. Branch prediction is a technique borrowed from RISC designs.
6. Pipelining allows instructions to execute in parallel.
7. Cache decreases the average memory access time.

Multiple Choice

Select the best answer.

8. The 80286 accesses memory in protected mode through the use of
   a. An on-board coprocessor.
   b. Segment descriptors.
   c. 8086 instructions.
9. The extended registers on the 80386 are able to store
   a. 8 bits.
   b. 16 bits.
   c. 32 bits.
   d. 1 byte.
10. The 80486 uses a special high-speed memory called a
    a. Cache.
    b. Page directory.
    c. RAM.
11. A pipelined processor is capable of executing one instruction
    a. Every three clock cycles.
    b. Every other clock cycle.
    c. Every clock cycle.
12. CISC stands for
    a. Creative instruction scheduler.
    b. Complex instruction set computer.
    c. Computing index scalar.
13. Which was the first processor to provide protected mode?
    a. 8086
    b. 80286
    c. 80386
14. To cope with instructions like JMP and CALL, the Pentium relies on a technique called
    a. Branch prediction.
    b. Jump management.
    c. Dynamic IP loading.

Completion

Fill in the blanks with the best answers.

15. Running more than one program at a time is called _____.
16. Physical memory is divided into pages in a(n) _____ memory system.
17. The Pentium uses two integer _____.
18. Pipeline operation is improved through the use of branch _____.
19. The first processor with an on-chip FPU was the _____.
20. RISC stands for _____ instruction set computer.

1. If possible, prepare identical AUTOEXEC.BAT and CONFIG.SYS files for two different computers that have different processor speeds or different processors. Run the same program (Windows, for example) on both machines and measure the system response time for each. System response time can be gauged by how long it takes for the Windows desktop to appear, for each machine to spell check the same file within Microsoft Word, or for some other similar type of operation.
2. Compare the two response times. Are they related to the clock speeds?
3. If possible, use a machine that has a TURBO button on the front panel that switches the processor between two different execution speeds. Repeat the response-time measurement described in step 1 and compare the results. Why would the lower processor speed be necessary once the faster one is available?

Go to a library and search through back issues of *Computer Design*, *Electronic Design*, or *BYTE* magazine for articles on computer architecture. Read one article and share your thoughts with your instructor.

Under the supervision of your instructor

1. Discuss the improvements offered by each new 80x86 processor.
2. Explain the basic operation of an instruction pipeline.

54

A Detailed Look at the System BIOS

INTRODUCTION

Joe Tekk was working on a 386-based system. The BIOS setup parameters, necessary for proper operation of the computer, were constantly being lost, requiring them to be reentered every time the computer booted up.

Joe had seen this problem once before, at the college where he studied engineering. One of the computers in the department was always forgetting its BIOS parameters as well. His professor, Alonzo Dixon, told him that the battery in the CMOS RAM might be bad, causing it to lose data when the computer is turned off.

Joe took the CMOS RAM chip out of a spare motherboard and replaced the suspect chip in the 386 machine. Since he had no idea what the BIOS parameters were in the spare CMOS RAM, he once again set them to their proper values during power-on. To test the new RAM, he ran several power-down/power-up cycles, until he was convinced that the CMOS RAM was working properly.

Old Al knew what he was talking about, he said to himself.

PERFORMANCE OBJECTIVES

Upon completion of this exercise, you will be able to

1. Describe the purpose of the system BIOS.
2. Explain why CMOS RAM replaced motherboard DIP switches for specifying system configuration parameters.
3. Discuss how to upgrade BIOS on a computer.

KEY TERMS

BIOS
CMOS RAM
Autodetect

Chip set
BIOS ID
FLASH BIOS

BACKGROUND INFORMATION

As we have seen in previous exercises, the system **BIOS** has a great deal to do with the overall operation of the personal computer. At power-on, the system BIOS program is responsible for testing all hardware in the computer, as well as starting the boot sequence. This process is illustrated in Figure 54.1. In this exercise we will examine the operation of the

FIGURE 54.1 BIOS flowchart

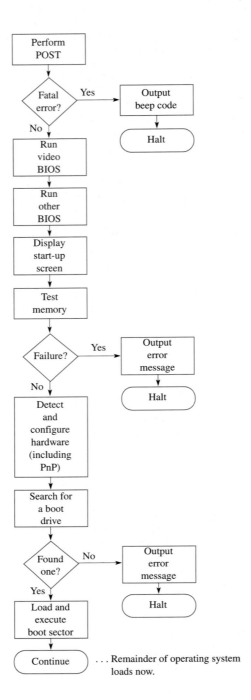

BIOS setup program provided by AMIBIOS. Other BIOS programs (e.g., Award Modular BIOS software) have similar features.

GETTING INTO THE SYSTEM BIOS

The AMIBIOS setup program is started when the machine is booting up if the user presses the Del key (or Ctrl-Alt-Esc for Award Modular BIOS). If the password option is enabled, the user must enter the correct password to gain access to the setup program.

The main menu of the AMIBIOS setup program looks like this:

STANDARD CMOS SETUP
ADVANCED CMOS SETUP
ADVANCED CHIP SET SETUP

AUTO CONFIGURATION WITH BIOS DEFAULTS
AUTO CONFIGURATION WITH POWER-ON DEFAULTS
CHANGE PASSWORD
HARD DISK UTILITIES
WRITE TO CMOS AND EXIT
DO NOT WRITE TO CMOS AND EXIT

The arrow keys are used to highlight a menu item, which is selected when the user presses Enter. Notice the numerous references to CMOS.

THE CMOS RAM

Beginning with the AT model computer, the old way of configuring the system with motherboard-mounted DIP switches was eliminated in favor of a **CMOS RAM** that stored system parameters. There were simply too many options to be set with switches. The CMOS RAM stores 64 bytes of data and uses a battery backup circuit so that it retains its information when the computer is turned off. During the boot sequence, the CMOS RAM is read by the BIOS software to establish the required configuration. The BIOS setup program allows you to modify the CMOS RAM and thereby reconfigure your system.

STANDARD CMOS SETUP

This menu option allows you to change the system time and date, the types and numbers of hard and floppy drives, the type of display being used, and the keyboard test option.

When specifying the hard drive parameters, you must enter either a predefined type number (provided by the BIOS manufacturer) or your own set of parameters. For example, the parameters for drive C: may look like this:

	Cyln	Head	WPcom	LZone	Sect	Size
Hard Drive C:	5248	16	0	0	63	2.7GB

All the numbers shown may be altered by the user. Note that some BIOS setup programs have an **autodetect** feature that will interrogate the hard drive and read the setup parameters automatically.

In addition, this option also shows how much base and extended RAM is available in the system.

ADVANCED CMOS SETUP

This menu option allows you to alter the following settings:

- Keyboard typematic delay (milliseconds)
- Keyboard typematic rate (characters per second)
- Memory test sequence
- Boot sequence (C then A, or A then C)
- Internal/external cache (enable/disable)
- NUM-LOCK indicator at boot time (on/off)
- Password checking (enable/disable)
- Video ROM shadow (enable/disable)

The Video ROM *shadow* is a clever technique used to speed up the video ROM BIOS software. Newer display cards come equipped with their own video ROM BIOS chips that contain the software required to operate the display electronics. When a video ROM is shadowed, its contents are copied into RAM (which has faster access). The system then uses the RAM copy during video operations instead of the ROM copy. Thus, intensive video applications will run faster with ROM shadowing enabled.

ADVANCED CHIP SET SETUP

This menu option is used to set up the *chip set* that controls data transfers over the EISA bus. Recall that the EISA bus grew out of a need to expand the older ISA bus so that 32-bit processors could communicate with adapter cards. Thus, the ways in which 8-, 16-, and 32-bit transfers take place, and the clock speeds for the transfers, need to be specified. For example, an 80486DX2-66 is an 80486 processor running at 66 MHz *internally* and 33 MHz *externally*, which means that the bus operates at a speed of 33 MHz (one-half the internal processor speed). The EISA chip set needs to be configured to operate under these speed conditions. Normally, only experienced technicians modify the chip set parameters. New chip sets that control operations on the local bus are configured here as well.

AUTO CONFIGURATION WITH BIOS DEFAULTS

This option sets all parameters to defaults stored in the system BIOS ROM. The default parameters are chosen to represent a broad range of typical EISA systems.

AUTO CONFIGURATION WITH POWER-ON DEFAULTS

This option is similar to the previous option, except that fewer parameters are enabled. This option is included to help diagnose hardware problems.

CHANGE PASSWORD

This option allows the BIOS password to be changed. In order to change the password, the current password must be entered first.

HARD DISK UTILITIES

This option allows the hard disk to be low-level formatted (not a good idea for IDE and SCSI drives). Normally, the hard disk utilities available with DOS and Windows (and supplied by the hard drive manufacturer) would be used instead of this menu option.

WRITE TO CMOS AND EXIT

After the BIOS parameters have been modified, they can be saved in the CMOS RAM before the BIOS setup program is exited. This menu option is used to save the new BIOS parameters.

DO NOT WRITE TO CMOS AND EXIT

This option allows the user to exit the BIOS setup program without changing any of the BIOS parameters. This is a good menu option to have, especially if the user has forgotten which parameters have been changed. The changes made in the BIOS parameters during a setup session may be ignored when this option is used.

THE BIOS DATA AREA

Once the system BIOS has completed booting the system, a data table of BIOS parameters will have been written into system RAM, beginning at address 0040:0000. These BIOS parameters are used by the BIOS interrupt service routines to both control and reflect the

TABLE 54.1 Selected BIOS data area locations

Address	Data
0040:0000	I/O addresses for synchronous communication
0040:0008	I/O addresses for printers
0040:0010	Installed devices
0040:0013	Installed memory in K
0040:0017	Keyboard shift flags
0040:001E	Keyboard buffer
0040:003F	Disk motor status
0040:0042	Disk controller status
0040:0049	Current video mode
0040:0072	Reset flag
0040:0076	Fixed disk control

status of all system hardware. Table 54.1 lists some of the predefined RAM locations and their associated parameters.

Experienced programmers are able to make use of the BIOS data area and control the system by manipulating selected parameters. For example, only a few instructions are needed to access and change the NUM-LOCK status bit stored in the byte at address 0040:0017.

EXAMINING THE BIOS DATA AREA

The DEBUG utility provides an easy way to examine the contents of the BIOS data area. Simply use the following command from inside DEBUG:

```
-D 40:0
```

The resulting display will be similar to the following:

```
0040:0000   F8 03 F8 02 00 00 00 00-78 03 00 00 00 00 00 00   ........x.......
0040:0010   63 C4 00 80 02 80 00 00-00 00 22 00 22 00 30 0B   c........."."..0.
0040:0020   0D 1C 61 1E 74 14 61 1E-2E 34 74 14 78 2D 74 14   ..a.t.a..4t.x-t.
0040:0030   0D 1C 64 20 20 39 30 0B-34 05 30 0B 3A 27 00 00   ..d  90.4.0.:'..
0040:0040   D6 00 C0 00 00 00 00 00-00 03 50 00 00 10 00 00   ..........P.....
0040:0050   00 18 00 00 00 00 00 00-00 00 00 00 00 00 00 00   ................
0040:0060   0E 0D 00 D4 03 29 20 03-00 00 C0 20 C1 60 0C 00   .....)  ... .'..
0040:0070   00 00 00 00 00 02 00 00-14 14 14 34 01 01 01 01   ...........4....
```

Once again, only experienced programmers should attempt to work directly with BIOS data.

BIOS MANUFACTURERS

Unlike motherboard manufacturers, there are very few BIOS manufacturers. Thus, even though there are millions of personal computers in use, many of them run the same BIOS program (or an appropriate upgrade). Three BIOS manufacturers to look at first are

- American Megatrends, maker of AMIBIOS. Visit them on the Web at www.ami.com.
- Award Software and Phoenix Technologies. These two companies merged in September 1998. Visit them on the Web at www.award.com or www.phoenix.com. Upgrades to Award BIOS are provided by Unicore Software, Inc. (www.unicore.com). Upgrades to Phoenix BIOS are provided by Micro Firmwave, Inc. (www.firmware.com).
- Mr. BIOS. Visit them on the Web at www.mrbios.com.

TABLE 54.2 A small sample of motherboard manufacturers

Motherboard Manufacturer	Manufacturer ID	Web Address
Epox	PA	www.epox.com
Gigabyte	G0	www.gigabyte.com.tw
Micro Star	M4	www.msi.com.tw
NEC	N5	www.neccomp.com
PC-Chips	P1	www.pcchips.com
Shuttle	H2	www.spacewalker.com

UPGRADING BIOS

Why is it necessary to upgrade your BIOS? There are many reasons. BIOS upgrades allow you to utilize new hardware in your computer (larger hard drives, processor upgrades), fix bugs discovered in the BIOS code (remember the Y2K phenomenon?), and take advantage of other enhancements to motherboard operation.

To upgrade your BIOS, you must first locate your motherboard manufacturer. Table 54.2 shows a partial list of manufacturers and their associated Web addresses. A detailed list of motherboard manufacturers can be found at www.motherboards.org.

The *BIOS ID* is displayed at power-on, while memory is being tested. Pressing the Pause key will freeze the display and allow you to copy the BIOS ID down. You can also download and run one of the many BIOS ID utilities available for free on the Web. Try www.ping.be/bios for information on Award and AMIBIOS. Figure 54.2 shows an Award BIOS display, with its BIOS ID.

If you examine the BIOS ID, the motherboard manufacturer can be determined by locating its manufacturer ID, which is found within the BIOS ID as indicated in Figure 54.3.

FIGURE 54.2 Award BIOS start-up screen

```
Award Modular BIOS v4.51PG, An Energy Star Ally
Copyright (C) 1984-98, Award Software, Inc.

06/03/1999 For Apollo MVP3 AGP/PCIset

AMD-K6(tm)-2/400 (100*4) CPU Found
Memory Test : 131072K OK

Award Plug and Play BIOS Extension v1.0A
Copyright (C) 1998, Award Software, Inc.
  Detecting Primary Master    ... Maxtor 72700 AP
  Detecting Primary Slave     ... Maxtor 85120A8
  Detecting Secondary Master  ... Hewlett-Packard CD-Writer Plus 8100
  Detecting Secondary Slave   ... SAMSUNG CD-ROM SC-148F


Press DEL to enter SETUP
06/03/1999-MVP3-596-W877-2A5LEPAEC-00
```

FIGURE 54.3 Motherboard information within BIOS ID

684

**TABLE 54.3   Chip set
information**

Manufacturer	ID	Name
Intel	2A59C	Triton FX
VIA	2A5LE	Apollo MVP3
SiS	2A5IJ	5120 Mobile
OPTi	2A5UN	Viper
ALi	2A5KK	Aladdin V

Note that the chip set used on the motherboard is also identified. Table 54.3 lists some common chip sets.

The BIOS ID for AMIBIOS contains manufacturer information as well, in the form of an ID number. For example, Epox computer is number 1519. This number is found in the third group of numbers in the AMIBIOS ID. Visit the www.ping.de/bios for a list of AMIs manufacturer codes.

On older computers (before the PCI bus) the system BIOS was stored in a ROM. To upgrade the BIOS, the ROM has to be replaced with a specific upgrade ROM from the motherboard manufacturer. Newer motherboards utilize EEPROM (electrically erasable PROM), or *FLASH ROM*, which allows the BIOS to be modified by running a special upgrade program. The upgrade program can be downloaded from the manufacturer's Web site. For example, American Megatrends provides AMIFLASH, a ZIP file that contains a program to flash the BIOS ROM on one of their motherboards. Note that FLASH ROMs come in 1-MB and 2-MB sizes and require different flash programs to be upgraded correctly.

If you determine that it is necessary to upgrade your BIOS, spend some time making a backup of all the current BIOS settings. This can be done by carefully stepping through all the setup screens and recording the value of each setting.

TROUBLESHOOTING TECHNIQUES

The BIOS software is a very important part of the PC's operating system. New BIOS programs, called "Plug-and-Play" BIOS, work together with add-on peripherals (sound cards, modems, etc.) to automatically recognize new hardware when it is added to the machine. The user does not have to fool around with DIP switch settings or tiny option jumpers. Windows 95/98 contains built-in support for Plug-and-Play BIOS, and does a nice job of detecting and configuring new plug-and-play hardware.

SUMMARY

In this exercise we discovered that

- BIOS is an important component of the microcomputer system.
- BIOS parameters are stored in CMOS RAM.
- The BIOS setup program allows you to configure BIOS for your particular system.
- FLASH BIOS ROMs can be updated by software.
- The BIOS ID indicates the manufacturer of the motherboard.

SELF-TEST

This self-test is designed to help you check your understanding of the background information presented in this exercise.

True/False

Answer *true* or *false*.

1. The BIOS setup program controls only the video display and hard drive.
2. Password protection is provided with the BIOS setup program.
3. AMIBIOS is the only BIOS available for the PC.

4. The BIOS data area and the CMOS RAM data are stored in two different places.
5. It is not possible to update your BIOS.
6. All BIOS errors are fatal.

Multiple Choice

Select the best answer.

7. The AMIBIOS setup program is started by
   a. Entering BIOS at the DOS prompt.
   b. Pressing Del when the machine is booting.
   c. Pressing both mouse buttons at once.
8. Using a video ROM shadow
   a. Slows down video BIOS routines.
   b. Increases the graphics resolution.
   c. Speeds up video BIOS routines.
9. The data stored in the CMOS RAM is
   a. Lost when the computer's power is turned off.
   b. Backed up by a battery.
   c. Loaded every time the machine boots.
10. A BIOS ROM that can be programmed using software is called a
    a. Static ROM.
    b. Dynamic ROM.
    c. EEPROM (or FLASH ROM).
11. In the BIOS manufacturer ID, the first five characters specify the
    a. Chip set.
    b. Processor.
    c. Motherboard manufacturer.

Completion

Fill in the blank or blanks with the best answers.

12. Data transfers on the EISA bus are controlled by a(n) _____ _____.
13. The BIOS program is stored in a(n) _____.
14. The starting address of the BIOS RAM data is _____.
15. EEPROM stands for _____ erasable PROM.
16. The VIA Apollo MVP3 is an example of a(n) _____.

FAMILIARIZATION ACTIVITY

1. Reboot your computer by turning the power off and on. Enter the setup program as indicated by your machine.
2. If your system is equipped with two floppy drives, change the setting on the second drive (for drive B:) to NONE (or the equivalent). If your system has a single floppy, change the hard drive type to NONE (be sure to check with your instructor before doing this).
3. Save your changes and exit the BIOS program.
4. When DOS boots, attempt to get a directory of drive B: (or drive C: on the single-floppy system). What error message, if any, do you get?
5. Reboot the computer using Ctrl-Alt-Del. Are you able to enter the BIOS setup program from a warm boot?
6. Change the drive parameters back to their original settings, save your changes, and exit the setup program.
7. Repeat the DIR command. Is everything back to normal?

QUESTIONS/ACTIVITIES

1. Examine the motherboard of your computer. Can you find the BIOS ROMs? Write down the manufacturer's name and other information.

2. Use DEBUG to view the date of your BIOS ROM. This is done by displaying the contents of memory beginning at F000:FFF0. Use the following DEBUG command:

```
-D F000:FFF0
```

Your display will be similar to the following:

```
F000:FFF0  EA F4 04 A6 02 30 34 2F-33 30 2F 39 30 00 FC EF   .....04/30/90...
```

The bytes highlighted in boldface contain the date stamp of the installed BIOS.

3. Using information from the BIOS ID of your computer, identify the manufacturer of the motherboard.

REVIEW QUIZ

Under the supervision of your instructor

1. Describe the purpose of the system BIOS.
2. Explain why CMOS RAM replaced motherboard DIP switches for specifying system configuration parameters.
3. Discuss how to upgrade BIOS on a computer.

55

Windows Internal Architecture

INTRODUCTION

Joe Tekk was browsing through the books in the computer section of a local bookstore. He found a book that illustrated the technical details of the internal operation of the Windows 98 operating system. The book was intended for developers writing application software. As Joe leafed through it, he found answers to many of the little questions he had in his mind about how Windows 98 did its job. For example, how does Windows 98 handle old hardware devices that are unsupported today?

Joe bought the book to put on his shelf as a reference.

PERFORMANCE OBJECTIVES

Upon completion of this exercise, you will be able to

1 Discuss some of the architectural features of Windows 95/98 and Windows NT.
2. Explain why real-mode device drivers reduce performance.

KEY TERMS

Application programming interface
Kernel

Graphical device interface
Virtual machine

BACKGROUND INFORMATION

The internal operation of Windows is a far more advanced topic than we are able to delve into here. But a short look at some of the major architectural components of Windows should provide you with a general picture of how things work in the operating system. You may never need to use this information during day-to-day computing, but it may be valuable when a problem crops up that you are unable to resolve using the standard help mechanism.

A LOOK INSIDE WINDOWS 95/98

Figure 55.1 illustrates a simplified view of the Windows 95/98 architecture. Three main components, the API layer (*application programming interface*), the system virtual machine, and the MS-DOS virtual machine all communicate with the base system through a set of protection rings (provided by 80x86 protected mode). Let's take a look at each component.

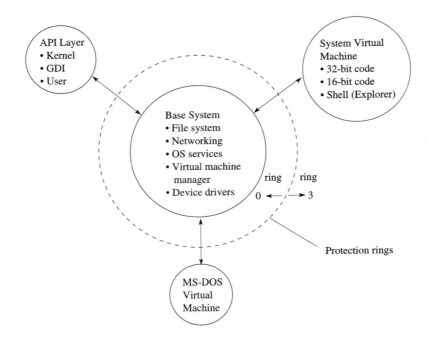

FIGURE 55.1   Windows 95/98 internal architecture

THE WINDOWS API LAYER

This layer provides system services to both 16-bit and 32-bit applications. The 16-bit API is used for old Windows 3.1 programs. The 32-bit API provides similar services, plus many more, for newer 32-bit applications.

The API layer is composed of three components: the *kernel*, the GDI (*graphical device interface*), and the user portion. The kernel provides services such as memory management. The GDI controls what appears on the display screen, managing all graphical output. The user portion provides services to applications such as supplying icons for buttons.

THE SYSTEM VIRTUAL MACHINE

The system *virtual machine* is responsible for handling 32-bit and 16-bit applications. The 32-bit applications run in a preemptive multitasking mode, meaning they can be interrupted by another application and restarted at any time. The 16-bit applications (old Windows 3.1 programs) are executed in a different memory space, to simulate the Windows 3.1 environment and help protect the rest of Windows 95/98 from an out-of-control 16-bit application.

The shell portion of the system virtual machine is typically the Windows Explorer program. Users new to Windows 95/98 have the option of running the Program Manager from Windows 3.1 to maintain a sense of familiarity.

THE MS-DOS VIRTUAL MACHINE

Running a program in a DOS shell uses the MS-DOS virtual machine. This service is provided by the virtual-8086 mode of the advanced 80x86 processors. Each virtual machine contains 1MB of memory space, a copy of DOS (configured by CONFIG.SYS and AUTOEXEC.BAT at boot time), and operates as a unique, individual machine.

THE BASE SYSTEM

The base system contains several components. Unlike Windows 3.1, which ran on top of DOS and used DOS to manage files, Windows 95/98 contains its own file system, managed by the base system. Included is support for long file names and larger hard drives (up to 2GB partitions).

The base system also contains built-in networking support, allowing Windows 95/98 machines to be easily networked via network interface card, modem, or serial port. The base system also provides operating system support (e.g., DirectX graphical hardware control), manages any virtual machines created, and controls the various installed device drivers. There are two types of device drivers: virtual device drivers (files with .VXD extensions) and real-mode device drivers. Virtual device drivers run in protected mode. Real-mode device drivers (typically used for old or unsupported hardware) force Windows 95/98 to switch from protected mode to real mode and back whenever they are used. This could lead to a loss in performance for a heavily used device driver.

WINDOWS NT ARCHITECTURE

The Windows NT operating system internal architecture is much more complex than Windows 95/98 architecture because of the internal modifications necessary to achieve a more stable, reliable, and secure environment. As such, the Windows NT operating system can accommodate any type and size of organization. Figure 55.2 illustrates all the various components in both the *user* and *kernel* modes of Windows NT.

FIGURE 55.2   Windows NT system architecture

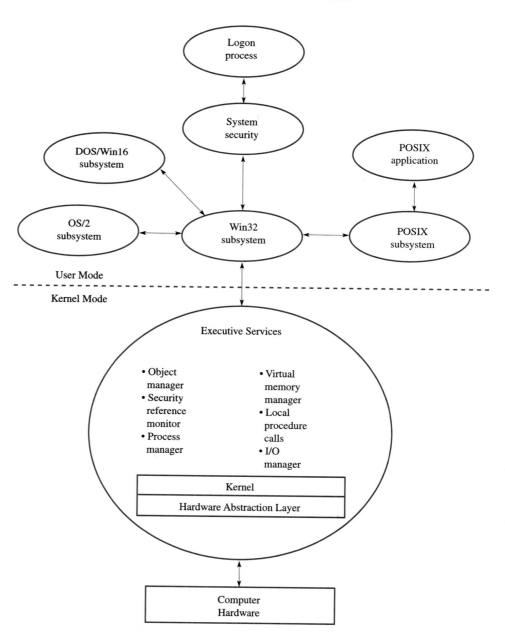

**TABLE 55.1   Windows NT
components**

Component	Description
DOS, Win16, and OS/2 subsystems	Support for applications written for earlier operating systems.
POSIX subsystem	Portable operating system interface for computing environments (POSIX) support.
Win32 subsystem	32-bit API support between application and operating system. Also manages keyboard, mouse, and display I/O for all subsystems.
Object manager	Responsible for creating, naming, protecting, and monitoring objects (operating system resources).
Security reference monitor	Manages security (access to objects).
Process manager	Manages processes and threads.
Virtual memory manager	Implements and manages virtual memory. All processes are allocated a 4-GB virtual address space.
Local procedure calls	Message-passing facility providing communication between client and server processes.
I/O manager	Interfaces with I/O drivers to provide all operating system I/O functions.
Kernel	Schedules tasks (processes of one or more threads) among multiple processors. Manages dispatcher objects and control objects.
Hardware abstraction layer	Provides a machine-independent interface by "virtualizing" the actual hardware of the base system.

Notice from Figure 55.2 that only the kernel mode has access to the physical hardware, thereby offering a high level of system protection. The executive services provided by the kernel are the foundation on which all other processing activities are performed in user mode. Table 55.1 shows a brief description for each of the components in each mode.

This modular approach to operating system design and development has allowed the Windows NT operating system to be ported to other hardware platforms, such as RISC-based microprocessors. You are encouraged to explore in more detail each of the internal components of the Windows NT operating system.

TROUBLESHOOTING TECHNIQUES

Knowing even a little bit of the internal operation of Windows may help you understand the reason behind a problem you encounter with an application or may serve as a starting point as you begin determining what is wrong. You are encouraged to learn more about Windows from books devoted to the subject, from information posted on the Web, and from other Windows users. The more information you have at your disposal, the better.

SUMMARY

In this exercise we discovered that

- The Windows 95/98 operating system contains three major parts: the API layer, the system virtual machine, and the MS-DOS virtual machine.
- The API layer contains the GDI, kernel, and user portions.
- The system virtual machine handles 16-bit and 32-bit code.
- Windows NT has a different internal architecture than Windows 95/98.
- The Windows NT operating system contains support for security, POSIX, and OS/2.

This self-test is designed to help you check your understanding of the background information presented in this exercise.

True/False

Answer *true* or *false*.

1. Windows 95/98 offers no type of protection mechanism.
2. The GDI is a component of the system virtual machine.
3. Only 32-bit code can be executed in the API layer.
4. You can use Program Manager to run applications under Windows 95/98.
5. All programs are executed using preemptive multitasking.
6. Windows NT runs on RISC-based microprocessors.
7. Windows 98 and Windows NT have the same internal architecture.
8. An MS-DOS virtual machine contains its own 1-MB addressing space.

Multiple Choice

Select the best answer.

9. Device drivers are maintained by the
   a. API layer.
   b. Base system.
   c. System virtual machine.
10. GDI stands for
   a. General device indicator.
   b. Graphical device interface.
   c. Graphical data interface.
11. Windows NT Executive Services run in
   a. User mode.
   b. POSIX subsystem.
   c. Kernel mode.
12. Which operating systems contain support for security?
   a. Windows 95/98.
   b. Windows NT.
   c. Both a and b.
13. Virtual device drivers run in
   a. Real mode.
   b. Protected mode.
   c. Both a and b.

Completion

Fill in the blank or blanks with the best answers.

14. The three components of the API layer are the _____, _____, and _____.
15. All 32-bit code is executed in the _____ virtual machine.
16. The typical shell used by the system virtual machine is _____.
17. The extension on a virtual device driver file is _____.
18. The _____ _____ _____ in Windows NT "virtualizes" the actual hardware of the base system.
19. Access to operating system features is controlled by four rings of _____.
20. 32-bit code is executed using _____ multitasking.

1. Experiment with various options with CONFIG.SYS and AUTOEXEC.BAT. What does the DOS environment look like for each configuration?
2. Determine which RISC processors can run Windows NT.
3. Read Appendix C, then search the Web for RAID products.

1. What real-mode device drivers are being used on your Windows 95/98 system, if any?
2. How many MS-DOS windows can you open?
3. Why is Windows NT considered more reliable than Windows 95/98?

Under the supervision of your instructor

1. Discuss some of the architectural features of Windows 95/98 and Windows NT.
2. Explain why real-mode device drivers reduce performance.

56

Computer Viruses and Security

INTRODUCTION

Joe Tekk was puzzled. First, the time displayed on his Windows 98 taskbar was incorrect. Next, for no apparent reason, some applications would launch with a single-click instead of a double-click.

Then the printer began working erratically. Joe checked the cable, which was fine. He connected the printer to a different computer, where it also worked fine.

The problem must be inside my computer, he thought to himself. Deciding to look inside, he shut Windows 98 down and powered off. He took the case off and gave the motherboard, plug-in cards, and connecting cables a good visual. Seeing nothing out of the ordinary, he powered his machine back on and waited while it booted up.

A warning tone from his computer took Joe by surprise. One look at the screen told him what was wrong. His computer had a virus.

PERFORMANCE OBJECTIVES

Upon completion of this exercise, you will be able to

1. Describe the operation of a typical file-infecting virus.
2. Discuss how viruses are transmitted between computers and files.
3. Use a virus scanner to scan a system for viruses.
4. Explain how firewalls and encryption help provide security.

KEY TERMS

Computer virus
Interrupt hook
Boot sector virus
Worm
Trojan horse
Macro virus

Checksum
Signature
Denial of service attacks
Firewall
Public key encryption

BACKGROUND INFORMATION

There are literally thousands of programs available for the personal computer that provide meaningful and constructive service to the user. Unfortunately, there is a growing group of destructive programs, called ***computer viruses***, as well. These virus programs are written by clever but dangerous programmers with the intent of doing some kind of damage to the computers of others. This damage can be as simple as a message on the display that reads "You are infected!" or as devastating as a destructive hard drive format. In this exercise, we will examine the operation of a computer virus, its method of infection and replication, its classification, and methods of preventing and eliminating virus infections.

OPERATION OF A TYPICAL FILE-INFECTING VIRUS

A virus is a computer program designed to place a copy of itself into another program. Programs are stored on floppy and hard disks as .COM and .EXE files. A virus program intercepts the .COM or .EXE file as DOS begins to load it into memory for execution. The virus checks the .COM or .EXE file to see if it already contains an infection. If the file is not infected, the virus inserts a copy of itself into the .COM or .EXE file and makes whatever other changes are necessary to the file so that the virus, and not the original .COM or .EXE program, executes first the next time the file is executed.

If the file is already infected, the virus does nothing, and allows the program to load normally. This is a clever way of avoiding detection, and it makes no sense to reinfect an already-infected file. This process is illustrated in Figure 56.1. Notice that control in the computer switches from DOS to the virus program, and back to DOS. This implies that the virus is already in memory, watching what is going on in the computer. How did the virus get there in the first place? The answer is given in the next section.

FIGURE 56.1   Typical virus infection sequence

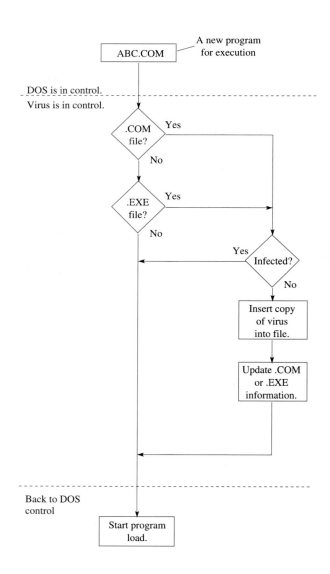

Virus infections come from a limited number of sources. First, you may get an infection by copying or running a program from someone else's floppy disk (or a swapped hard drive). If you copy an infected program (e.g., a game program, a popular hiding place for viruses), you must run the copied program to activate the virus. If you run an infected program, the virus code gets control first and loads itself into memory to reinfect more programs later. The virus is active until you turn your computer off. It is not good enough to just reboot your machine using Ctrl-Alt-Del when you think you have an infection, because some viruses take over the reboot code and keep themselves in memory during a warm boot.

The second way to get an infection is to download an infected program from a computer bulletin board or network site. Most system operators in control of these types of installations already scan their software to help guarantee it is virus-free.

It is rare to find a virus hiding in a newly purchased software product or a box of pre-formatted disks, but anything is possible. It only takes one infected program to begin the spread of a virus, as you can see in Figure 56.2. The user in this example has unknowingly infected two programs. What happens the next time the STARGATE or CHESS program is executed? The virus will be back in business, resident in memory, waiting to infect any other programs that are run that day. So the user will spread the virus, once again unknowingly, to more and more files.

FIGURE 56.2   How a virus spreads

Floppy with infected file

STARGATE.EXE

(a) User puts friend's floppy into disk drive.

> Copy STARGATE.EXE C:\Games

(b) User copies the infected program.

> CD\Games
> STARGATE

(c) User runs the infected game program. The virus installs itself into memory.

> CHESS

(d) User gets tired of game and starts a new one.

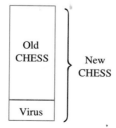

Old CHESS

Virus

New CHESS

(e) Virus infects the CHESS program prior to execution.

THE ANATOMY OF A VIRUS

For a self-replicating virus to be able to survive, it must be capable of the following:

- Operating as a memory-resident program
- Interfacing with the disk I/O routines
- Duplicating itself

You may think, from our brief introduction to assembly language, that the code for a virus that does all of this might be substantial. But the whole trick to writing a virus is to make it as small as possible, because large chunks of virus code might be more easily spotted. Many self-replicating viruses are written with fewer than 1000 bytes of machine code, but are still capable of mass destruction of system files.

For the virus to install itself as a memory-resident program, it requires a method of getting control of the system. This is usually done through the use of an ***interrupt hook***. Recall that BIOS and DOS initialize the interrupt vector table at boot time with the addresses for each interrupt service routine. All the virus has to do is pick an interrupt frequently used by BIOS and DOS (e.g., the keyboard interrupt or INT 21H) and make the interrupt vector point to itself, instead of to the BIOS or DOS service routine. To give the impression that everything is fine, the virus always completes its job by running the original interrupt service routine. This concept of hooking an interrupt is illustrated in Figure 56.3. As you can see, the virus makes a copy of the original INT 21H service routine address and uses it upon exit.

The process of hooking the interrupt and installation as a memory-resident program is called *initialization*.

FIGURE 56.3   Hooking an interrupt

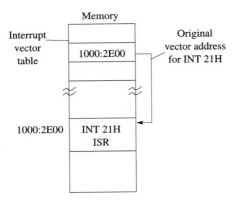

(a) Original vector.

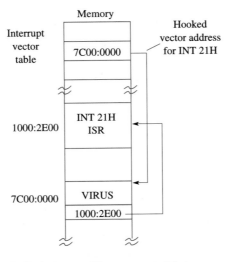

(b) Hooked vector. Virus gets control first.

698

When a virus detects that a file should be infected, it does the following:

1. Positions the file pointer at the end of the file. This is accomplished with calls to BIOS or DOS file I/O routines.
2. Writes a copy of itself (directly from memory) to the end of the file being infected.
3. Modifies the initial code of a .COM file (usually a JMP instruction), or the header information in an .EXE file, so that the virus code is executed first whenever the file loads.
4. Tells DOS to update the file information. This is done with another call to BIOS/DOS file routines.
5. Resumes the process of loading the file into memory, so that the user thinks everything is normal.

The duplication (or replication) phase of the virus continues either forever or until the virus has determined that it has reached a preset number of duplications. In the latter case, the virus takes whatever destructive action it was designed to take. Some viruses merely slow down the system response time when they are activated. For example, files that get sent to the printer take longer to print, the mouse gets sluggish, the display scrolls at a slower rate—anything annoying to the user. Other viruses are more troublesome, scrambling the internal key codes so that the keyboard becomes impossible to use (and Ctrl-Alt-Del does not work anymore). A destructive virus might swap numbers around in a spreadsheet file, encrypt a file so that it is impossible to decipher, format a few random tracks on a floppy disk, or delete important system files such as AUTOEXEC.BAT, CONFIG.SYS, and COMMAND.COM. Some viruses are mutated into *strains* or *variants* by other programmers who make a few changes in the virus code so that it does slightly different things when activated. File-infecting viruses are the most plentiful of the known viruses.

BOOT SECTOR VIRUSES

The viruses discussed so far infect actual files residing on floppy or hard disks. A ***boot sector virus*** is a virus that takes over the boot sector of a floppy or hard drive. Recall that the boot sector is the *first* piece of code read in from the disk and is in control of what happens next during the boot process. Boot sector viruses are very sophisticated and require good programming skills, which is why there are fewer boot sector viruses than file viruses.

The operation of a typical boot sector virus is detailed in Figure 56.4. The boot sector virus moves the original boot sector to a new sector on the disk and marks the new sector as bad so that DOS does not try to access it. The portion of the boot sector virus code that does not fit into the boot sector is stored in sectors marked bad also, to help avoid detection.

Any system booted from an infected disk begins running the virus immediately. Once again, the virus operates transparently to the user as a memory-resident program, possibly infecting the boot sector of any disk placed in the floppy drives.

WORMS

A ***worm*** is essentially a virus that does not replicate by infecting other programs. When a worm-infected program is loaded into memory, the worm gets control first and does its damage (possibly transparently).

TROJAN HORSES

A ***Trojan horse*** is a virus disguised as a normal program. For example, a chess program with sophisticated 3-D graphics and sound could be a Trojan horse program containing a file-infecting virus. So, while the user is having fun playing chess, the virus is secretly running, too, examining directories on the hard drive for files to infect. The hard drive activity seems normal to the user because of the 3-D effects on the screen. But the effects are a mask for the file infection happening under the user's nose.

FIGURE 56.4 Typical boot sector virus operation

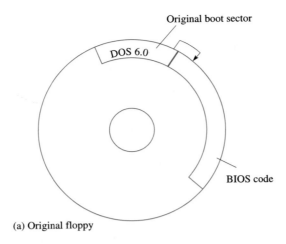

Original boot sector

(a) Original floppy

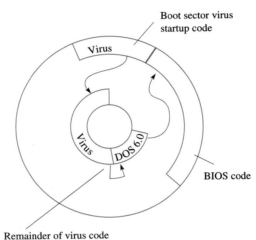

Boot sector virus
startup code

BIOS code

Remainder of virus code
and original boot sector
all marked bad

(b) Infected floppy

MACRO VIRUSES

Relative newcomers to the world of viruses are the ***macro viruses***. Typically, a macro virus is written in a scripting language, such as Visual BASIC, and contains commands that can be processed from within a word-processing or spreadsheet application. These commands, stored as a macro for the particular application, can be very destructive, for instance by deleting files or randomly rearranging data within a spreadsheet when the macro is executed. Putting a simple macro virus together does not require the programming skills of an assembly language or C/C++ programmer, making them easier to write. Word-processing and spreadsheet documents should be included in the list of files routinely scanned by your virus scanner.

VIRUS DETECTION

Fortunately, there are methods that can easily detect the presence of a virus hiding in *any* file. One method involves the use of a file ***checksum***. A checksum is a numerical sum based on every byte contained in a file. For example, an 8-bit checksum is made by adding all the bytes of the file, ignoring any carries out of the most significant bit. A file infected by a virus will practically always have a different checksum, thereby making detection a simple matter of comparing the current checksum of a file with a previous one.

A second method of virus detection involves searching a file for a known virus *signature*. A virus signature is an encoded string of characters that represents a portion of the actual virus code for a specific virus. A file is scanned for an entire collection of virus signatures to help guarantee that it is free of infection. The only problem with this technique is that a brand-new virus does not have a signature that the virus detection software will recognize, and it will escape detection.

There are many good virus scanning programs on the market. Many are also available as shareware, or through the computer centers at many colleges, which have an interest in keeping the number of virus infections on students' disks to a minimum. Students migrate from machine to machine on a large campus and will rapidly spread a virus before it is detected.

When DOS 6.0 was released, a virus scanning program called MSAV.EXE (Microsoft AntiVirus) was included; it runs under DOS and is very easy to use. MSAV contains a detailed list of all the viruses it scans for and is capable of removing a virus from an infected file. When MSAV is first started, it scans the current drive for all possible directories. A menu allows the user to choose one of the following operations:

1. Detect
2. Detect and clean
3. Select new drive
4. Options
5. Exit

Selections are made by using the arrows keys followed by Enter, by typing in specific letters, or by using the mouse. MSAV also contains an information menu, activated when F9 is pressed, which allows the user to view details of a specific virus chosen from MSAV's internal list of viruses.

VIRUS DETECTION IN WINDOWS

A number of companies make virus detection products for Windows. One popular virus scanner package is McAfee VirusScan. VirusScan's VShield runs at boot time, installing itself so that it can watch everything that is going on. VirusScan is used to scan entire drives for infected files. VShield scans files when they are accessed. When a floppy disk is inserted into the drive, it is automatically scanned for viruses. VirusScan is also able to automatically update its virus information database over the Web (for registered users). A screen shot of VirusScan's control window is shown in Figure 56.5. As shown, the hard

FIGURE 56.5   McAfee VirusScan control window

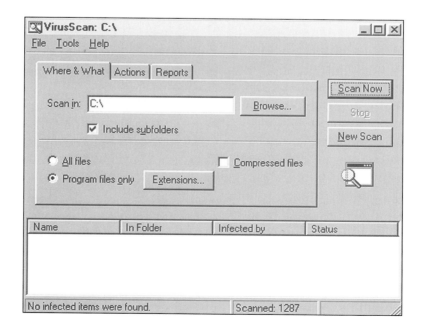

FIGURE 56.6   Detection menu

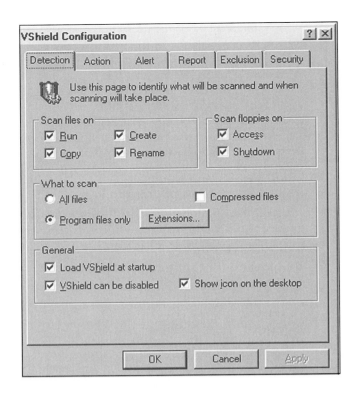

FIGURE 56.7   Action menu

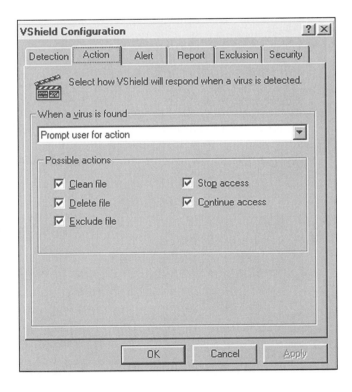

drive C: has been scanned and no viruses were found. Several configuration options allow you to control what types of files are scanned and what to do when a virus is found. Figures 56.6 and 56.7 illustrate VShield's Detection and Action option menus, respectively.

It is well worth the investment (money and scanning time) to use a good virus scanner. You will appreciate its value the first time it finds a virus.

VIRUS PREVENTION

The simple rules that follow should help eliminate the threat of infection:

- Never share your floppy disks with another person.
- Make sure your floppy disks are write protected.
- Never copy or run software from another person's floppy before it has been scanned for viruses.
- Never execute downloaded software (from a computer bulletin board or network site) before it has been scanned for viruses.
- Never run a program from your disk on another person's computer.
- Run a virus scanning program such as MSAV periodically on your computer to ensure that there are no spreading infections.
- Examine e-mail attachments before opening them. Beware of executable programs and Visual BASIC scripts sent as attachments.

Keeping your disks and software to yourself is the best method of avoiding a virus infection.

SECURITY

Security, on stand-alone and networked computers, begins with the security measures in place on a host computer. Protection of files and limited access to resources are simple measures that can be instituted. For example, you may decide not to share any files or your printer. This is a good start, but it is only the beginning of a successful effort to provide security.

THREATS

The threats to a networked computer environment are many. Essentially, the goal for networked computers is to transmit information from a source location to a destination. In practice, the communication may be encrypted for added security, or IP tunneling may be employed to further restrict access to sensitive information. Figure 56.8 shows several different scenarios that are common problems associated with exchanging information between computers. Note that any interference with the exchanged information, even capturing it with a packet sniffer, could constitute a security breach.

Very sophisticated attacks called *denial of service attacks* have now become a significant threat as well. In this attack, special programs secretly installed on selected hosts around the Internet participate in a coordinated attack against one site by flooding it with network requests. The result is to deny service to anyone attempting to use the site.

Two methods we will examine to help cope with ongoing threats are the use of firewalls and encryption.

FIREWALLS

A *firewall* is a program that will examine packets of information between two network hosts to determine whether or not to allow the communication. Figure 56.9 shows how a firewall is used to protect an intranet from external access.

Note that the firewall is selective about how information may be exchanged between the public and private networks. The firewall may be implemented through hardware as a stand-alone component of your network or through software on a designated gateway machine.

When a host computer is "behind a firewall," it is not visible to the outside network. Host computers that are not protected by a firewall stand the chance of being infiltrated by an attacker, so it is a good idea to utilize a personal firewall on your computer, in addition to the one provided by your ISP.

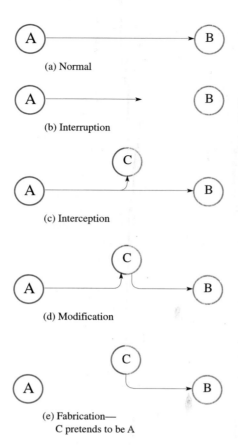

FIGURE 56.8 Typical information exchange scenarios between A and B

(a) Normal

(b) Interruption

(c) Interception

(d) Modification

(e) Fabrication—
C pretends to be A

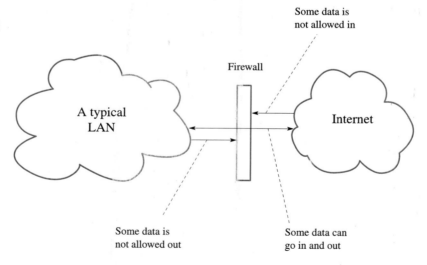

FIGURE 56.9 Communication networks connected to a firewall

Some data is
not allowed in

Firewall

A typical
LAN

Internet

Some data is
not allowed out

Some data can
go in and out

Several manufacturers provide trial versions of their firewall utilities, which block and report unwanted access to your networked computer. Visit www.zonelabs.com to download and experiment with the ZoneAlarm firewall product.

PUBLIC KEY ENCRYPTION

As the name indicates, *public key encryption* uses a public key. Public key encryption uses two keys: one public key and one private key. The public key is used to encrypt the data to be transmitted. Once the data is encrypted it bears little resemblance to the original data. For example, the phrase

```
Hello Ken, how are you?
```

might be encrypted as

```
G6YpomCW2a1JcxiBks
```

The public key cannot be used to decrypt the data. Instead, the private key is used to decrypt the data. This is a more secure method to encrypt and decrypt the data since the public key can be posted for public access. The private key must be guarded very carefully and protected from disclosure.

Using public key encryption, the public keys for individuals are stored on a public key ring. As messages are created, the public key ring can be accessed and the appropriate public key used to encrypt the message so that the only receiver of the message (with the corresponding private key) can decode the message and read it. This procedure is shown in Figure 56.10. Notice that Jim's key ring contains entries for Ken and Jeff. When a message is composed and sent to Ken, Ken's public key is used to encrypt the message. Although the message is transmitted on the network, the message contents cannot be examined. Only Ken, the receiver of the message, can decrypt and read the message text.

PGP

PGP stands for Pretty Good Privacy, a security application produced by Phil Zimmerman. PGP provides confidentiality and authentication service that can be used with electronic

FIGURE 56.10 Information exchange using public key encryption

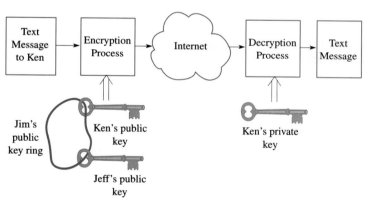

(a) Encryption

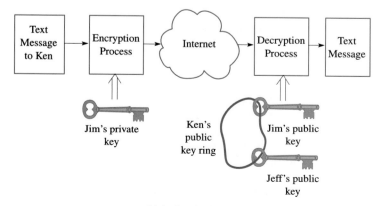

(b) Authentication

messages as well as file storage applications. PGP has gained significant popularity due to the following important elements:

- The best cryptographic algorithms as basic building blocks.
- Unlimited distribution of source code, and documentation.
- Not controlled by a government or other standards organization.

Specifically, PGP provides supports mechanisms for digital signatures, message encryption, compression, and transparent compatibility with many application programs.

PGP provides an easy method to begin corresponding with a person who prefers to use encrypted messaging.

The PGP freeware can be downloaded from the MIT Web server using the following address:

```
http://web.mit.edu/network/pgp.html
```

TROUBLESHOOTING TECHNIQUES

Four simple words say it all: When in doubt, scan.

SUMMARY

In this exercise we discovered that

- There are many different types of computer viruses.
- File-infecting viruses insert copies of themselves into other files, which in turn infect even more files.
- Viruses can be sent through e-mail as attachments.
- Firewalls are used to control the exchange of information to/from a private network.
- Encryption can be used to prevent access to sensitive information.

SELF-TEST

This self-test is designed to help you check your understanding of the background information presented in this exercise.

True/False

Answer *true* or *false*.

1. Only .COM files can be infected.
2. File infections come only from other infected files.
3. The hard drive is safe from infection.
4. Viruses are easy to find because they require huge amounts of code and cannot easily hide in a file.
5. A firewall only prevents access into a private network.
6. A PGP is an encryption tool.

Multiple Choice

Select the best answer.

7. In order to reproduce, a virus must be able to
   a. Install itself as a memory-resident program.
   b. Use BIOS/DOS disk I/O routines.
   c. Duplicate itself from its own image in memory.
   d. All of the above.
8. The best way to stop a virus from spreading is to
   a. Reboot the computer.
   b. Run as many programs as possible to tire the virus out.
   c. Eliminate the virus with a virus scanning program.
   d. Format the hard drive.

9. Viruses are written
   a. By beginning programmers.
   b. With the intent of doing damage to PCs.
   c. For amusement at computer trade shows.
10. When sending data between two network hosts, the data may be
    a. Blocked.
    b. Intercepted and examined.
    c. Changed.
    d. All of the above.
11. Public key encryption uses two keys: a public key and
    a. Master public key.
    b. Private key.
    c. Universal key.

Completion

Fill in the blank or blanks with the best answers.

12. A virus disguised as a normal program is called a(n) _____ _____.
13. Viruses are detected by comparing _____ or _____.
14. A virus may also hide inside the _____ _____ of a floppy or hard disk.
15. Security may become a problem when you _____ files or other resources.
16. ZoneAlarm from www.zonelabs.com is an example of a(n) _____.

FAMILIARIZATION ACTIVITY

1. Start up the MSAV program from hard drive C:.
2. When MSAV finishes checking directories, choose Detect from the menu. Detect will scan system RAM for memory-resident viruses and then begin scanning the hard drive. The current file (and its associated directory path) is displayed in the upper-left corner of the display. Warning! If MSAV finds a virus, the speaker will emit a short tone, and a message will be displayed such as:

   Virus Friday 13th was found in: SHIP.COM

   MSAV will give you the option of cleaning the file, continuing, or stopping the scan. Consult with your instructor if this happens.

   You may also get a *verify* error for a particular file. This indicates that MSAV found a different time/date stamp on the file, or that the file's length or checksum has changed. MSAV allows you to update the file's information or ignore the error and resume scanning. Consult with your instructor if this happens (in case you have discovered a brand-new virus).
3. Use the Select option to choose a new drive. Select the drive appropriate for the floppy disk you are going to scan.
4. Use Detect to scan your floppy disk. Once again, consult with your instructor if a virus is detected.
5. Press the F9 key to get into MSAV's information menu. An alphabetical list of viruses should appear, beginning with the As.
6. Enter the three characters "F," "R," and "I." MSAV should update the virus list so that the selected virus begins with the name "Friday." Use the arrow keys to select the Friday the 13th virus.
7. Press Tab until the Info box is highlighted. Then press Enter. You should get a screen of information on the virus that shows what types of files are infected, how long the virus code is, and what the side effects of the virus are.
8. Exit MSAV.
9. Restart MSAV with the following command line:

```
C> MSAV  C:\DOS
```

MSAV will now scan only the \DOS directory, not the entire hard drive. Individual files can also be scanned in this way by using their full path names after MSAV, as in

```
C> MSAV C:\DOS\PRINT.EXE
```

QUESTIONS/ACTIVITIES

1. Use MSAV to make a list of 10 boot sector viruses. Compare their relative sizes and infection information. Include the Pakistani Brain, Michelangelo, and Alameda viruses in your list.
2. Repeat step 1 for the file-infecting viruses. Include the Friday the 13th and Columbus Day viruses in your list.
3. Search the Web for firewall products and compare three different utilities.

REVIEW QUIZ

Under the supervision of your instructor

1. Describe the operation of a typical file-infecting virus.
2. Discuss how viruses are transmitted between computers and files.
3. Use a virus scanner to scan a system for viruses.
4. Explain how firewalls and encryption help provide security.

57 Performance and Diagnostic Software

INTRODUCTION

Joe Tekk watched the image on his computer monitor. A 3-D spaceship flew through a mine-field of subatomic particles, accompanied by the typical sounds of outer space.

After 30 seconds, the display blanked out, and then the Desktop reappeared with a Result window, displaying the message Video tests completed. 46 fps, CPU 92%, Memory 67%.

Joe frowned. He was running a program to evaluate the performance of the video system in his personal computer, and did not like the results.

"It looks like I'll be buying more memory for you," he said to his computer. "67% is not good enough."

After adding another 64MB of RAM, Joe ran the test program again. The memory per-formance increased to 87%.

PERFORMANCE OBJECTIVES

Upon completion of this exercise, you will be able to

1. Discuss how system performance is measured and the steps that can be taken to improve it.
2. Explain how a system may be diagnosed using software.
3. Describe the features of the MSINFO32 utility.

KEY TERMS

Bottleneck
Benchmark

Resource
Resource conflict

BACKGROUND INFORMATION

The reasons people have for purchasing and using their computers are unique for each indi-vidual. However, there is a large number of users who demand all the power available from a computer system. They run one or more applications that require every cycle of CPU time. Some applications that demand fast processing include

- Games (real-time 3-D graphics)
- Modeling
- Computer-aided design
- Simulations
- Image and video processing

Hand-in-hand with performance are the diagnostics that help identify trouble spots that may lead to **bottlenecks**, software or hardware problems that reduce performance.

In this exercise we will discuss how to evaluate the performance of a computer system, general methods used to improve performance, and additional ways to diagnose system problems.

PERFORMANCE EVALUATION

What affects performance in a computer system? Both the hardware in the system and the software that is controlling the hardware can affect the performance. The major hardware components to consider are

1. CPU
2. Motherboard buses
3. RAM
4. Display adapter
5. Hard drives

It is not just the individual components you need to consider, but also the way the components interact. For example, consider the following scenario:

- Processor data bus rate: 800 Mbps
- I/O bus data rate: 400 Mbps
- Video adapter data bus rate: 600 Mbps

In this example, the I/O bus is the bottleneck, hampering the performance of the video adapter.

The software controlling the hardware has an effect on performance as well, since it may not use the hardware effectively. For example, an application may not take advantage of MMX or 3D Now! instructions, ignore the cache, access the hard drive too often, or not take advantage of other aspects of the system.

How is the performance of a computer system evaluated? What should be measured? How can it be measured? How can software be used to evaluate the characteristics of a complex hardware architecture?

One way to rate the performance of a system is through a **benchmark**. A benchmark is a set of software tools designed to put the system through a series of typical operations found in high-performance computing. Many different parameters are measured by the benchmark (disk transfer speed, memory bandwidth, computational speed, etc.). These parameters can be measured by moving blocks of data around on the system (between hard disk, RAM, and display), performing calculations on groups of numbers, or running actual applications (video game, word processor, browser, MPEG video player), to name a few techniques.

The ZDNet eTesting Labs Web site provides many different benchmark applications for the PC. The Winbench application, for example, tests the performance of the disk and video/graphics systems. Winbench comes in two sizes: 10MB (no video files) and 132MB (video files included). Visit ZDNet at

http://www.zdnet.com/etestinglabs/filters/benchmarks/

Another place to get performance evaluation software is at Intel's Evaluation Software Center Web site, located at

http://developer.intel.com/software/products/eval/

Specifically, look for their VTune application. VTune can be used to analyze an applications performance and determine where the software bottlenecks are. This allows the application programmer to redesign the code and eliminate the bottleneck.

INCREASING PERFORMANCE

To increase performance, you may need to do one or more of the following:

- Adjust BIOS parameters and other system settings.
- Run one or more diagnostic programs.
- Replace/upgrade hardware.
- Get some new software.

It may not be wise to take the BIOS settings for granted. Just because a computer boots up properly does not mean the BIOS parameters are set to their optimal values. For example, the DRAM, cache, or video parameters may not be set to their most appropriate values (more wait states on DRAM than necessary, cache disabled). Refer to the system board manual for your computer, or locate information on your BIOS and motherboard chip set on the Web.

It has been mentioned a number of times that a fragmented hard disk will lower performance due to the increased amount of time required to read and write files. Periodic defragmentation should help keep disk performance at a high level. Some users may think it is not necessary to defragment a hard disk that is only 5% fragmented. But what if that small 5% contains the programs the user needs to run? In general, it never hurts to defragment a hard disk.

Replacing hardware one component at a time has its benefits. For example, simply upgrading the CPU from 200 MHz to 400 MHz will make a big difference. Replacing an old ISA video adapter with a newer PCI video adapter will improve graphics performance. But the new component still has to interface with the old components. Maybe the system bus is the bottleneck and will rob some of the performance gained by the new video adapter.

Sometimes, no matter what you do, it is not possible to get the performance you need out of your old system. In this case, it is best to get a new system (possibly joining your old system on a network). This way, everything will be better and faster: the CPU, motherboard buses, RAM, display adapter, and hard drives.

Last, you may need to get new software to take full advantage of your system. For example, your old CAD program that "works just fine" most likely has a new version that will benefit from the multimedia improvements now found in microprocessors.

Right-clicking on My Computer and selecting the Performance tab results in the window shown in Figure 57.1. Notice that Windows has evaluated itself and found 86% of its resources free. Naturally, as the percentage of free resources gets smaller, performance will suffer. To keep that from happening, it may be necessary to adjust the File System, Graphics, or Virtual Memory settings. Figure 57.2 shows the windows associated with these three performance areas. Note that you have a good deal of control over the settings associated with each resource. Caution should be used when changing the settings, as system performance can be severely affected if the settings are incorrect.

SOFTWARE DIAGNOSTICS

There are many software utilities designed to help diagnose and repair system problems.

One built-in Windows diagnostic tool is MSINFO32. Launch MSINFO32 from the Run menu. Figure 57.3 shows the initial MSINFO32 screen. Specific system information is displayed, such as the Windows version, CPU type, and the number of hard drives and their associated file system and free space. This information can be saved in a file or printed using the File pull-down menu. It is often necessary to provide information like this to a service technician when diagnosing problems over the telephone.

Let's take a quick tour of the four System Information sections. The first is Hardware Resources. The IRQ's Hardware Resource entry is illustrated in Figure 57.4.

Resources are any hardware entities used by the system. These include interrupts, memory, I/O locations, and DMA channels. Sometimes when installing new hardware, a *resource conflict* occurs. For example, adding a network interface card causes a conflict with interrupt 5, which is currently in use by the sound card hardware. The Hardware Resources section is a good place to get information on hardware-related problems.

Next is the Components section. Figure 57.5 shows the information in the Display entry. MSINFO32 provides details about the interrupt, memory and I/O locations, and drivers used by each hardware component, as well as any updates to the components.

FIGURE 57.1   System performance parameters

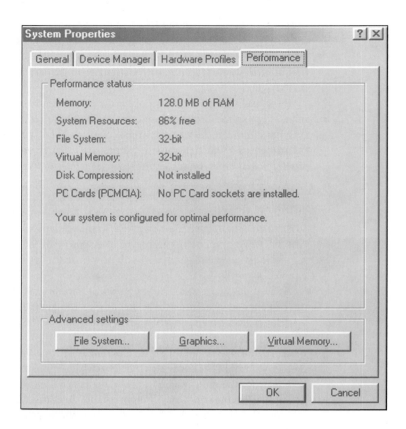

FIGURE 57.2   (a) File System Properties window

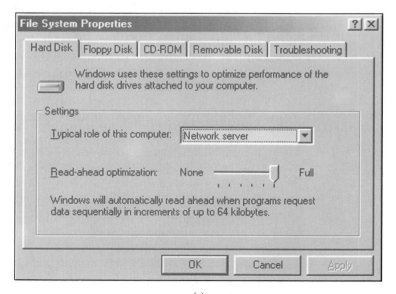

(a)

(b)

(c)

The Software Environment section contains information about software drivers, 16- and 32-bit code modules, start-up programs, and other system software items. The Drivers (Kernel, MS-DOS, and User-mode) entry can be used to determine what drivers are active. Note that MS-DOS drivers require switching from protected mode to real mode, and should be eliminated if possible to increase performance.

Figure 57.6 shows the Running Tasks information, a list of processes currently running on system. Many of these processes can be disabled through the MSCONFIG utility Start-up menu. It is interesting to note that this information helped a user discover that several applications he thought had been deleted were still present and running.

The last section contains information about Internet Explorer. Here you can check file versions, cache and security settings, and other Internet-related properties.

A number of diagnostic tools are available through the Tools pull-down menu, as indicated in Figure 57.7. These tools evaluate the integrity of the operating system files and the Registry, scan the hard disks for errors, and perform other types of diagnostics. Figure 57.8(a) shows the System File Checker tool at work. It takes only a few minutes to verify that all system files (more than 2000 files in 111 directories) are intact. Figure 57.8(b) indicates that no files needed to be restored.

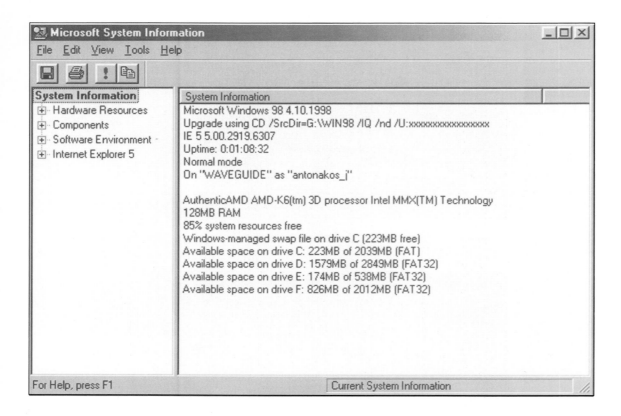

FIGURE 57.3   Initial Microsoft System Information screen

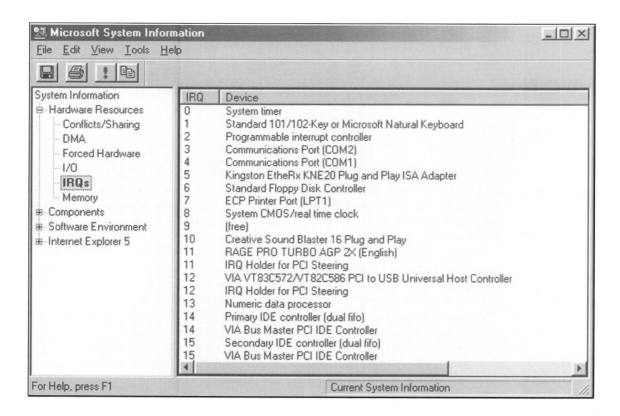

FIGURE 57.4   Interrupt assignments in Hardware Resources

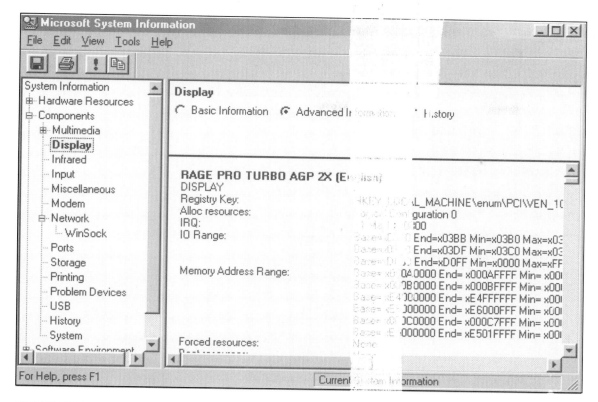

FIGURE 57.5   Advanced Display Information

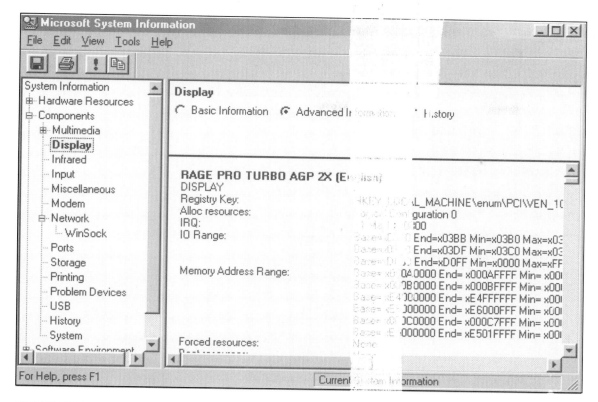

FIGURE 57.6   Tasks running on a system

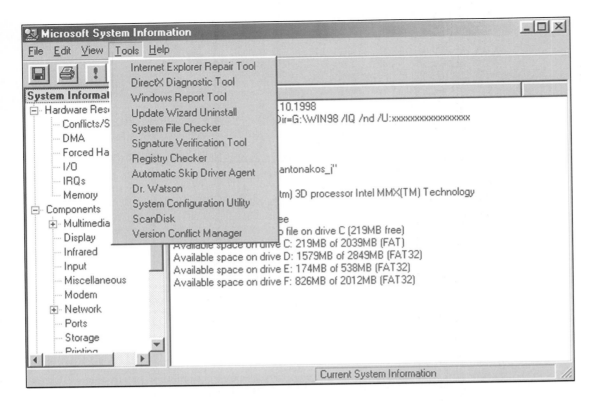

FIGURE 57.7 The Tools pull-down menu

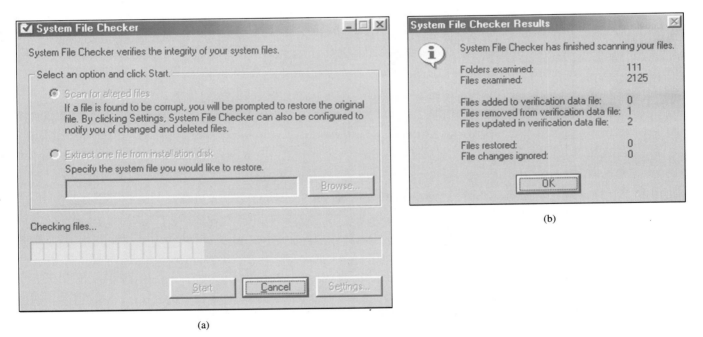

FIGURE 57.8 (a) Checking files with the System File Checker and (b) Results of the system file check

It would be worth your time to get acquainted with the features of MSINFO32. You will learn new things about your system and how all of its components work together.

Many companies offer their own diagnostic utilities. Two examples are

- Norton Utilities (www.symantec.com)
- McAfee Utilities (www.mcafee.com)

These utility packages provide the means to repair disks, scan for viruses, and create disaster recovery disks, and they support many other diagnostic and performance-based operations. Anyone seriously involved in microcomputer servicing and repair should invest in one or more diagnostic packages.

TROUBLESHOOTING TECHNIQUES

It is very difficult to get a grip on intermittent problems that plague a system. For example, a system that has run fine for years suddenly begins booting up occasionally with the error message "Windows Protection Error." Restarting the computer solves the problem, but it is aggravating not knowing why the error message comes up in the first place. Intermittent errors force us to be patient and to explore avenues of troubleshooting that we had not considered before. So, we become better troubleshooters in our elusive hunt for the cause of the problem.

SUMMARY

In this exercise we discovered that

- There are many tools for analyzing system performance and diagnosing system problems.
- Performance can be improved in many ways, including replacing hardware and upgrading software.

SELF-TEST

This self-test is designed to help you check your understanding of the background information presented in this exercise.

True/False

Answer *true* or *false*.

1. Software bottlenecks increase performance.
2. A benchmark is used to rate the performance of a system.
3. Adjusting the BIOS parameters may increase system performance.
4. Only disks with 25% or more fragmentation should be defragmented.
5. Software always uses the MMX or 3D Now! instructions.

Multiple Choice

Select the best answer.

6. Which application does not require a powerful machine?
   a. Simulating an electronic circuit.
   b. Sending e-mail.
   c. Watching a DVD movie in an on-screen window.
7. VTune is an example of a
   a. Benchmark.
   b. Firewall.
   c. Hard disk optimizer.
8. A bottleneck may be caused by
   a. Hardware.
   b. Software.
   c. Both a and b.
9. Which will not improve performance?
   a. Increasing the fragmentation on the hard disk.
   b. Adding more memory.
   c. Adding a faster processor.
10. Problems that occur randomly are called _____ problems.
    a. Dangling.
    b. Intermittent.
    c. Nonperiodic.

Completion

Fill in the blanks with the best answers.

11. Software and/or hardware problems that reduce performance are called _____.
12. DRAM wait states are ___ _____ parameters that affect system performance.
13. _____ ___ are any hardware entities used by the system.
14. A resource _____ may occur when installing new hardware.
15. Some of the hardest problems to troubleshoot are called _____ problems.

1. Download one of the performance evaluation tools described earlier (such as Winbench) and use it to evaluate your system.
2. Examine all the categories contained in the four System Information sections of the MSINFO32 utility. Print out several items from different categories.
3. Use the MSINFO32 Tools pull-down menu to check the system files and the Registry. Inform your instructor of any problems.

1. Explain how MSINFO32 can help diagnose a problem.
2. Search the Web for performance and diagnostic software. How do the features of one test program compare to another?

Under the supervision of your instructor
1. Discuss how system performance is measured and the steps that can be taken to improve it.
2. Explain how a system may be diagnosed using software.
3. Describe the features of the MSINFO32 utility.

58

Setting Up a Repair Shop

Joe Tekk was eating lunch with Don, the senior technician. Don had just told him privately that he was resigning his position at RWA Software.

"What are you doing that for?" Joe asked, shocked and surprised.

"Well," Don began, "I live in a small town. There are several businesses that consult me from time to time, and I have many friends who seem to always need some kind of help with their computers. But there isn't even an electronics store in town. So, I've decided to start my own repair shop."

Joe thought that was a great idea. "Are you going to sell computers, or anything like that?"

"Right now I'm concentrating on repairs, so I'll just be selling peripheral stuff such as RAM, hard drives, cables, and things like that. If there is a demand, I'll begin building complete systems for individuals."

"When are you doing this?"

Don winked at Joe as he replied, "As soon as I can hire a technician to help me."

PERFORMANCE OBJECTIVES

Upon completion of this exercise, you will be able to

1. Describe the tools, spare parts, equipment, and software required to repair computers.
2. Explain the importance of keeping a problem/solution journal.

KEY TERMS

Tools
Spare parts
Equipment

Software
References
Problem/solution journal

BACKGROUND INFORMATION

In this last exercise we will cover some of the items necessary to begin repairing computers. It is not the author's intention that you go out and start your own business after reading this exercise, or even after having read this entire book. Rather, this exercise provides a starting point for your own investigation into what is required in a computer-repair shop.

TOOLS

You will need tools to perform repairs and upgrades to your customers' computers. You may need to replace or install disk drives, motherboards, power supplies, adapter cards, co-processors, and/or RAM. The following list should include most of the necessary tools:

- Screwdriver assortment (regular and Phillips)
- Plier assortment (large and long-nose)
- Wrench assortment
- Nut drivers
- Wire cutter strippers
- IC extractor
- Soldering iron
- Files
- Magnifying glass

The wire cutter and stripper and the soldering iron are included so that worn or broken cables can be repaired. It is not recommended that you use a soldering iron to remove or replace any nonsocketed ICs.

NETWORKING TOOLS

Installing, upgrading, and repairing networking components (cards, cables, hubs) requires a special set of tools, possibly kept in a separate tool case so that network jobs and computer jobs can be easily prepared for. This set should include the following tools:

- Coaxial cable and UTP cable crimpers
- Wire cutter strippers
- Spare cable (with and without connectors)
- Spare connectors (BNC, RJ45)
- T-connector (for tapping into coax)
- 10Base2-to-10baseT transceiver
- Cable tester or DMM
- Spare hub and several NICs
- Protocol analyzer (hardware, software, or both)

SPARE PARTS

If you find a bad or broken part in a machine, you will save a great deal of time and effort if you have a replacement handy. Driving to a hardware store or a computer store to get a replacement will cost you more in the long run by virtue of lost repair or installation time.

Spare parts for a typical microcomputer system are numerous. Some of the most important ones are:

- Nut, washer, and screw assortment
- Metal and insulated standoffs
- Power cords
- Printer cables
- Null-modem cable
- SIMMs and DIMMs
- Cache RAM
- Resistors and capacitors
- Common semiconductor devices (diodes, transistors)
- I/O connectors (9- and 25-pin male and female)
- I/O adapter cables (male-male, female-female, male-female)
- Ribbon cable with crimp-on connectors
- Power splitter adapter cable
- LEDs (for front panel indicators)
- Push buttons (for RESET and TURBO buttons)

- Metal slot guards
- Fuses
- Plastic face plates
- Muffin fans
- Processor fans
- Processors
- Motherboards
- Floppy drives
- IDE and SCSI hard drives
- An assortment of adapter cards
- Printer cartridge and paper
- Blank disks

The size of your spare parts inventory will determine what kind of service you can provide to your customers. At the very least, customers should be able to have additional floppy disk drives installed, have CD-ROMs or larger hard drives installed, or upgrade the sizes of their system RAM.

EQUIPMENT

Although you will encourage your customers to bring their entire systems in for repair or upgrading, you will still need equipment of your own to handle some jobs. The most important piece of equipment your repair shop should contain is one or more computer systems, complete with mouse, sound card, CD-ROM, lots of RAM, and the most current processor. This will allow you to test both hardware and software from a customer's system in your own system (as a last resort). In addition, the following equipment may also come in handy:

- Multimeter
- Logic probe
- Oscilloscope

The multimeter will allow you to find broken wires in cables and missing power supply voltages. The logic probe and the oscilloscope will allow you to monitor the activity of a signal or a pin on an IC, with the scope providing much more information than the logic probe.

SOFTWARE

It would be asking for trouble for you to run a customer's software on your own machine before checking it for viruses. Or, if a portion of a customer's Windows directory was accidentally deleted, what happens if the customer does not have the installation disks? These are two examples of why you should also maintain a software library in your repair shop. The library should be strictly for repairing damaged software, and not for doing free installations of popular programs (unless they are shareware).

Some of the more useful software packages to have in your library might include the following:

- Virus scanner
- DOS (versions 3.0 and up)
- Windows (all versions)
- CD-ROMs (for testing CD-ROM installations/problems)
- Typical DOS and Windows applications
- 3.5- and 5.25-inch system disks
- Modem software
- Data exchange software (such as FastLynx)
- Specialized diagnostic software (such as Norton Disk Doctor)
- Hard drive software (such as Partition Magic)

The system disks are for booting a machine whose hard drive has failed for some reason.

REFERENCES

Eventually, even the best technician reaches a dead end and requires an external source of information to help solve a problem. Most bookstores now have entire sections filled with computer books on practically every computer subject, from building your own computer to advanced Windows applications. You may benefit greatly from owning one or more references. At the very least, you will pick up valuable information just reading the books. Here are a few easy-to-find types of references:

- PC hardware references
- PC software references
- Manuals for printers, modems, hard drives, and other hardware devices
- BIOS and DOS references
- PC interrupt reference

The PC interrupt reference is a comprehensive listing of how all of the software interrupts are used by various manufacturers. This is a good reference to have when you suspect that an interrupt conflict is the cause of a system problem.

THE BUSINESS ASPECT

Just having the technical skills to diagnose and repair computers is not enough to open your own business. There are many aspects of running a business that must be examined and provided for. Some of the required office supplies include

- Office space (and the associated rent and utilities)
- Office supplies (paper, pens, etc.)
- Fax machine
- Postal supplies
- Cash register and credit card reader

In addition, someone will need to "do the books," keep track of sales and expenses, fill out tax forms, manage the business account at the bank, and generally do everything associated with running a business except for the repair part. If your interest in repairing computers does not include the business aspects, you have already grown to a business of two people, since you will need to hire an office manager.

The alternative to all of this is running the business out of your own home, which reduces or eliminates many of the costs. Then you simply may use an accountant to help with the business end of things.

Clearly, starting your own business is a major step and must be considered and planned very carefully.

ONE FINAL COMMENT

Whether or not you open your own repair shop, as a user of computers you will eventually find yourself staring at a new 8-GB hard drive that you would like to install, or at an error message that reads

```
Bad or missing command interpreter
```

and you will have to figure out what to do. It would be a good idea to write down the problem and its eventual solution in a *problem/solution journal*. Keep track of all the strange things that happen to your computer. Often, a problem comes back again, and you may not remember exactly what you did to fix it. Windows problems are especially notorious in this regard and also difficult to diagnose. Keeping a journal of your repair (and installation) efforts will be rewarding in the long run.

TROUBLESHOOTING TECHNIQUES

When you begin a new repair job, it is important to ask yourself a few questions. Did the computer or network ever work? If it was working, your job should require less effort to repair than a system that never worked. For example, if a customer's laser printer suddenly quits, there are several obvious things to check, such as cabling, fuses, the power cord, and toner level. If, however, a customer complains that the 32-node network they just installed themselves does not work, your job is much different. It is possible that the network may never work, no matter what you do. Or it could be as simple as bad crimps on the connectors.

In general, it pays to think about these things before you even begin a repair job. You may save yourself time and effort.

SUMMARY

In this exercise we discovered that

- Starting your own computer repair business is a major undertaking.
- A computer repair shop requires spare parts, tools, equipment, software, and references.
- In addition to technical skill, you must also understand the basic business practices.
- You should use a problem/solution journal to keep track of your experiences.

SELF-TEST

There is no self-test for this exercise. Each new computer problem that you solve will be a self-test.

FAMILIARIZATION ACTIVITY

1. Have your lab partner create a problem on your lab machine. This can be done by deleting (or renaming) an important file, such as a mouse driver. Your lab partner will then tell you that something is wrong with the machine (e.g., the mouse does not work, the system does not boot, or Windows will not run).
2. Diagnose the cause of the problem and fix it. Your lab partner should not assist you during this step, except to answer questions you might ask of a customer with the same problem, such as "What was the last thing you did before the problem showed up?" or "Did it ever work?"
3. Write the problem and solution down in a notebook.
4. When the entire class has completed the activity, add your problem/solution to a common journal for future classes to use as a guide.

QUESTIONS/ACTIVITIES

1. Visit a local computer-repair shop and ask for a tour. Keep track of everything you see. Ask the technicians what their most common service problems are.
2. Price out the cost of a small parts inventory. What are the most expensive items?
3. Visit a local bookstore and examine the computer books. What books offer the most comprehensive information on the computer?

REVIEW QUIZ

Under the supervision of your instructor

1. Describe the tools, spare parts, equipment, and software required to repair computers.
2. Explain the importance of keeping a problem/solution journal.

Glossary

A

Across-the-wire migration A method of performing an upgrade from an older version to a newer version of NetWare that transfers all NetWare files from the current server to a new machine attached to the network that is already running NetWare 4.x or above. This method allows the original server to continue running during the upgrade.

Additive color mixing The process of mixing three colors of light, red, green, and blue, to produce many more colors as well as white light.

Add New Hardware Wizard An assistant available in Windows 95/98 that leads the user step-by-step through the process of configuring the software necessary to communicate with a new hardware device.

Addressing mode A method used by a programmer to refer to a memory location.

AGP adapter Accelerated graphics port adapter, an interface specification developed by Intel that allows the graphics controller to access the system's main memory directly.

Alphanumeric mode *See* Text mode.

Ambient temperature The temperature of the air surrounding an object, such as a computer system.

Ampere A unit of measurement used to measure electrical *current*.

Antistatic wrist strap A device used to avoid damage from static electricity when handling integrated circuits such as CPU and memory devices. A typical antistatic wrist strap consists of a strap that wraps around the wrist, with a cable attached to it that connects to a ground terminal. The cable provides a path to ground for static electricity.

Application programming interface The "layer" of Windows 95/98 internal architecture that contains the *kernel*, the *graphical device interface* (GDI), and the user portion.

ASCII code The American Standard Code for Information Interchange, which is the code used to transmit information to a printer to enable it to print characters. Appendix A includes a complete list of ASCII codes.

Aspect ratio The ratio of the horizontal measurement to the vertical measurement of a graphic or a graphic display, such as a video monitor. Monitors are usually rectangular, with an aspect ratio of the horizontal to the vertical measurements of 4:3.

Assembly language The unique language used to program a microprocessor.

AT command set A standard set of commands that can be sent to a *modem* to configure, test, and control it.

ATAPI The AT Attachment Packet Interface, an improved version of the IDE hard drive interface that uses packets of data during transfers. ATAPI specifications support CD-ROM drives, hard drives, tape backup units, and plug-and-play adapters.

Attachment Any file attached to an e-mail message. An attachment can be a document, a graphic image, or even an executable program.

Autodetect A feature of some *BIOS* setup programs that checks the hard drive and reads the setup parameters automatically.

B

Backbone An RG-11 *thickwire* cable used to distribute *Ethernet* signals throughout a building, office complex, or other large installation.

Background image   A centered or tiled picture that appears behind the *icons* in the Windows *desktop*.

Backup   A copy of the information on a disk; hard disks are backed up on floppy disks or magnetic tape, and backups can be full (all the information on the disk) or incremental (the information that has changed since the last full backup).

Bar code reader   A common multimedia device used to read a variety of types of bar codes, such as those used for product information, to track mail and packages, and to control the circulation of library books.

Baseband   A type of communications network that has a single information carrier that is modulated with the digital network data. Baseband systems are less expensive than *broadband* systems and are generally used for computer networks.

Base memory   The 1MB of memory that is addressable by the 8088, 8086, and new microprocessors running in real mode. All 80x86 microprocessors have this base memory.

BASIC interpreter   A program that is used to automatically convert code written in the high-level BASIC language into low-level machine language that can be understood by the microprocessor.

Basic rate interface   The simplest Integrated Services Data Network (ISDN) connection, which consists of two 64Kbps B channels for data and one 16Kbps D channel (for signaling).

Baud   A unit of measurement, named for J. M. E. Baudot, that indicates the number of discrete signal changes per second when transmitting data. The popular usage of the term *baud* has become the same as bits per second (bps).

Benchmark   A set of software tools designed to put a computer system through a series of typical operations found in high-performance computing to measure parameters such as disk transfer speed, memory bandwidth, and computational speed.

Best effort delivery   Provided by IP (Internet Protocol), this is considered an unreliable type of delivery because when an *IP datagram* runs into trouble on the network, it is simply discarded as undeliverable, and an error message may or may not be returned to the sender.

Binary number   A number composed of digits (called *bits*) from a set of only two digits, 0 and 1.

BIOS   Stands for Basic Input/Output System, the instructions that test all the hardware in a computer and start the boot sequence when the computer is turned on.

BIOS ID   An identification that appears on a computer screen at power-on that indicates the software manufacturer of the system's *BIOS*.

BIOS upgrade   An upgrade performed on the Basic Input/Output System of a processor, which was stored in the ROM of older *motherboards*, and was thus unchangeable,

but can be updated in many newer motherboards that have Flash memory.

Bit   A unit of computer information represented by one of two digits, 0 or 1, which represents the result of a choice between two alternatives, such as *yes* or *no*, *true* or *false*, or *on* or *off*.

Bitmap file   A standard Windows file format used for graphic images. A bitmap file can be used as a *background image* for the Windows desktop.

Boot sector virus   A virus that takes over the boot sector, the first piece of code read, of a floppy or hard drive, and begins running the virus immediately.

Bottleneck   A hardware or software problem that reduces the performance of a computer system.

Branch prediction   A technique that allows copies of instructions that may be needed by a processor to execute a program to be held in a branch target buffer, from which it quickly can be fed to the instruction cache. Branch prediction allows a steady flow of instructions to the Pentium microprocessor via *pipelining*.

Bridge   A hardware connection used to separate a large network into smaller groups in order to isolate traffic and improve performance. A bridge keeps traffic from one *segment* of the network separate from another segment.

Broadband   A type of communications network, such as a television cable system, that uses many different carriers and supports multiple channels of data. Broadband systems require more expensive hardware than *baseband* systems and are typically used for high-bandwidth applications such as broadcast video and FM audio.

Broadcast storm   A flooding of a computer network with so many packets that the network's switches are forced to drop packets to try to maintain their buffers, resulting in interruption of service for network users.

Browser   A software application, such as Windows Internet Explorer or Netscape Navigator, used to navigate the *Internet* by selecting links on a Web page or specifying a *Universal Resource Locator (URL)* address to point to a specific page of information.

Bubble-jet printer   A type of printer that uses heat to form bubbles of ink. As the bubbles cool, they form droplets applied to paper to create the printed image.

Button   A graphical element in a Windows screen or dialog box that allows a user to choose an action by clicking on it.

Bus   A group of wires dedicated to a specific task in a microprocessor; for example, there are data buses for handling data and address buses for getting and placing data in different locations.

Bus network   A computer network *topology* in which all the nodes are connected to the same communication link.

Byte   One 8-bit binary number.

Byte-swapping   Refers to a way of storing 16-bit numbers in memory, so that the lower byte is always written first and the higher byte second. For example, if the 16-bit hexadecimal number 2055 is written into locations 18000 and 18001, the low-order 8 bits (55) are written into the first location (18000) and the high-order 8 bits (20) are written into the second location (18001). This method is also called *little endian format.*

C

C++ compiler   A program that is used to automatically convert code written in the high-level C++ language into low-level machine language that can be understood by the microprocessor.

.CAB file   Files on the Windows CD-ROM that provide the Windows operating system components in compressed form for distribution.

Cable modem   A high-speed network device connected to a local cable television provider. The cable company allocates two channels (one for transmitting and one for receiving) on the cable system for data communication. A traditional Internet service provider service is established at the cable company to service the network clients.

Cable tester   A tool used to check electronic cables that can identify problems, measure cable length, and provide profile information about the cable.

Cache   A very high speed memory, with an access time usually 10 times faster than conventional RAM.

Cache hit/miss   A cache hit indicates that data or an instruction has been found in the cache memory; a cache miss means that the data or instruction was not found, and the slower external memory must be accessed.

Caching   A technique used to increase the data transfer rate of a microprocessor by using a hardware or software *cache* to store frequently accessed instructions or data.

Capacitor   An electronic component that is used to store an electrical charge. Capacitors are used in computers to help maintain a steady supply of voltage to circuits.

CAV   Constant angular velocity, one of two methods used for spinning a CD in a CD-ROM drive that spins the CD at a constant rate. Thus, the data is transferred slower from the inner tracks and faster from the outer tracks.

Cell relay   A method of data transfer, also called asynchronous transfer mode (ATM), that uses fixed-size cells of data and supports voice, data or video at either 155.52 Mbps or 622.08 Mbps.

Channel bar   An element of the Windows 98 *desktop* that provides one-click access to Internet links.

Character   The letter, number, or other symbol that appears on a key on the computer *keyboard*, or the same letter,

number, or symbol stored in computer *memory* or displayed on the *video monitor.*

Character code   The *ASCII code* associated with each key on a keyboard.

Charge-coupled device (CCD)   An imaging device used instead of CMOS chips in more expensive *digital cameras.*

Checksum   A numerical sum based on every byte contained in a file. Checksum is used to detect the presence of a *computer virus*, because a file infected with a virus will almost always have a different checksum value than the previous file.

Chip set   The microprocessor hardware component that controls data transfers over a bus.

Circuit switching   A mechanical switching system originally used in telephone communication, employing rotary switches to make necessary connections to allow end-to-end communication. The switches completed a circuit, so the system was called *circuit switching.*

CISC   One of two major types of computer architecture; it stands for Complex Instruction Set Computer. All 80x86 machines prior to the Pentium are considered CISC machines.

Clipboard   A Windows accessory that serves as a temporary holding place for information being exchanged between applications. For example, if a graphic is *cut* from one location, and then the *Paste* command is used to place it in another file, between the cut and paste actions, the graphic image will temporarily be in the Clipboard.

Cloud   A graphic symbol used to describe a network without specifying the nature of the connections.

Cluster   A set of contiguous disk *sectors.*

CLV   Constant linear velocity, one of two methods used to spin a CD in a CD-ROM drive. In CLV, the disk is rotated at different speeds so the data transfer rate remains constant; that is, the CD rotates slower when the read head is over the outer tracks and faster when it is over the inner tracks.

CMOS   Complementary Metal-Oxide Semiconductor, a low-power logic family that uses field-effect transistors instead of bipolar junction transistors.

CMOS RAM   The memory that stores the system parameters for a computer. CMOS RAM stores 64 bytes of data and uses a battery backup circuit to retain the information when the computer is turned off.

Cold solder joint   A bad solder joint that results in an unreliable electrical connection.

Collision   A condition that occurs when two or more nodes in a computer network transmit data at the same time, requiring each node to stop and wait before retransmitting.

Computer virus   One of several types of destructive computer programs written with the intent of doing damage to the computers of others.

Concentrator   *See* Hub.

Connection   A connection is established between two nodes on a network using TCP (Transmission Control Protocol) and is used, with error checking, to acknowledge received packets.

Context-sensitive menu   A pop-up menu that appears when a user right-clicks a program icon, a toolbar, a taskbar button, or a random location in the desktop, as examples. The context-sensitive menu lists commands that pertain specifically to that object or screen region.

Control code   *See* Printer-control code.

Controller   A device used to control *MIDI* messages that communicates with a *sequencer* connected to a MIDI device.

Conventional memory   The memory located between 0K and 1MB, with 640K of memory that is usable by DOS-based programs in any microprocessor and 340K reserved. Conventional memory uses the *real mode* of processor operation.

Coprocessor   A companion chip that was used with Intel microprocessors before the 80486 to help the processor do arithmetic calculations. Some software programs such as CAD (computer-aided design) programs require so many math calculations that the coprocessor was required to run them on the older microprocessors. The 80486 and higher processors have built-in coprocessors.

Copy   The command used to make a copy of a specific object. In Windows Explorer, the Copy command is used to make another copy of a file or folder; when the copy is pasted to a new location, the original file or folder remains in its previous location as well.

Corrosion   A chemical change to a metal component that can eat away critical parts of metal pin connectors, wires, interface cards, and other microprocessor components. Corrosion is caused by chemicals in the air and contact between computer parts and human hands.

CPU   The central processing unit or "brain" of a computer.

CSMA/CD   Carrier Sense Multiple Access with Collision Detection; refers to the technique of nodes in a network sharing a common bus and is the basis of the *Ethernet* communication system.

Current   The amount of movement of an electrical charge, measured in *amperes*, or amps. The letter symbol for current is *I*.

Cut   The command used to cut an object from its current location. In Windows Explorer, the Cut command generally is used with the *Paste* command to cut and paste the file or folder to a new location.

Cylinder   A combination of tracks on different *platters* in a hard disk system that are used to organize data on the hard drive.

D

Data communications equipment   Any equipment used to transmit data, such as a *modem*.

Data migration   A method used by NetWare to move data from one location to another to maintain effective use of all available hard disk space. For example, large files are moved to a secondary storage system and are copied back (demi-grated) to the hard drive when needed.

Data terminal equipment   Any equipment used to process data, such as a computer.

Defect mapping   The process of locating any defects on a hard disk. All new hard disk drives have the location of these defects printed on a label on the drive itself and/or in the accompanying instruction manual. The information is also recorded during the *low-level formatting* process in order to have the formatting program mark these sections so that no data is ever written to them.

Denial of service attacks   A breach of a computer system's security that involves a special program, which is secretly installed on selected host computers, and that allows a coordinated attack against an Internet site by flooding it with network requests. The resulting overload denies service to anyone attempting to use the site.

Desktop   The Windows start-up screen, which, depending on the version of Windows in use, can include *icons* representing *program groups*, *shortcut* icons, pull-down menus, a *taskbar*, the current time and date, and open folders.

Desoldering   The process of removing a soldered component or unwanted solder.

Diagonal cutters   A hand tool used to cut wire.

Dialog box   A Windows element used to communicate with the user, a dialog box usually contains a brief description of what Windows is doing and one or more buttons that the user can click to choose an action.

Digital camera   A camera that stores images digitally as opposed to recording them on photographic film.

Digital signal processor   A custom chip included on a sound card that is responsible for digital/analog signal processing.

Digital-to-analog converter   A device used by a sound card that translates the binary patterns of computer data into corresponding analog voltages, which are then passed to a *low-pass filter* to create a continuous analog waveform that we hear as a sound.

Digital versatile disk (DVD)   A type of CD that takes advantage of advancements in laser technology to significantly increase the storage capacity of a disk. An ordinary CD may contain 650MB of data, whereas a DVD can hold 4.7GB to 17GB of information. A DVD is often comprised of two disks bonded together.

Diode   An electrical device that allows current to flow in only one direction.

Direct cable connection   A method of connecting, or networking, computers that is simple and inexpensive. Direct cable connection can use either serial cable or parallel cable; this method works only for computers that are near each other.

Direct memory address   Control lines that provide direct access to memory without having to go through the microprocessor.

Disk geometry   The parameters of a disk, specifying the number of heads, cylinders, and sectors per track.

Dithering   A method used by a printer to represent a color by using one or more similar colors.

Domain   A category into which computer servers on the *Internet* or other networks are divided to make the large number of nodes in some networks more manageable.

DOS shell   A copy of the DOS environment maintained by Windows that can be accessed from the MS-DOS Prompt icon in the Main window in Windows 3.x or the MS-DOS Prompt choice in the Programs submenu of the Start menu in Windows 95/98. Once inside the DOS shell, you can run a DOS program and issue DOS commands.

Dot-matrix printer   The most common type of *impact printer,* which prints characters by means of a series of tiny mechanical pins that move in and out to form the various characters printed on the paper.

Dot pitch   The size of a *pixel*, which is a function of the number of pixels on a scan line and the distance across the display screen.

Double word   A data type that consists of a 32-bit number.

Driver   Software used to manipulate the input and/or output of hardware devices connected to a computer.

Dual inline memory module (DIMM)   A component used to organize memory in Pentium-class processors containing 64-bit data buses. A DIMM is like having two *single inline memory modules (SIMMs)* side by side.

Dual-scan twisted nematic (DSTN)   A type of *liquid crystal display* technology, also known as passive matrix display. DSTN displays are not as bright as *thin-film transistor (TFT)* displays and have a smaller viewing angle.

Duplex   A term used to describe a communications channel that allows for two-way communication between two systems.

E

Echo   The method used to display characters on the monitor screen; that is, the character that is typed is *echoed* on the display.

ECP printer port   An *extended capabilities port*, a parallel port that allows 8-bit bidirectional data flow.

EDO DRAM   Extended Data Out RAM, which is used with bus speeds at or below 66 MHz and is capable of starting a new access while the previous one is being completed, which works well with the pipelined Pentium processors.

Electrical environment   Aspects of the location of a computer system, how it is supplied with power, and its proximity to magnetic fields and sources of electrical interference, that can affect the operation of the system.

Electronic mail   A personal communication tool that can be used to electronically send a message to one or several recipients, send a message that contains text, graphics, or multimedia files, and/or send a message that a computer program will respond to, such as a mailing list program or mail exploder.

Electrical noise   Any undesirable and sometimes unpredictable random changes caused by sources of electricity.

Electrostatic discharge   Static electricity that can accumulate in objects or in people and, when discharged, can amount to several thousand volts. Care should be taken to avoid electrostatic discharge into or around a computer system, because the integrated circuits in the computer can be severely damaged by it. *See also* Static discharge.

Environmental checklist   A guide to possible physical and electrical environmental hazards at a computer station, with items to check such as *ambient temperature*, power cycling, presence of dust or smoke, proximity of *magnetic fields*, and other potential problems.

Error code   A code, which can be a number or a series of beeps, that indicates there is a problem with a component of a computer system.

Escape code   A code preceded by the ESC (escape) character that is used as a printer command. The ESC character doesn't do anything by itself, but it alerts the printer that the character or set of characters that follows should be handled in a special way; for example, <ESC>E means to begin bold text.

Ethernet   A popular communication network developed in 1980 that operates as a baseband system, which means that a single digital signal is used to transmit data.

Expanded memory   A type of memory that requires special hardware and software that allows chunks of memory to be switched in and out of *conventional memory*.

Expansion slot   An element of a computer motherboard that allows for the connection of electronic cards that expand and enhance the operation of a computer.

Extended character set   The group of characters, including all those on a computer keyboard plus many more, that are available for display when a *video monitor* is in *text mode*.

Extended memory   Memory that cannot be used by DOS-based programs, although it can be used by a virtual disk in DOS systems. Extended memory uses the *protected mode* of processor operation and is the type of memory used with 80386, 80486, and Pentium chips.

Extraction tool   A hand tool used to remove integrated circuit packages from a computer.

F

FAT   Stands for file allocation table, the method an operating system uses to organize the data stored on a disk.

FAT32   A file allocation system that allows the use of larger disks and disk partitions by using 32-bit FAT entries, permitting very large hard drives without having to use large *cluster* sizes.

Fiber   Refers to fiber optic cable, which relies on pulses of light to carry information. Fiber can be made of plastic or glass, and is constructed of an inner core and an outer cladding. The beam of light travels by reflecting off the boundary between the core and the cladding.

Firewall   A program that examines packets of information moving between two network hosts to determine whether or not to allow the communication in order to protect the computer system of one of the hosts.

Flag register   A 16-bit register used to indicate the results of arithmetic and logical instructions.

Flash EPROM   A type of read-only memory that can be electrically reprogrammed, similar to *FLASH ROM*.

FLASH ROM   A type of read-only memory, also called EEPROM (electrically erasable PROM) found on newer motherboards, which allows the *BIOS* to be modified by running a special upgrade program.

Flat addressing   An addressing scheme used with *extended memory*, in which Windows 95/98 runs most applications above the first 1MB of RAM.

Flat-bladed screwdriver   A screwdriver with a blade designed to remove or tighten a slot screw, which has one long slot in its head. This type of screwdriver comes in many sizes and may have a carbide tip for strength and/or an insulated handle for work on electrical components.

Flat panel display   An alternative to a CRT (cathode-ray tube) monitor. Instead of controlling the intensity of each *pixel*, as in a CRT monitor, the flat panel display controls the pixels by individually turning them on or off, using *liquid crystal display* technology to develop an image.

Floppy disk drive   A device that enables a computer to read and write information from a floppy disk.

Flux   A chemical, also called *rosin*, used to form the core of solder.

Folder   A graphical representation of a directory or subdirectory; in general, Windows 95/98 uses the term *folder* instead of *directory*, which is used in Windows 3.x.

Form factor   A label used to describe the physical size, layout, and features of a particular *motherboard*.

Fragmented   Describes a file that is spread out over many different areas of a disk rather than being stored in one big block. Fragmentation increases the time required to read or write to the file.

Frame relay   An improved method of error checking, relying on fewer acknowledgments during a transfer of data. *Frame relay* relies on improvements in communication technology so that only the receiving station needs to send an acknowledgment, while still providing a low error rate.

Full duplex   A term describing a communication link in which data can be passed in two directions at the same time.

Full-mesh network   *See* Mesh network.

G

Galvanic corrosion   A type of *corrosion* that creates the effect of a small battery, causing a tiny current to flow between pieces of metal via an oil residue, such as natural skin oil left when a component is touched by human hands. Galvanic corrosion gradually eats away the metal, causing computer failure.

General-purpose register   Refers to one of six data or index registers that are 16 bits wide. These registers are referred to as AX, BX, CX, DX, SI, and DI. The first four, which are data registers, can be split into two halves of 8 bits each by the programmer.

Gigabyte   A measurement of computer information equal to more than 1 billion *bytes*. It is exactly 1,073,741,824 bytes.

Graphical device interface   The part of the *application programming interface* in Windows 95/98 internal architecture that controls what appears on the display screen, managing all graphical output.

Graphical user interface   Part of an operating system such as Windows that provides an easy-to-use computer environment with buttons, icons, pop-up boxes, and pull-down menus.

Graphics accelerator   A *video adapter card* containing a microprocessor designed specifically to handle the graphics processing workload, which allows the system processor to process other, nongraphic instructions.

Graphics mode   One of two modes in which the *video monitor* communicates with the computer. In graphics mode, the display receive information from RAM because the application being used has complete control over the intensity and the color of each *pixel*. The more pixels that are available on the screen, the more detailed the display can be, but each added pixel requires more memory in the system.

H

Half duplex   A term describing a communication link that allows data to pass in both directions, one at a time. An example of a half duplex system would be two people communicating with walkie-talkies.

Handheld PC (HPC) A small computing device with a miniature keyboard and display (640 × 240 pixels); a stylus is used to select items on the screen. HPCs use the Windows CE operating system to perform many common computer tasks, but they are not as powerful as a microcomputer.

Hand tools Instruments used with the hands to extend the hands' working capabilities. In a microcomputer-repair lab, hand tools are used to disassemble and reassemble computer equipment and parts.

Hardware interrupt A hardware event that causes a response from the software. For example, when a key is pressed on the computer keyboard, its scan code is transmitted to keyboard logic on the motherboard, which generates a *hardware interrupt* to signal the processor that the key code needs to be read.

Hayes compatible Refers to a standard for *modems* for the use of a set of commands, the *AT command set*, to configure, test, and control a modem.

Header The first part of an *electronic mail* message, which consists of a keyword followed by a colon and additional information, such as To: Recipient's Name.

Heat sink In the soldering process, a tool placed between the heat source and the device to help conduct heat away from delicate electronic components.

Hexadecimal number A number in base 16, which is made up of the digits 0 through 9 and the letters A through F.

Hierarchy A layered organization applied to a computer network to make it run more efficiently.

High-level language A computer programming language, such as BASIC, that uses symbols and letters that the operator can read, rather than pure machine language. Any program written in a high-level language must be converted to machine language before it can be executed.

Hot swapping The capability of a PCMCIA (Personal Computer Memory Card International Association) card to be removed or inserted with the power on. The standard was developed to expand the memory available on early laptop computers and has since expanded to include almost any kind of peripheral.

Hub The central node, also called a *concentrator*, in a *star network*.

Hypertext markup language (HTML) The source code for a Web page that contains commands (called tags) that are understood and interpreted by a Web *browser* to create specific formats and graphics.

Hypertext Transport Protocol (HTTP) The *protocol* used on the Internet. HTTP is also called the World Wide Web (WWW), and it allows hypermedia information to be exchanged, such as text, video, audio, animation, images, and more.

I

IC extractor *See* Extraction tool.

Icon A graphical representation of an object that a user can click on to open the corresponding drive, disk, *folder*, file, document, or program.

IC socket A platform used to provide a solid mechanical and electrical connection for an integrated circuit.

IEEE 802 Standards Standards for computer networking established by the Institute of Electrical and Electronic Engineers (IEEE). Companies entering the network marketplace must manufacture hardware that complies with the published standards to ensure compatibility between systems.

Impact printer A type of printer that uses a mechanical device to impart an impression to the paper through an inked ribbon.

Inductor An electrical component that opposes a rapid change in current. Inductors are used to help maintain a steady source of current to supply electrical energy to computer circuits.

Ink-jet printer A type of printer that uses electrostatically charged plates to direct jets of ink onto paper to create a printed image.

In-place migration A method of performing an upgrade from an older version to a newer version of NetWare that involves shutting down the NetWare server to perform the upgrade directly on the machine. *See also* Across-the-wire migration.

Input device A device that can only provide information to a microcomputer.

Insertion tool A hand tool used to put integrated circuit packages in a computer.

Instant messaging A method of sending an electronic message instantly, without the intermediate steps and additional time required for a conventional e-mail message.

Integrated circuit A tiny complex of electronic components and their connections that is produced in or on a small chip or wafer of material such as silicon.

Internal temperature The temperature inside an electronic component, such as a microprocessor. The internal temperature of a microprocessor is usually much higher than the outside, or *ambient temperature*.

Internet The largest computer network in the world, connecting all types of computers in different *domains*, or categories, and allowing someone with a computer and an Internet service provider to connect to virtually any other computer anywhere in the world at any time.

Internet service provider (ISP) Any facility that contains its own direct connection to the Internet. A school, business, or organization may have their own dedicated high-speed connection, or a company may provide the service to individuals who dial up the connection with a modem or connect via a cable modem or a dedicated telephone line.

Interrupt handler   A routine that handles the action needed by a *hardware interrupt*, which allows the software to interact with the hardware.

Interrupt hook   A method used by a *computer virus* program to get control of a computer system by picking an *interrupt* frequently used by BIOS and DOS and making the interrupt vector point to itself instead of the BIOS or DOS service routine.

Interrupt service routine   *See* Interrupt handler.

Interrupt vector table   A collection of 4-byte addresses (2 for the CS, or code segment, and 2 for the IP, or instruction pointer) that indicate where the 80x86 processor should jump to execute the associated *interrupt handler.*

I/O (input/output) device   A peripheral device that is capable of both getting information from and providing information to a computer.

IP address   The unique identification of each machine on the Internet. It consists of a 4-byte number, commonly represented in dotted-decimal notation, for example, 204.210.133.51.

IP datagram   A unit used to package TCP/IP (Transmission Control Protocol/Internet Protocol) data for delivery on a network.

ISO/OSI network model   *See* OSI reference model.

J

Jaz drive   A Jaz drive operates like a hard disk drive, but has a removable cartridge that holds up to 1GB of data.

Jukebox   A secondary storage system used in a process called *data migration* to store large files temporarily before they are copied back (demigrated) to the hard drive when they are needed.

K

Kernel   An element of the *application programming interface* in Windows 95/98 internal architecture that provides services such as memory management.

Keyboard   A computer input device that consists of an array of separate keys representing letters, numbers, and other *characters* that, when pressed, send specific codes to the computer.

Keyboard interface connector   The cable that connects the *keyboard* to the computer.

L

LAN (local area network)   A collection of computers and other devices connected in a network so they can share

information. Local area networks are geographically small, connecting offices in a building or classrooms in a school, for example.

Laser printer   A type of printer that uses a six-step electrophotographic process to produced high-quality text and graphics.

Liquid crystal display   A screen composed of pixel cells containing liquid crystals that change orientation when a small voltage is applied, affecting the way light passes through the crystal.

List file   A file that contains all the text from a *source file*, plus some additional information, including the set of memory locations in which the instructions are stored.

Little endian format   *See* Byte-swapping.

Local bus   A type of bus architecture that provides the fastest communication possible between a plug-in expansion card and the processor by bypassing the EISA (extended industry standard architecture) chip and connecting directly to the CPU. Local bus video cards and hard drive controllers offer high-speed transfer capability.

Local echo   The term used to describe the situation in which a sending modem immediately returns or *echoes* each character back to the screen as it is entered into the keyboard.

Local printer   A printer that is local, or attached only to a particular computer. Only that computer can print to a local printer, even if the computer is networked.

Log file   A record of all the files that have been created in a *backup.*

Logical block addressing (LBA)   A method used to access IDE (Integrated Drive Electronics) hard disk drives. Using LBA, disks larger than 504MB can be *partitioned* using the FDISK command or a special program designed for disk partitioning.

Logical topology   The path a packet of data takes through a computer network, to be distinguished from the *physical topology* of the network.

Long file name   A file name that can contain up to 255 characters, allowable in the Windows 95/98 operating system. File names in DOS were limited to eight characters with a three-character extension, called 8.3 notation. Long file names allow them to be more descriptive of the file's contents; they are also downwardly compatible with operating systems that use 8.3 file name notation, in which they appear truncated to eight characters.

Long-nose pliers   A type of pliers specifically designed to bend wire. Long-nose pliers come in various sizes, but they are delicate instruments and should never be used to remove nuts or bolts.

Lossless compression   A method of compressing the data in a file that uses a digital data compression algorithm to compact the amount of storage space used for a file or group of files. Specialty programs such as PKZIP or WINZIP create lossless compression, in which files must be identical to their original contents when they are uncompressed.

Low-level format   The process of writing the outline of the tracks and sectors on a disk, the first step in preparing it to store data. Low-level formatting is performed on disks at the factory.

Low-pass filter   A filter used to smooth out the analog voltages received from a *digital-to-analog converter* to create a continuous analog waveform we hear as a sound.

M

MAC address   The address of the destination of a packet of data to be transmitted to another computer via a network. MAC stands for media access control, and the number is a 48-bit binary number that uniquely identifies one machine from every other machine on the network. Every *NIC* (network interface card) manufactured responds to a unique pre-assigned MAC address.

Machine language   The binary language of a computer microprocessor, which is made up of hexadecimal code.

Macro assembler   A program that takes an assembly language *source file* as input and determines the machine language for each source statement.

Macro virus   A virus written in a scripting language such as Visual BASIC that contains commands that can be processed from within an application program.

Magnetic field   The space around a magnetic object or an object carrying current in which magnetic forces can be detected.

Magneticware   Refers to floppy and hard disks and other magnetic storage material such as magnetic tape.

Mailbox   An electronic mailbox is a user's unique e-mail address, which includes two parts: a mailbox name and a computer host name, separated by an "at" sign (@), such as mailbox@rwa.com.

Mass storage   *See* Secondary storage.

Matched memory expansion   A section of a *micro-channel expansion slot* used when a higher memory transfer rate can be used by the expansion card, allowing data to be transferred at a 25% increase in speed.

Math chip   *See* Coprocessor.

Megabyte   A measurement of computer information equal to approximately 1 million *bytes*. It is exactly 1,048,576 bytes.

Memory location   A place that a microprocessor uses for getting and placing data; the number of lines used for the address *bus* determines how many different places the microprocessor can use for this operation.

Memory map   A simple way of graphically showing what is located at different addresses in memory.

Mesh network   A computer network *topology* in which every node is connected to every other node in the network; this type of network is also called a fully connected network or a full-mesh network.

Micro-channel architecture (MCA)   An approach developed by IBM to handle the requirements of newer microprocessors for an *expansion slot* that could accommodate a 32-bit bus.

Microcomputer   A small (usually desktop-size) computer that uses a microprocessor to handle information.

Microprocessor   The central processing unit (CPU) or the "brains" of a computer.

MIDI (Musical Instrument Digital Interface)   A MIDI-capable device, such as an electronic keyboard, uses a MIDI-in and MIDI-out serial connection to send messages between a *controller* and a *sequencer*, specifying the type of note and how to play it, as well as other information about the sounds.

Mirror   An arrangement that supports multiple hard drives (up to 32) through fault-tolerant RAID (Redundant Array of Inexpensive Disks) technology.

MMX technology   Intel's hardware and software extensions to the 80x86 architecture designed to support multimedia operations involving audio, video, and graphical data.

Mode   The path used by a beam of light inside a *fiber* optic cable.

Modem   A hardware device that connects one computer to another, remote computer using a telephone line. Modem stands for <u>mo</u>dulator/<u>dem</u>odulator.

Motherboard   The main system board of a computer.

Mouse   The computer input device used to move the cursor around the screen, click on icons and buttons, highlight and select text, and otherwise communicate with applications.

MPC compliant   Refers to a computer being multimedia compliant, that is, capable of using applications that include graphic images and sounds, as well as having a CD-ROM drive. There are several levels of multimedia compliancy applied to computer systems.

Multisession CD-ROM   A CD-ROM that can be written to more than once; CD-ROMs that can be written to use a layer of gold instead of aluminum.

Multitasking   The ability of an operating system to allow many programs or tasks to run seemingly simultaneously; that is, you can have a word-processing application and a spreadsheet application "open" on the desktop at the same time, and switch back and forth between them.

N

Nanosecond   A measurement equal to 0.000000001 second, used to measure memory chip speed in a microprocessor; abbreviated as ns.

NetBEUI   NetBIOS Extended User Interface, a network protocol that provides limited networking capabilities in the Windows operating system. It allows printer and file sharing in small peer-to-peer networks.

Network access point (NAP)   A facility that provides access to national and global computer network traffic. An experimental network was originally established with just four major data processing facilities, in Chicago, New York, San Francisco, and Washington, DC, but today NAP facilities are located all across the country and around the world.

Network administrator   The person who determines how a computer network is set up and how each of the components of the network is configured.

Network diameter   The total distance that can be spanned by a computer network, which varies based on different types of cable and signal speeds.

Network drive   A disk drive that is accessible from a computer through a network connection. A network drive can be mapped by Windows Explorer using the Map Network Drive tool.

Network printer   A printer that is connected to a network of computers. A network printer is also a *local printer* to the computer that hosts it, but it can be accessed for printing by anyone on the network who has made a connection to that printer.

NIC   Network interface card, the interface between the PC (or other networked device) and the physical network connection.

Nonimpact printer   A type of printer that uses heat, a jet of ink, electrostatic discharge, or laser light to create a printed image on paper without making physical contact with the paper.

Nonmaskable interrupt   A line used primarily when a parity check error occurs in a system; so called because it cannot be "masked," or switched off, by software.

NTFS   A more advanced file system than FAT 16/32; NTFS is a feature of Windows NT.

Nut driver   A hand tool used to remove or tighten a nut. Nut drivers are not adjustable, so a microcomputer-repair tool kit should have a variety of sizes for different-size nuts.

O

Object file   A file that contains only the machine language created when a *source file* has been translated by a *macro assembler.*

Object-oriented programming   A programming language, such as C++, that supports objects, which contain both code and data and interact with the language environment in certain, predetermined ways. Object-oriented programming allows many elements of data management in a computer program to be performed automatically.

Optical character recognition   A process that examines a scanned image that contains text, in which specialty software compares the scanned data to match characters with the graphical representation and then converts the scanned image to *ASCII* characters.

OSI reference model   The Open Systems Interconnection reference model, one of the accepted standards governing the use of *protocols* in computer networks. The OSI reference model defines seven layers required to establish communication between two nodes in a network.

Output device   A device that can only get information from a microcomputer.

P

Packet switching   An electronic switching system used in computer networks that allows packets of data to be transmitted out of order and then reassembled in the correct order at the receiving station by including a sequence number within the packet.

Page fault   In a virtual memory computer environment, physical memory is divided into several fixed-size pages, which are loaded with a portion of an executing program or other type of data. If the correct page needed is not loaded into memory, a page fault is generated that instructs the processor to load the missing page from memory.

Palm-size PC (PPC)   A computing device so small it can be held in the palm of one hand. A PPC has a display that is 240×320 pixels, one-fourth the size of the Windows 98 desktop, and requires a stylus to select items on the screen. PPCs use the Windows CE operating system to perform many common computer tasks, but they are not as powerful as a microcomputer.

Parity bit   An extra bit used by microprocessors to check parity, a method of detecting errors in data.

Parity checking   A method used by microprocessors to check errors when working with computer bits. Parity checking uses an extra bit called a *parity bit.*

Partitioning   A method of formatting a hard disk so that it can act as two or more independent systems. For example, a hard disk can be partitioned so that one computer can operate under two different operating systems, such as Windows and UNIX.

Password   A sequence of characters that must be entered to gain access to a computer system.

Paste   The command used to paste an object that has been cut or copied. In Windows Explorer, the Paste command generally is used to place a file or folder in a new location.

Peering agreements   Contracts or agreements between companies that connect to *network access points (NAPs)* that allow them to establish conditions under which they can exchange traffic.

Peripheral device   Part of a microcomputer system that is separate from the microcomputer processor itself. Examples include a printer, mouse, modem, or keyboard.

PGA (pin grid array)   A type of integrated circuit or computer chip that has its pins arranged in a two-dimensional structure, protruding from the bottom of a square ceramic housing.

Phillips screwdriver   A screwdriver with a blade designed to remove or tighten a Phillips screw, one with a cross-shaped slot in its head. This type of screwdriver comes in many sizes and may have a carbide tip for strength and/or an insulated handle for work on electrical components.

Physical address   A notation created by an addressing mode that refers to information that resides in a specific register location in the 00000 to FFFFF real-mode addressing space of the computer processor.

Physical environment   Aspects of the physical location of a computer system that can affect the operation of the system.

Physical topology   The type of hardware used to construct a computer network, to be distinguished from the *logical topology* of the network.

Pipelining   A process that allows instructions to be executed faster because the Fetch, Decode, and Execute operations overlap, as opposed to being handled as separate actions. For example, one instruction, which has already been fetched and decoded, is being executed at the same time another instruction, just fetched, is being decoded, and a new, third instruction is being fetched. This reduces the total number of clock cycles needed by the processor to execute a program.

Pits   Microscopic indentations on the surface of a CD-ROM that are used to store binary information. The pits are so small that a standard CD-ROM can store more than 650MB of data.

Pixel   The smallest area on a video monitor screen that can have its intensity controlled individually. The more pixels on a screen, the greater the detail that can be displayed.

Planar   *See* Motherboard.

Platter   The physical disk in a hard disk system. Most hard disk systems now consist of more than one platter.

PLCC (plastic leaded chip carrier)   A type of integrated circuit or computer chip that has spring pins mounted on all four sides of a square-shaped housing.

Pliers   A hand tool used for holding small objects or bending or cutting wire. Pliers can damage computer surfaces and should not be used to remove or tighten nuts on computer equipment.

Point-of-presence (POP)   A connection to the Internet.

Port   A predefined connection used for two computers to communicate together over interconnected networks.

Power   The amount of electrical energy. Power can be calculated as $P = IE$, where P is the power in *watts*, I is the *current* in *amperes*, and E is the *voltage* in *volts*.

Power-on self-tests (POSTs)   A series of internal tests that are run automatically when you turn on a computer. The POSTs, which are contained in the computer's *ROM*, check the major sections of the motherboards as well as peripherals such as disk drives, the display, and the keyboard.

Power supply   In a microcomputer, the power supply is the component that converts the 120V AC to a low, steady DC voltage that can be used by the computer circuits.

Preemptive multitasking   A method used by Windows 95/98 to run multiple 32-bit applications while still running older 16-bit Windows 3.x applications in cooperative mode. In preemptive multitasking, the current application can be interrupted and another application can be started or switched to, allowing a fair allocation to each of the applications competing for processor time.

Primary partition   On a disk that has multiple partitions, the primary partition is the one from which the computer boots.

Primary storage   The hardware memory of a computer. Primary storage is immediately accessible to the processor but has limited capacity and is used for short-term storage.

Print spooling   A technique used to speed up the time required by an application to send data to the printer. In print spooling, the print job is first sent to a temporary spooling file on the hard disk, and then Windows prints the spooled file in background, freeing the user to work on other computer tasks.

Printer-control code   The *ASCII codes* with number values less than 32, which are used for controlling the operations of a printer. Appendix A includes a complete list of ASCII codes.

Problem/solution journal   A notebook or other record of problems encountered in a microcomputer-repair lab or shop, along with the solution to each problem, that can be used as a reference when similar problems are encountered.

Program groups   A group of one or more software programs represented by an *icon* in the Windows 3.x *desktop*. Clicking on a program group icon will open a window that shows icons for the programs in the group.

Protected mode   A mode of internal organization for microprocessor operation that was introduced with the 80286 microprocessor, and which allows *virtual addressing*, *multitasking*, and protected memory accesses.

Protocol   The type of format used to communicate between two nodes in a computer network. The type of protocol used in a network must be firmly defined before information can be transferred.

Protocol analyzer   A software tool that is used to monitor all traffic transmitted on a local network.

Public key encryption   A type of program encryption that uses two keys, a public key and a private key. The public key is used to encrypt the data, but only the private key can be used to decrypt it.

R

RAID technology   A method of using multiple hard drives (up to 32) together in arrangements called *mirrors* or *stripe sets*. RAID stands for "Redundant Array of Inexpensive Disks."

Random access memory (RAM)   The type of memory that a computer can store information (write) to as well as get information (read) from. RAM is also known as *volatile memory*, because it is not permanent, and any information in RAM is lost when the computer is turned off.

Read-only memory (ROM)   The type of memory that a computer can get information (read) from but cannot store information (write) to. ROM usually contains the programs called *BIOS* that are needed by the computer when it boots and is permanent memory, usually contained in a chip or chips that are programmed at the factory.

Read/write memory   *See* Random access memory.

Real mode   A mode of internal organization for microprocessor operation that is used in the original 8086 and 8088 microprocessors. The internal organization includes number of registers, number of bits in the registers, names of the registers, and what they are used for.

Real-mode device driver   A software *driver* that operates in real mode; a real-mode driver is needed for a device that is being used in a DOS environment.

Rectification   In electronics, the conversion of alternating current (electrical current that repeatedly changes its direction) to direct current.

Recycle Bin   In Windows 95/98, a temporary storage place for deleted files. You can retrieve deleted files from the Recycle Bin if needed; files are not actually removed until the Recycle Bin is emptied.

Registry   The Windows 95/98 replacement for the SYSTEM.INI and WIN.INI configuration files used by Windows 3.x; an internal operating system file.

Relay   An electrically operated switch.

Remote echo   The term used to describe the situation in which the remote modem (the one being transmitted to) sends back or *echoes* each character it receives, so the characters that display on your screen are the result of a transmission from the remote unit, as a method of verifying what you are sending.

Repeater   A connector used to connect two network *segments* and broadcasts packets between them. Using a repeater helps amplify the signal and extend the usable length of a network.

Resistor   An electronic component used to resist or limit the flow of current in a circuit. The value of a resistor is measured in ohms; the more ohms it has, the more a resistor will limit the flow of electrical current.

Resolution   The size of the Windows desktop. The resolution of the desktop can be adjusted in the Settings tab of the Display Properties window, accessed through the Settings option on the Start menu.

Resource   Any hardware entity used by a computer system, such as interrupts, memory, I/O locations, and DMA channels.

Resource conflict   A situation in which two (or more) hardware peripherals attempt unsuccessfully to use the same interrupt or other hardware entity.

Ring network   A computer network *topology* in which the nodes are connected in a circular communication path.

RISC   One of two major types of computer architecture; it stands for Reduced Instruction Set Computer. Pentium microprocessors are a mixture of RISC and *CISC* architecture, which allows for upward compatibility with 80x86 machines while providing for additional performance improvements.

Rosin   A chemical, also called *flux*, used to form the core of solder.

Router   A hardware connector that connects two or more networks together by providing an interface for each network.

Run-time error   An error in an executable computer program that appears when the program runs. They are typically caused by incorrect sequences of instructions and incomplete or faulty logical thinking.

S

Safety glasses   Protective eye covering used when dealing with tools or procedures that could cause heat, chemicals, or small particles of material to fly up or splash into the user's eyes.

ScanDisk   A utility provided with Windows 95/98 that allows the user to check the integrity of a disk, the organization of the file allocation table *(FAT)*, the directory areas, bad sectors, and every file on the disk.

Scanner   A common multimedia device used to convert text and/or graphical images into binary data that can be stored and manipulated using various software applications. Many different types of scanners are available, including flatbed, image, and hand held scanners.

Secondary storage   The *magneticware* used for long-term storage by a computer, such as floppy and hard disks and magnetic tape.

Sector   A structural element of a floppy or hard disk, essentially an arc-shaped area on the disk. A single-sided 5.25-inch floppy disk has eight sectors in each of its 40 tracks, for a total of 320 sectors; a double-sided 3.5-inch floppy disk has nine sectors per track, for a total of 720 sectors.

Security accounts manager (SAM) database   Maintains user profiles in a *domain.*

Segment   In a memory map of the 80x86 addressing space, a *segment* refers to one 64K block that contains all the memory locations that can be reached when the particular segment register is used.

Segment descriptor   A device that actually makes it possible for a computer operating system to control and protect memory. Segment descriptors are stored in memory and certain bits within them are used to grant or deny access to memory.

Segment register   One of four registers used by the computer processor to control all accesses to memory and input/output. Segment registers must be maintained by the programmer.

Sequencer   A microcomputer used with a *MIDI* device, such as an electronic keyboard, which instructs the device about the type of note to play and how to play it, along with other information.

Setup Wizard   A Windows assistant that guides the user through the steps involved in installing new software.

Shortcut   A user-defined link to frequently used programs or other files.

Signature   An encoded string of characters that represents a portion of the actual virus code for a specific virus. Viruses can be detected by a program that scans and compares files against a list of virus signatures.

SIMD   Single instruction multiple data, a method used by *MMX technology* that allows one MMX instruction to operate on 64 bits of data stored in an FPU register.

Simplex   A term that refers to a communications channel in which information flows in one direction only, such as a radio or television station.

Single inline memory module (SIMM)   A small board with several memory chips soldered to it that can be inserted into a memory slot in a motherboard. A SIMM is used to physically organize memory in a computer and was developed to solve the problem of chip creep, which is what happens when a chip works its way out of its socket due to thermal expansion and contraction.

Socket   A TCP/IP connection. *See also* Port.

Software interrupt   A software event that is used to perform a special task, such as file I/O.

Solder bridge   An undesirable condition caused by using too much solder, which forms an accidental connection between two adjoining parts of a printed circuit board.

Soldering   The process used to secure the wire connections of electrical components by heating a metal alloy (solder) around the components to be joined.

Soldering iron   A tool used in soldering.

Source disk   A disk that is to be backed up or copied.

Source file   A file consisting of the assembly language code used to create a computer program.

Speculative execution   A technique used by Pentium Pro and Pentium II/III microprocessors that allows multiple instructions to be fetched and executed, possibly out of order, to keep the *pipeline* busy.

Stackable   Refers to a *hub* that can use one of its ports to connect to existing network wiring. One port is designed to operate in either straight-through or crossover mode, selected by a switch on the hub.

Star network   A computer network *topology* in which there is one centralized node connecting all the other nodes.

Start button   The button at the far left of the *taskbar* (in its default position) in Windows 95/98. Everything it is possible to do in Windows is accessible through the Start button, which generates the Start menu. The Start menu includes Shut Down, Run, and Help choices as well as submenus for Programs, Favorites, Documents, and Settings.

Static discharge   A discharge of electricity that consists of isolated motionless charges, such as those produced by friction. Static electricity discharges should be avoided in a microcomputer-repair lab because they can harm many types of electronic equipment and integrated circuits. *See also* Electrostatic discharge.

Store and forward   A common method used to route and direct traffic in a computer network via a *switch.* In the store and forward method, the received packet is stored before it is examined to check for errors, and if no errors are found, the packet is retransmitted. Bad packets are not forwarded.

Stripe set   An arrangement that supports multiple hard drives (up to 32) through fault-tolerant *RAID technology.*

SuperDisk   A type of *floppy disk drive* capable of reading and writing both 120-MB SuperDisk floppy disks and 1.44/2.88-MB 3.5-inch disks.

Superpipelined   A 14-stage *pipeline* that allows less logic to be used in each stage, which permits the pipeline to be clocked at a higher speed.

Surface-mount component   A type of electronic component such as a *capacitor* that is much smaller than through-hole-mounted components, allowing more components to be placed on a board, or the same number of parts to be placed on a smaller board.

Switch   An electronic component that regulates the flow of electric current. Also, in a computer network, a switch is a hardware connection similar to a *bridge*, but with multiple ports that can direct traffic to several different *segments* of the network.

Switching power supply   The type of *power supply* used in personal computers, which converts the 60-Hz power line frequency into a 20,000-Hz frequency, before rectification and voltage regulation.

Synchronous DRAM (SDRAM)   Very fast *RAM* (up to 100-MHz operation) designed to synchronize with the system clock to provide high-speed data transfers.

System board   *See* Motherboard.

System settings area   The area at the lower right-hand corner of the *taskbar* (in its default position) that includes icons for background tasks such as virus protection, speaker volume, and the clock.

T

T1 carrier   Basic carrier in time-division multiplexing (TDM), a technique used in telephone communications to combine multiple digitized voice channels over a single wire. T1 provides 1.544Mbps multiplexed data for twenty-four 64,000 bps channels.

Target disk   A disk that will contain backed-up or copied files.

Taskbar   The bar that appears by default at the bottom of the Windows 95/98 *desktop* and contains the Start button, icons for applications that are currently running or suspended, open desktop folders, and the current time.

Text mode   One of two modes in which the *video monitor* communicates with the computer. In text mode, the display receives information from a built-in ROM chip that contains all the characters on the keyboard, plus many more. Simple graphics created from special characters in the *extended character* set can be displayed, and text mode requires very little memory compared to *graphics mode*. Text mode is sometimes also called *alphanumeric mode*.

Thermal shock   A adverse effect that occurs when a rapid increase or decrease in temperature causes an undue mechanical stress on an electrical component.

Thickwire   A specific type of cable in computer networks, RG-11 coaxial cable, used for 10base5 operation.

Thin-film transistor (TFT)   A type of *liquid crystal display* technology, also known as active matrix display. TFT displays are brighter than *dual-scan twisted nematic* displays and have a larger viewing angle and better resolution.

Thinwire   A specific type of cable in computer networks, RG-58 coaxial cable, used for 10base2 operation.

Tinning   A process performed on a soldering iron tip or a piece of stranded wire. Tinning a piece of stranded wire results in the strands being soldered together to make it easier

to solder the wire to other parts of a circuit. Tinning a soldering iron tip by melting solder on it and then wiping the tip clean with a damp cloth or sponge makes a soldering iron tip heat more efficiently by providing a bright, shiny surface.

Tip   The working part of a soldering iron, the part that actually melts the solder.

Token-ring network   A computer network *topology* in which a special token (a specific binary pattern) is circulated between nodes in a ring. The token is used by a specific node to gain access to the network and transmit data.

Topology   The way in which computers are connected in a network.

Trackball   A device that is similar to a mouse, except that the device does not move. Instead, the user pushes the trackball around inside its case, allowing the same movement of the pointer on the screen as with a mouse, but without requiring a mousepad or a large surface for movement. Many laptop and notebook computers have a built-in trackball instead of a mouse.

Transceiver   A hardware device that converts from one media type to another, so that 10base2 coaxial cable and fiber optic cable can be connected, for example. It is common to use more than one media type in a network, so many different kinds of transceiver are available.

Trojan horse   A virus that is disguised as a normal program. For example, a game program can contain a file-infecting virus that works while the user is playing the game.

Trust relationship   A network relationship in which one *domain* either provides or receives services from an external domain, permitting users in one domain to use the resources of another domain.

Tunneling   The method in which data is transferred in an encapsulated form via a *virtual private network (VPN)* that uses public network connections and encryption as well as encapsulation.

Tweezers   A hand tool used to hold a small part in place. Tweezers can also be used as a heat sink in soldering.

U

Undo   The command used to back up and erase the results of the last operation performed. In Windows Explorer, if you accidentally move a file or folder to the wrong location, you can use the Undo command to move it back.

Unicode character set   An international standard for representing text characters from many different languages. *Handheld* and *palm-size PCs* use programs that utilize the Unicode character set.

Uninstall   A Windows feature that removes a software application, as well as all of its associated files, folders, and program registry entries, from a disk.

Uninterruptible power source (UPS)   An electrical generator designed to provide protection against voltage spikes and electrical noise on the power line as well as to maintain an uninterrupted source of power for a limited time in the event of a loss of outside electric power.

Universal Resource Locator (URL)   A unique sequence of letters and/or numbers that directs a Web *browser* to a specific page of information on the Internet; for example, www.rwasoftware.com is the URL of a fictitious software company.

Universal serial bus   A peripheral bus designed to make it easier to connect many different types of devices, such as audio players, joysticks, digital cameras, data gloves, and printers to a computer.

Universal service   Connection to virtually any computer, anywhere in the world, any time, via the Internet.

Upper memory   The memory above the conventional 640K of memory in a microprocessor.

V

Vacuum pickup   A small tool that uses a suction cup and vacuum to pick up small components without harming them.

Video adapter card   The interface between the *motherboard* and the *video monitor* that allows you to see the output of the microprocessor to be displayed on the monitor screen.

Video coprocessor   A microprocessor that is dedicated to the display, relieving the main processor of responsibility for this element of the computer system, and providing more detailed and quicker video display for graphics and animation.

Video monitor   The part of the computer system that displays output as text or graphics on a screen.

Virtual addressing   A mode of addressing that allows a large program to execute in a smaller physical memory.

Virtual channel   Connections used by asynchronous transfer mode (ATM) or *cell relay*, which sets up a *virtual channel* between the end-to-end stations on the network and sends fixed-size cells back and forth.

Virtual circuit   A prearranged path through a computer network that all packets will travel for a particular session between machines. A virtual circuit is established to transmit a large amount of data more effectively.

Virtual device driver   A software driver for use with devices operating in protected mode. Virtual drivers have .VXD extensions in Windows 95/98 operating systems and exist for all types of hardware.

Virtual 8086 mode   A mode of operation that allows a single 80386, 80486, or Pentium processor to divide its memory into many "virtual" computers that can run programs in total isolation from each other. This means that more than one DOS program can run on the same computer at the same time.

Virtual machine   The part of the internal architecture of Windows 95/98 that handles 32-bit and 16-bit applications. The system virtual machine runs 32-bit applications in a preemptive multitasking mode, and it runs the 16-bit applications (old Windows 3.1 programs) in a different memory space that simulates the Windows 3.1 environment and protects the Windows 95/95 environment from out-of-control 16-bit applications.

Virtual memory   A type of memory made up of mass storage devices such as disks. The processor senses when its usable real memory is used up, and then stores some data onto a disk (usually a hard disk). If it needs the data stored on the disk again, it frees up some more real memory by placing other data on the hard disk, and then reads what it needs from the disk back to the real memory.

Virtual private network (VPN)   A type of virtual connection that uses public network connections (such as the Internet or a telephone system) to establish private communication by encrypting the data.

Virus   *See* Computer virus.

Volatile memory   Computer memory that is not permanent; the data contained in volatile memory is lost when the computer is powered off. *See also* Random access memory.

Volt   A unit of measurement used to measure electrical potential.

Voltage   The amount of electrical potential that can cause a current to flow; voltage is measured in *volts*. The letter symbol for voltage is *E*.

W

Wallpaper   An optional *background image* in the Windows desktop that can be chosen and modified in the Background tab of the Display Properties window, accessed through the Settings option on the Start menu.

WAN (wide area network)   A collection of computers and other devices connected in a network so they can share information. Wide area networks are geographically large, connecting computers in separate buildings or other widely scattered locations.

Watt   A unit of measurement used to measure *power*; 1 watt is the power produced by a *current* of 1 *ampere* through a potential difference of 1 *volt*.

Wave table synthesis   An alternative to frequency modulation used by a computer's sound card to control the sound output using prerecorded samples of notes stored in a data table.

Window   A screen in which an application can run, a document or Web page can be viewed, or the DOS shell can be accessed. Multiple windows can be open on the *desktop* at the same time, and a user can switch back and forth between windows, but only one window can be active at any time.

Windows desktop   The screen you see when a microcomputer using the Windows operating system boots up, providing easy access to the programs and information in the computer system.

Wireless network   A computer network *topology* in which the nodes of the network are connected to a base station that broadcasts data into the air in the form of a high-frequency RF signal or through a line-of-sight infrared laser.

Wire stripper cutters   A hand tool used to strip the insulation from wire to prepare the wire for use in an electrical connection.

Word   One 16-bit binary number.

Workgroup   A number of networked computers that share a common protocol. Also called a *domain*.

Worm   A virus that does not infect files. When a worm-infected program is loaded, the worm gets control first and does its damage.

WYSIWYG   Stands for "what you see is what you get," referring to a graphical, user-friendly computer environment, a big difference from the plain text interface used by earlier operating systems such as DOS.

Z

ZIF (zero insertion force)   Describes a type of integrated circuit socket with a small handle that, when lifted, releases an integrated circuit by removing the pressure on its pins.

Zip drive   A device similar to a *floppy disk drive* but with a removable cartridge that can hold much more data than a floppy disk, from 100MB up to 250MB. A Zip drives is a proprietary device produced by Iomega and it connects to a computer's printer port or USB port.

Appendix A
ASCII Character Set

DECIMAL VALUE →		0	16	32	48	64	80	96	112	128	144	160	176	192	208	224	240
↓	HEXA-DECIMAL VALUE	0	1	2	3	4	5	6	7	8	9	A	B	C	D	E	F
0	0	BLANK (NULL)	►	BLANK (SPACE)	0	@	P	`	p	Ç	É	á	▒	└	╨	∝	≡
1	1	☺	◄	!	1	A	Q	a	q	ü	Æ	í	▒	┴	╤	β	±
2	2	☻	↕	"	2	B	R	b	r	é	FE	ó	▓	┬	╥	γ	≥
3	3	♥	‼	#	3	C	S	c	s	â	ô	ú	│	├	╙	π	≤
4	4	♦	¶	$	4	D	T	d	t	ä	ö	ñ	┤	─	╘	Σ	∫
5	5	♣	§	%	5	E	U	e	u	à	ò	Ñ	╡	┼	╒	σ	∫
6	6	♠	▬	&	6	F	V	f	v	å	û	ª	╢	╞	╓	µ	÷
7	7	•	↨	'	7	G	W	g	w	ç	ù	º	╖	╟	╫	τ	≈
8	8	◘	↑	(	8	H	X	h	x	ê	ÿ	¿	╕	╚	╪	Φ	°
9	9	○	↓	)	9	I	Y	i	y	ë	Ö	⌐	╣	╔	┘	Θ	•
10	A	◙	→	*	:	J	Z	j	z	è	Ü	¬	║	╩	┌	Ω	•
11	B	♂	←	+	;	K	[	k	{	ï	¢	½	╗	╦	█	δ	√
12	C	♀	∟	,	<	L	\	l	¦	î	£	¼	╝	╠	▄	∞	η
13	D	♪	↔	-	=	M	]	m	}	ì	¥	¡	╜	═	▌	∅	²
14	E	♫	▲	.	>	N	^	n	~	Ä	Pts	«	╛	╬	▐	∈	■
15	F	☼	▼	/	?	O	_	o	△	Å	ƒ	»	┐	╧	▀	∩	BLANK 'FF'

Appendix B
DOS Reference

INTRODUCTION

The information in this appendix is included to provide details on the operation of DOS for Windows users performing work within the MS-DOS Prompt window.

BACKGROUND INFORMATION

THE DOS PROMPT

A typical DOS prompt looks like this:

C>

The C means that disk drive C: is active (remember that the colon is used to indicate a disk drive). The "greater than" sign (>) is the symbol used by DOS to indicate that the control of the computer has been turned over to you. To change the DOS prompt—that is, to make a different disk drive active—simply type the drive letter followed by the colon:

B:

This will cause drive B: to be the active drive, and the DOS prompt will change to B:. Similarly, typing in A: will change the active drive to the A: drive. This assumes that these drives are on your system.

CHANGING THE DATE AND TIME

Table B.1 shows two DOS commands for changing the date and the time. Figure B.1 shows the actions of these two DOS commands.

COPYING A DISK

There are many times when you will find it necessary to make backup copies of programs on a disk. Remember that purchased software is copyrighted, and the legal owner of the software is usually allowed to make one backup copy to keep in case the original is damaged. It is illegal to make copies of purchased software for any other reason.

To make a copy of a disk on a single-drive system, follow the procedure shown in Figure B.2.

TABLE B.1   DOS DATE and TIME commands

Command	Action
DATE	Allows you to change the date.
TIME	Allows you to change the time.

FIGURE B.1   Actions of DATE and TIME commands

```
Current date is Sun 01-14-2001
Enter new date (mm-dd-yy):
```

(a) DATE

```
Current time is 4:27:36.14p
Enter new time:
```

(b) TIME

FIGURE B.2   Procedure for copying a disk on a single-drive system

```
C>DISKCOPY A: A:
```

STEP 1
Type in DOS command. Press the Enter key.

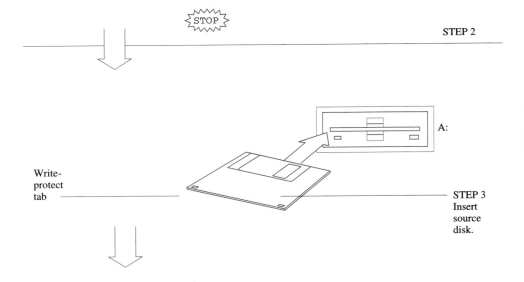

```
C>DISKCOPY A: A:

Insert source diskette into drive A:

Press any key when ready
```

STOP

STEP 2

A:

Write-protect tab

STEP 3
Insert source disk.

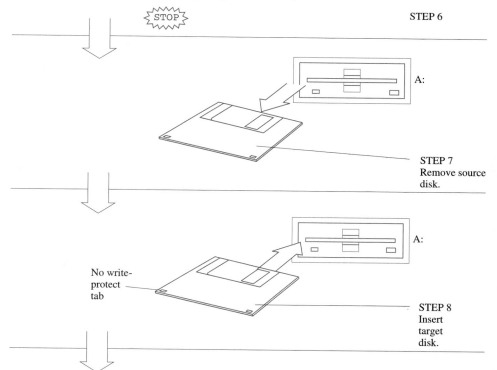

```
C>DISKCOPY A: A:

Insert source disk into drive A:

Press any key when ready
```

STEP 4
Press the
Enter
key.

```
C>DISKCOPY A: A:

Insert source disk into drive A:

Press any key when ready
Copying 80 tracks
9 sectors/track, 2 side(s)
```

STEP 5

```
C>DISKCOPY A: A:

Insert source disk into drive A:

Press any key when ready
Copying 80 tracks
9 sectors/track, 2 side(s)

Insert target disk into drive A:

Press any key when ready
```
STOP

STEP 6

A:

STEP 7
Remove source
disk.

A:

No write-
protect
tab

STEP 8
Insert
target
disk.

(continued on the next page)

```
C>DISKCOPY A: A:

Insert source disk into drive A:

Press any key when ready

Copying 80 tracks
9 sectors/track, 2 side(s)

Insert target disk into drive A:

Press any key when ready
```

STEP 9
Press the
Enter key.

```
C>DISKCOPY A: A:

Insert source disk into drive A:

Press any key when ready

Copying 80 tracks
9 sectors/track, 2 side(s)

Insert target disk into drive A:

Press any key when ready
Copy another disk (Y/N?)
```

STEP 10
After programs are
copied to
target disk,
this message
appears.

STEP 11
Copying process
completed.

The DISKCOPY command is used to make backup copies of original disks. You should not use original disks on a daily basis; instead, you should run your programs from the backup copies.

Note: It is not possible to use DISKCOPY to copy a 5.25-inch disk to a 3.5-inch disk, or vice versa.

DEFINITION OF DOS FILES

The purpose of a disk is for storing programs. When you store information on a disk, DOS places this information in what is called a *file*, much as you might place any information you have in a filing cabinet. You can change the contents of disk files, give them names, change their names, transfer them to other places within the computer system (to other disks or to the printer or monitor), or erase them completely. To start with, you should know how DOS names a file. Every DOS file has a name. The DOS file name consists of two important parts: the file name and the extension.

FILENAME.EXT

The following rules apply:

FILENAME: Must contain from one to eight characters, starting with a number or letter of the alphabet and then any letter or number, including the following symbols: ~, {, }, _, -, @, #, %, ^, &. No blanks are allowed.

EXT: Optional; may contain up to three characters or numbers.

As an example, the following are all legal DOS file names:

```
COMMAND.COM
PLAYIT.EXE
DO_12
WOW.
FILE.05
12345.678
```

The following are not legal DOS file names and will not be accepted by DOS:

TOOMANYLETTERS	More than eight characters
.WOW	No file name
BIG.EXTENSION	More than three letters in the extension
HEY YOU.	Blank used in file name

As you will see later, the file extension can be a very powerful tool for keeping, maintaining, and using files.

DOS has some file names that it reserves for its own special use. They are listed here so that you will not use them for now:

```
AUX    CON    COM1    COM2    NUL
LPT1   LPT2   LPT3    PRN     CLOCK$
```

Note: Windows 95/98/NT allow the use of long file names, such as "My Documents," that appear in truncated form in DOS (MYDOCU~1). See Unit III for more details.

HOW DOS NAMES FILES

Table B.2 lists the file-naming conventions used by DOS. These conventions consist of file extensions.

TABLE B.2   Common DOS file extensions

Extension	Meaning
.COM	DOS commands or programs you can directly execute. To execute the program, simply enter the file name of the program (without the .COM extension), press Enter, and the program will be loaded into your computer and will begin running. Application programs may have this extension.
.BAT	DOS batch files. This extension provides a way of storing several commands in a file and having them done automatically just by entering the name of the file.
.DAT	Data files used by programs.
.CPI	Code page information containing foreign character sets.
.EXE	Executable programs. To execute the program, simply enter the file name of the program (without the .EXE extension) and press Enter, and the program will be loaded into your computer and will begin running. Some application programs may have this extension.
.SYS	For installing device drivers. These are programs that allow DOS to communicate with hardware devices attached to your computer, such as the keyboard, printer, and monitor.
.TXT or .DOC	Files used for word processing.
.BAS	File is a BASIC program.
.ASM	File is an assembly language source program.
.C or .CPP	File is a C/C++ source program

Note: The main difference between the .COM and .EXE extensions is in how the program uses memory when loaded into the computer.

In addition, special programs (such as business application programs) use their own extensions. You need to consult the application program documentation in order to know what these extensions are.

DISPLAYING DOS FILES

DOS places files on your disk in a list called a *directory*. You will learn more about directories, but for now you should know how to display a list of file names on the computer screen. The command for displaying the list of file names is

```
DIR
```

This command causes DOS to list the disk files, as shown in Figure B.3.

Another useful modifier is

```
DIR/W
```

Think of the /W as standing for *wide*. The results of issuing this command are shown in Figure B.4. As shown in the figure, the DIR/W command allows the display of more files at the same time. What is lost here is the individual file size and the creation or modification date and time.

COPYING THE SCREEN TO THE PRINTER

Your PC makes it easy to copy the contents of the screen to the printer. This means that whatever text is displayed on the monitor can be copied to your printer, giving you a hard copy of what is on the screen. To do this, use the sequence given in Table B.3.

FIGURE B.3 Typical DOS directory

```
Volume in drive C is FIREBALLXL5
Volume Serial Number is 3729-11D6
Directory of C:\DOS

.                  <DIR>         01-18-95  10:22a
..                 <DIR>         01-18-95  10:22a
COMMAND  COM      47,845 11-11-91   5:00a
FORMAT   COM      33,087 11-11-91   5:00a
COUNTRY  SYS      17,069 11-11-91   5:00a
KEYB     COM      14,986 11-11-91   5:00a
KEYBOARD SYS      34,697 11-11-91   5:00a
ANSI     SYS       9,029 11-11-91   5:00a
ATTRIB   EXE      15,796 11-11-91   5:00a
CHKDSK   EXE      16,200 11-11-91   5:00a
EDIT     COM         413 11-11-91   5:00a
MORE     COM       2,618 11-11-91   5:00a
SYS      COM      13,440 11-11-91   5:00a
DEBUG    EXE      20,634 11-11-91   5:00a
FDISK    EXE      57,224 11-11-91   5:00a
MODE     COM      23,537 11-11-91   5:00a
EGA      CPI      58,873 11-11-91   5:00a
EGA      SYS       4,885 11-11-91   5:00a
HIMEM    SYS      11,616 11-11-91   5:00a
XCOPY    EXE      15,820 11-11-91   5:00a
EDIT     HLP      17,898 11-11-91   5:00a
HELP     EXE      11,473 11-11-91   5:00a
PRINT    EXE      15,656 11-11-91   5:00a
SETVER   EXE      12,015 11-11-91   5:00a
APPEND   EXE      10,774 11-11-91   5:00a
DISKCOPY COM      11,879 11-11-91   5:00a
GRAPHICS COM      19,694 11-11-91   5:00a
LABEL    EXE       9,390 11-11-91   5:00a
SORT     EXE       6,938 11-11-91   5:00a
TREE     COM       6,901 11-11-91   5:00a
DOSKEY   COM       5,883 11-11-91   5:00a
        31 file(s)       526,270 bytes
                      22,773,760 bytes free
```

FIGURE B.4   Result of the DIR/W command

```
Volume in drive C is FIREBALLXL5
Volume Serial Number is 3729-11D6
Directory of C:\DOS

[.]             [..]            COMMAND.COM     FORMAT.COM      COUNTRY.SYS
KEYB.COM        KEYBOARD.SYS    ANSI.SYS        ATTRIB.EXE      CHKDSK.EXE
EDIT.COM        MORE.COM        SYS.COM         DEBUG.EXE       FDISK.EXE
MODE.COM        EGA.CPI         EGA.SYS         HIMEM.SYS       XCOPY.EXE
EDIT.HLP        HELP.EXE        PRINT.EXE       SETVER.EXE      APPEND.EXE
DISKCOPY.COM    GRAPHICS.COM    LABEL.EXE       SORT.EXE        TREE.COM
DOSKEY.COM
        31 file(s)          526,270 bytes
                         22,773,760 bytes free
```

TABLE B.3   Copying the screen to the printer

Step	Process
1	On the monitor, view the text that you want to copy to the printer.
2	Make sure the printer is attached to the computer, the printer is on, and paper is in the printer.
3	Hold down the Shift key and press the key marked <Print Scrn> at the same time. (The cursor will now move across the screen from top to bottom and left to right, transferring the screen's contents to the printer.)
4	Remove the copy from the printer and turn the printer off.

TRANSFERRING A DOS FILE TO ANOTHER DISK

There are times when you will need to transfer just a single file from one disk to another. The command that lets you do this is

COPY source_file target_file [/V][/A][/B]

where   source_file = the complete file name of the file to be copied.
        target_file = the name of the destination file. *Note:* If a file with the same name exists, it will be overwritten by the new copy.
        /V = disk verification. It does a double check to make sure that the file was copied exactly the same as the original. It takes a little more copying time to do this.
        /A = a modifier that lets the COPY command know that the file is an ASCII file (a text file).
        /B = a modifier that lets the COPY command know that the file is a binary image file (a file that contains computer code instead of text).

Table B.4 shows three examples of the use of the COPY command. In all the examples, the active drive is drive A:.

A special form of the COPY command allows you to copy several files and merge them all into one single file. As an example, suppose you have three text files on the disk, as shown:

```
MYFILE1.TXT
MYFILE2.TXT
MYFILE3.TXT
```

You can now use the COPY command to copy all three files into one large file called ALLFILES.TXT, as follows:

```
A> COPY MYFILE1.TXT + MYFILE2.TXT + MYFILE3.TXT ALLFILES.TXT
```

749

DOS Command	Resulting Action
A>COPY MYFILE.TXT B:	Copies the file MYFILE.TXT from the A: drive to the disk on the B: drive and uses the same name.
A>COPY MYFILE.TXT MYFILE.BAK	Copies the file MYFILE.TXT into a new file on the same drive called MYFILE.BAK (the extension .BAK is normally used to indicate a backup copy of a file).
A>COPY MYFILE.TXT B:NEW.TXT	Copies the file MYFILE.TXT from the A: drive to the disk on the B: drive and uses the name NEW.TXT, which will now be the file's name on the B: drive.

DISPLAYING ONE FILE

There may be times when you want to find out if a particular file is on the disk at all. One way to do this is, of course, to use the DIR command and then look through all the files on the screen. Another, more professional, way is to use the name of the file with the DIR command. Suppose you want to see if a file by the name of THISONE.TXT is in your disk directory. You can enter the DIR command as follows:

DIR THISONE.TXT

If the file is there, it will be the only one displayed. If it isn't there, a message telling you so will be displayed.

WILD CARD CHARACTERS

There may be times when you want to display a listing of all the files on a disk that have the extension .EXE and none of the others. For example, suppose your disk contains the following files:

TEXT1.TXT   GAME1.EXE   TEXT2.TXT   WONKERS.EXE
CONFIG.SYS   TEXT1.BAK   WONKERS.BAK   WONKINS.EXE

You can use the wild card symbol * and enter the following command:

DIR *.EXE

The only files that will be displayed are

GAME1.EXE   WONKERS.EXE   WONKINS.EXE

The * symbol tells the system to use any group of characters in its search for a file. As an example, if you instead enter the command

DIR TEXT1.*

then the only files you will get are

TEXT1.TXT   TEXT1.BAK

Of course, if you enter

DIR *.*

you will get all the files.

Another wild card character is the question mark, ?. The ? allows substitution for only one letter. For example, if you enter the command

DIR WONK???.EXE

you will get the files

WONKERS.EXE   WONKINS.EXE

COPYING A GROUP OF FILES

The wild card characters are useful when you want to copy a group of files. For example, suppose you want to copy all the .TXT extension files from the hard drive (drive C:) to the A: drive. To do this (assuming that your active drive is the C: drive), all you need to enter is

```
COPY *.TXT A:
```

All the files with the .TXT extension will be copied to the A: drive.

DISPLAYING FILE CONTENTS

You can display the contents of any file, but only the text files will make any sense. To do this, use the command

$$\text{TYPE file\_specification}$$

where   file_specification = the complete name of the file

To display the contents of the text file MYTEXT.TXT, make sure that the file is on the active drive and enter

```
TYPE MYTEXT.TXT
```

and the contents of the file will be displayed on the monitor. A better method involves the use of a text editor (such as EDIT) to view the file.

Many software vendors have a file on their disk called README.TXT. This file usually contains important last-minute information about the software on the disk. In order to see the contents of this file, you simply enter

```
TYPE README.TXT
```

REDIRECTING FILES

If you want the contents of a file to be displayed on the printer (meaning printed on the paper in the printer) rather than on the screen, you can issue the following command:

```
TYPE README.TXT > PRN
```

This command sends the text of the file README.TXT directly to the printer. The > symbol is the redirection operator, and it means that the file is to be directed elsewhere than the monitor. PRN is the device name for the printer. So, to transfer the contents of any text file to the printer, use

```
TYPE FILENAME.EXTENSION > PRN
```

Using the redirection operator, you can also place a copy of the directory listing into a file:

```
DIR > DIRFILE.TXT
```

This command instructs the system to copy the contents of the directory listing into a file on the active drive and name the file DIRFILE.TXT. You can then send the contents of that file to the printer:

```
TYPE DIRFILE.TXT > PRN
```

You can also transfer the disk directory to the printer with

```
DIR/W > PRN
```

or

```
DIR > PRN
```

In either case, the directory output is redirected from the monitor to the printer.

USING EDIT

EDIT is a command that allows *screen* editing. EDIT allows you to use arrow keys to move the cursor around on the screen. This makes it much easier to correct typos, enter new text, and select other text for deletion.

STARTING EDIT

EDIT allows you to specify the name of the text file you wish to edit:

```
A> EDIT MYBATCH.BAT
```

If the MYBATCH.BAT file does not exist, EDIT creates it. If the file does exist, EDIT opens the file and shows you the initial screen of file text. Figure B.5 shows what the EDIT start-up screen looks like when the file does not exist. At the top of the screen are *pull-down menu items* File, Edit, Search, Options, and Help. The Help menu is especially helpful for new users. The large box with arrows on the right side and bottom is called the *editing window* and is where your text file is displayed. Horizontal and vertical scroll bars provide a graphic display of your current position within the text file. The four arrow keys can be used to navigate around inside the editing window.

At the lower-right corner are two numbers that display the current line and column position of the cursor in the text file. The cursor controls where text is inserted and deleted.

Since MYBATCH.BAT does not yet exist, EDIT creates the file and waits for you to begin entering text. Enter the following three lines right now:

```
CLS
VOL
VER
```

Those three lines should now appear in the upper-left-hand corner of the editing window.

FIGURE B.5   Initial EDIT screen

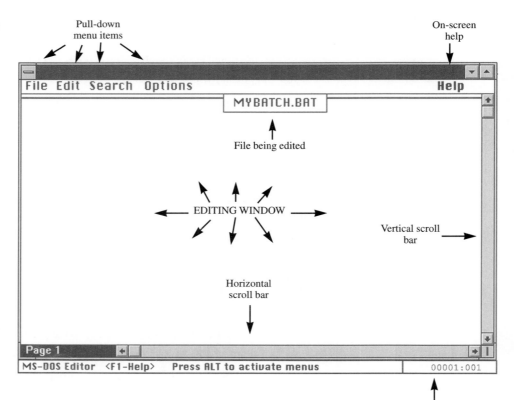

```
New
Open...
Save
Save As...

Print...

Exit
```

To save the new text file, it is necessary to access the File menu. This is accomplished by pressing Alt and then the F key. When you do this, EDIT displays the File menu, which is shown in Figure B.6. Choosing a command from the File menu can be done in several ways:

- Enter the letter of the command you want to use (N for New, O for open, etc.).
- Use the up and down arrows to highlight the desired command, and then press Enter.
- Use the mouse to select a command and click the left mouse button.

To save the file, use the SAVE command from the File menu (press S). Then exit EDIT by choosing the EXIT command (Alt-F-X). EDIT writes the new text into MYBATCH.BAT and exits.

EDITING AN EXISTING FILE

Now that we have a text file to work with, let us edit it again. First, enter

```
A> EDIT MYBATCH.BAT
```

EDIT opens the file and displays the three lines of text. The cursor will be at line 1, column 1. Use the arrow keys to position the cursor so that it is underneath the V in the VER command on line 3. Now press the Del key on your keyboard. EDIT erases the V character and shifts the remaining two characters (ER) to the left. Press Del again. Now only the R from VER should remain. Enter the characters D and I now so that the third line reads DIR. This is one way we can modify an existing file. The Backspace key can also be used to delete single characters, and there are other EDIT commands that allow entire blocks of text to be moved, deleted, and copied.

OTHER EDIT FEATURES

Large text files can be navigated quickly through the use of the Page Down and Page Up keys, which replace the entire editing window with a new set of lines. Ctrl-End takes you to the last page of the text file, and Ctrl-Home takes you to the beginning.

Also, EDIT allows the use of user-selectable screen colors. This is a great advantage over monochrome (black and white) editors, because it allows the screen colors to be adjusted for a visually pleasing effect. And, as stated before, EDIT can be controlled with the mouse, adding another dimension of ease to the task of editing.

Finally, choices within the Search menu allow the user to search for a word, phrase, or group of symbols, and to replace it or them with a different character or set of characters, if necessary. This is very useful for programmers, writers, and individuals who handle large data files.

RENAMING FILES

The command for renaming any file (or group of files using wild card characters) is

```
RENAME file_name newfilename[.ext]
```

or

```
REN file_name newfilename[.ext]
```

For example, if you want to rename the file MYFILE.TXT to YOURFILE.TXT, type

```
RENAME MYFILE.TXT YOURFILE.TXT
```

and the name of MYFILE.TXT will be changed to YOURFILE.TXT.

Entering REN HELLO.* BYE.* uses the wild card on the extension to rename all files beginning with HELLO to BYE.

ERASING FILES

Because you can create files, you can also get rid of them; however, use this command with caution. The DOS command is

DEL filename

For example, if you want to delete the file OLDSTUF.TXT, you enter

```
DEL OLDSTUF.TXT
```

and the file is deleted. You can also use wild card characters to delete files. Be very cautious when you do this so you don't delete more files than you intend to delete.

ORGANIZING YOUR FILES

An important part of using a computer system is understanding the different methods by which files are kept on the disk. Up to this point, when you needed to store information on a disk, you simply gave this information a file name and used the proper DOS command to store it on the disk. Storing files in this way is a rather simple and direct approach. However, you will find that this storage method is not the most practical one actually used in the field. Figure B.7 shows the direct approach to storing disk files.

The problem with this storage method is that any time you want to access a file or get information about it, you must be in the environment of all the other files on the disk. In practical applications, there are usually stored files that contain related kinds of information or have some other kind of unique relationship, as illustrated in Figure B.8.

It would be nice to be able to deal with just one file type at a time. One method of doing this is to use a separate disk for each type of file. This, however, may not be practical, especially if a single hard disk is used to store a variety of disk files.

Fortunately, DOS provides a way of making a single disk appear as if it were many smaller, separate disks, as shown in Figure B.9. DOS does this by creating *directories*. A directory is nothing more than a part of a disk that looks like a smaller disk, while the rest of the disk is invisible to the user. This important concept is illustrated in Figure B.10.

Theoretically, you can break your disk into as many smaller disks as you want. You can even break up your directories into even smaller parts, called *subdirectories*. A block diagram representing the use of subdirectories is shown in Figure B.11.

CREATING DIRECTORIES

The command for creating a directory on your disk is

MKDIR [drive:]directory

where   drive = the name of the drive (optional)
         directory = the name you want to give the directory

A directory must be named following the same DOS rules used for naming a file. This means that each name can consist of from one to eight characters, with a three-character extension. If you do not specify a drive, DOS will use the active drive.

Everything in one stack!

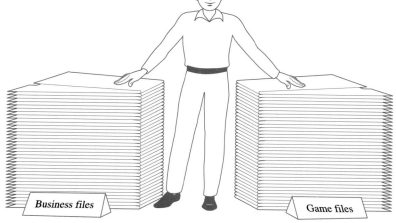

Text files

Business files

Game files

FIGURE B.8   Files with related information

**FIGURE B.9   Breaking a disk
into smaller parts**

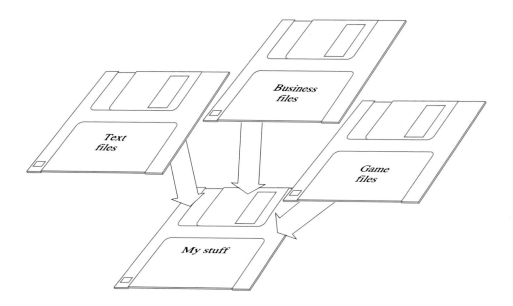

Text files

Business files

Game files

My stuff

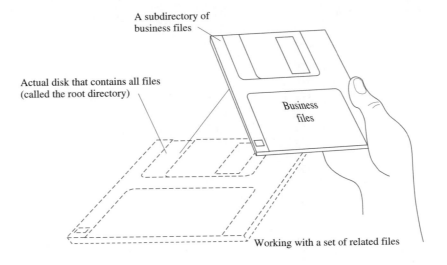

A subdirectory of
business files

Actual disk that contains all files
(called the root directory)

Business
files

Working with a set of related files

**FIGURE B.11   Block diagram
representing the use of
subdirectories**

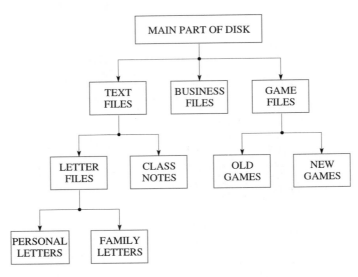

**FIGURE B.12   Disk listing of a
directory**

```
A:\> DIR

Volume in drive A is MYDISK
Directory of A:\

DIR_1          <DIR>        12-23-98  2:32a  dir_1
   0 file(s)         0 bytes
   1 dir(s)      1,456,640 bytes free

A:\>
```

For example

```
MKDIR DIR_1
```

will cause the directory DIR_1 to be created on the disk in the active drive.

When you use the DIR command to list the disk directory, if there is nothing else on the disk, your output will be as shown in Figure B.12.

Because the MKDIR command is frequently used, the abbreviated form MD is provided. For example, using the same disk, you could create two other directories:

```
MD DIR_2
MD DIR_3
```

You have effectively created three new disks called DIR_1, DIR_2, and DIR_3. You still have the main disk as well. The main disk is called the *root directory*, meaning that it is the one with which you always start.

Now that you know how to create a directory, the next step is to discover how to get inside one.

GETTING INTO DIRECTORIES

The DOS command for getting into a directory is

CHDIR [drive:]directory

where   drive = the name of the drive that contains the directory
        directory = the name of the directory

If you don't use a drive name, DOS will look for the directory on the active drive. If it isn't there, DOS will respond with an error message, telling you that the directory cannot be found.

As an example, to get inside the DIR_1 directory, the command

```
CHDIR DIR_1
```

will activate the DIR_1 directory. If you now do a listing with the DOS DIR command, you will see what is shown in Figure B.13. As you can see from the figure, all the other information on the disk (such as the other directories) is no longer displayed, just as if you were now on another disk.

Note that the top part of the directory display shows the drive the disk is in and the directory:

```
Directory of A:\DIR_1
```

As shown in Figure B.13, there are two other listings:

```
.    <DIR>    12-23-98    2:33a    .
..   <DIR>    12-23-98    2:33a    ..
```

The single period is the abbreviation for the current directory. The double period is the abbreviation for the directory immediately above the current directory. For example, entering DIR . will produce exactly what is shown in Figure B.13, whereas entering DIR .. will produce the directory immediately above DIR_1, which is, in this case, the root directory.

Since the CHDIR command is so common, DOS allows you to abbreviate it CD. Thus, CD DIR_1 works the same as CHDIR DIR_1.

GETTING OUT OF DIRECTORIES

To get back to the root directory (the one you start with when the system first boots up), use

```
CD \
```

FIGURE B.13   Inside a DOS directory

```
A:\> CHDIR DIR_1

A:\DIR_1> DIR

Volume in drive A is MYDISK
Directory of A:\DIR_1

.             <DIR>         12-23-98      2:33a  .
..            <DIR>         12-23-98      2:33a  ..
        0 file(s)          0 bytes
        2 dir(s)     1,456,640 bytes free
A:\DIR_1>
```

This will automatically put you back into the root directory. The directory that is just above a current directory is called the *parent directory*. Thus, the root directory is the parent directory for DIR_1. If there were a directory called SUB_1 inside DIR_1, then DIR_1 would be the parent directory for SUB_1. Using CD .. you move up one directory to the parent directory of the current directory.

SOME DIRECTORY EXAMPLES

Suppose you had a disk with the directory structure shown in Figure B.14. Note from the figure that there are two subdirectories (the root directory is the main directory). These sub-directories are called DIR_1 and DIR_2. Also note that there are three files, each called MYFILE.01. Each of these files, even though they all have the same name (MYFILE.01), can be entirely different *because they are in different directories.* Remember to think of a directory as if it were a separate minidisk. The directory structure of Figure B.14 is similar to having three different disks. Of course, you could have three entirely different files using exactly the same name on three different disks.

Normally you wouldn't give the files inside different directories the same name; we did this here just to illustrate that the files inside directories are indeed isolated from files inside other directories. Figure B.15 shows another example of two directories, each with some files contained in them.

Table B.5 shows some of the various combinations that you could achieve with the directory structure of Figure B.15.

CREATING OTHER DIRECTORIES (PATHNAMES)

As a computer user you will encounter directories contained inside directories. Directories inside directories are usually found on hard disk drives (because hard drives can hold so

FIGURE B.14 Example of directory structure

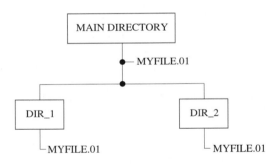

FIGURE B.15 Directories with different files

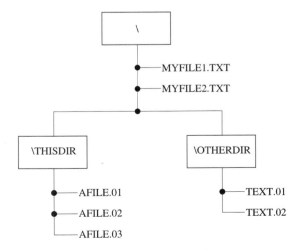

TABLE B.5 Various directory combinations from Figure B.15

Active Directory	Directory Listing Display				Comments
\(root)	MYFILE1.TXT	128	3-5-98	9:30a	Active files:
	MYFILE2.TXT	250	3-6-98	10:05a	MYFILE1.TXT,
	THISDIR	\<DIR>	3-6-98	10:15a	MYFILE2.TXT.
	OTHERDIR	\<DIR>	3-7-98	9:03a	Files not accessible: all the files in the other two subdirectories.
Change to another directory with CD\THISDIR					
\THISDIR	AFILE.01	356	3-9-98	1:08p	Active files:
	AFILE.02	550	3-9-98	1:15p	AFILE.01,
	AFILE.03	128	3-9-98	1:25p	AFILE.02, AFILE.03. Files not accessible: all the files in the other directories (the root directory and OTHERDIR).
Change to the other directory with CD\OTHERDIR					
\OTHERDIR	TEXT.01	256	3-7-98	9:20a	Active files:
	TEXT.02	256	3-8-98	11:46a	TEXT.01, TEXT.02. Files not accessible: all the files in the other directories (the root directory and THISDIR).

much information), but they are possible on floppy disks as well. Figure B.16 shows an example of what you might encounter on a hard disk drive directory used by a small business.

As shown in Figure B.16, there are three main subdirectories, called TEXT, GAMES, and ACCOUNTS. Each of these subdirectories has subdirectories within it. For example, the subdirectory TEXT contains the subdirectories LETTERS and MAILINGS. If you start at the root directory and you want to play the game ZAP.EXE, you first enter

CD\GAMES\ZAPPER

This DOS command puts you into the ZAPPER subdirectory, where you can access the game ZAP.EXE.

In DOS, a *pathname* is simply a series of subdirectory names DOS must combine to give you access to the files you want. Thus, in the DOS command CD\GAMES\ZAPPER, \GAMES\ZAPPER is the pathname.

FIGURE B.16 Typical directory tree used by a small business

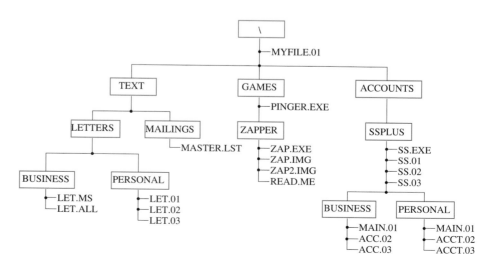

If the Current Directory Is	To Gain Access to File	You Must Use DOS Command (Pathname)
\	LET.MS	CD\TEXT\LETTERS\BUSINESS
TEXT\LETTERS\BUSINESS	LET.01	CD\TEXT\LETTERS\PERSONAL
TEXT\LETTERS\PERSONAL	ACC.02	CD\ACCOUNTS\SSPLUS\BUSINESS
ACCOUNTS\SSPLUS\BUSINESS	SS.01	CD..
ACCOUNTS\SSPLUS	MYFILE.01	CD\

TABLE B.7 Summary of DOS directory rules

Main Point	Comments
Directory names conform to DOS file name standards.	A file name can consist of eight characters with a three-character extension.
Do not create a subdirectory called \DEV.	DOS uses a *hidden* internal subdirectory called \DEV to perform I/O operations to devices such as the printer.
DOS allows the creation of any number of subdirectories according to the amount of disk space.	Disk Space Maximum Subdirectories 160/180K 64 320/360K 112 1.2MB 224 (The more disk space, the more subdirectories the disk can contain.)
DOS pathnames cannot exceed 63 characters.	

To get to the game PINGER.EXE, you enter CD\GAMES (the pathname here is \GAMES). This allows you to access the game PINGER.EXE, because you are in the GAMES directory.

Table B.6 gives some examples of how to access various files in the directory structure given in Figure B.16.

Table B.7 summarizes the directory rules.

REMOVING DIRECTORIES

The command for removing a directory is

RMDIR [drive:]pathname

or

RD[drive:]pathname

where   drive = the drive containing the subdirectory (If no drive is specified, DOS will look for the subdirectory in the active drive.)
pathname = the subdirectory names required to get to the directory

For example, in order to remove the directory called MYGRADES, use RMDIR MYGRADES or RD MYGRADES, assuming that this directory is on the active drive. To remove the subdirectory LETTERS in the subdirectory TEXT, use RMDIR\TEXT\LETTERS or RD\TEXT\LETTERS.

A directory must be empty (not contain any files or other directories) before it can be removed. Thus, if you want to remove a directory that contains files or other subdirectories, you must first get rid of all the files and subdirectories.

DELETING A NONEMPTY DIRECTORY

The RMDIR command requires an empty directory for the command to proceed. The DELTREE command can be used to delete entire directories (and any subdirectories) quickly and easily. For example, the "punchin" directory is deleted as follows:

```
C:\> deltree  punchin
Delete directory "punchin" and all its subdirectories? [y/n] y
Deleting punchin...
```

Wild cards are allowed in the directory name, which allows multiple directories to be deleted with a single command. Use the DELTREE command with caution, especially when using wild cards.

TRANSFERRING FILES

There are times when you need to copy a file from one directory to another. To do this, you use the DOS COPY command just as before, except now you include the pathname. Thus, to copy the file MYFILE.01 in the root directory to a subdirectory called TEXTFILE, you enter

```
COPY MYFILE.01 \TEXTFILE
```

In this example both the root directory and the subdirectory are on the same disk in the same active drive.

If you want to copy a file from one directory to another on a different drive (assuming the active drive is A:), you use

```
COPY MYFILE.01 B:\TEXTFILE
```

SETTING THE PATH

Figure B.17 shows the structure of the programs in a typical hard drive configuration (usually drive C:). As you can see from the figure, a typical hard drive configuration has many subdirectories, including a subdirectory for DOS files. DOS provides a command that makes it easy to tell DOS where to look for external commands. For example, every time you execute an external command, DOS first searches the current directory for the command file. If this file exists (such as a .COM, .EXE, or .BAT file), it is executed. Otherwise, DOS checks to see if you have defined a *path* or other subdirectories or disk drives to use in searching for the command.

The DOS command used for this purpose is PATH; it has the form

PATH[drive:][pathname][;drive][pathname] . . .

where   drive = the drive with a disk that may contain a required file
          pathname = the name of the path that may contain a required file

FIGURE B.17   **Typical hard drive configuration**

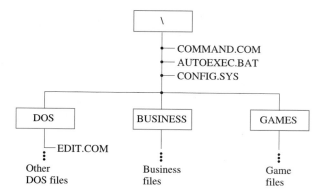

For example, the command

$$PATH = C:\;C:\DOS;C:\MYSTUFF;C:\WORDS;C:\GAMES$$

tells DOS to search these five directories for external commands. The DOS path command is normally placed in the AUTOEXEC.BAT file so that it is automatically activated every time the system is booted.

SYSTEM CONFIGURATION

The file called **CONFIG.SYS** is a special file for which DOS looks every time the system is booted. CONFIG.SYS is a *system configuration file*. If the file exists, DOS reads its contents and configures itself according to the entries provided. If the file does not exist, DOS leaves the system in the default condition. Table B.8 lists the DOS configuration parameters.

There will be times when you will have to create a CONFIG.SYS file to meet certain specifications. Typical CONFIG.SYS commands are illustrated in the following sections.

BREAK Command

The BREAK command has the form

$$BREAK = Condition$$

where   Condition = the word **on** or the word **off**

Normally the Ctrl-C key combination can be used to stop operation of most programs. In normal operation, DOS checks for a Ctrl-C only when it is reading from the keyboard or writing to the printer or screen. If you want DOS to check for a Ctrl-C at other times (such as during a disk read or write), you must set BREAK ON in the CONFIG.SYS file:

```
BREAK = ON
```

To set BREAK back to its normal default condition, use

TABLE B.8   DOS configuration parameters

Parameter	Meaning
BREAK	May be set ON or OFF. Determines how often DOS checks for a Ctrl-C.
BUFFERS	Allows you to set the number of buffers (memory locations) used to hold data when it reads from or writes to a disk.
COUNTRY	Lets you change specific items (such as the currency format) used by certain countries.
DEVICE	Allows you to let DOS control specific devices such as plotters or create a spot in memory that looks like an extra disk drive.
DRIVEPARM	Sets parameters for floppy disk drives.
FCBS	Used with early versions of DOS for file control blocks.
FILES	Introduced with DOS 3.0. Specifies the number of files that DOS can open at a time.
INSTALL	Invokes a special **fastopen** command from processing of the CONFIG.SYS file. Available with DOS 4.01 and later.
LASTDRIVE	Lets you specify the last disk drive to which DOS can refer.
REM	Allows you to add remarks to the CONFIG.SYS file. Available with DOS 4.01 and later.
SHELL	Lets you specify an alternate command process other than COMMAND.COM.
STACKS	Designates how many places in memory DOS can use to store what it was doing while servicing an interrupt.

```
BREAK = OFF
```

If BREAK is not specified in the CONFIG.SYS file, DOS uses the default condition, BREAK = OFF.

The disadvantage of setting BREAK = ON is that it may slow down the system somewhat because DOS has to check for the Ctrl-C more often.

BUFFERS Command

The BUFFERS command has the form

$$BUFFERS = n[,m][/x]$$

where   n = the number of disk buffers, from 1 to 99

        m = the maximum number of sectors that can be read or written in one I/O operation (1–8); the default setting is 1

        /x = a qualifier that, when used, allows the maximum number of disk buffers to be 10,000 or the largest number of buffers to fit into computer memory, whichever value is less

A *buffer* can be thought of as a special storage place in memory that holds data when it is being written to or read from a disk. The size of a buffer is 528 bytes. Whenever an application program is required to read data from a disk, it first checks the buffers to see if the data is there. If the required data is not in any of the buffers, the program performs a read operation from the disk. This operation is illustrated in Figure B.18.

Some application programs (such as database programs) access the disk many times to get required information. Doing this can slow down the system. The system can be speeded up by increasing the number of buffers, normally from 10 to 25. As an example, to increase the number of buffers to 20, you enter the following in the CONFIG.SYS file:

```
BUFFERS = 20
```

The use of too many buffers can result in a waste of computer memory, with a resulting slowdown in system operation. The slowdown occurs because every time DOS performs a disk read, it must first search through all the buffers. If you have allocated 60 or more buffers, for instance, this reading operation can take as much time as a simple disk I/O.

DEVICE Command

Your computer needs to know what *devices* it is required to operate. By default, DOS supplies *device drivers* (programs), which instruct it on how to read the keyboard (standard input), place information on the monitor (standard output), and use the printer, the system clock, and the disk drives. However, if you add devices such as a mouse or CD-ROM to the system, the computer needs to know this. This information is placed again in the CONFIG.SYS file.

The format for the DEVICE command is

$$DEVICE = [drive]:[path]filename[argument]$$

where   drive = the drive containing the device driver

FIGURE B.18 Example of reading buffers

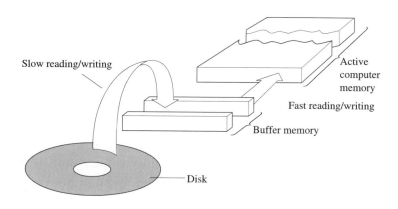

```
DEVICE=C:\HIMEM.SYS
DEVICE=C:\DOS\ANSI.SYS
DEVICE=C:\CCARD\CCDRIVER.SYS
DEVICE=C:\MOUSE\MOUSE.SYS
DEVICE=C:\WINDOWS\SMARTDRV.SYS 2048
```

path = the path of the device driver
filename = the device driver file name
argument = any information needed with the file name

There may also be other requirements for the DEVICE settings. Some of these are shown in Figure B.19. DOS 5.0 and later versions allow the use of the DEVICEHIGH command to load a device driver into high memory. This may be necessary if RAM below 640K must be conserved.

FILES

The FILES entry in the CONFIG.SYS file lets DOS know the number of disk files that DOS can have open at the same time. Normally DOS allows up to eight files to be opened at the same time. However, some application programs require more than eight files to be opened. When this is the case, you need to use the FILES entry in the CONFIG.SYS file. The syntax is:

FILES = NumberFiles

where   NumberFiles = the number of files to be opened (The minimum value is 8—the default value—and the maximum is 255.)

For example, if you have a database program that requires a maximum of 20 files to be opened, you add

```
FILES = 20
```

to your CONFIG.SYS file.

STACKS

Some computer systems must operate many different devices, and some computer programs have a lot of interaction with these devices. When this happens, DOS is interrupted many times. Each time DOS is interrupted, it must find a place in memory to store what it was doing, so that when it completes its interrupt it can retrieve this information and continue. Memory spaces used in this fashion are referred to as *stacks*. If there are too many interrupts, DOS may run out of stack space, and the following error message will appear:

```
Fatal: Internal Stack Failure, System Halted
```

When this message appears, you need to increase the number of stacks in your CONFIG.SYS file. The syntax is

STACKS = Number,Size

where   Number = the number of stacks to be used (The values range from 0 to 64.)
Size = the size of each stack (The values range from 0 to 512 bytes.)

For example, the entry

```
STACKS = 8,512
```

in your CONFIG.SYS file results in eight stacks of 512 bytes each.

DOS 6.0 and later versions allow function keys F5 and F8 to skip or single-step through CONFIG.SYS and AUTOEXEC.BAT. This is very helpful when you are troubleshooting a computer, because the specific hardware or software problem can be detected quickly through the process of elimination.

DEFINITION OF BATCH PROCESSING

Up to this point, when you entered DOS commands, you did so by entering one command at a time, pressing the Enter key, waiting for the command to act, and following up with an action of your own. This is called *interactive processing*. Simply stated, interactive processing consists of the user typing a command at the system prompt and then waiting for the command to complete before proceeding.

DOS allows you to develop a file that will contain all the DOS commands you need. All you have to do is activate the file (by typing its name); the file will then process each DOS command you entered into it, just as if you had entered these commands from the keyboard. For example, you could have a file that you named DOIT that would automatically clear the screen and display the directory. This type of processing, where the commands are obtained from a file rather than from the user at the keyboard, is called *batch processing*. The difference between interactive and batch processing is shown in Figure B.20.

BATCH FILES

A DOS batch file allows you to group DOS commands in a file. This file must have the extension .BAT (meaning "batch"). When you key the name of this file (without the extension), DOS automatically opens the file and executes each DOS command just as if it were entered by you from the keyboard. As an example, you could create a batch file called TEST.BAT that would clear the screen and display the current directory, as follows:

1. First, create the batch file using your preferred editor. Include these two commands in TEST.BAT:

   ```
   CLS
   DIR
   ```

2. The new file now exists on your disk:

   ```
   TEST.BAT
   ```

3. To activate the file, simply enter the file name (without the .BAT extension):

   ```
   A> TEST
   ```

4. DOS automatically opens the file, looks at the first DOS command (CLS), executes it, looks for the next DOS command (DIR), and executes it.

Effectively, you have created a new external DOS command. That is, every time you enter the word TEST, the screen will automatically be cleared and the directory will be displayed. This DOS feature gives you tremendous power to customize any computer system for a variety of different needs.

FIGURE B.20   Difference between interactive and batch processing

```
A>PROMPT $T

17:01:08:39
```

Figure B.21 shows the execution of a batch file that clears the screen and changes the DOS prompt to display the current time. Note from the figure that, by default, DOS displays each of the commands it uses. Do you know why the CLS command is not shown?

BATCHES WITHIN BATCHES

You can evoke a batch file from another batch file. For example, suppose you have a batch file called FIRST.BAT that contains the following:

```
TIME
DATE
```

This batch file displays the time (TIME) and the date (DATE). You can create another batch file, called DOIT.BAT, that first clears the screen, displays the directory, and then calls the batch file FIRST.BAT. DOIT.BAT should contain the following commands:

```
CLS
DIR
FIRST
```

If the second batch file is not the last DOS command, you must use a special DOS command to activate it. For example, if you want the file DOIT.BAT to activate FIRST.BAT before showing the directory, you have to write DOIT.BAT as follows:

```
CLS
CALL FIRST
DIR
```

The CALL command allows the original batch file (DOIT) to resume processing when the second batch file (FIRST) completes.

THE REM COMMAND

As you will soon discover, you can create very powerful batch files that can perform many complex tasks. Because of this, you may want to get into the habit of using the REM command to put *remarks*, or comments, in your batch files that explain what the file is to do. You will find this useful, especially if you come back to a file after a time and have forgotten your original intention in creating it. This command can also be used to display messages to the screen. The DOS command for doing this is

REM [message]

where   message = an optional character string that can contain up to 123 characters

A batch file that activates several .EXE files may use remarks like this:

```
CLS
REM Activating ACCOUNTS
ACCOUNTS.EXE
REM Activating PROCESS
PROCESS.EXE
```

THE ECHO COMMAND

DOS provides a command that will turn on or turn off characters being displayed on the screen as a batch file is processed. The command has the form

ECHO[ON/OFF/message]

where ON = turn on screen displays
OFF = turn off screen displays
message = an optional string of characters that will appear on the screen

Note: ECHO is ON by default.

The ECHO command may also be used to display information directly to the screen. For example, the following batch program displays information about the program developer:

```
ECHO OFF
CLS
ECHO          ***************************
ECHO          *       BATCH PROGRAM      *
ECHO          *  Brad's Repair Service   *
ECHO          *       Phone: 339-3115    *
ECHO          ***************************
```

THE PAUSE COMMAND

There may be times when it is necessary to halt the batch process temporarily. A halt may be needed to give the user a chance to insert or change a disk or to remind the user that something is about to happen. This feature is provided for by the PAUSE command:

PAUSE [message]

where message = an optional string of characters to be displayed on the screen

An example of the use of the DOS PAUSE command in a batch file (called DOACCT.BAT) is as follows:

```
CLS
REM Activating ACCOUNTS
ACCOUNTS.EXE
PAUSE Be sure to insert disk in drive B:
REM Activating PROCESS
PROCESS.EXE
```

Activation of the program produces the output shown in Figure B.22.

The user may continue batch processing by pressing any key or not continue it by a Ctrl-Break (holding down the Ctrl key at the same time as the Break key). When you do a Ctrl-Break, the following message will appear:

```
Terminate batch job (Y/N)
```

To terminate the batch process, press Y; otherwise, press N.

FIGURE B.22 Using the PAUSE command

```
A>REM Activating ACCOUNTS

A>PAUSE Be sure to insert disk in drive B:
Strike a key when ready...
```

(Press the RETURN key.)

```
A>REM Activating ACCOUNTS

A>PAUSE Be sure to insert disk in drive B:
Strike a key when ready...

A>REM Activating PROCESS

A>
```

BATCH PARAMETERS

DOS provides a way of allowing the user to decide how the batch file will operate without having to know much about DOS or anything about the batch process. This is done by making a batch file to which a user can pass instructions. To give you an idea of how this works, consider a batch file called WATCH.BAT, which displays the user's instructions:

```
A> WATCH Testword
```

The batch program would now display

```
This is an example of
a batch parameter.
The parameter you entered
was Testword for the
batch file called WATCH.
```

The actual contents of the batch file that do this are as follows:

```
ECHO OFF
ECHO This is an example of
ECHO a batch parameter.
ECHO The parameter you entered
ECHO was %1 for the
ECHO batch file called %0.
```

A *parameter* is nothing more than something that is passed on to the batch program itself. You may pass several parameters (the %0 parameter will always contain the name of the batch file). These parameters are identified by DOS as %1, %2, and %3, and so on. Consider the following batch file, called ALLTHREE.BAT:

```
ECHO OFF
ECHO The first parameter is %1
ECHO The second parameter is %2
ECHO The third parameter is %3
```

If you activated this batch file as

```
A> ALLTHREE Look at this
```

the result would be

```
The first parameter is Look
The second parameter is at
The third parameter is this
```

The ability to pass parameters to a batch file is a very powerful feature, as you will soon see.

THE IF COMMAND

DOS provides a method of allowing decision making in batch files by the use of the IF command. For example, one of the common uses of the IF command is in the following form:

IF EXIST file.ext DOS Command

where   file.ext = the name and extension of a file
DOS Command = a legal DOS command

As an example of the usefulness of this command, suppose you want to have the batch file (called CHECK.BAT) check to see if a specified program (called MYTEXT.TXT) is available in the current directory and, if it is, to have its contents displayed on the screen. The batch file CHECK.BAT would contain

```
ECHO OFF
REM Batch file to check for program
REM MYTEXT.TXT and display its
```

```
REM contents to the screen if it
REM exists.
IF EXIST MYTEXT.TXT TYPE MYTEXT.TXT
```

When executed, the program displays the contents of the file MYTEXT.TXT. If the file does not exist, the contents are not displayed. You will learn more about the IF command as other commands are presented.

THE NOT COMMAND

You can modify a DOS command to do the reverse of an IF condition by using the NOT command. As an example, suppose you have a batch file called REDO.BAT that checks to see if a file exists and, if it does exist, renames it. The batch file REDO.BAT would contain

```
ECHO OFF
REM Batch file to check if a
REM program exists and, if it
REM does, to rename it.
IF NOT EXIST %1 ECHO File %1 does not exist.
IF EXIST %1 RENAME %1 %2
```

To use the batch file, proceed as follows. Suppose the user is in the root directory and wants to change the name of a file called OLD.TXT to NEW.TXT. All that is necessary is

```
A> REDO OLD.TXT NEW.TXT
```

The program first checks whether the file OLD.TXT exists and then automatically renames it NEW.TXT, all without the user ever having to change the directory.

THE GOTO COMMAND

DOS provides a method of allowing a batch file to take a different set of instructions or to repeat a set of instructions. The action of the GOTO command is shown in Figure B.23.

The form of the GOTO command is

GOTO Labelname

where   Labelname = the place where commands will continue (It is distinguished from other DOS commands by starting with a colon.)

An example using labels is shown here for the batch file FINDIT.BAT. This batch program looks for the existence of a file and lets the user know if the file exists.

```
ECHO OFF
REM File Locator Program
IF EXIST %1 GOTO OK
IF NOT EXIST %1 GOTO NOGOOD
:OK
ECHO File %1 exists.
GOTO EXIT
:NOGOOD
ECHO File %1 does not exist.
:EXIT
```

Figure B.24 illustrates the action of this batch program.

FIGURE B.23   Action of the GOTO command

```
┌─── GOTO Label
│
│    REM HELLO ──────── This is skipped
│                       because of the
└──► :Label             GOTO.

     REM GOODBY
```

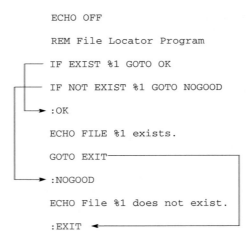

```
ECHO OFF

REM File Locator Program

IF EXIST %1 GOTO OK

IF NOT EXIST %1 GOTO NOGOOD

:OK

ECHO FILE %1 exists.

GOTO EXIT

:NOGOOD

ECHO File %1 does not exist.

:EXIT
```

COMBINING THE IF AND GOTO COMMANDS

Using the GOTO makes the IF even more powerful. The IF can be used to compare two strings, as shown:

IF String1 == String2 DOS Command

Note: Be sure to use the double equal sign (==) for this command.

What this means is that if String1 is identical to String2, the DOS command that follows will be executed. As an example, consider the batch file called YOURID.BAT that checks for a user ID. If the user gives the correct ID, the contents of a file are displayed; if not, the screen is cleared.

```
ECHO OFF
REM ID check program,
IF '%1' == 'HOWDY' GOTO YES
CLS
ECHO You are an imposter!
GOTO EXIT
:YES
TYPE SECRET.TXT
:EXIT
```

Observe that strings (a *string* is a sequence of keyboard characters *strung* together) must be enclosed by single quotation marks. To implement this program, enter

```
A> YOURID HOWDY
```

This command causes the contents of the text file SECRET.TXT to be displayed on the screen (assuming that the file exists).

AUTOEXEC.BAT FILES

The operating system provides a method of automatically activating a batch file when the system is first booted. What happens is that each time the system boots, the root directory is examined for a file named

```
AUTOEXEC.BAT
```

If a file by the name of AUTOEXEC.BAT is found, all the commands the file contains are processed.

An AUTOEXEC.BAT file is exactly the same as any other batch file, with the exception that it is automatically executed each time the system is booted. All the rules you have discovered about batch files apply to the AUTOEXEC.BAT file as well.

An AUTOEXEC.BAT file might clear the screen, change the DOS prompt so it displays the root directory, and then display a directory listing. You could create such a file with the following commands:

```
ECHO OFF
REM Batch file for creating
REM the DOS Prompt and listing
REM the directory.
PROMPT $N$G
CLS
DIR
```

Frequently users want particular application programs to be activated automatically every time they turn on their computers. For example, a user may want an accounting program, a word-processing program, or a listing of computer games to appear. This is the place to use an AUTOEXEC.BAT file. The following AUTOEXEC.BAT file sets the DOS prompt to the current date and activates an accounting program in the file ACCOUNT.EXE.

```
ECHO OFF
PROMPT $D
ACCOUNT.EXE
```

DOS 6.0 and later versions allow the use of two function keys to control what happens during boot time. If F5 is pressed at the beginning of the boot process, DOS skips activating the AUTOEXEC.BAT file entirely (as well as CONFIG.SYS). Booting up with the F5 key is useful for times when a new software package must be installed into a machine that does not contain any *extras* (such as a CD-ROM or sound card), and for certain diagnostic programs.

Another feature DOS 6.0 provides at boot time is activated when the F8 key is pressed. This feature allows the user to *single-step* through the CONFIG.SYS and AUTOEXEC.BAT files one line at a time. DOS prompts the user with

```
[Y/N]
```

after each line, allowing custom boot sequences depending on how the user answers each prompt.

Furthermore, DOS 6.0 provides a mechanism that allows the CONFIG.SYS and AUTOEXEC.BAT files to be written in such a way that a *menu* of boot options is provided at boot time. This allows the user to pick a particular machine configuration (for example, CD-ROM only, Network only, or CD-ROM and Network) as needed, without having to constantly modify AUTOEXEC.BAT for each configuration.

CHANGING THE DOS PROMPT

Normally, the DOS prompt is just a small flashing line. It indicates the position on the screen where the next keyboard entry will appear. However, DOS allows you to modify the prompt to display useful information. To change the DOS prompt, use the command

PROMPT PromptString

where the meaning of PromptString is given in Table B.9. (*Note:* The PromptString letters may be either upper- or lowercase.)

Using the command

```
A:\> PROMPT $P
```

results in the following DOS prompt (assuming you are in the directory called MYDIRECT):

```
A:\MYDIRECT
```

The command

```
A:\> PROMPT [$P]
```

PromptString Letter	Causes the DOS Prompt to Display
$b	:
$d	The current date
$e	The Esc character
$h	Backspace
$g	>
$l	<
$n	Current disk drive
$p	Current directory
$q	=
$t	Current time
$v	DOS version number
$$	$
$_	Carriage-return line-feed combination

causes the prompt to display

```
[A:\MYDIRECT]
```

Entering

```
A:\> PROMPT $P $G
```

produces

```
A:\MYDIRECT>
```

To restore the prompt back to its original condition, simply enter

```
PROMPT
```

This produces the *standard* DOS prompt

```
A>
```

It is common to have the DOS prompt modified in order to display the current directory (as used throughout this appendix).

STANDARD INPUT AND OUTPUT

Figure B.25 shows a computer system and what is referred to as *standard input* and *standard output*. As shown in the figure, the computer normally gets its input from the keyboard and normally sends its output to the monitor. You can cause the computer to change its standard input and get its instructions from some place other than the keyboard. You can also instruct the computer to direct its output to places other than the monitor. You have already done this with disk files, as shown in Figure B.26.

DOS comes with built-in *redirection operators*. You have already used one of them (>) to redirect output to the printer.

DOS REDIRECTION OPERATORS

A redirection operator allows you to to change its standard input and/or standard output to something else. The redirection operators are listed in Table B.10.

Figure B.27 shows some of the uses of the redirection operators.

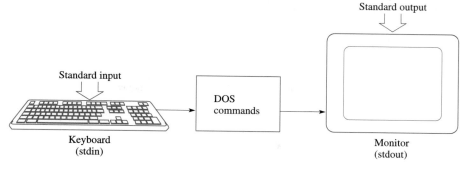

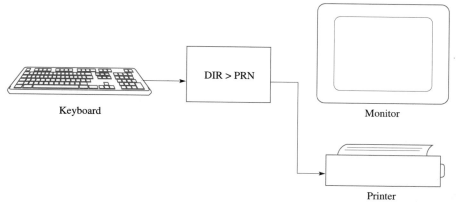

TABLE B.10 Standard redirection operators

Name	DOS Symbol	Action
Redirect Output	>	Redirects output from monitor to somewhere else.
Append Output	>>	Redirects output from monitor and appends it to an existing file.
Redirect Input	<	Redirects input from the keyboard to input from somewhere else.
DOS Pipe	|	Redirects both output and input at the same time.

DOS OUTPUT REDIRECTION

As indicated in Table B.10, the redirect output operator > instructs the system to redirect its output from the monitor to somewhere else. As an example, you can use the > operator to direct output to a DOS file rather than to the monitor. You can do this by entering

```
A> DIR > DIRFILE.TXT
```

Recall that you can also use the redirect output operator to send output to the printer:

```
A> DIR > PRN
```

This causes the directory to be printed on the printer instead of being displayed on the monitor. PRN is a standard device name meaning *printer.*

By using the append output operator >>, you can append information to an existing file. Figure B.28 shows the difference between the > and >> redirection operators.

As an example of the use of the >> operator, suppose you need to merge a group of separate text files into one text file. Given that the original group of text files is

TEXT.01
TEXT.02
TEXT.03

A> DIR > DIR.TXT

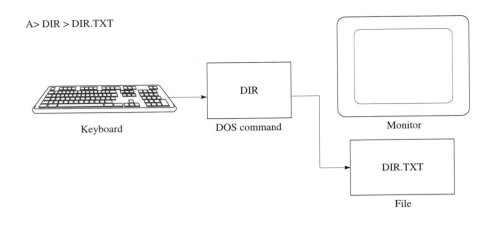

A> SORT < NAMES.TXT

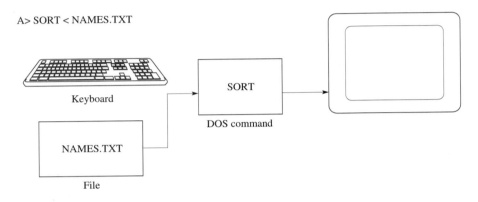

A> SORT < NAMES.TXT> SORTED.TXT

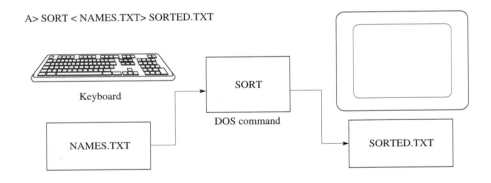

you can place all the text in these three separate files into one text file as follows:

```
A> TYPE TEXT.01 > ALLTEXT.TXT
```

This command creates a new file called ALLTEXT.TXT and copies the contents of TEXT.01 into it. Next, to merge the contents of TEXT.02, enter

```
A> TYPE TEXT.02 >> ALLTEXT.TXT
```

This command places the contents of TEXT.02 into ALLTEXT.TXT without removing the previous contents there from TEXT.01. Next issue the following:

```
A> TYPE TEXT.03 >> ALLTEXT.TXT
```

Now the contents of all three text files, TEXT.01, TEXT.02, and TEXT.03, are contained in one file called ALLTEXT.TXT.

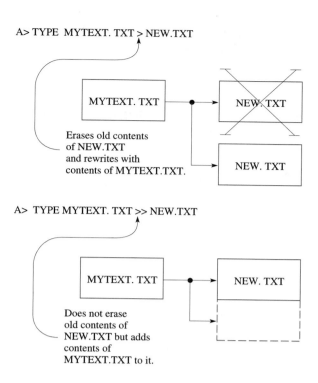

A> TYPE MYTEXT. TXT > NEW.TXT

MYTEXT. TXT

NEW. TXT

NEW. TXT

Erases old contents of NEW.TXT and rewrites with contents of MYTEXT.TXT.

A> TYPE MYTEXT. TXT >> NEW.TXT

MYTEXT. TXT

NEW. TXT

Does not erase old contents of NEW.TXT but adds contents of MYTEXT.TXT to it.

FIGURE B.29 Concept of the MORE command

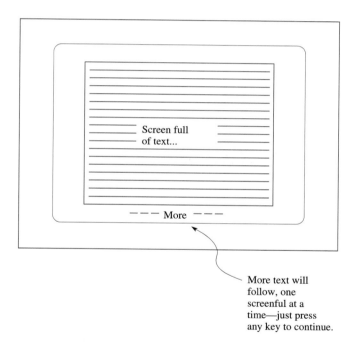

Screen full of text...

– – – More – – –

More text will follow, one screenful at a time—just press any key to continue.

THE MORE COMMAND

The MORE command gets input from standard input and displays it, one screen at a time, to standard output, as shown in Figure B.29. This utility is a program called MORE.COM. The syntax of the MORE command is

MORE filename

or

DIR | MORE

where   filename = the name of the file to be viewed one screen at a time

DIR | MORE = the directive allowing the directory to be viewed one screen at a time. This routes the output of the directory to the input of the MORE command through a "pipe" (indicted by the | symbol).

An alternative way of viewing a text file with MORE is to use this command:

A> TYPE FILENAME | MORE

which causes the output of the TYPE command to be piped to the MORE utility.

SETTING FILE ATTRIBUTES USING ATTRIB

There are several important characteristics of a file. These characteristics have applications when it comes to write-protecting files. Write-protecting a file means fixing it so that its contents may be read but not changed. Another characteristic of a file is its ability to remember whether it has been changed since a backup copy of it was last made. All of these file characteristics have important applications in microcomputer systems. As a computer user, you will be the one responsible for knowing what these special file characteristics mean and how to apply them.

ATTRIBUTE OF A FILE

An *attribute* of a file is a software assignment to that file that instructs it to behave in a specified manner. Table B.11 lists the attributes of a file.

You can set the attribute of a file from the following command:

ATTRIB [± R][± A][± H][± S][drive:]pathname[/S]

where   +R = sets the file to read only

−R = removes the read-only status of the file

+A = sets the file to archive status

−A = removes the archive status of the file

+H = sets the file to hidden

−H = removes the hidden status of the file

+S = sets the system attribute of the file

−S = removes the system attribute of the file

/S = directs ATTRIB to process all files that reside in a given directory

READ-ONLY FILES

Suppose you have an important text file that you want to protect from modification or erasure. Using the ATTRIB command, you can change the file to a ***read-only file***. If the name of the file is MYFILE.TXT, you enter

A> ATTRIB +R MYFILE.TXT

TABLE B.11 DOS file attributes

Attribute	Meaning
Read Only	The contents of the file can be read but cannot be modified, nor can the file be erased from the disk.
Archive	The file has not been backed up since it was last modified.
Hidden	The file is hidden and will not show up in a DIR listing.
System	The file is a system file. System files are hidden by default.

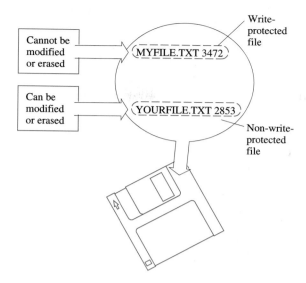

FIGURE B.30 Characteristics of write-protected and non-write-protected files

Cannot be modified or erased

Can be modified or erased

MYFILE.TXT 3472 — Write-protected file

YOURFILE.TXT 2853 — Non-write-protected file

The file MYFILE.TXT will now appear exactly the same on the directory as before. You can look at its contents with the TYPE command, but you cannot change its contents with the EDIT or any other text processor; you also cannot erase the file. Figure B.30 illustrates the characteristics of write-protected and non-write-protected files.

If later you decide that you want to change the file contents (or erase the file), you remove its read-only status:

```
A> ATTRIB -R MYFILE.TXT
```

ARCHIVE FILES

When a file is first created, the system automatically labels it an *archive* file. Anytime a file is modified, the system also automatically labels it an archive file. When a file is copied from one place to another, the copied file is automatically labeled an archive file. This is done as an aid in backing up files from one disk to another and backing up only those files on the original disk that either are new or have been recently modified.

To set a file (say, MYFILE.TXT) to be an archive file, use

```
A> ATTRIB +A MYFILE.TXT
```

The file is now an archive file.

To remove the archive status of an archive file (again, MYFILE.TXT), use

```
A> ATTRIB -A MYFILE.TXT
```

Now the file no longer has the status of an archive file.

HIDDEN FILES

Often, for reasons of security or privacy, you may want to prevent a file or files from showing up in the output of a DIR command. DOS allows you to protect files in this manner by setting its *hidden attribute*. To turn MYFILE.TXT into a hidden file, you use

```
A> ATTRIB +H MYFILE.TXT
```

After the hidden attribute is set, the file will no longer show up in directory listings. It is also impossible to copy or delete a hidden file. This is a valuable feature to keep in mind, and helps provide a measure of security to the file system.

To change a hidden file back to its normal, visible status, use the command

```
A> ATTRIB -H MYFILE.TXT
```

Note that a directory may also be hidden through the use of ATTRIB.

SYSTEM FILES

The last file attribute is reserved for *system files*. A system file is a file used exclusively by the system, such as IO.SYS and MSDOS.SYS, and is hidden by default. System files cannot be copied or deleted. Normally, you have no reason to set the system attribute of a file, but you can do so by means of the command

```
A> ATTRIB +S MYFILE.TXT
```

CHECKING THE FILE STATUS

You can check the status of any DOS file by issuing the following command:

```
A> ATTRIB [filename]
```

As an example, to check the attribute of the file MYFILE.TXT, you would enter, from the prompt,

```
A> ATTRIB MYFILE.TXT
```

Figure B.31 illustrates the different results from this command and their meanings. As you can see from Figure B.31, the attributes of a file can easily be determined.

THE XCOPY COMMAND

The XCOPY command is a powerful copy command. Unlike the COPY command or the DISKCOPY command, XCOPY has many special features. The syntax for this command is

XCOPY[Filespec1][Filespec2][/A][/D:mm-dd-yy][/E][/M][/P][/S][/V][/W]

where   Filespec1 = the source file(s) to copy
           Filespec2 = the destination file name(s) for the copied file(s)
                /A = copies only archive files (Does not change the archive status of the original files.)
                /D:mm-dd-yy = copies only files that have the date mm-dd-yy or later
                /E = creates subdirectories on the target disk that have the same contents as those on the source disk
                /M = copies only archive files and changes the status of the copied files (This modifier allows the XCOPY command to be used as a file backup command.)
                /P = causes XCOPY to present a "(Y/N?)" prompt, letting you decide whether you want to create each target file

FIGURE B.31 Possible results of the ATTRIB command

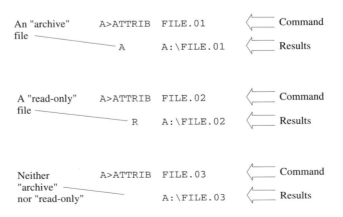

FIGURE B.32 XCOPY command examples

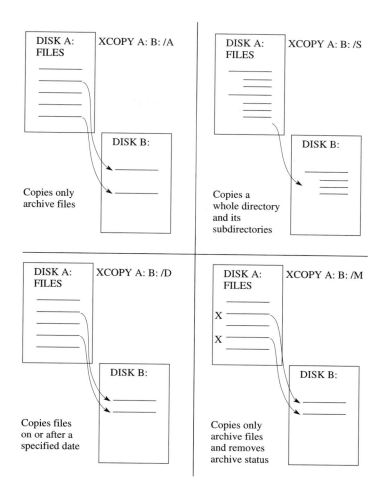

/S = causes XCOPY to copy directories and lower-level subdirectories (When this modifier is omitted, only the contents of the current directory are accessed.)

/V = forces XCOPY to verify that each file that is copied is an exact match of the original file

/W = makes XCOPY wait before starting the copying process. The following message is

Press any key when ready to start copying files.

(This gives you a chance to change disks if necessary.)

Figure B.32 illustrates the effects of the XCOPY command with the use of various modifiers.

COPYING A SET OF FILES FROM A HARD DISK

The XCOPY command is a convenient command to use for copying the contents of the hard disk (C: drive) or a large subdirectory to a floppy disk. To do this, modify the attributes of the files on your hard drive with

```
A> ATTRIB +A C:*.* /S
```

Next, use the XCOPY command:

```
A> XCOPY C:*.* B: /M
```

The files will then be copied from the hard drive to the floppy disk in drive B:. When the floppy disk becomes full, the following message is displayed:

```
Insufficient disk space
        nn files copied
A>
```

Now all you need do is remove the disk from drive B:, insert another formatted disk, and issue the same command:

```
A> XCOPY C:*.* B: /M
```

XCOPY will now pick up where it left off. It will copy only those files that it has not yet completely copied to the previous backup disk (in drive B:). These are the files that still have the archive attribute. Those files already copied have had their archive attributes turned off (because of the /M modifier used in XCOPY).

Appendix C
Fault Tolerance in Windows NT

Fault tolerance is the ability to tolerate failures in system hardware. Consider a networked environment in which many users access files on a shared server. What would happen if the server loses a hard drive, or even just one or more files? The impact on the network could be substantial. Windows NT provides technology called *RAID* that can help prevent this type of hardware failure.

RAID stands for Redundant Array of Inexpensive Disks. A system using RAID uses two or more hard drives to implement fault tolerance. Windows NT supports three levels of RAID technology:

- *RAID Level 0:* Stripe Sets
- *RAID Level 1:* Disk Mirroring
- *RAID Level 5:* Stripe Sets with Parity

Let's examine the features of these technologies.

RAID LEVEL 0: STRIPE SETS

In this technique, files are read/written in 64K chunks simultaneously using from 2 to 32 physical drives. The data is not duplicated. Figure C.1 shows how two drives are used to store a large file.

It is important to note that using a stripe set does not provide fault tolerance, since data is not duplicated on the drives. Performance is improved, however, because of the parallelism available during reads and writes. For example, the large file in Figure C.1 only requires two reads from each disk to access the entire file. Each read brings in 128K (64K from each disk at the same time). Write operations are similarly improved.

A disadvantage to using stripe sets is that system and boot partitions may not be stored on them.

RAID LEVEL 1: DISK MIRRORING

Disk mirroring is used to make an *exact* copy of data on two drives. Data is written to both drives simultaneously. If one drive fails, the second drive still has a good copy of the data, so the system is not affected. Both system and boot partitions may be mirrored. Figure C.2 shows how a file is mirrored.

FIGURE C.1   RAID level 0: Stripe sets

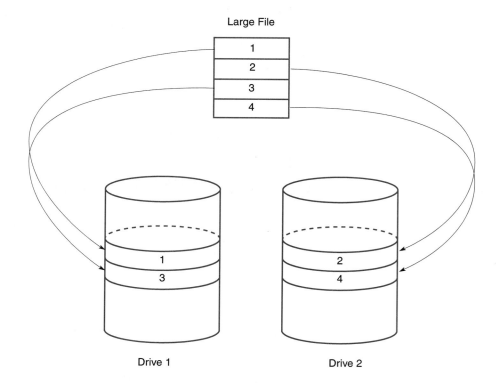

FIGURE C.2   RAID level 1: Disk mirroring

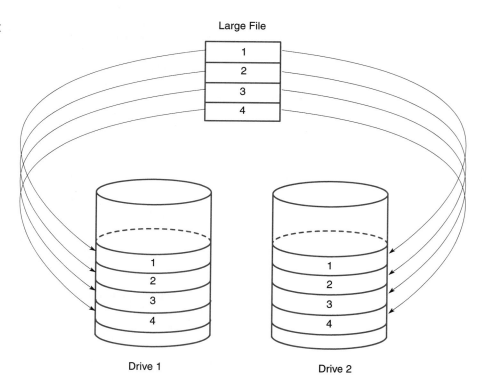

Typically, one controller is used to control both drives. A special variation of disk mirroring is *disk duplexing*, in which each drive has its own controller. This provides additional fault tolerance, since now a drive or a controller may fail without affecting the system.

A disadvantage to using disk mirroring is that you only get 50% of the hard drive space you pay for. For example, using two 8-GB drives only provides 8GB of storage capacity. Using the same drives in a stripe set would provide 16GB of capacity (but no fault tolerance).

RAID LEVEL 5: STRIPE SETS WITH PARITY

This technique is similar to ordinary stripe sets, except parity information is also written to each disk, as indicated in Figure C.3. If one of the drives in the stripe set fails, the parity data stored on each drive can be used to reconstruct the missing data.

A minimum of three drives must be used to employ RAID level 5. A maximum of 32 drives is allowed. The equivalent of one drive is used for parity information, even though parity is distributed across all drives. The available data capacity can be found by the following equation:

$$\text{Capacity} \quad = \quad \frac{\text{Drives} - 1}{\text{Drives}} \times 100\%$$

Therefore, with three drives, the available storage capacity for data is 66%. For four drives, the capacity becomes 75%.

A disadvantage to using stripe sets with parity is that extra system memory is required for the parity calculations. File servers should be equipped with enough RAM to handle the additional workload.

FIGURE C.3   RAID level 5: Stripe sets with parity

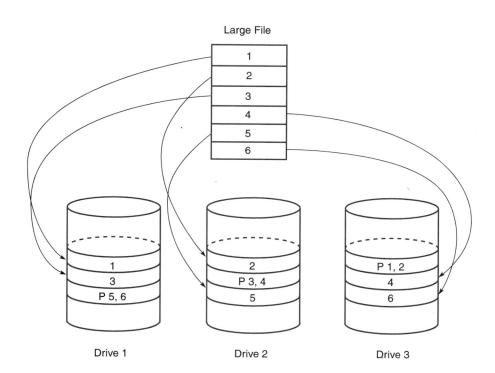

Appendix D
Installing and Upgrading Windows

This appendix provides information on installing and upgrading Windows. Note that it is extremely important to back up all critical files before attempting a change to the operating system. This will allow recovery from an installation that does not complete successfully.

INSTALLING WINDOWS

With the exception of Windows 3.x, which ran on top of DOS (thus making DOS a required component of the overall operating system), the newer Windows operating systems are stand-alone. This means Windows 95, 98, ME, NT, and 2000 can all be installed on a newly formatted hard disk, without having to first install DOS. In many cases, you also can upgrade an older version of Windows to a newer one. This avenue is explored in the next section.

Each operating system has a minimum hardware configuration that must be met, such as processor type and speed, the amount of system RAM, and the amount of free space on the hard disk. These requirements are indicated in Table D.1. Failure to meet the minimum requirements may prevent the installation from completing, or cause the operating system to run poorly or not at all.

Let's examine several installation scenarios.

Installing Windows on an Old Computer

The meaning of "old" is a matter of opinion. One person may think the 386 is an old processor, whereas another may think the Pentium II is old. One factor that helps define the

TABLE D.1 Minimum installation requirements

Operating System	Processor	Minimum RAM	Maximum Hard Disk Space
95	386DX	4MB	55MB
98	486DX-66	16MB	295MB
ME	Pentium-150	32MB	645MB
NT Server	486DX-25	16MB	124MB
2000 Server	Pentium-133	32MB	650MB

age of a system is the BIOS version used on the motherboard. Older BIOS ROMs do not support the newer, high-capacity hard drives used today, and limit the operation of the hardware in other ways (no plug-and-play support, for example). Imagine an older system whose BIOS does not support hard drives larger than 528MB. It may not even be possible to install Windows ME or 2000 with all desired options without filling up the hard disk. Commonly, disk drive manufacturers provide special software that translates the disk parameters (track, sector, and surface) to allow for large disk partitions to be created using older versions of BIOS.

If the hard disk has enough capacity, it will be necessary to boot DOS with the appropriate device drivers for the CD-ROM drive in your system. This includes the driver file in CONFIG.SYS and MSCDEX in AUTOEXEC.BAT. DOS can be booted from floppy, or from the hard disk (assuming the hard disk was formatted as a system disk). Note that Windows NT and Windows 2000 provide their own boot floppies to begin the installation process.

After the system has booted, insert the Windows installation CD into the CD-ROM drive and run the SETUP program if the installer does not automatically start up.

Installing Windows on a New Computer

The BIOS available on new computers typically supports using the CD-ROM drive as a boot device. This makes installation much simpler, since the installation CD can be used to boot the system, via the bootable CD-ROM drive. The hard disk is not involved in the boot sequence.

Windows 2000 takes advantage of this feature. In fact, the hard disk does not even have to be formatted. The Windows 2000 installation procedure will format the hard disk automatically, allowing you to choose the size of the partition and other options.

Note that some manufacturers provide a CD that contains the operating system and all prepackaged software for their system. Booting the system with the manufacturers' CD allows a complete system restore in the event of a serious hard disk failure.

If the BIOS does not support booting from the CD-ROM drive, or if the installation media is not bootable, the system must be prepared as it is for an old computer, with DOS and an operational CD-ROM drive.

Dual-Boot Systems

A dual-boot machine is one that contains more than one operating system. Typically, the hard disk is divided into two or more partitions, which are made bootable when an operating system is installed on them. One partition is chosen as the active partition, the one that will actually take over during the boot sequence and bring up the operating system. A partition manager, such as PartitionMagic or even FDISK, is required to create, specify the partition size, and activate a new partition.

The new operating system may also share the same partition as the old operating system. For example, Windows NT can be installed on the same partition as Windows 98 as long as the disk is formatted using FAT16. In this case, Windows NT will also install a boot manager that allows the operating system to be selected at boot time.

UPGRADING WINDOWS

Table D.2 indicates how one version of Windows may or may not be upgraded to another. For example, Windows 98 can be upgraded to Windows ME and Windows 2000, but

TABLE D.2   Upgrade paths for Windows operating systems

From	To 95	To 98	To ME	To NT	To 2000
3.x	✔	✔			
95		✔	✔		✔
98			✔		✔
NT					✔

Windows NT can only be upgraded to Windows 2000. Note that no operating systems can be upgraded to Windows NT. Also, different versions of the same operating system (for example, Windows 98 and Windows 98 SE, or Windows NT 3.51 and Windows 4.0) are not considered in upgrade paths.

To upgrade your current operating system, place the installation CD for the new operating system into the CD-ROM drive and run the SETUP program (if it does not automatically start by itself once the CD is inserted). Typically, the new installation will perform a check on your current system, copy files to the hard drive, reboot several times, and possibly lock up during hardware detection. Perhaps most importantly, you are also given the option to back up all of your old operating system files, in case you want to uninstall the new operating system and return to your old one. To offer the highest level of protection against data loss, it is recommended that a complete system backup be created prior to beginning any upgrade process.

Appendix E
Installing and Using McAfee Utilities

In this appendix we briefly examine the installation and operation of a popular software package called McAfee Utilities. Note that this software is a trial version that will expire 60 days after it has been installed. *Do not install* this product until required by your instructor.

To begin the installation, insert the McAfee Utilities companion CD into your CD-ROM drive. The installation program should automatically start. If it does not, navigate to the root directory of the CD-ROM and run the SETUP.EXE program.

The first screen displayed during the installation process is shown in Figure E.1. Left-click on the Install McAfee Utilities button to begin the installation. Bear in mind that your system will have to be rebooted one or more times during the installation. To avoid unexpected problems, choose the Complete installation option when prompted. Advanced users may choose the Custom option, which will allow the destination directory to be specified, as well as the choice of utilities to be installed. A license agreement will require you to answer *Yes* to continue with the installation. Note that a previous version of McAfee Utilities may have to be removed before the installation can continue.

After successfully completing the installation, an icon for McAfee Utilities Central is placed on the desktop. Left double-clicking the icon brings up the main window shown in Figure E.2.

The Status section indicates the state of several options that are particularly important. The four buttons on the left provide access to the various utilities that can be used to diagnose and tune up your system. The Update button is used to connect to the McAfee Web site and upgrade the product (for a small fee).

The utilities provided by each of the four buttons are as follow:

Repair and Recover

- DiskMinder
- McAfee Image
- Rescue Disk
- First Aid
- Discover Pro
- Undelete

Clean and Optimize

- QuickClean Lite
- Disk Tune
- Registry Wizard

FIGURE E.1   Initial McAfee Utilities installation screen

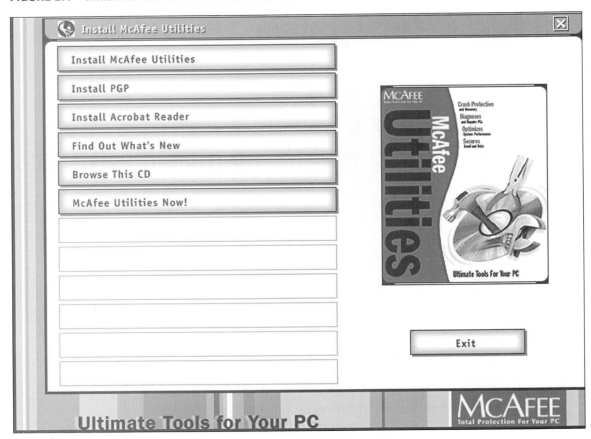

FIGURE E.2   Main McAfee Utilities main window

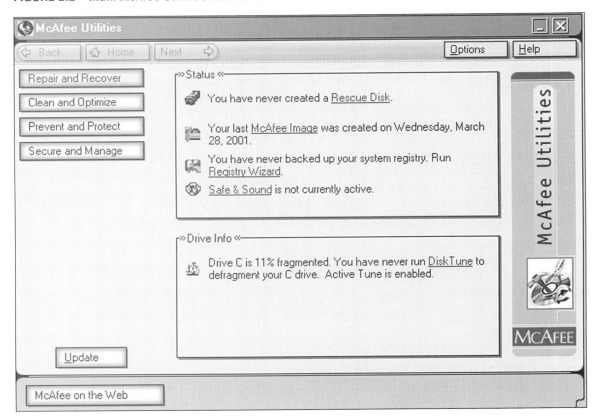

Prevent and Protect

- Crash Protector
- WinGuage
- Trash Guard
- Safe & Sound

Secure and Manage

- EZSetup
- PGP
- McAfee Shredder
- Zip Manager
- Registry Pro
- TaskMaster

You are encouraged to examine the operation of each utility under the direction and supervision of your instructor. You will find many useful and powerful tools for maintaining your computer. For example, the Discover Pro utility contains a diagnostic tool that can be used to test all of the hardware in your system. Figure E.3 shows the diagnostics display during a RAM test.

McAfee Utilities provide beginning and experienced users the tools required to diagnose, repair, and optimize their systems with a minimum of fuss. The first disk problem fixed or invalid Registry entry found will prove the value of this important collection of utilities.

FIGURE E.3   RAM Diagnostics display

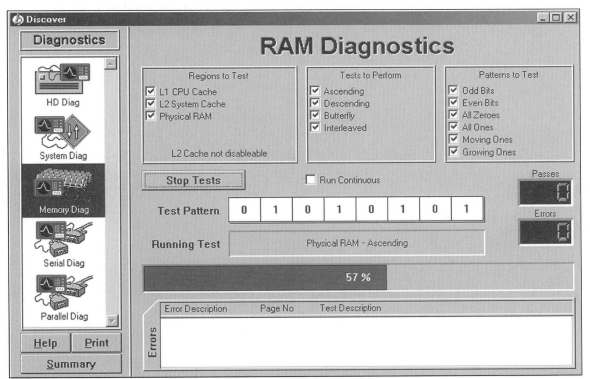

Answers to Odd-Numbered Self-Test Questions

EXERCISE 1

None.

EXERCISE 2

Answers to questions will be specific to your micro-computer repair lab.

EXERCISE 3

1. c   3. b   5. b   7. d

9.   The 10 safety rules presented in this exercise are (1) no horseplay is allowed in the lab; (2) always get instructor approval; (3) immediately report any injuries; (4) use safety glasses; (5) use tools correctly; (6) use equipment correctly; (7) do not distract others; (8) use correct lifting techniques; (9) remove jewelry; and (10) avoid static discharge.

EXERCISE 4

1. True   3. False   5. True   7. True   9. b   11. a
15. c   17. d   19. e   21. tweezers   23. pliers
25. plastic

EXERCISE 5

1. True   3. False   5. True   7. False   9. b   11. b   13. d   15. h, i   17. f   19. a   21. output
23. mouse   25. engine

EXERCISE 6

1. True   3. True   5. True   7. True   9. b   11. a
13. b   15. c   17. d   19. e   21. dual inline
package   23. photodiode   25. electrolytic

EXERCISE 7

1. True   3. False   5. False   7. True   9. d
11. b   13. c   15. b   17. d   19. b   21. bent
23. exact   25. grid

EXERCISE 8

1. True   3. True   5. False   7. True   9. b   11. b
13. b   15. b   17. f   19. a   21.   bridge
23. mechanical   25. surface

EXERCISE 9

1. False   3. True   5. False   7. b   9. c
11. protected   13. dialog box   15. bars

EXERCISE 10

1. False   3. True   5. False   7. False   9. True
11. c   13. a   15. b   17. TCP/IP   19. desktop
21. Registry   23. shutdown

EXERCISE 11

1. True   3. True   5. False   7. False   9. b
11. b   13. b   15. Shut Down   17. Applications
19. Diagnostics   21. NT

EXERCISE 12

1. True   3. False   5. False   7. True   9. False
11. False   13. b   15. c   17. b   19. b   21. b
23. b   25. Taskbar   27. tiled   29. Briefcase
31. screen saver   33. Settings   35. Start

EXERCISE 13

1. False   3. True   5. True   7. True   9. c   11. d
13. c   15. MPEG   17. Properties   19. .WAV
21. Network

EXERCISE 14

1. False   3. False   5. False   7. True   9. d
11. d   13. c   15. My Computer, Network
Neighborhood, Recycle Bin   17. pull-down   19. Find
Now   21. network

EXERCISE 15

1. False 3. False 5. False 7. False 9. b 11. c
13. b 15. a 17. dithering 19. local 21. default
23. NetBEUI

EXERCISE 16

1. False 3. False 5. True 7. False 9. False
11. d 13. b 15. a 17. d 19. a 21. fragmented
23. WordPad 25. slower

EXERCISE 17

1. False 3. True 5. False 7. True 9. a 11. a
13. a 15. Extended User Interface 17. Accessories
19. Internet service provider 21. Application

EXERCISE 18

1. False 3. True 5. False 7. True 9. a
11. a 13. a 15. a 17. Installation
19. Application setting 21. Wizard

EXERCISE 19

1. False 3. True 5. False 7. True 9. c
11. a 13. b 15. virtual device driver
17. interrupts 19. lower 21. NT

EXERCISE 20

1. False 3. True 5. False 7. b 9. c
11. devices 13. WIN32 15. logo

EXERCISE 21

1. True 3. False 5. False 7. a 9. b
11. NDS 13. IPX 15. Time

EXERCISE 22

1. True 3. False 5. False 7. b 9. b 11. c
13. a 15. c 17. fully connected 19. data-link
21. wireless

EXERCISE 23

1. False 3. False 5. False 7. True 9. a 11. b
13. c 15. c 17. b 19. virtual 21. Carrier
Sense Multiple Access with Collision Detection
23. virtual private

EXERCISE 24

1. False 3. False 5. False 7. False 9. a 11. a
13. a 15. c 17. d 19. e 21. combo 23. five
25. mode

EXERCISE 25

1. False 3. True 5. False 7. False 9. d 11. c
13. c 15. c 17. e 19. b 21. electronic mail
(e-mail) 23. dotted-decimal notation
25. connectionless

EXERCISE 26

1. False 3. True 5. True 7. True 9. c 11. b
13. c 15. NetBEUI 17. host 19. TRACERT
21. leased

EXERCISE 27

1. False 3. True 5. False 7. True 9. c 11. a
13. c 15. HTML 17. active 19. HTTP
21. instant

EXERCISE 28

1. False 3. True 5. True 7. False 9. c 11. b
13. a 15. Multimedia Internet Mail Extensions
17. Cc 19. Server 21. client

EXERCISE 29

1. False 3. False 5. True 7. True 9. b 11. b
13. b 15. least 17. hierarchy 19. environmental
21. straight-through

EXERCISE 30

1. False 3. True 5. False 7. False 9. c 11. c
13. c 15. d 17. a 19. e 21. workgroup
23. one 25. wizards

EXERCISE 31

1. True 3. True 5. True 7. b 9. a 11. T1
13. 13572468 15. cell

EXERCISE 32

1. True 3. False 5. True 7. c 9. b 11. c
13. b 15. UPS 17. electro

EXERCISE 33

1. False 3. True 5. True 7. a 9. b 11. b
13. c 15. fan 17. antistatic

EXERCISE 34

1. True 3. True 5. True 7. a 9. c 11. b
13. c 15. current 17. regulator

EXERCISE 35

1. True 3. False 5. False 7. False 9. a
11. d 13. c 15. b, c 17. a 19. d 21. disk
drive 23. power supply 25. 34 27. Answer
depends on your system. 29. Answer depends on
your system 31. Yes, you must have an
operating system (DOS or Windows).

EXERCISE 36

1. True 3. True 5. True 7. c 9. c 11. a
13. b 15. coprocessor 17. software 19. 3DNow!

EXERCISE 37

1. True 3. False 5. True 7. d 9. b 11. a
13. b 15. c 17. Hard; virtual 19. level-2
21. ported

EXERCISE 38

1. True 3. True 5. False 7. True 9. c 11. a
13. c 15. d 17. a, c 19. b 21. PC-card
23. hot swapping 25. bridge
27. Interrupt request lines are used to temporarily get
the attention of the microprocessor.

29. The EISA expansion slot is made compatible with the ISA expansion card by having two rows of connectors; the top row is compatible with the ISA card. Since the card isn't notched, it will go down only as far as the first row of connectors.

31. The distinguishing feature of micro-channel architecture is the use of smaller-sized expansion slots that will not accommodate earlier IBM PC, XT, or AT expansion cards but will accommodate 32-bit microprocessors such as the 80386 and up.

EXERCISE 39

1. True 3. False 5. True 7. b 9. b 11. c
13. a 15. math coprocessor 17. speaker

EXERCISE 40

1. False 3. True 5. True 7. False 9. d
11. b 13. a 15. unbootable 17. NLX
19. primary, secondary 21. flash

EXERCISE 41

1. False 3. True 5. False 7. True 9. b
11. d 13. a 15. c 17. a 19. root 21. Small Computer System Interface 23. stripe

EXERCISE 42

1. False 3. False 5. True 7. True 9. d 11. b 13. b 15. a 17. b 19. BIOS 21. partition
23. disk geometry

EXERCISE 43

1. False 3. True 5. False 7. False 9. d
11. a 13. a 15. b 17. sectors
19. FILE000N.REC 21. slave

EXERCISE 44

1. False 3. False 5. True 7. True 9. a 11. d 13. a 15. c 17. d 19. b 21. resolution
23. 640 × 480 25. Twisted
27. *Multisync* refers to a monitor capable of generating many different sync frequencies for various display resolutions. 29. 1280 × 1024
31. The monitor type is changed when you want to take advantage of the features of your new monitor.

EXERCISE 45

1. True 3. False 5. False 7. True 9. c 11. d
13. b 15. b 17. c 19. escape 21. portrait, landscape 23. exposure

EXERCISE 46

1. True 3. False 5. False 7. c 9. character
11. serial 13. trackball 15. driver

EXERCISE 47

1. False 3. True 5. False 7. b 9. a 11. b
13. a 15. line 17. protocol 19. Digital

EXERCISE 48

1. True 3. True 5. False 7. False 9. b 11. c 13. b 15. 300K/sec 17. advanced wave effects
19. sequencer

EXERCISE 49

1. False 3. True 5. False 7. True 9. c 11. a 13. a 15. c 17. e 19. a 21. wireless
23. device driver 25. GIF

EXERCISE 50

1. False 3. True 5. True 7. False 9. c
11. b 13. segment 15. byte-swapping 17. 17

EXERCISE 51

1. False 3. False 5. False 7. False 9. d
11. b 13. b 15. object 17. compiler
19. interpreted

EXERCISE 52

1. False 3. False 5. True 7. c 9. b 11. c
13. BIOS; the operating system 15. programmable interrupt controller 17. CS

EXERCISE 53

1. True 3. False 5. True 7. True 9. c
11. c 13. b 15. multitasking 17. pipelines
19. 80486

EXERCISE 54

1. False 3. False 5. False 7. b 9. b 11. a
13. ROM 15. electrically

EXERCISE 55

1. False 3. False 5. True 7. False 9. b
11. c 13. b 15. system 17. .VXD
19. protection

EXERCISE 56

1. False 3. False 5. False 7. d 9. b 11. b
13. checksums; signatures 15. share

EXERCISE 57

1. False 3. True 5. False 7. a 9. a
11. bottlenecks 13. resources 15. intermittent

EXERCISE 58

None.

Index